Statistical Computing

Statistical Computing

An Introduction to Data Analysis using S-Plus

Michael J. Crawley

Imperial College of Science, Technology and Medicine, UK

WILEY

Copyright © 2002   John Wiley & Sons Ltd, The Atrium, Southern Gate, Chichester, West Sussex PO19 8SQ, England

Telephone (+44) 1243 779777

Email (for orders and customer service enquiries): cs-books@wiley.co.uk
Visit our Home Page on www.wileyeurope.com or www.wiley.com

Reprinted with corrections March 2003, March 2004

Other Wiley Editorial Offices

John Wiley & Sons Inc., 111 River Street, Hoboken, NJ 07030, USA

Jossey-Bass, 989 Market Street, San Francisco, CA 94103-1741, USA

Wiley-VCH Verlag GmbH, Boschstr. 12, D-69469 Weinheim, Germany

John Wiley & Sons Australia Ltd, 33 Park Road, Milton, Queensland 4064, Australia

John Wiley & Sons (Asia) Pte Ltd, 2 Clementi Loop #02-01, Jin Xing Distripark, Singapore 129809

John Wiley & Sons Canada Ltd, 22 Worcester Road, Etobicoke, Ontario, Canada M9W 1L1

Wiley also publishes its books in a variety of electronic formats. Some content that appears in print may not be available in electronic books.

British Library Cataloguing in Publication Data

A catalogue record for this book is available from the British Library

ISBN 0-471-56040-5

Typeset in 10/12pt Times by Laserwords Private Limited, Chennai, India
Printed and bound in Great Britain by Antony Rowe Ltd, Chippenham, Wiltshire
This book is printed on acid-free paper responsibly manufactured from sustainable forestry
in which at least two trees are planted for each one used for paper production.

Contents

Preface

Learning S-Plus will change the way you do statistics. Most scientists are familiar with regression, analysis of variance and the statistics of contingency tables, but the problem is that they tend to use these techniques whether or not the models they embody are correct, or the error structure of the data is appropriate. Part of the problem has been that elementary statistics courses have tended to encourage rote learning of a small set of stereotyped tests:

- without critical attention to the underlying model involved

- without due regard to the precise distribution of sampling errors

- with little concern for the scale of measurement

- careless of dimensional homogeneity

- without considering the ideal transformation

- without any attempt at model simplification

- with too much emphasis on hypothesis testing and too little emphasis on parameter estimation.

S-Plus changes all that. It forces the user to think about all these problems and, in so doing, fosters a generally more critical approach to statistical analysis. It can handle most of the analyses that you are likely to carry out: regression, analysis of variance, log-linear models of counts, models using logits or probits for the analysis of proportions, models in which the variance increases with the mean, models of survival, and so on. All these models are defined, fitted and interpreted in essentially the same way, so that once the initially unfamiliar and rather daunting output from S-Plus has been mastered, all the different model structures are handled in exactly the same way. S-Plus is also useful in encouraging good statistical habits by:

- providing a first-rate graphics environment, allowing a thorough inspection of data before statistical analysis is begun

- permitting the choice of an appropriate error structure, rather than forcing the implicit assumption of normality of errors

- encouraging a conscious decision about the ideal transformation for linearising the relationship between the response variable and the explanatory variables

- demanding a precise specification of the model to be fitted to the data

- allowing tests of hypotheses and the construction of confidence intervals, using the correct standard errors, especially when there are missing data, offsets or unequal weights

- encouraging model simplification, and the search for a minimal adequate model

- focusing on model criticism, and on a comparison of different models for the same data (e.g. determining whether the assumptions made about the error structure, transformation and model definition are appropriate to the question in hand)

- allowing the identification of data values that are particularly influential to the parameterisation of a given model, and encouraging a willingness to present a range of models and to discuss the implications of the influential points (e.g. repeating the analysis with and without the influential data points included)

- providing a wide range of modern robust techniques in which outliers have much less influence

- introducing state-of-the-art techniques in mixed effects models that deal with repeated measurements and account properly for temporal and spatial pseudoreplication

- highlighting the need for extra data collection and the performance of new experiments

One of the objectives of statistical analysis is to distil a long and complicated set of data into a small number of meaningful descriptive statistics. Many of the modern computer statistics packages, however, do exactly the opposite of this. They generate literally pages of output from the most meagre sets of data. This copious output has several major shortcomings: it is open to uncritical acceptance; it can lead to over interpretation of data; and it encourages the bad habit of data trawling (dredging through the output looking for significant results without any prior notion of a testable hypothesis).

S-Plus, on the other hand, tells you nothing unless you explicitly ask for it. But this strength of S-Plus can also be a major stumbling block for beginners. Because you need to know what you are doing when you use S-Plus, and S-Plus demands that you tell it everything that you want to be done, the language can appear somewhat unfriendly on first acquaintance. Again, because the output is minimal, its interpretation takes a lot of getting used to. Without investing a certain amount of time in learning to understand the output from S-Plus, the exercise will be futile. On the other hand, the investment will be amply rewarded. There is no point investing masses of effort in collecting data and then not analysing them properly.

This book is intended as both an introduction to and a reference manual for statistics and computing. It assumes nothing by way of background in either subject, and starts from absolute basics. All it takes for granted is an enthusiasm to learn. It covers everything from the simplest non-parametric techniques (e.g. the runs test), up to the most advanced modern methods (e.g. mixed effects modelling). It covers generalised linear models and generalised additive models. It deals with tree models and non-parametric smoothing models. It covers time series analysis and multivariate techniques. You should be able to find the statistical test you need by using the dichotomous key on p. 2. Likewise, you should be able to find a comprehensively worked example to guide you through the computer analysis.

This is a statistics book for non-statisticians. In the past, such books have typically been aimed at relatively specialised niche markets: statistics for biologists, statistics for engineers, statistics for medics, statistics for economists, and so on. This book is different. It is based on the premise that effective data analysis requires the mastery of a relatively small core of central ideas and methods, and that these cut across the boundaries of academic disciplines. I have tried to use examples from as broad a range of disciplines as possible in order to minimise the risk that unfamiliarity with the subject matter of the example stands in the way of understanding the statistical analysis.

The intended audience comprises advanced undergraduate and postgraduate students as well as experienced researchers in science, medicine, engineering, economics and the social sciences. The material is developed from first principles in small steps, with practice and examples at every stage. No experience in statistics or computing is assumed at the outset. All of the classical methods are covered (regression, analysis of variance, contingency tables), plus generalised linear models (logistic regression, log-linear models, survival analysis), multivariate statistics, time series analysis, random effects (variance components analysis, mixed effects models, computationally intensive statistics (bootstrap, jackknife, simulation), and robust modern methods. The emphasis is on graphical data inspection, parameter estimation and model criticism rather than on classical hypothesis testing.

The book can be used in three ways. An introductory course might involve Chapters 1 to 11. If time permitted, adding regression (Chapter 14) and analysis of variance (Chapter 15) would be useful. An intermediate course (say for master's degree students) might start with Chapter 13 and cover regression, analysis of variance and analysis of covariance (Chapters 14 to 16) then cover the main classes of generalised linear models (log-linear models for count data and logistic regression for proportion data; Chapters 28 and 29). An advanced course, for PhD students or postdocs might begin with GLMs (Chapter 27) and go on to more advanced aspects of analysis of variance (Chapters 18 to 20) if time allowed.

Nobody ever learned statistics by reading a book about it. The way to get the most out of this book is to work through the examples while sitting at your computer, and to check the calculations by hand, so that you can see exactly what the program is doing, and precisely where all the output comes from.

The web site for the book is on http://www.bio.ic.ac.uk/research/mjcraw/statcomp/ This contains all of the data frames used in the book; you can download them as a batch, or inspect the contents of files individually. All of the executable code is available for you to copy and paste rather than type at the command line. Again, you can download these as a batch, or inspect files individually. The files are named by the Chapter number to which they apply. There are extra chapters on material for which there was no space in the printed version.

The computing is presented in S-Plus, but all the examples will also work in the freeware program called R, which can be downloaded from the web, free of charge, anywhere in the world.

There are a few differences between S-Plus and R, so that one or two of the examples in the book do not work in R. If you encounter any difficulties of incompatibility, you should visit the web site and look in the section called Updates, where all of the R equivalents are explained in full. If you find any differences that I have missed, then please let me know, and I'll post the fixes in Updates.

The comments of successive generations of Silwood students on the annual statistical computing course have greatly improved the clarity of the presentation, and have helped me to understand which bits of statistical modelling and computing are particularly daunting for beginners. Learning S-Plus will not be easy, but you won't regret making the effort.

M.J. CRAWLEY

1

Statistical methods

The hardest part of any statistical work is getting started. And one of the hardest things about getting started is choosing the right kind of statistical analysis for your data and the particular question you are trying to answer. The truth is that there is no substitute for experience: the way to know what to do is to have done it properly lots of times before. But here are some useful guidelines. It is essential, for example, that you ask:

- Which of your variables is the response variable?

- Which are the explanatory variables?

- Are the explanatory variables continuous or categorical, or a mixture of both?

- What kind of response variable have I got? Is it a continuous measurement, a count, a proportion, a time at death, or a category?

The answers to these questions should lead you quickly to the appropriate choice of method. The next thing to decide is whether you need to use one of the so-called classical tests, or carry out more sophisticated statistical modelling. *There is no point doing an analysis that is more elaborate than necessary for the question in hand*. The classical tests include:

- one-sample tests on means (e.g. Student's *t* test)

- two-sample tests on means (e.g. Wilcoxon rank sum test)

- two-sample tests on variances (e.g. Fisher's *F* ratio)

- tests of correlation (e.g. Spearman's rank test)

- analysis of count data using contingency tables (e.g. Fisher's exact test)

- comparisons of distributions (e.g. Kolmogorov–Smirnov test)

The classical tests are described in full in Chapter 11. If you decide that your analysis requires a more complex treatment, then you should use the following dichotomous key to find out what kind of statistical model is most appropriate. The key begins by asking questions about the nature of your *explanatory* variables, and ends by asking about what kind of *response* variable you have got. The response variable is the thing you are working on—the variable whose variation you are attempting to understand. It is the variable that goes on the *y* axis of the graph (the ordinate). The explanatory variable goes on the *x* axis of the graph (the abscissa); you are interested in the extent to which variation in the response variable is associated with variation in the explanatory variable. A continuous measurement is a variable like height or weight that can take any real numbered value. A categorical variable is a *factor* with two or more *levels*: sex is a factor with two levels (male and female), and rainbow might be a factor with seven levels (red, orange, yellow, green, blue, indigo, violet).

There is a small core of key ideas that needs to be understood from the outset. We cover these here before we get into any detail about the classical tests or the different kinds of statistical model.

Everything varies

If you measure the same thing twice, you will get two different answers. If you measure the same thing on different occasions, you will get different answers because the thing will have aged. If you measure different individuals, they will differ for both genetic and environmental reasons (nature and nurture). Heterogeneity is universal: spatial heterogeneity means that places always differ, and temporal heterogeneity means that times always differ.

Because everything varies, finding that things vary is simply not interesting. We need a way of discriminating between variation that is scientifically interesting, and variation that just reflects background heterogeneity. That is why you need statistics. It is what this whole book is about.

The key concept is the amount of variation that we would expect to occur by chance alone, when nothing scientifically interesting was going on. If we measure bigger differences than we would expect by chance, we say that the result is statistically significant. If we measure no more variation than we might reasonably expect to occur by chance alone, then we say that our result is not statistically significant. This is not to say that the result is not important. Non-significant differences in human life span between two drug treatments may be massively important (especially if it is your loved ones that are involved). Non-significance is not the same as 'not different'. It may be simply that our replication is too low to interpret the measured difference as being statistically significant.

On the other hand, when nothing really *is* going on, then we want to know this. It makes life much simpler if we can be reasonably sure that there is no relationship between y and x. Some students think that 'the only good result is a significant result'. They feel that their study has somehow failed if it shows that 'A has no significant effect on B'. This is an understandable failing of human nature, but it is not good science. The point is that we want to know the truth, one way or the other. We should try not to care too much about the way things turn out. This is not an amoral stance, it just happens to be the way that science works best. Of course, it is hopelessly idealistic to pretend that this is the way that scientists really behave. Scientists often want passionately that a particular experimental result will turn out to be statistically significant, so that they can get a *Nature* paper and get promoted. But that doesn't make it right.

Significance

What do we mean when we say that a result is significant? The normal dictionary definitions of significant are 'having or conveying a meaning' or 'expressive; suggesting or implying deeper or unstated meaning'. But in statistics we mean something very specific indeed. We mean that 'a result was unlikely to have occurred by chance'. In particular, we mean 'unlikely to have occurred by chance if the null hypothesis was true'. So there are two elements to it: we need to be clear about what we mean by 'unlikely', and also what exactly we mean by the 'null hypotheses'. Statisticians have an agreed convention about what constitutes 'unlikely'; they say an event is unlikely if it occurs less than 5% of the time. In general, the null hypothesis says 'nothing's happening' and the alternative says 'something *is* happening'.

Good and bad hypotheses

Karl Popper was the first to point out that a good hypothesis was one that was capable of **rejection**. He argued that *a good hypothesis is a falsifiable hypothesis*. Consider the following two assertions:

A.   There are vultures in the local park

B.   There are no vultures in the local park

   Both involve the same essential idea, but one is refutable and the other is not. Ask yourself how you would refute option A. You go out into the park and you look for vultures. But you don't see any. Of course, this doesn't mean that there aren't any. They could have seen you coming, and hidden behind you. No matter how long or how hard you look, you cannot refute the hypothesis. All you can say is, 'I went out and I didn't see any vultures'. One of the most important scientific notions is that *absence of evidence is not evidence of absence.*
   Option B is fundamentally different. You reject hypothesis B the first time that you see a vulture in the park. Until the time that you *do* see your first vulture in the park, you work on the assumption that the hypothesis is true. But if you see a vulture, the hypothesis is clearly false, so you reject it.

p values

A *p* value is an estimate of the probability that a particular result could have occurred by chance, *if the null hypothesis were true*. If something is very unlikely to have occurred by chance, we say that it is statistically significant e.g. $p < 0.001$. For example, if we are comparing two sample means, the null hypothesis is that the means are the same. A low *p* value means this hypothesis is unlikely to be true. A large *p* value (e.g. $p = 0.23$) means there is no compelling evidence on which to reject the null hypothesis. Of course, saying 'we do not reject the null hypothesis' and 'the null hypothesis is true' are two quite different things. For instance, we may have failed to reject a false null hypothesis because our sample size was too low, or because our measurement error was too large.

Interpretation

It should be clear by this point that we can make two kinds of mistakes in the interpretation of our statistical models:

- We can reject the null hypothesis when it is true

- We can accept the null hypothesis when it is false

These are referred to as **Type I** and **Type II** errors respectively. Supposing we knew the true state of affairs (which, of course, we seldom do), then we could put it in tabular form.

Null hypothesis	Actual	situation
	True	False
Accept	Correct decision	Type II
Reject	Type I	Correct decision

Statistical modelling

The object is to determine the values of the parameters in a specific model that lead to *the best fit of the model to the data*. The data are sacrosanct, and they tell us what actually happened under a given set of circumstances. It is a common mistake to say 'the data were fitted to the model' as if the data were something flexible, and we had a clear picture of the structure of the model. On the contrary, what we are looking for is the minimal adequate model to describe the data. *The model is fitted to data, not the other way around.* The best model is the model that produces the minimal residual deviance, subject to the constraint that all the parameters in the model should be statistically significant.

You have to specify the model. It embodies your best hypothesis about the factors involved, and about the way they are related to the response variable. We want the model to be *minimal* because of the principle of parsimony, and *adequate* because there is no point in retaining an inadequate model that does not describe a significant fraction of the variation in the data. It is very important to understand that *there is not one model*; this is one of the common implicit errors involved in traditional regression and Anova, where the same models are used, often uncritically, over and over again. In most circumstances there will be a large number of different, more or less plausible models that might be fitted to any given set of data. Part of the job of data analysis is to determine which, if any, of the possible models are adequate, and then, out of the set of adequate models, which is the minimal adequate model. In some cases there may be no single best model and a set of different models may all describe the data equally well.

Maximum likelihood

What exactly do we mean when we say that the parameter values should afford the 'best fit of the model to the data'? The convention we adopt is that our techniques should lead to *unbiased, variance-minimising estimators*. We define 'best' in terms of **maximum likelihood**. This notion is likely to be unfamiliar, so it is worth investing some time to get a feel for it. This is how it works:

- given the data
- and given our choice of model
- what values of the parameters of that model
- make the observed data most likely?

Here are the data: y is the response variable and x is the explanatory variable. Because both x and y are continuous variables, the appropriate model is regression.

```
x<-c(1,3,4,6,8,9,12)
y<-c(5,8,6,10,9,13,12)
plot(x,y)
```

Now we need to select a regression model to describe these data from the vast range of possible models available. Let's choose the simplest model, the straight line

$$y = a + bx$$

This is a two-parameter model: the first parameter, a, is the intercept (the value of y when x is 0) and the second, b, is the slope (the change in y associated with unit change in x). The response variable, y, is a linear function of the explanatory variable x. Now suppose that we knew that the slope was 0.68, then the maximum likelihood question can be applied to the intercept, a. If the intercept were 0 (left graph), would the data be likely? The answer of course is no. If the intercept were 8 (right graph) would the data be likely? Again, the answer is obviously no. The maximum likelihood estimate of the intercept is shown in the central graph (its value turns out to be 4.827).

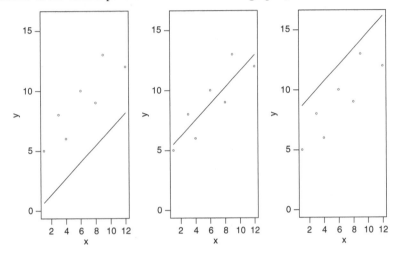

We could have a similar debate about the slope. Suppose we knew that the intercept was 4.827, then would the data be likely if the graph had a slope of 1.5 (left graph)?

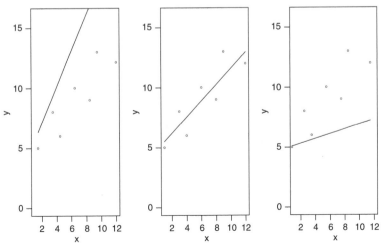

The answer of course is no. What about a slope of 0.2 (right graph)? Again, the data are not at all likely if the graph has such a gentle slope. The maximum likelihood of the data is obtained with a slope of 0.679 (centre graph). This is not how the procedure is actually carried out, but it makes the point that we judge the model on the basis *how likely the data would be if the model were correct*. In practice both parameters are estimated simultaneously.

Data frames

One of the fundamental concepts to grasp about statistical computing is the notion of the data frame. Tables of data come in all shapes and sizes, but only one kind of table is a data frame. In a data frame *all of the values of a given variable are in a single column.*

Data frames are the fundamental data structure used by most of the S-Plus modelling software. They are tightly coupled collections of variables that share many of the properties of matrices. The main difference is that the columns of a data frame may be of differing types (numeric, factor, logical or character), whereas a matrix has to have all of its columns of the same type. This is what a data frame looks like:

Field Name	Area	Slope	Vegetation	Soil pH	Damp	Worm density
Nash's Field	3.6	11	Grassland	4.1	F	4
Silwood Bottom	5.1	2	Arable	5.2	F	7
Nursery Field	2.8	3	Grassland	4.3	F	2
Rush Meadow	2.4	5	Meadow	4.9	T	5
Gunness' Thicket	3.8	0	Scrub	4.2	F	6
Oak Mead	3.1	2	Grassland	3.9	F	2
Church Field	3.5	3	Grassland	4.2	F	3
Ashurst	2.1	0	Arable	4.8	F	4
The Orchard	1.9	0	Orchard	5.7	F	9
Rookery Slope	1.5	4	Grassland	5	T	7
Garden Wood	2.9	10	Scrub	5.2	F	8
North Gravel	3.3	1	Grassland	4.1	F	1
South Gravel	3.7	2	Grassland	4	F	2
Observatory Ridge	1.8	6	Grassland	3.8	F	0
Pond Field	4.1	0	Meadow	5	T	6
Water Meadow	3.9	0	Meadow	4.9	T	8
Cheapside	2.2	8	Scrub	4.7	T	4
Pound Hill	4.4	2	Arable	4.5	F	5
Gravel Pit	2.9	1	Grassland	3.5	F	1
Farm Wood	0.8	10	Scrub	5.1	T	3

- The columns represent variables (like Soil pH or Vegetation)

- The first row contains the names of the variables

- The columns can represent variables of various kinds

— Soil pH is a numeric variable

— Vegetation is a factor variable

— Damp is a logical variable

- All the values of one variable are in a single column

- The rows are labelled in column 1 (in this case by field names, but by row numbers if there are no row names) and represent the cases (or replicates) of the data frame

One important detail needs to be mentioned at this point. The names of the variables must not contain blank spaces (like Soil pH, above); if they do, then the computer thinks you have an extra variable (called pH in this example). In S-Plus your variable names can be joined together using the decimal (full stop) character, e.g. Soil.pH or Worm.density. Don't use the underline symbol to make composite variable names, because in S-Plus this symbol means 'allocate' (see p. 18).

Here is another data frame:

Area	Slope	Vegetation	Soil.pH	Damp	Worm.density
3.6	11	1	4.1	0	4
5.1	2	2	5.2	0	7
2.8	3	1	4.3	0	2
2.4	5	3	4.9	1	5
3.8	0	4	4.2	0	6
3.1	2	1	3.9	0	2
3.5	3	1	4.2	0	3
2.1	0	2	4.8	0	4
1.9	0	5	5.7	0	9
1.5	4	1	5	1	7
2.9	10	4	5.2	0	8
3.3	1	1	4.1	0	1
3.7	2	1	4	0	2
1.8	6	1	3.8	0	0
4.1	0	3	5	1	6
3.9	0	3	4.9	1	8
2.2	8	4	4.7	1	4
4.4	2	2	4.5	0	5
2.9	1	1	3.5	0	1
0.8	10	4	5.1	1	3

The difference here is that the values of the categorical variables (the *factors* of the data frame) have been replaced by numbers (the *levels* of the factors). Notice that there are no level 0 factors; factor numbering always starts at 1 not at 0. Also, the logical variable Damp has been replaced by numbers: by convention, FALSE = 0 and TRUE = 1. The row names have been omitted, and the default row names = row number will therefore be used.

Here is a table that is *not* a data frame:

Temp Grass	Growth	Temp Wood	Growth	Temp Heath	Growth	Temp Arable	Growth
10	2.1	10	2.7	10	1.9	10	3.3
12	3.1	12	2.7	12	1.5	12	3.1
14	2.9	14	3.4	14	1.7	14	2.9
16	2.7	16	3.6	16	1.5	16	3.0
18	3.2	18	2.9	18	2.1	18	2.8
20	2.8	20	3.2	20	1.3	20	2.7

It is not a data frame because values of the same variable (e.g. Temp) occur in several different columns. Because of this, each row of the table is not uniquely associated with one value of the other explanatory variable (Land Type: Grass, Wood, Heath or Arable). To turn this table into a data frame you would need to cut and paste the Temp data into a single column, then do the same for the Growth data, then create a *new* column for Land Type in which 'Grass' appears in the first six rows, 'Wood' in the second six rows, 'Heath' in the next six rows, and 'Arable' in the last six rows.

Preparing data frames

It is good practice to *create* the data frame and *analyse* the data frame as two quite separate exercises. Data entry and error checking are the first step. Only when the data frame is error-free should it be read into S-Plus and analysed. This approach is likely to save you a lot of time that you might otherwise waste by analysing data frames that subsequently turn out to have errors in them, in which case the whole of the analysis needs to be started over again, once the errors have been corrected. Most people create and edit their data frames in a spreadsheet like Excel. The error-free spreadsheet is then imported to S-Plus and converted directly into a data frame.

Reading data into S-Plus from a file

There are two options. You can use the graphic user interface (GUI) by pointing the mouse and clicking, to import the Excel file directly. From the main menu, choose File > Import Data > From File. The Import Data dialog opens, and you specify the directory, type of file and file name you want to import. Then click on Open. The name of your spreadsheet becomes the name of the data frame within your current session of S-Plus.

Alternatively, you can save your Excel file as a tab-delimited text file. Then you can read the file and turn it into a data frame from the command line. If you are using early versions of the software R (with no GUI) this is your only option for reading data from a file. The relevant function is **read.table**.

Give the file a memorable and informative name. You will be amazed how quickly you forget what is inside a file called data1! Let's call this file Silwood.worms.txt.

There are three things to remember:

- The whole path and file name needs to be enclosed in double quotes, "c:\\abc.txt"

- header = T says that the first row contains the variable names

- Always use double backslash \\ rather than \ in the file path definition

Think of a name for the data frame (say worms in this case). Now use the **gets arrow** <-, which is made up of the two characters < (less than) and - (minus) like this:

worms<-read.table("c:\\data\\Silwood.worms.txt",header=T)

This creates a data frame called worms. To see the content of the object called worms just type worms then the return key:

```
worms
         Field.Name Area Slope Vegetation Soil.pH Damp Worm.density
1        Nash's.Field  3.6   11  Grassland     4.1    F            4
2      Silwood.Bottom  5.1    2     Arable     5.2    F            7
3       Nursery.Field  2.8    3  Grassland     4.3    F            2
4         Rush.Meadow  2.4    5     Meadow     4.9    T            5
5     Gunness.Thicket  3.8    0      Scrub     4.2    F            6
6            Oak.Mead  3.1    2  Grassland     3.9    F            2
7        Church.Field  3.5    3  Grassland     4.2    F            3
8             Ashurst  2.1    0     Arable     4.8    F            4
9         The.Orchard  1.9    0    Orchard     5.7    F            9
10      Rookery.Slope  1.5    4  Grassland     5.0    T            7
11        Garden.Wood  2.9   10      Scrub     5.2    F            8
12       North.Gravel  3.3    1  Grassland     4.1    F            1
13       South.Gravel  3.7    2  Grassland     4.0    F            2
14   Observatory.Ridge  1.8    6  Grassland     3.8    F            0
15         Pond.Field  4.1    0     Meadow     5.0    T            6
16       Water.Meadow  3.9    0     Meadow     4.9    T            8
17          Cheapside  2.2    8      Scrub     4.7    T            4
18         Pound.Hill  4.4    2     Arable     4.5    F            5
19         Gravel.Pit  2.9    1  Grassland     3.5    F            1
20          Farm.Wood  0.8   10      Scrub     5.1    T            3
```

You will see that S-Plus has automatically added a set of row numbers, and that these start in row 2 (not in row 1 because we said header = T). An important thing to remember is that you are *not* allowed to have multiple-word variable names or factor levels in which the words are separated by blanks.

Attaching a data frame

Once the file has been imported to S-Plus we want to do two things:

- Use **attach** to make the variables accessible by name within the S-Plus session

- Use **names** to get a list of the variable names

Typically, the two commands are issued in sequence, whenever a new data frame is imported from file:

```
attach(worms)
names(worms)

[1] "Field.Name"    "Area"       "Slope"         "Vegetation"
[5] "Soil.pH"       "Damp"       "Worm.density"
```

For the examples in this book, the **read.data** command is omitted, because most of you will probably use the dialog box to import spreadsheets. If you forget to attach the data frame, you will get an error message the first time you try to refer to one of the variables by name. Suppose you wanted to plot worms against pH:

```
plot(Soil.pH,Worm.density)
```

produces the message

```
Error: Object "Soil.pH" not found
```

Once the data frame is successfully imported and attached, you can begin your analysis. Read on.

2

Introduction to S-Plus

S-Plus was first developed by AT&T's Bell Laboratories to provide a software tool for professional statisticians who wanted to combine state-of-the-art graphics with powerful model-fitting capability. This does not mean that you have to be an expert statistician or a mathematical prodigy to use S-Plus effectively. On the contrary, you can come to S-Plus as an absolute beginner, and develop your computing skills as your knowledge of statistics increases, and your experience in applying them develops. The strength of S-Plus lies in its generality and its flexibility.

S-Plus is made up of three components. First and foremost, it is a powerful tool for statistical modelling. It enables you to specify and fit statistical models to your data, assess the goodness of fit and display the estimates, standard errors and predicted values derived from the model. It provides you with the means to define and manipulate your data, but the way you go about the job of modelling is not predetermined, and the user is left with maximum control over the model-fitting process. Second, S-Plus can be used for data exploration, in tabulating and sorting data, in drawing scatterplots to look for trends in your data, or to check visually for the presence of outliers. Third, it can be used as a sophisticated calculator to evaluate complex arithmetic expressions, and a very flexible and general object-oriented programming language to perform more extensive data manipulation. One of its great strengths is in the way in which it deals with vectors (lists of numbers). These may be combined in general expressions, involving arithmetic, relational and transformational operators such as sums, greater-than tests, logarithms or probability integrals. The ability to combine frequently used sequences of commands into functions makes S-Plus a powerful programming language, ideally suited for tailoring one's specific statistical requirements.

S-Plus is especially useful in handling difficult or unusual data sets, because its flexibility enables it to cope with such problems as unequal replication, missing values, non-orthogonal designs, and so on. Furthermore, the open-ended style of S-Plus is particularly appropriate for following through original ideas and developing new concepts.

One of the great advantages of learning S-Plus is that the simple concepts that underlie it provide a unified framework for learning about statistical ideas in general. By viewing particular models in a general context, S-Plus highlights the fundamental similarities between statistical techniques and helps play down their superficial differences.

The S-Plus system is described in detail in the S-Plus 6 Manual, produced by Insightful Corporation, Seattle, Washington, USA.

There are two ways of driving S-Plus. If you enjoy wasting time, you can pull down the menus and click in the dialog boxes to your heart's content. However, this takes about 5 to 10 times as long as writing the command line. Life is short. Use the command line.

S-Plus operates by executing commands that the user enters at the keyboard; these commands are called functions and they are shown in **bold** in the text. Each command consists of a function name specifying the action required, followed by items indicating how the action is to be performed (contained within round brackets like these).

A useful feature of S-Plus is that past command lines can be retrieved using the Up Arrow, the history can be scrolled back, and Copy and Paste can be used to move material back to the command line. Multiple graphics windows can be opened, and each graphics window can be divided up into as many panels as you like.

There is a vast network of S-Plus users worldwide, exchanging functions with one another, and a vast resource of libraries containing data and programs.

A completely free version of software that emulates S-Plus is called R. It is available from the web at http://www.mirror.ac.uk/sites/lib.stat.cmu.edu/R/CRAN/ then run Rsetup.exe.

This is excellent for teaching and learning S-Plus, and although it does not have all the functionality of its parent product, it is becoming more comprehensive all the time.

Calculator

The screen prompt is an excellent calculator. All of the usual calculations can be done directly, and the standard order of precedence applies: powers and roots are done first, then multiplication and division, then finally addition and subtraction.

Arithmetic

The arithmetic operations are addition, subtraction, multiplication, division, exponentiation, integer division, and modulo. All but the latter two accept complex as well as numeric arguments. Addition and subtraction are carried out in the obvious way:

```
> 3+5
[1]   8
```

Notice the conventions adopted: > is the input prompt for the command line. From here onwards, this will be omitted. The material that you type at the command line is shown in Arial font. Just press the Return key to see the answer. The output from S-Plus is in Courier font. The advantage of Courier is that it has absolute rather than proportional spacing, so columns of numbers remain neatly aligned on the page or on the screen.

Multiplication uses the asterisk * operator (not × or x):

```
3*9
[1]   27
```

Division uses the forward slash / operator (not ÷ or $\frac{a}{b}$):

```
27 / 9
[1]   3
```

Exponentiation (powers) uses the hat ^ operator (not ** as in some older computing languages like Fortran or GLIM):

```
3^2
[1]   9
```

Fractional powers are roots. For example, cube root is power 1/3:

```
8^(1/3)
[1]   2
```

Negative powers are reciprocals. For example $3^{-2} = 1/3^2 = 1/9 = 0.11111$:

```
3^-2
```

```
[1]   0.1111111
```

Integer divide

Sometimes we want to know how many times one number goes into another; we are not concerned about the size of the remainder. The arithmetic of this is carried out by the integer divide operator % / % (percent forward slash percent). So if we want to know how many times 19 goes into 137, we type

```
137 % / % 19
```

```
[1]   7
```

The remainder (0.210 526) is dropped. Technically, the result is *floor*(a/b)—the largest integer that is smaller than a/b—so long as the denominator, b, is not 0 (and 0 if the denominator is 0; see below).

Modulo

Sometimes we are more interested in the remainder than the result of the integer divide. Suppose we want to test whether a number is a multiple of 19. If it is, then the remainder will be zero. If it is not, then the remainder will be greater than zero. We can test to see if 137 is a multiple of 19 like this:

```
137 % % 19
```

```
[1]   4
```

The remainder is 4, so 137 is not a multiple of 19. But 133 is. If we ask 133 %%19 then we should get zero:

```
133 % % 19
```

```
[1]   0
```

The answer you get from a modulo operation a %% b is $a - b \times$ floor (a/b) so long as b is not zero. If b is zero then the answer returned is a.

Fractional powers of negative numbers

The square root of −1 is a very special number in mathematics—it is the fundamental concept that lies behind complex numbers. Generally speaking, however, fractional powers of negative numbers are not allowed.

Suppose we try to work out the cube root of −4:

```
(-4)^(1/3)
```

```
[1]   NA
```

Note the use of brackets. The NA output means 'not available' and is the S-Plus missing value symbol. This is very different from

```
-4^(1/3)
```

```
[1]   -1.587401
```

where S-Plus has assumed that you want the negative of the cube root of +4. If you try to take an integer power of a negative number

```
(-4)^2
```

```
[1]   16
```

it will work, but the following will not work, even though the power is very close to 2:

```
(-4)^2.00000000000001
```

```
[1]   NA
```

A machine-dependent test is performed to decide if a number is exactly or nearly an integer, but you should not count on its behaviour in doubtful cases.

Continuation of multiline expressions

Each line can have at most 128 characters, so if you have a complicated expression, you may need to continue it on one or more lines. A typed expression may be continued on further lines by ending the line at a place where the line is obviously incomplete (e.g. with a trailing comma, operator, or with more left parentheses than right parentheses implying that more right parentheses will follow). The default prompt character is >, and when continuation is expected the default prompt is +:

```
> 5+6+3+6+4+2+4+8+
+      3+2+7
```

```
[1]   50
```

Note that the + continuation prompt does not carry out arithmetic plus. Two or more expressions can be placed on a single line if they are separated by a semicolon ;.

Built-in Functions

All the mathematical functions you could ever want are here (see Table 2.1). The **log** function gives logs to base e (e = 2.718282), for which the antilog function is **exp**:

```
log(10)
```

```
[1]   2.302585
```

```
exp(1)
```

```
[1]   2.718282
```

If you are old-fashioned, and want logs to base 10, then there is a separate function

```
log10(6)
```

```
[1]   0.7781513
```

Logs to other bases are possible by providing the log function with two arguments: the second argument is the base to which you want the logs to be taken. Suppose we want log to base 3 of 9:

```
log(9,3)
```

```
[1]   2
```

Table 2.1 Mathematical functions in S-Plus

Function	Meaning
log(x)	log to base e of x
exp(x)	antilog of x (2.7818^x)
log(x,n)	log to base n of x
log10(x)	log to base 10 of x
sqrt(x)	square root of x
factorial(x)	$x!$
choose(n,x)	binomial coefficients $n!/(x!(n-x)!)$
gamma(x)	$\Gamma\ x$
	$(x-1)!$ for integer x
lgamma(x)	natural log of $\Gamma(x)$
floor(x)	greatest integer $< x$
ceiling(x)	next integer $> x$
trunc(x)	closest integer to x between x and 0
	$\mathrm{trunc}(1.5)=1$, $\mathrm{trunc}(-1.5)=-1$
	trunc is like floor for positive values and like ceiling for negative values
round(x, digits=0)	rounds the value of x to an integer
signif(x, digits=6)	gives x to 6 digits in scientific notation
runif(n)	generates n random numbers between 0 and 1 from a uniform distribution
cos(x)	cosine of x in radians
sin(x)	sine of x in radians
tan(x)	tangent of x in radians
acos(x), asin(x), atan(x)	inverse trigonometric transformations of real or complex numbers
acosh(x), asinh(x), atanh(x)	inverse hyperbolic trigonometric transformations on real or complex numbers
abs(x)	the absolute value of x, ignoring the minus sign if there is one

The trigonometric functions measure angles in radians. A circle is 2π radians, and this is $360°$, so a right angle ($90°$) is $\pi/2$ radians. S-Plus knows the value of π as pi:

```
sin(pi/2)

[1]    1

cos(pi/2)

[1]    6.123032e-017
```

Notice that the cosine of a right angle does not come out as exactly zero, even though the sine came out as exactly 1. The e-017 means 'times 10^{-17}'. While this is a very small number it is not exactly zero.

Numbers with exponents

For very big numbers or very small fractions it is useful to adopt the following scheme:

1.2e3 means twelve hundred, or 1200, because the e3 means 'move the decimal point three places to the right'

1.2e-2 means twelve-thousandths, or 0.012, because the e-2 means 'move the decimal point 2 places to the left'

3.9+4.5i is a complex number because *i* is the square root of -1

Logical arithmetic

In logical arithmetic an expression evaluates to either true (T) or false (F). The 'double equals' operator == asks the equality question:

cos(pi/2)==0

[1]   F

to which S-Plus gives the answer false F. But sin(pi/2) does exactly equal 1 (see above):

sin(pi/2)==1

[1]   T

You need to be careful with this asymmetry.

Whole numbers (integers)

Various sorts of rounding (rounding up, rounding down, rounding to the nearest integer) can be done easily. Take 5.7 as an example. The 'greatest integer' function is **floor**:

floor(5.7)

[1]   5

and the 'next integer' function is **ceiling**:

ceiling(5.7)

[1]   6

You can round to the nearest integer by adding 0.5 to the number then using **floor**. There is a built-in function for this, but we can easily write one of our own to introduce the notion of function-writing. Call it rounded, then define it as a function like this:

rounded <- function (x) floor(x + 0.5)

Now we can use the new function:

rounded(5.7)

[1]   6

Generating sequences

The simplest way to generate a regularly spaced sequence of numbers is to use the colon operator : like this:

1:7

[1]   1   2   3   4   5   6   7

For more complicated sequences, use the **seq** function in which the third argument is the step length (the increment or decrement you want to use). Going up:

```
seq(0, 1, 0.1)
```

```
[1]   0.0   0.1   0.2   0.3   0.4   0.5   0.6   0.7   0.8   0.9   1.0
```

or coming down:

```
seq(5, -5, -1)
```

```
[1]   5   4   3   2   1   0   -1   -2   -3   -4   -5
```

The **along** option allows you to map a sequence onto an existing vector (to ensure equal lengths). For example, you may know how many numbers you want, but you can't be bothered to work out the final value of a series. Suppose we want a series of 20 numbers, starting at 5 and going down in steps of -1:

```
seq(from = 5, by = -1, along = 1:20)
```

```
[1]   5   4   3   2   1   0   -1   -2   -3   -4   -5   -6   -7   -8   -9   -10   -11  -12   -13   -14
```

Generating repeats

The **rep** function replicates the input either a certain number of times, or to a certain length. In the unlikely event that you wanted to get 5.3 repeated 14 times, you would type

```
rep(5.3, 14)
```

```
[1]   5.3   5.3   5.3   5.3   5.3   5.3   5.3   5.3   5.3   5.3   5.3   5.3   5.3   5.3
```

The first parameter is the object that you want to be repeated, and the second parameter is the number of times you want it to be repeated. The general syntax is like this: rep(x, times). If times consists of a single integer, the result consists of the whole object x repeated this many times.

```
rep(1:6, 2)
```

```
[1]   1   2   3   4   5   6   1   2   3   4   5   6
```

This says repeat the whole series 1 to 6, twice. That's fine, but what if we wanted each of the numbers 1 to 6 repeated, say, three times. In S-Plus we could write

```
rep(1:6,each=3)
```

```
[1]   1   1   1   2   2   2   3   3   3   4   4   4   5   5   5   6   6   6
```

If times is a vector of the same length as x, the result of **rep** consists of x[1] repeated times[1] times, x[2] repeated times[2] times, and so on. The thing to remember is that the *length* of the second argument (the number of each repeat) has to match the length of the first:

```
rep(1:6,rep(3,6))
```

```
[1]   1   1   1   2   2   2   3   3   3   4   4   4   5   5   5   6   6   6
```

This emulates the old 'generate levels' function for those of you who grew up with GLIM. We can write a simple function to do this; the first argument is the maximum and the second argument is the repeat:

```
gl<-function(x,y) rep(1:x,rep(y,x))
```

Now we can use the function **gl** on the example to obtain three repeats of each number up to 6:

```
gl(6,3)
[1]  1  1  1  2  2  2  3  3  3  4  4  4  5  5  5  6  6  6
```

The most complex case arises when we want to repeat each element of the first list a different number of times. One very symmetric case might be if we wanted one 1, two 2s, three 3s, and so on. This is

```
rep(1:6,1:6)
[1]  1  2  2  3  3  3  4  4  4  4  5  5  5  5  5  6  6  6  6  6  6
```

but more generally we would like to be able to specify each repeat separately. We do that by using the concatenate function, **c**, to make a list of the individual repeats for each element of the first series. The list of numbers inside **c**(...) must match the length of the initial sequence:

```
rep(1:6,c(1,2,3,3,2,1))
[1]  1  2  2  3  3  3  4  4  4  5  5  6
```

Finally, we might want to customise both the sequence and the repeat. This is especially useful in generating repeated factor levels that are text:

```
rep(c("monoecious","dioecious","hermaphrodite","agamic"),c(3,2,7,3))
[1]  "monoecious"     "monoecious"     "monoecious"     "dioecious"      "dioecious"      "hermaphrodite"
[7]  "hermaphrodite"  "hermaphrodite"  "hermaphrodite"  "hermaphrodite"  "hermaphrodite"  "hermaphrodite"
[13] "agamic"         "agamic"         "agamic"
```

Allocation

In S-Plus **allocation** is done using 'gets' rather than 'equals'. 'Gets' is a composite arrow symbol <- made up from a less than symbol < and a minus symbol -. Note that allocation destroys existing objects with the same name. Thus, to make a vector called *x* that contains the numbers 1 through 10, we just type

```
x<- 1:10
```

which is read as saying '*x* gets the values 1 to 10 in series'. This overwrites any existing objects called *x*. Notice that this command does not generate any output; if you want to see the contents of the vector, you need to type x:

```
x
[1]  1  2  3  4  5  6  7  8  9  10
```

Variable names

- Variable names are case sensitive, so x is not the same as X

- Variable names should not begin with numbers (e.g. 1x) or symbols (e.g. %x)

- Variable names should not contain breaks: use back.pay (not back_pay or back pay)

The reason that you cannot use the underline symbol as a character in a name is that it is a single-character alternative for the composite 'gets' operator; in the example above, it would mean that variable called back gets the contents of the variable called pay.

Vectors

A vector is a variable with more than one value. A variable with just one value is called a *scalar*. The real power of the calculator comes to the fore in its abilities for vector calculations. Suppose that *y* contains the following 10 values:

> y<-c(7,5,7,2,4,6,1,6,2,3)

where **c** is the concatenation function. This is a long name for a simple idea; **c** combines objects from left to right into a vector (as here) or a list, tagging the next item onto the bottom of the list so far.

Here are some vector functions in action. Perhaps the most commonly used vector functions are **sum** and **mean**:

> sum(y)

> [1]   43

> mean(y)

> [1]   4.3

We can also find the extreme values in the vector, either separately using **max** and **min**:

> max(y)

> [1]   7

> min(y)

> [1] 1

or together, using **range,** which gives the minimum first, then the maximum:

> range(y)

> [1]   1   7

Other useful statistical vector functions are listed in Table 2.2.

Subscripts

The use of subscripts in S-Plus is superb. It is so simple, yet so powerful. There are basically two things you would want to do with subscripts:

- Select values from a vector according to some logical criterion (i.e. questions involving the **contents** of the vector), e.g. the value of the fourth element of *y* is 2

- Discover which elements of a vector (i.e. which subscripts) contain certain values (i.e. questions involving the **addresses** of items within the vector), e.g. the elements of *y* that contain 2s are 4 and 9

In working with vectors, it is important to understand the distinction between:

- the logical status of a number (T or F)
- the value of a number (0.64 or 2541)

Table 2.2   Vector functions in S-Plus

Operation	Meaning
max(x)	maximum value in x
min(x)	minimum value in x
sum(x)	total of all the values in x
mean(x)	arithmetic average of the values in x
median(x)	median value in x
range(x)	vector of min(x) and max(x)
var(x)	sample variance of x, with degrees of freedom $=$ length$(x) - 1$
cor(x,y)	correlation between vectors x and y
sort(x)	a sorted version of x
rank(x)	vector of the ranks of the values in x
order(x)	an integer vector containing the permutation to sort x into ascending order
quantile(x)	vector containing the minimum, lower quartile, median, upper quartile, and maximum of x
cumsum(x)	vector containing the sum of all the elements to that point
cumprod(x)	vector containing the product of all the elements to that point
cummax(x)	vector of non-decreasing numbers which are the cumulative maxima of the values in x to that point
cummin(x)	vector of non-increasing numbers which are the cumulative minima of the values in x to that point
pmax(x,y,z)	vector, of length equal to the longest of x, y or z containing the maximum of x, y or z for the ith position in each
pmin(x,y,z)	vector, of length equal to the longest of x, y or z containing the minimum of x, y or z for the ith position in each
colMeans(x)	column means of data frame x
colSums(x)	column totals of data frame x
colVars(x)	column variances of data frame x
rowMean(x)	row means of data frame x
rowSums(x)	row totals of data frame x
rowVars(x)	row variances of data frame x
peaks(x)	logical vector of length(x) with T for peak values bigger than both their neighbours

Take the example of a vector containing the 11 numbers 0 to 10:

```
x<-0:10
```

There are two quite different things we might want to do with this. We might want to *add up* the values of the elements:

```
sum(x)
```

```
[1]   55
```

Alternatively we might want to *count* the elements that passed some logical criterion. Suppose we wanted to know how many of the values were less than 5:

```
sum(x<5)
```

```
[1]   5
```

You see the distinction. We use the vector function **sum** in both cases. But sum(x) adds up the *values* of the element of *x* and sum(x<5) counts up the number of *cases* that pass the logical condition '*x* is less than 5'.

That is all well and good, but how do you add up the values of just some of the elements of *x*? We specify a logical condition, but we don't want to count the number of cases that pass the condition, we want to add up all the values of the cases that pass. This is the final piece of the jigsaw, and involves the use of *logical subscripts*. Note that when we counted the number of cases, the counting was applied to the entire vector, using sum(x<5). To find the sum of the elements of *x* which have a value less than 5, we write

```
sum(x[x<5])
```

```
[1]   10
```

This is *so* important that it is worth working through the logic that underpins it. The logical condition x<5 is either true or false:

```
x<5
```

```
[1]   T   T   T   T   T   F   F   F   F   F   F
```

You can imagine false as being numeric zero and true as being numeric one. Then the vector of subscripts [x<5] is five 1s followed by six 0s. *Multiplying logical values by numeric values produces numeric values*:

```
1*(x<5)
```

```
[1]   1   1   1   1   1   0   0   0   0   0   0
```

Now imagine multiplying the values of *x* by the values of this logical vector:

```
x*(x<5)
```

```
[1]   0   1   2   3   4   0   0   0   0   0   0
```

When the function **sum** is applied, it gives us the answer we want—the sum of the values of the numbers $0 + 1 + 2 + 3 + 4 = 10$.

```
sum(x*(x<5))
```

```
[1]   10
```

This gives the same answer as sum(x[x<5]) but is rather less elegant. It is essential that you understand the distinction here between round brackets (functions) and square brackets [subscripts].

Suppose we want to work out the sum of the three largest values in a vector. There are two steps: first sort the vector into descending order. Then add up the values of the first three elements of the sorted array. Let's do this in stages. First, the values of *y*:

```
y<-c(8,3,5,7,6,6,8,9,2,3,9,4,10,4,11)
```

Now if you apply **sort** to this, the numbers will be in ascending sequence, and this makes life a bit harder for the present problem:

```
sort(y)
```

```
[1]   2   3   3   4   4   5   6   6   7   8   8   9   9   10   11
```

We can use **rev**, the reverse function, like this (use the Up Arrow key to save typing):

```
rev(sort(y))
```

```
[1]   11   10   9   9   8   8   7   6   6   5   4   4   3   3   2
```

So the answer to our problem is $11 + 10 + 9 = 30$. But how to compute this? We can use specific subscripts to discover the contents of any element of a vector. We can see that 10 is the second element of the sorted array. To compute this we just specify the subscript [2]:

```
rev(sort(y))[2]
```

```
[1]  10
```

A range of subscripts is simply a series generated using the colon operator. We want the subscripts 1 to 3, so this is

```
rev(sort(y))[1:3]
```

```
[1]   11   10    9
```

And the answer to the exercise is just

```
sum(rev(sort(y))[1:3])
```

```
[1]   30
```

Note that we have not changed the vector y in any way, nor have we created any new space-consuming vectors during intermediate computational steps.

Ifelse

An important element of statistical computing involves doing one thing in one case, but a different thing in another case. This is called *conditional evaluation*. Suppose that you wanted to replace all the negative values in an array by zeros. In the old days, you might have written something like this:

```
for (i in 1:length(y))  if(y[i] < 0) y[i] <- 0
```

Now you would use logical subscripts like this:

```
y [y< 0] <- 0
```

Sometimes you want to do one thing if a condition is true and a different thing if the condition is false (rather than do nothing, as in the last example). The **ifelse** function allows you to do this for vectors without using **for** loops. We might want to replace any negative numbers by -1 and any positive values and zero by $+1$:

```
ypos <- ifelse (y < 0, -1, 1)
```

The next example avoids the warning message that arises from an attempt to take the log of a negative number, by replacing the problem entries (where the function would attempt to take the log of a negative number) by missing values NA:

```
log(ifelse(y>0, y, NA))
```

Trimming vectors using negative subscripts

Individual subscripts are referred to in square brackets. So if x is like this:

```
x<- c(5,8,6,7,1,5,3)
```

we can find the fourth element of the vector just by typing

x[4]

[1] 7

An extremely useful facility is the ability to use negative subscripts to drop terms from a vector. Suppose we wanted a new vector, *y*, to contain everything but the first element of *x*:

y <- x [-1]
y

[1] 8 6 7 1 5 3

This facility allows some extremely useful things to be done. Suppose we want to calculate a trimmed mean of *x* which ignores both the smallest and largest values (i.e. we want to leave out the 1 and the 8). There are two steps to this. First we **sort** the vector *x*. Then we remove the first element using x[-1] and the last using x[-length(x)]. We can do both drops at the same time by concatenating both instructions like this: -c(1,length(x)). Then we use the built-in function **mean**:

trim.mean <- function (x)   mean(sort(x) [- c(1,length(x))])

Now try it out. The answer should be the mean of $\{5, 6, 7, 5, 3\} = 26/5 = 5.2$:

trim.mean(x)

[1] 5.2

Addresses within vectors

There are two important functions for finding addresses with arrays. The function **which** is very easy to understand. Suppose we want to know which elements of *y* contained values bigger than 5, we type

which(y>5)

[1]   1   4   5   6   7   8   11   13   15

Notice that the answer to this enquiry is *a set of subscripts*. We *don't* use subscripts in the **which** function itself. The function is applied to the whole array. To see the *values of y* that are larger than 5, we just type

y[y>5]

[1]   8   7   6   6   8   9   9   10   11

If we want to know whether or not each element of *y* is bigger than 5, we just type

y>5

[1]   T   F   F   T   T   T   T   T   F   F   T   F   T   F   T

and the answer is a logical vector of Ts (true) and Fs (false). It is important that you work through these examples until you understand the distinction between these three operations:

● which subscripts?

● which values?

● which elements?

Sorting, ranking and ordering

Sorting, ranking and ordering are related concepts; all three but important and ordering is difficult to understand on first acquaintance. Let's take a simple example:

```
data<-c(7,4,6,8,9,1,0,3,2,5,0)
```

Now we apply the three different functions to the vector called data:

```
ranks<-rank(data)
sorted<-sort(data)
ordered<-order(data)
```

We make a data frame out of the four vectors like this:

```
view<-data.frame(data,ranks,sorted,ordered)
view
```

	data	ranks	sorted	ordered
1	7	9.0	0	7
2	4	6.0	0	11
3	6	8.0	1	6
4	8	10.0	2	9
5	9	11.0	3	8
6	1	3.0	4	2
7	0	1.5	5	10
8	3	5.0	6	3
9	2	4.0	7	1
10	5	7.0	8	4
11	0	1.5	9	5

Rank

The data themselves are in no particular sequence. Ranks contains the value that is the rank, out of length (data), of the particular data point. So the first element data = 7 is the ninth highest value in data. You should check that there are eight values smaller than 7 in the vector called data. Fractional ranks indicate ties. There are two zeros in data and their ranks are 1 and 2. Because they are tied, each gets the average of the sum of their ranks, $(1 + 2)/2 = 1.5$.

Sort

The sorted vector is very straightforward. It contains the values of data sorted into ascending order. If you want to sort into descending order, use the reverse order function **rev** like this: y<-rev(sort(x)). Note that *sort is potentially very dangerous*, because it uncouples values that might need to be in the same row of the data frame (e.g. because they are the explanatory variables associated with a particular value of the response variable).

Order

This returns an integer vector containing the permutation that will sort the input into ascending order. You will need to think about this one. Look at the data frame and ask yourself, What is the subscript in the original vector called data of the value in the first element of the sorted vector? The first zero is in location 7 and this is order[1]. Where is the second value in the sorted vector found within the original data? It is in position 11, so this is order[2]. The only 1 is found in position 6 within data, so this is order[3]. And so on.

This function is particularly useful in conjunction with subscripting for sorting several parallel arrays. Suppose we want to sort the whole of the data frame called view on the basis of the calculated ranks, and store the result in a new data frame called sortview. Note that the *square brackets signify subscripts*, i.e. particular rows within view:

```
sortview<-view[order(view[,2]),1:4]
sortview
```

	data	ranks	sorted	ordered
7	0	1.5	5	10
11	0	1.5	9	5
6	1	3.0	4	2
9	2	4.0	7	1
8	3	5.0	6	3
2	4	6.0	0	11
10	5	7.0	8	4
3	6	8.0	1	6
1	7	9.0	0	7
4	8	10.0	2	9
5	9	11.0	3	8

All the rows remain intact, but their sequence has been changed. In general, this is a much safer option than using **sort**, because with **sort** the values of the response variable and the explanatory variables can be uncoupled with potentially disastrous results if this is not realised at the time that modelling is carried out. Note the use of 1:4 to apply this ordering to all four columns within view. Here is a more familiar example:

```
attach(houses)
houses
```

	Location	Price
1	Ascot	325
2	Sunninghill	201
3	Bracknell	157
4	Camberley	162
5	Bagshot	164
6	Staines	101
7	Windsor	211
8	Maidenhead	188
9	Reading	95
10	Winkfield	117
11	Warfield	188
12	Newbury	121

To see how this works, let's inspect the order of the house prices:

```
order(Price)
```

[1] 9 6 10 12 3 4 5 8 11 2 7 1

This says that the lowest price is in subscript 9 (where Price = 95 and Location = Reading). Next cheapest is in location 6 (Price = 101) where Location = Staines, and so on. The beauty of it is that we can use order (Price) as a subscript for Location to obtain the price-ranked list of locations:

Location[order(Price)]

```
[1]   Reading       Staines       Winkfield    Newbury
[5]   Bracknell     Camberley     Bagshot      Maidenhead
[9]   Warfield      Sunninghill   Windsor      Ascot
```

When you see it used like this, you can see exactly why the function is called 'order'. If you want to reverse the order, just use the **rev** function like this:

Location[rev(order(Price))]

```
[1]   Ascot         Windsor      Sunninghill   Warfield
[5]   Maidenhead    Bagshot      Camberley     Bracknell
[9]   Newbury       Winkfield    Staines       Reading
```

Sample

This function *shuffles the contents of a vector* into a random sequence while maintaining all the numerical values intact. It is extremely useful in simulation and Monte Carlo techniques of computationally intensive hypothesis testing. Here is the original *y* again:

y

```
[1]   8   3   5   7   6   6   8   9   2   3   9   4   10   4   11
```

and here are two samples of y:

sample(y)

```
[1]   8   8   9   9   2   10   6   7   3   11   5   4   6   3   4
```

sample(y)

```
[1]   9   3   9   8   8   6   5   11   4   6   4   7   3   2   10
```

The option replace=T allows for sampling *with replacement*, which is the basis of bootstrapping (Chapter 12). The vector produced by the **sample** function is the same length as the vector sampled, but some values are left out at random and other values, again at random, appear two or more times. The values selected will be different each time that **sample** is invoked. In this sample, 10 has been left out and there are now three 9s.

sample(y,replace=T)

```
[1]   9   6   11   2   9   4   6   8   8   4   4   4   3   9   3
```

In this sample there are two 10s and only one 9

sample(y,replace=T)

```
[1]   3   7   10   6   8   2   5   11   4   6   3   9   10   7   4
```

More advanced options in **sample** include taking fewer than (length(x)) numbers (size=), and specifying different probabilities with which each element of *x* is to be sampled (prob=). For example, if we want to take four numbers at random from the sequence 1:10 where the probability of selection is greater for the middle numbers ($p = 1, 2, 3, 4, 5, 5, 4, 3, 2, 1$) and we want to do this five times, we could write

```
p <- c(1, 2, 3, 4, 5, 5, 4, 3, 2, 1)
x<-1:10
for (i in 1: 5){
v<-sample(x, 4, prob=p)
cat(v,"\n")}
```

```
6    4    1    5
4    3    9    2
9    5    7    1
3    5    7    6
5    8    6    4
```

Logical arithmetic

A really useful computing skill involves the use of logical statements in arithmetic calculations. This takes advantage of the fact that logical 'true' evaluates to 1.0 and logical 'false' evaluates to 0.0. When we make a command that involves only logic, we get back a logical statement, T or F

Note that 'equals' is handled rather oddly in logical statements. We need to use = = (double equals) to mean 'exactly equal to'. Suppose we want to add 1 to the value of y if $y = 3$, but otherwise leave y unaltered:

```
y
```

```
[1]   8   3   5   7   6   6   8   9   2   3   9   4   10   4   11
```

```
y+(y==3)
```

```
[1]   8   4   5   7   6   6   8   9   2   4   9   4   10   4   11
```

It is often important to make one or more shorter versions of an array, and logical subscripting is a superb way of doing this. Let's make an array *ys* containing the small values of *y*, and *yb* containing the big values:

```
ys <- y[y<5]
ys
```

```
[1]   3   2   3   4   4
```

```
yb <- y[y>=5]
yb
```

```
[1]   8   5   7   6   6   8   9   9   10   11
```

We can **count** the number of elements in an array that have various (potentially quite complicated) properties by using the **sum** function applied to some logical condition: the true conditions evaluate to 1, and **sum** counts these simply by adding them up. For example, we can count the number of extreme values in y, defining extreme values as those more than 2 above the mean, and those less than 2 below the mean (there are nine of them in this case):

```
sum(y> mean(y)+2 | y < mean(y)-2)
```

```
[1]   9
```

Table 2.3   Logical and arithmetic operations in S-Plus

Symbol	Meaning
!=	not equal
%%	the remainder of a division, modulo
%/%	the integer part
*	multiplication
+	addition
-	subtraction
^	to the power
/	division
<	less than
<=	less than or equal to
>	greater than
>=	greater than or equal to
==	logical equals (double =)

Note the use of | (the OR operator) to separate the two parts of the expression. If we want the total of the y values that are extreme in one direction or the other, we type

```
sum(y[y> mean(y)+2 | y < mean(y)-2])
```

```
[1]    55
```

Do you see the distinction between these two operations? The first one evaluates as TRUE or FALSE whereas the second evaluates as numerical values of y. See Table 2.3.

Elements of two vectors

You are often interested to find the elements that two vectors have in common. On other occasions you might want to know which elements are in one vector but *not* in another. S-Plus has an excellent set of built-in functions to perform these tasks.

Suppose that our two vectors are 3,5,5,6,8,9 and 1,3,3,3,6,7,8,8,9

```
list1<-c(3,5,5,6,8,9)
list2<-c(1,3,3,3,6,7,8,8,9)
```

What about the numbers that the two lists have in common? This is found by **intersect**:

```
intersect(list1,list2)
```

```
[1]   3   6   8   9
```

Note that **intersect** returns *unique* values. To find the elements that occur in one vector or the other, you use **union**:

```
union(list1,list2)
```

```
[1]   3   5   6   8   9   1   7
```

Notice that the numbers are not sorted, and again, the union does not include repeats. To find out the values in one list that are *not* in another list you use **setdiff**:

```
setdiff(list1,list2)
```

```
[1]   5   5
```

This gives you all the elements (including repeats) that are in list1 but not in list2. Note also that this function is *order dependent*:

setdiff(list2,list1)

[1] 1 7

If you want to select all of the elements that are duplicates, you need to use the logical vector produced by the function **duplicated**; this takes the value T (true) for the second and subsequent appearances of an element in a vector. We use this as a subscript (square brackets) to list the duplicates that appear in list2 (i.e. for which the subscript duplicated(list2) is TRUE):

list2[duplicated(list2)]

[1] 3 3 8

It might help to understand how this works if we list the contents of the two vectors side by side:

duplicated(list2)
list2

[1] F F T T F F F T F
[1] 1 3 3 3 6 7 8 8 9

We can make one list out of two lists using the concatenation function **c**:

list3<-c(list1,list2)

We might want to know the **unique** values in the combined list (i.e. ignoring duplicates):

unique(list3)

[1] 3 5 6 8 9 1 7

Note that the numbers are not sorted; they appear in the order that they appear in the concatenated list. To get the ordered unique values, we would write

sort(unique(list3))

[1] 1 3 5 6 7 8 9

Loops

Although you can often avoid the use of explicit loops in S-Plus, by using subscripts and vector functions to the full, there are occasions when you want to repeat an operation. The controls for looping and branching in S-Plus expressions are as follows:

- for (name in values) expression
- while (test) expression
- repeat expression
- break
- next

The **for** loop works like this. You specify the name of an index for the loop, and say what values you want the index to take. Suppose you want to loop round with the index i taking the values 1 to 5 and print the cube of i:

```
for (i in 1:5) print (i^3)
```

```
[1]    1
[1]    8
[1]    27
[1]    64
[1]    125
```

When there are several expressions to be executed within the loop, these run over several lines, and are enclosed within curly brackets { } like this:

```
ss<-0; total<-0

for (x in c(12,44,31,2,19)){
total<-total+x
ss<-ss+x^2
}
```

This is a case where the loop could easily be avoided using vector arithmetic:

```
x<-c(12,44,31,2,19)
total<-sum(x)
ss<-sum(x^2)
```

Loops controlled by **while** are equally straightforward. The key point is that *the logical variable controlling the while operation must be recalculated inside the loop*:

```
x<-0
test<-1
while (test>0){
x<-x+1
test<-(x<6)
print (x^2)}
```

```
[1]    1
[1]    4
[1]    9
[1]    16
[1]    25
[1]    36
```

Can you see why this printed 6^2 but not 7^2?

If

In loops we typically use **if** whereas for operations on entire vectors we use **ifelse** (see p. 22). We can use **if** on its own, or in a pair with **else**. Control of loops often involves conditions of one sort or another. We can use **if** to control the three important loop controls: **next, break** and **return**. Suppose we are in a loop controlled by the index i, then

```
for (i in 1:10){
     ...
     if (test1) next
```

goes immediately to the next iteration of the loop, if **test1** is TRUE, and

> if (test2) break

exits the for loop if **test2** is TRUE, and

> if (test3) return(x)

if **test3** is TRUE, exits the program in which the loop is running (see below), returning the value x to the point at which the function was invoked.

Simple functions

We begin with a very simple, but highly practical function to compute the standard error of a mean when provided with a vector of numbers, x. The first thing to decide upon is a name for the function: **se** would be ideal in this case (it is memorable and quick to type). We need to check that it hasn't been used before, either by us (and we have forgotten) or by the machine: just type se then the enter key:

> se

```
Error:  Object  "se"  not found
```

So that's okay then. The syntax is that we write 'function-name gets function of x' like this:

> se <- function (x)

and then we define the function (i.e. provide the computer code for what we want the function to do). If the function extends over more than one line, we enclose the relevant rows in curly brackets {} immediately after the brackets () enclosing the function's parameters (x). Now we need to think about how to write the code. The formula for the standard error of a mean (p. 97) is

$$\text{standard error} = \sqrt{\frac{s^2}{n}}$$

where s^2 is the variance and n is the sample size. We are provided only with a vector of numbers, x, so we shall have to calculate both of the quantities we need. Variance is easy because there is a built-in function **var** to work it out, so $s^2 = \text{var}(x)$. Sample size is slightly more tricky, because we don't know in advance how many numbers are going to be in the vector. This does not matter because we can use the built-in function **length** to work how many numbers there are, so $n = \text{length}(x)$. That is all we need to know before writing the function in full:

> se<-function(x) sqrt(var(x)/length(x))

To see the contents of a function, just type its name:

> se

```
function(x)
sqrt(var(x)/length(x))
```

That looks fine, but we should test that it works properly by providing it with some data and making sure that it gives the right answer. Inside the function, the vector is called x, but when we use the function it should work with a vector with any name. Let's create a vector of data and call it y, then work out $se(y)$:

```
y<-c(5,8,3,5,7,2,6,6,4)
se(y)
```

```
[1]   0.6334308
```

The length of this vector was $n = 9$ and its variance was 3.6111 (hand calculator) so the function seems to work as we intended. Testing and debugging functions by trying them out with strange numbers and empty vectors is a vital part of the process of function writing, but we gloss over this issue here.

A function for printing measures of central tendency

This function computes the arithmetic, geometric and harmonic means of a vector. The harmonic and geometric means are home-made functions as follows (Chapter 4):

```
harmonic <- function(x) 1/mean(1/x)
geometric <- function(x) exp(sum(log(x))/length(x))
```

The function called **central** is very straightforward except for the output printing, using **cat** (described later):

```
central<-function(x){
    c1<-mean(x)
    c2<-geometric(x)
    c3<-harmonic(x)
    c4<-median(x)
    c5<-range(x)

    cat("Arithmetic mean = ",round(c1,3),"\n")
    cat("Geometric mean = ",round(c2,3),"\n")
    cat("Harmonic mean = ",round(c3,3),"\n")
    cat("Median     = ",round(c4,3),"\n")
    cat("Range     from", round(c5[1],3), "to", round(c5[2],3), "\n")
}
```

The function is used like this:

```
w<-c(12,11,15,14,10,9,13,12)
central(w)
```

and produces output like this:

```
Arithmetic mean    =    12
Geometric mean     =    11.851
Harmonic mean      =    11.701
Median             =    12
Range            from   9   to 15
```

If you want to print more than one thing from a function, you should use **cat** because this allows you to specify the format of numbers (using **round** to specify the number of decimal places printed), the creation of new lines (with "\n"), and to interpose text with the output numbers.

Functions with several arguments

A general function to calculate weighted average (see p. 107) would require two vectors as arguments: the response variable x and the weights w. It might look like this:

```
weighted.mean<-function(x,w) sum(x*w)/sum(w)
```

Functions involving loops

This function uses **while** to generate the Fibonacci series 1, 1, 2, 3, 5, 8, etc., in which each term is the sum of its two predecessors. The key point about while loops is that the logical variable controlling their operation *is altered inside the loop*. In this example we alter n, the number whose Fibonacci number we want, starting at n, reducing the value of n by 1 each time around the loop, and ending when n gets down to 0. Here is the code:

```
fibonacci<-function(n){
    a<-1
    b<-0
    while(n>0)
        {swap<-a
        a<-a+b
        b<-swap
        n<-n-1}
    b  }
```

An important general computing point involves the use of swap. When we replace a by a+b on line 6 we lose the original value of a. If we had not stored this value in swap, we could not set the new value of b to the old value of a on line 7.

Now test the function by generating the Fibonacci numbers 1 to 10:

```
for (i in 1:10) print(fibonacci(i))

[1]   1   1   2   3   5   8   13   21   34   55
```

Multiple cases: switch and if else

Many computer programs have a *case* function, but S-Plus does not. Instead, it uses one of two powerful alternatives: the **if, else if** construction or the **switch** function.

We want to write a general function to calculate central tendency of data in a vector, y, with the option of specifying the arithmetic average, geometric mean, harmonic mean or median as our measure of central tendency. Here are two ways to write it. First, with **if else**:

```
central<-function(y, measure){
    if (measure=="Mean") mean(y) else
    if (measure=="Geometric") exp(mean(log(y))) else
    if (measure=="Harmonic") 1/mean(1/y) else
    if (measure=="Median") median(y) else
        stop("Measure not included")}
```

We can try it out using the vector

```
x<- c(5,8,6,7,1,5,3)
```

```
central(x,"Mean")
```

```
[1]   5
```

```
central(x,"Geometric")
```

```
[1]   4.253746
```

```
central(x,"Harmonic")
```

```
[1]   3.228995
```

```
central(x,"Median")
```

```
[1]   5
```

The arguments to the function are case sensitive, so this fails:

```
central(x,"median")
```

```
Error in central(x, "median"): Measure not included Dumped
```

Here is the same thing written using **switch**:

```
central<-function(y, measure){
     switch(measure,
     Mean = mean(y),
     Geometric = exp(mean(log(y))),
     Harmonic = 1/mean(1/y),
     Median = median(y),
       stop("Measure not included") )}
```

I think this is more elegant and easier to follow than the **if else** construction, but you may have different tastes. Both work equally well. Note that you have to include the character strings in quotes inside the function in the **if else** format, but they must *not* be in quotes in **switch**. They must always be in quotes in the call to the function, otherwise an 'object not found' error will occur.

Flexible handling of arguments to functions

Because of the *lazy evaluation* practised by S-Plus, it is very simple to deal with missing arguments in function calls, giving the user the opportunity to specify the absolute minimum number of arguments, but to override the default arguments if they want to do so.

As a simple example, take a function **plotx2** that we want to work when provided with either one or two-arguments. In the one-argument case, we want it to plot z^2 against z for $z = 1$ to x. In the case when y is supplied, we want it to plot y against z for $z = 1$ to x.

Here is the preferred form:

```
plotx2 <- function (x, y=z^2){
     z<-1:x
     plot(z,y,type="l")}
```

In other languages, the first line would fail because z is not defined at this point. S-Plus does not evaluate an expression until the body of the function actually calls for it to be evaluated (i.e. never, in the case where y is supplied as a second argument).

Below we see the less efficient version. This requires z^2 to be evaluated, even when y is supplied as the second argument and the calculation need never be carried out:

```
plotx2 <- function (x, y){
     z<-1:x
```

```
                    if (missing(y)) plot(z, z^2,type="l")
                    else plot(z,y,type="l")}
```

Variable numbers of arguments

Some applications are much more straightforward if the number of arguments need not be specified in advance. There is a special formal name **...** (dot dot dot) which is used in the argument list to specify that an arbitrary number of arguments are to be passed to the function within the body of the function.

Here is a function that takes any number of vectors and calculates their means and variances:

```
        many.means <- function (...){
            data <- list(...)
            n<- length(data)
            means <- numeric(n)
            vars <- numeric(n)
            for (i in 1:n){
                    means[i]<-mean(data[[i]])
                    vars[i]<-var(data[[i]])
            }
            print(means)
            print(vars)
            invisible(NULL)
        }
```

Note that the function definition has ... as its only argument. The first thing done inside the function is to create an object called data out of the list of vectors that are actually supplied in any particular case. The length of this list is the number of vectors, not the lengths of the vectors themselves (these could differ from one vector to another; see the example below). Then the two output variables (means and vars) are defined to have as many elements as there are vectors in the parameter list. The loop goes from 1 to the number of vectors, and for each vector uses the built-in functions **mean** and **var** to compute the answers we require. It is important to note that because data is a *list*, we use double subscripts [[ ]] in addressing its elements.

Now try it out. To make things difficult, we shall give it three vectors of different lengths. All come from the standard Normal distribution (mean = 0, variance = 1) but x is 100 in length, y is 200 and z is 300 numbers long:

```
        x<-rnorm(100)
        y<-rnorm(200)
        z<-rnorm(300)
```

Now we invoke the function:

```
        many.means(x,y,z)
        [1]   -0.039181830    0.003613744    0.050997841
        [1]    1.146587       0.989700       0.999505
```

As expected, all three means are close to zero and all three variances are close to 1.

Loop avoidance

You often want to apply the same function (e.g. **mean** or **sort**) repeatedly to a series of vectors. Suppose we wanted to test whether the central limit theorem worked for 100 replicated samples of size 4 drawn from a uniform distribution using runif(4). The temptation might be to do something like this:

```
average<-numeric(100)
for(i in 1:100) average[i]<-mean(runif(4))
par(mfrow=c(1,2))
hist(average)
qqnorm(average)
qqline(average)
```

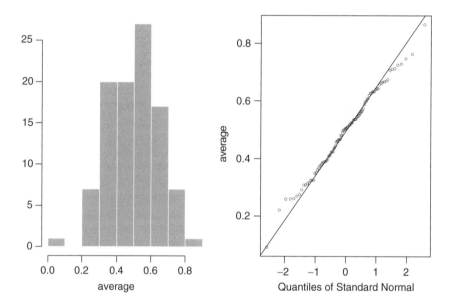

This shows that averaging as few as four uniformly distributed random numbers gives a distribution of means that is remarkably close to Normal. It is much more efficient, however, to create a matrix of the raw data, then use **apply** to repeatedly apply the function to either the rows (1) or the columns (2) of the entire matrix. First work out how many random numbers are required in total ($100 \times 4 = 400$). Now create a two-dimensional matrix containing 400 random numbers from a uniform distribution in 100 rows of 4 columns, like this:

```
numbers<-matrix(runif(400),100)
```

then use **apply** across the rows (second argument $= 1$ means rows; second argument $= 2$ means columns) to get the 100 averages directly:

```
averages<-apply(numbers,1,mean)
```

In a different application we might want 8 sets of 10 sorted random Normal numbers with mean $= 12$ and standard deviation $= 2$. First work out how many numbers you need in total ($8 \times 10 = 80$). Now generate all 80 numbers in a matrix called vector with 8 rows:

```
vector<-matrix(rnorm(80,12,2),8)
```

Now use **apply** across the rows (second argument $= 1$) to apply **sort** to each separate column of the data:

apply(vector,1,sort)

```
         [,1]        [,2]        [,3]        [,4]        [,5]        [,6]         [7]      [,8]
 [1,] 11.18062    8.381093    8.814371   10.16823    7.873338    9.368173    7.655775 10.55234
 [2,] 11.86850    9.655026    9.282948   10.64336    7.892675   10.534404    9.374074 11.21825
 [3,] 12.41992   10.337677    9.409168   11.12605   10.165011   11.639672   10.747172 12.10577
 [4,] 12.65146   11.227749    9.923746   11.54249   11.411768   11.893076   12.965538 12.16883
 [5,] 12.75655   11.328600   10.840820   12.22796   11.903638   12.337142   13.049084 12.36900
 [6,] 13.01849   12.678201   12.160369   12.37488   13.281493   12.400143   13.099213 12.85560
 [7,] 13.19379   12.682163   12.332475   12.67797   13.407532   12.550686   13.660919 13.25083
 [8,] 13.55015   13.240490   13.486250   12.82203   13.566356   13.123813   13.744214 13.61454
 [9,] 14.74835   13.658402   13.547243   13.47489   14.388554   13.385318   14.344051 13.69659
[10,] 15.31573   14.054028   13.982437   14.01373   15.649079   15.640584   15.608319 13.75310
```

The matrix has been sorted on the basis of its row values, separately for each column.

Character strings

Character strings are objects contained within double quotes. A categorical variable called sex, containing the names of the two sexes, could be made like this:

sex <- c ("male", "female")

which is a concatenation **c** of two character strings. There are four things you might want to do with character strings:

- find a specific subset of the characters they contain
- abbreviate them
- form composite character strings by joining them together
- check to see if they contain particular character sequences

Substrings

The function **substring** returns a vector of character strings that are substrings of the input string. Let's create a character string containing all the letters of the alphabet:

string<-"the quick brown fox jumps over the lazy dog"

Now we can extract different kinds of substrings. The first argument is essential and is the address of the first character of the substring. Suppose we want to extract "the quick brown" from string. The first character we want is 1 and the last is 15:

substring(string,1,15)

[1] "the quick brown"

To extract "lazy dog" we start at the 36th character and extract up to the 43rd character:

substring(string,36,43)

[1]   "lazy dog"

If **substring** is provided with just a single number, it extracts from this character up to the end of the character string:

```
substring(string,17)

[1]  "fox jumps over the lazy dog"
```

Vectors of numbers supplied for the second and third arguments produce vectors of substrings like this (notice that blanks are trimmed):

```
substring(string,1:18,35:18)
```

```
[1] "the quick brown fox jumps over the"  "he quick brown fox jumps over the"
[3] "e quick brown fox jumps over th"      "quick brown fox jumps over t"
[5] "quick brown fox jumps over"           "uick brown fox jumps over"
[7] "ick brown fox jumps ove"              "ck brown fox jumps ov"
[9] "k brown fox jumps o"                  "brown fox jumps "
[11] "brown fox jumps"                     "rown fox jump"
[13] "own fox jum"                         "wn fox ju"
[15] "n fox j"                             "fox"
[17] "fox"                                 "o"
```

Abbreviation

Sometimes it is useful to abbreviate character strings. The syntax is straightforward. By default, the function **abbreviate** takes the first four characters of the string.

```
abbreviate("alphabet")

"alph"
```

Matching characters

Character matching uses the functions **charmatch, pmatch, match** and **grep**. The function **charmatch** is useful for processing the arguments to functions. It is very similar to **pmatch** but **pmatch** does not allow a distinction between no match and an ambiguous match, nor does it allow a match to the empty string. The function **charmatch** returns a vector of the indices of target (the second argument) that are partially matched by input (the first argument):

```
charmatch("med", c("mean", "median", "mode"))
```

This looks for the string "med" in each of the elements of the target list c("mean", "median", "mode"), finds it only in element 2, and so returns the index 2. It returns 0 if there is more than one match:

```
charmatch("m", c("mean", "median", "mode"))

[1]  0
```

You can say what to do if there is no match in the **nomatch** option:

```
charmatch("my", c("mean", "median", "mode"), nomatch = -1)

[1]  -1
```

The first argument can itself be a list, in which case the output is a vector, so

```
charmatch(c("sin", "cot"), c("cos", "sin", "tan"), nomatch = -1)
```

produces

```
[1]   2  -1
```

The function **match**, used as match(x, table), produces an integer vector of the same length as *x* giving, for each element of *x*, the smallest *i* such that table[i] equals that element. If no value in table is equal to x[j], then the *j*th element of the result is nomatch. You can specify what to do with a nomatch in the nomatch option (e.g. nomatch = 0). It is used to find the location of particular values within vectors. Suppose we want to know where age = 1 first appears in the vector called age in the Parasite data frame:

```
attach(Parasite)
names(Parasite)

[1]  "infection"   "age"        "weight"    "sex"

match(1,age)

[1]  29
```

It is in position 29. We can check whether a value appears anywhere within the vector, and do something special if it doesn't:

```
match(235,age,nomatch=-9)

[1]  -9
```

You can use **match** as a logical subscript to select certain values of a vector: Suppose we want to find the sex of the first individual that was age = 1; i.e. sex[29] as we now know:

```
sex[match(1,age)]

[1]  male
Levels:
[1]  "female"  "male"
```

If we want to suppress the extra, unasked-for information on the set of factor levels from which male was chosen, we use **as.vector**:

```
as.vector(sex[match(1,age)])

[1]   "male"
```

It is important to understand the difference between this **match** function and the familiar use of logical subscripts:

```
as.vector(sex[age==1])

[1]   "male"   "male"   "male"   "male"   "male"
```

This gives us a vector of the sexes of *all* of the elements of the vector called age that were 1s. Function **match** gives us the location (the index) within the array called sex of the *first* instance of age = 1.

Searching for patterns in text

For logical subscripts to select subsets of data on the basis of information contained within names, we use the function **grep**. This searches for a text pattern in a vector of character strings, grep(pattern, text), where pattern is a character string that may include any of the following wildcards:

* matches any *sequence* of zero or more characters, irrespective of the previous character

? matches any *single* character

[ ] encloses a set of values and matches any character within the set. Any continuous subset of ([A-Z]) is allowed

! used within [ ] matches any character except those specified in the set (the exclamation mark ! is the 'not' symbol in S-Plus)

The result of **grep** is a numeric vector indicating the elements of text that matched the pattern. It returns numeric(0) if there are no matches. In all cases the return value can be used as a subscript to retrieve the matching elements of text.

To see this working at its best, we need a data frame with lots of text variables in it:

```
attach(worldfloras)
names(worldfloras)

[1]   "Country"   "Latitude"   "Area"   "Population"   "Flora"
      "Endemism"  "Continent"
```

Now we can use **grep** to select subsets of the values within variables, using logical subscripts [ ]. Character vectors are assumed by default to be **factors**, so if we want to work with them as character strings (as here), we need to prefix their names by **as.character**. For example, we can get all the countries ending in '*ia*' using the wildcard * like this:

```
as.vector(Country[grep("*ia",as.character(Country))])

[1]   "Albania"       "Algeria"          "Australia"
[4]   "Austria"       "Bolivia"          "Bulgaria"
[7]   "Colombia"      "Czechoslovakia"   "Ethiopia"
[10]  "Gambia"        "India"            "Indonesia"
[13]  "Liberia"       "Malaysia"         "Mauritania"
[16]  "Mongolia"      "Namibia"          "New Caledonia"
[19]  "Nigeria"       "Romania"          "Sardinia"
[22]  "Saudi Arabia"  "Somalia"          "Syria"
[25]  "Tanzania"      "Tunisia"          "Yugoslavia"
[28]  "Zambia"
```

or all the countries that consist of two or more words. Look for a blank with text on either side of it, using the string "* *" (two wildcards separated by a blank space):

```
as.vector(Country[grep("* *",as.character(Country))])
```

```
[1]    "Balearic Islands"    "Burkina Faso"        "Central African Republic"
[4]    "Costa Rica"          "Dominican Republic"  "El Salvador"
[7]    "French Guiana"       "Germany, East"       "Germany, West"
[10]   "Hong Kong"           "Ivory Coast"         "New Caledonia"
[13]   "New Zealand"         "Papua New Ginea"     "Puerto Rico"
[16]   "Saudi Arabia"        "Sierra Leone"        "Solomon Islands"
[19]   "South Africa"        "Sri Lanka"           "St Helena"
[22]   "Trinidad & Tobago"   "Tristan da Chuna"    "United Kingdom"
[25]   "Viet Nam"            "Yemen, North"        "Yemen, South"
```

This raises an important general point. You will notice that United States of America does not appear in this list. That is because its name is USA in this data frame. It is up to you to ensure the condition you have set really *does* capture all the elements that you want it to.

You can have endless fun with **grep**. For example, what condition would get you Burkina Faso and Canada into the same list?

For contiguous groups of characters, use the [A-D] convention. Countries beginning with letters in the range Q to R are obtained like this:

as.vector(Country[grep("[Q-R]*",as.character(Country))])

```
[1]    "Qatar"        "Reunion"        "Romania"        "Rwanda"
```

Conditions concerning a particular *position* can be handled with ?. Suppose we want all the countries with *y* as their second letter; this is ?y*:

as.vector(Country[grep("?y*",as.character(Country))])

```
[1]    "Cyprus"    "Syria"
```

Exclusion is handled with the ! operator. Countries ending other than with an *a* are obtained like this:

as.vector(Country[grep("*[!a]",as.character(Country))])

```
[1]    "Afghanistan"    "Bahrain"                     "BalearicIslands"
[4]    "Bangladesh"     "Belgium"                     "Belize"
[7]    "Benin"          "Bhutan"                      "Brazil"
[10]   "Brunei"         "Burkina Faso"                "Burundi"
[13]   "Cameroon"       "Central African Republic"    "Chad"
......
[91]
 "Turkey"             "United Kingdom"        "Uraguay"
[94]   "USA"             "USSR"                    "Vanuatu"
[97]   "Viet Nam"        "Yemen, North"            "Yemen, South"
[100]  "Zaire"           "Zimbabwe"
```

Care needs to be taken with special characters like / and .

Patterns matching a regular expression in character strings

The function **regexpr** uses the syntax [a-zA-Z] to match any letter in any case, and a plus sign means to match *one or more* of the preceding character or pattern, so that [0-9]+ matches any integer, so long as there is not a break or a comma in it. You can do quite complex tasks with these tools. Suppose that you wanted to extract the day numbers 10, 9, 2 and 4 from the following vector of dates in mixed formats:

```
dates<- c("10 Aug", "Oct 9th", "Jan 2", "4th of July")
```

There are two steps to this. First, find out where within each string the numbers are:

```
locations<-regexpr("[0-9]+", dates)
```

which produces the following two vectors on typing locations. The first vector gives the locations of the start character of the numbers within the string, and the second gives the length (in characters) of each number; only the first number (10) is longer than one digit:

```
locations

[1]   1   5   5   1
attr(,   "match.length"):
[1]   2   1   1   1
```

Now for the cunning bit:

```
as.numeric(substring(dates, locations, locations+attr(locations,"match.length")-1))

[1]    10   9   2   4
```

This has disentangled the four numbers from the jumbled mix of text and numbers. The substrings within each date start at locations and end at locations + attr(locations, "match.length")-1 (i.e. 1 to 2, 5 to 5, 5 to 5 and 1 to 1 for the four elements of dates). To extract the names of the months, the functions would be

```
months<- regexpr("[A-Z][a-z]*", dates)
months

[1]   4   1   1   8
attr(, "match.length"):
[1]   3   3   3   4

substring(dates, months, months+attr(months, "match.length")-1)

[1]   "Aug"      "Oct"           "Jan"       "July"
```

It is worth practising with these conventions because they are very useful in programming.

Paste

The **paste** function is good for making sets of variable (or model) names. Suppose we wanted to generate the names X1 to X10 in a vector called nameX. The long way would be like this:

```
nameX<-c("X1","X2","X3","X4","X5","X6","X7","X8","X9","X10")
```

The compact way is like this:

```
nameX<-paste("X",1:10,sep="")
nameX

[1]   "X1"   "X2"   "X3"   "X4"   "X5"   "X6"   "X7"   "X8"   "X9"   "X10"
```

Note the use of the separator "", which means 'leave nothing at all between the X and the number'. You can use this to insert any repeated pattern at all.

The **collapse** function allows you to make a single character string out of multiple strings and to specify the character to separate each element of the collapsed list. Suppose we wanted to make a model formula for a multiple regression using our 10 new variable names X1 to X10, and we wanted to fit them only as main effects, separated by + signs. This is what we do:

```
formula<-paste("X",1:10,sep="",collapse="+")
formula
[1]  "X1 + X2 + X3 + X4 + X5 + X6 + X7 + X8 + X9 + X10"
```

Matrices

Matrix arithmetic is handled in an intuitively obvious way. A matrix is defined with the **matrix** function. Suppose we want to turn a new vector y of length 15,

```
y<-c(8, 3, 5, 7, 6, 6, 8, 9, 2, 3, 9, 4, 10, 4, 11)
```

into a matrix called m consisting of 5 rows and 3 columns:

```
m<-matrix(y,nrow=5)
m
        [,1]    [,2]    [,3]
[1,]      8       6       9
[2,]      3       8       4
[3,]      5       9      10
[4,]      7       2       4
[5,]      6       3      11
```

Another way to make matrices is to bind vectors together into matrices, either row by row using **rbind**, or column by column using **cbind**:

```
cols.of.y<-cbind(y,y,y)
cols.of.y
         [,1]    [,2]    [,3]
[1,]       8       8       8
[2,]       3       3       3
[3,]       5       5       5
[4,]       7       7       7
[5,]       6       6       6
[6,]       6       6       6
[7,]       8       8       8
[8,]       9       9       9
[9,]       2       2       2
[10,]      3       3       3
[11,]      9       9       9
[12,]      4       4       4
[13,]     10      10      10
[14,]      4       4       4
[15,]     11      11      11
```

Matrix arithmetic

It is important to understand that the * operator works with matrices, but it does *not* carry out matrix multiplication. This requires the %*% operator. Consider this example where a Leslie matrix, **L**, is to be multiplied by a column matrix of population sizes, **n**:

```
L<-c(0,0.7,0,0,6,0,0.5,0,3,0,0,0.3,1,0,0,0)
L<-matrix(L,nrow=4)
```

Note that the elements of the matrix are entered column-wise, not row-wise. We make sure that the Leslie matrix is properly conformed:

```
L
```

```
        [,1]      [,2]      [,3]      [,4]
[1,]    0.0       6.0       3.0        1
[2,]    0.7       0.0       0.0        0
[3,]    0.0       0.5       0.0        0
[4,]    0.0       0.0       0.3        0
```

The top row contains the age-specific fecundities (0, 6, 3 and 1), and the subdiagonal contains the survivorships (0.7, 0.5 and 0.3). Now the population sizes at each age go in a column vector, **n**:

```
n<-c(45,20,17,3)
n<-matrix(n,ncol=1)
n
```

```
        [,1]
[1,]     45
[2,]     20
[3,]     17
[4,]      3
```

Population sizes next year in each of the four age classes are obtained by matrix multiplication:

```
L %*% n
```

```
        [,1]
[1,]    174.0
[2,]     31.5
[3,]     10.0
[4,]      5.1
```

We can check this out longhand. The number of juveniles next year (the first element of **n**) is the sum of all the babies born last year:

```
45*0+20*6+17*3+3*1
```

```
[1]    174
```

So that's okay. If you try to do ordinary multiplication with **L** and **n**, you will fail because their dimensions do not match:

```
L*n
```

```
Error in L * n: non-conformable arrays
```

To use the * operator, matrices don't have to be the same shape, but they do need to be the same total length. Suppose that **a** is a vector of length 6 and **b** is a 2×3 matrix:

```
a<-c(1,0,3,4,8,5)
b<-c(2,9,3,2,7,0)
```

```
b<-matrix(b,nrow=2)
b
```

	[,1]	[,2]	[,3]
[1,]	2	3	7
[2,]	9	2	0

then the command **a*b** does work and produces a matrix with the attributes of the most complex matrix involved (**b** in this case). The elements of the resulting matrix are the pairwise products calculated in column-wise sequence ($1 \times 2, 0 \times 9, 3 \times 3, 4 \times 2$, etc.):

```
a*b
```

	[,1]	[,2]	[,3]
[1,]	2	9	56
[2,]	0	8	0

Matrix multiplication **%*%** on vectors of the same length returns the *vector dot product* (the sum of the pairwise products) as a 1×1 matrix, as we can see by converting **b** back from a matrix into a vector, like this:

```
b<-as.vector(b)
```

then multiplying **a** by **b** using the matrix multiplication operator **%*%**,

```
a % * % b
```

	[,1]
[1,]	75

and comparing the answer with this:

```
sum(a*b)
```

```
[1]   75
```

When the vectors are of different lengths, the result takes on the length attribute of the longer vector, with the shorter vector cycled repeatedly to match the length of the longer one. This can have unexpected results. The length of **y** is 15 and the length of **a** is 6:

```
y/a
```

```
Warning in y/a : longer object length
     is not a multiple of shorter object length

[1]   8.0000000        Inf  1.6666667 1.7500000 0.7500000  1.2000000 8.0000000
[8]         Inf  0.6666667  0.7500000 1.1250000 0.8000000 10.0000000       Inf
[15] 3.6666667
```

Note the following points: the answer is the length of the longest vector (15), division by zero returns the answer `Inf`, and a warning is printed.

Solving systems of linear equations using matrices

Suppose we have two equations containing two unknown variables:

$$3x + 4y = 12$$

$$x + 2y = 8$$

We can use the function **solve** to find the values of the variables if we provide it with two matrices:

- a square matrix **A** containing the *coefficients* (3, 1, 4 and 2)

- a column vector **kv** containing the *known values* (12 and 8)

We set the two matrices up like this (column-wise, as usual):

```
A<-matrix(c(3,1,4,2),nrow=2)

A

          [,1]       [,2]
[1,]        3          4
[2,]        1          2

kv<-matrix(c(12,8),nrow=2)
kv

          [,1]
[1,]        12
[2,]         8
```

Now we can solve the simultaneous equations:

```
solve(A,kv)

          [,1]
[1,]        -4
[2,]         6
```

so $x = -4$ and $y = 6$ (which you can easily verify by hand). The function comes into its own when there are many simultaneous equations to be solved. More advanced applications of matrix methods are found in Chapter 24 on multiple regression. Matrix operations are listed in Table 2.4.

Table 2.4   Matrix operations in S-Plus

Operation	Meaning
solve(x)	inverse of matrix **x**
solve(x,y)	solution of simultaneous linear equations with coefficients **x** and known values **y**
backsolve(x,y)	solves a system of linear equations when the square matrix **x** is upper triangular, **y** is the matrix containing the right-hand sides to equations (the known values)
eigen(x)	eigenvalues and eigenvectors of a square matrix **x**
diag(x)	diagonal of the matrix **x**
sum(diag(x))	trace of square matrix **x**
prod(eigen(x)$values)	determinant of real-valued matrix **x**
svd(x)	a list containing the singular value decomposition of **x**
qr(x)	a representation of an orthogonal (or unitary) matrix and a triangular matrix whose product is the input **x**
chol(x)	an upper triangular matrix which is the Choleski decomposition of a (Hermitian) symmetric, positive definite (or positive semidefinite) matrix
kronecker(x,y)	A matrix with dimension the product of the dimensions of the input matrices **x** and **y**; each element of **x** is replaced by that element times the entire matrix **y**
t(x)	transpose of the matrix **x**

Using the help functions in S-Plus

Full details of all the S-Plus functions, with worked examples and literature references, can be found by clicking the Help icon and then typing in the function name. It is often quicker to get help from the command line by typing a question mark followed by the name of the function that you are interested in. For example, to find out about Choleski decomposition of a (Hermitian) symmetric matrix, just type

```
? chol
```

and you get the following detailed description:

chol   Choleski Decomposition of Symmetric Matrix

```
DESCRIPTION

Returns an upper-triangular matrix which is the Choleski
decomposition of a (hermitian) symmetric, positive definite
(or positive semi-definite) matrix.

USAGE

chol(x, pivot=F)
```

and much, much more besides.

In a Help script, the worked example is often the most useful part, and it is always found at the very bottom of the script. You can often copy the code of the example from Help, and paste it directly into your command line. If needs be, you can always edit it afterwards using the Up Arrow.

Function sapply and the avoidance of loops

Suppose you want to apply a function to all the elements of a list. For example, you could use **sapply** to compute the squares of the integers 1 to 10 like this:

```
sapply(1:10,function(x) x^2)

[1]   1   4   9   16   25   36   49   64   81   100
```

But you would never do this, because it is so much simpler to write

```
(1:10)^2

[1]   1   4   9   16   25   36   49   64   81   100
```

The function **sapply** comes into its own with complex iterative calculations. The data show decay of radioactive emissions over a 50 day period, and we intend to use non-linear least squares (see p. 413) to estimate the decay rate a in $y = \exp(-ax)$:

```
attach(sapdecay)
names(sapdecay)

[1]   "x"   "y"
```

We need to write a function to calculate the sum of the squares of the differences between the observed (y) and predicted (y_0) values of y, when provided with a specific value of the parameter a:

```
sumsq <- function(a,xv=x,yv=y)
    {y0 <- exp(-a*xv)
    sum((yv-y0)^2)}
```

The parameter a is the only argument that has to be passed to the function called **sumsq**; the data for xv and yv will be accessed from vectors called x and y unless different data are specified in the function call using =.

We can get a rough idea of the decay constant, a, for these data by linear regression of log y against x, like this:

```
lm(log(y)~x)
```

```
Coefficients:
(Intercept)                    x
0.04688374      -0.0584863
```

So our parameter a is somewhere close to 0.058. We generate a range of values for a spanning an interval on either side of 0.058:

```
a<-seq(0.01,0.2,.005)
```

Now we can use **sapply** to apply the sum of squares function for each of these values of a (without writing a loop), and plot the deviance against the parameter value for a:

```
plot(a,sapply(a,sumsq),type="l")
```

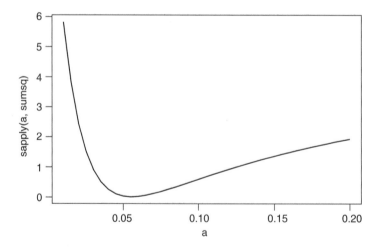

This shows the least squares estimate of a is indeed close to 0.06 (this is the value of a associated with the minimum deviance). To extract the minimum value of a, we use **min** with square brackets to extract the relevant value of a:

```
a[min(sapply(a,sumsq))==sapply(a,sumsq)]
```

```
[1]   0.055
```

Finally, we could use this value of a to generate a smooth exponential function to fit through our scatter of data points:

```
plot(x,y)
xv<-seq(0,50,0.1)
lines(xv,exp(-0.055*xv))
```

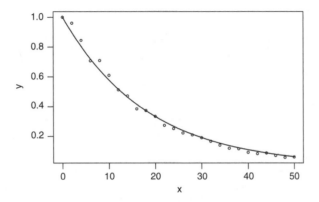

Using sapply to generate expected frequencies for a negative binomial distribution

Suppose that we know the two parameters of the negative binomial distribution to be μ and k (the mean and aggregation parameter, respectively; see p. 482). We want to use these to calculate the expected frequencies for comparison with a set of observed frequencies. The density function of the negative binomial in terms of μ and k is

$$P(x) = \left(1 + \frac{\mu}{k}\right)^{-k} \frac{\Gamma(k+x)}{x!\,\Gamma(k)} \left(\frac{\mu}{\mu+k}\right)^{x}$$

This is a classic case for using **sapply**. We want to do a relatively complicated calculation (the right-hand side) lots of times, once for each value of x (from 0 up to the largest observed count). First we write a function to work out the value of the right-hand side when provided with a particular value of x:

```
negbin<-function(x) (1+u/k)^(-k)*(u/(u+k))^x*gamma(k+x)/(factorial(x)*gamma(k))
```

Now we use **sapply** to apply this function to a sequence of x values 0:12 and present the result as a barplot like this:

```
u<-1.0042
k<-0.58
barplot(sapply(0:12,negbin),names=as.character(0:12))
```

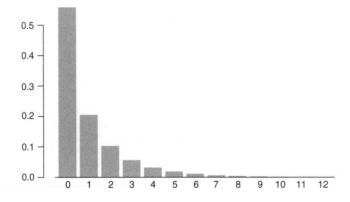

This example is developed further on p. 485.

The period-doubling route to chaos

We sometimes need to combine loops and **sapply**. In this example we use a loop to simulate the dynamics of a difference equation, then use **sapply** to simulate the time series for different values of the parameter, lambda (λ). The idea is to plot the population size at time 200 (once the transients have damped away). For small values of λ, the system has a stable point equilibrium. Larger values of λ produce stable two-point cycles, larger values four-point cycles, and so on, until the system becomes chaotic above $\lambda \approx 3.7$. The transitions between one behaviour and another are called Hopf bifurcations. First we write a function to create the last 10 points of the time series (see p. 494):

```
chaos<-function (lambda) {
x<-numeric(200)
x[1]<-0.6
for (t in 2: 200) x[t] <- lambda * x[t-1] * (1 - x[t-1])
x[191:200]}
```

Now we can use **sapply** to calculate the last 10 points in the time series for a range of different λ values between 2 and 4, and plot them as points:

```
plot(c(2,4),c(0,1),type="n",xlab="lambda",ylab="population")
for(lam in seq(2,4,0.01)) points(rep(lam,10),sapply(lam,chaos),pch=16,cex=0.5)
```

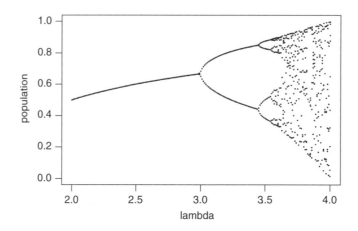

More advanced functions are described in several of the chapters.

Further reading

Chambers, J.M. and Hastie, T.J. (1992) *Statistical Models in S*. Pacific Grove, California, Wadsworth and Brooks Cole.

Everitt, B.S. (1994) *Handbook of Statistical Analyses Using S-PLUS*. New York, Chapman & Hall / CRC.

Krause, A. and Olson, M. (2000) *The Basics of S and S-Plus*. New York, Springer-Verlag.

Millard, S.P. and Krause, A. (2001) *Using S-PLUS in the Pharmaceutical Industry*. New York, Springer-Verlag.

Venables, W.N. and Ripley, B.D. (1997) *Modern Applied Statistics with S-Plus*. New York, Springer-Verlag.

3

Experimental design

Whole library shelves of books have been devoted to the subject of experimental design. Some valuable examples are Fisher (1954), Cochran and Cox (1957), Cox (1958), and Mead (1989). My sole object here is to provide a few notes that are relevant to using S-Plus. There are only two key concepts:

- replication

- randomisation

You replicate to increase reliability. You randomise to reduce bias. If you replicate thoroughly and randomise properly, you will not go far wrong. There are a number of other issues whose mastery will increase the likelihood that you analyse your data the right way rather than the wrong way:

- the principle of parsimony

- the power of a statistical test

- controls

- the pros and cons of different experimental designs

- spotting pseudoreplication and knowing what to do about it

- the difference between experimental and observational data (non-orthogonality)

It does not matter very much if you cannot do your own advanced statistical analysis. If your experiment is properly designed, you will often be able to find somebody to help you with the stats. But if your experiment is not properly designed, or not thoroughly randomised, or lacking adequate controls, then no matter how good you are at stats, some (or possibly even all) of your experimental effort will have been wasted. No amount of high-powered statistical analysis can turn a bad experiment into a good one. S-Plus is good, but not that good.

There is always a trade-off between including a wide range of conditions, in an attempt to make the experiment general; and restricting the set of conditions, so as to reduce variability and increase the likelihood of reaching firm conclusions.

The principle of parsimony (Occam's razor)

An important theme running through this book concerns model simplification. The principle of parsimony is attributed to the fourteenth-century English Nominalist philosopher William of Occam who insisted that, given a set of equally good explanations for a given phenomenon, then *the correct explanation is the simplest explanation*. It is called Occam's razor because he 'shaved' his explanations down to the bare minimum. In statistical modelling, the principle of parsimony means that:

- models should have as few parameters as possible

- linear models should be preferred to non-linear models

- experiments relying on few assumptions should be preferred to those relying on many

- models should be pared down until they are *minimal adequate*

- simple explanations should be preferred to complex explanation

The process of model simplification is an integral part of hypothesis testing in S-Plus. In general, a variable is retained in the model only *if it causes a significant increase in deviance when it is removed from the current model.* Seek simplicity, then distrust it.

In our zeal for model simplification, we must be careful not to throw the baby out with the bathwater. Einstein made a characteristically subtle modification to Occam's razor. He said, 'A model should be as simple as possible. But no simpler.'

Observation, theory and experiment

There is no doubt that the best way to solve scientific problems is through a thoughtful blend of observation, theory and experiment. In most real situations, however, there are constraints on what can be done, and on the way things can be done, which mean that one or more of the trilogy has to be sacrificed. There are lots of cases, for example, where it is ethically or logistically impossible to carry out manipulative experiments. In these cases it is doubly important to ensure that the statistical analysis leads to conclusions that are as critical and as unambiguous as possible.

Replication: it's the n's that justify the means

The requirement for replication arises because if we do the same thing to different individuals then we are likely to get different responses. The causes of this heterogeneity in response are many and varied (genotype, age, sex, condition, history, substrate, microclimate, etc.). The object of replication is to increase the reliability of parameter estimates, and to allow us to quantify the variability that is found within the same treatment. To qualify as replicates, the repeated measurements:

- must be independent

- must not form part of a time series (data collected from the same place on successive occasions are not independent)

- must not be grouped together in one place (aggregating the replicates means that they are not spatially independent)

- must be of an appropriate spatial scale

- ideally, one replicate from each treatment ought to be grouped together into a block, and each treatment repeated in many different blocks

- repeated measures (e.g. from the same individual or the same spatial location) are not replicates (this is probably the commonest cause of pseudoreplication in statistical work)

How many replicates?

The usual answer is as many as you can afford. An alternative answer is 30. A very useful rule of thumb is this: a sample of 30 or more is a big sample, but a sample of less than 30 is a small sample.

The rule doesn't always work: 30 would be derisively small as a sample in an opinion poll, for instance. In other circumstances it might be impossibly expensive to repeat an experiment as many as 30 times. But nevertheless, it is a rule of great practical utility, if only for giving you pause as you design your experiment with 300 replicates that perhaps this might really be a bit over the top. Or when you think you could get away with just 5 replicates this time.

There are ways of working out the replication necessary for testing a given hypothesis (see Chapter 8). Sometimes we know little or nothing about the variance or the response variable when we are planning an experiment. Experience is important. So are pilot studies. These should give an indication of the variance between initial units before the experimental treatments are applied, and also of the approximate magnitude of the responses to experimental treatment that are likely to occur. Sometimes it may be necessary to reduce the scope and complexity of the experiment, and to concentrate the inevitably limited resources of manpower and money on obtaining an unambiguous answer to a simpler question. It is immensely irritating to spend three years on a grand experiment, only to find at the end of it that the response is only significant at $p = 0.08$. A reduction in the number of treatments might well have allowed an increase in replication to the point where the same result would have been unambiguously significant.

Replicates or blocks?

There is always a trade-off between replication and blocking, because we always have limited resources (time, money, space, pairs of hands). The question always arises, therefore, as to whether it is better to have lots of replicates in a small number of blocks, or lots of blocks with no replication within each block. It is impossible to make a cast iron generalisation, but in many cases it is better to go for blocks rather than replicates, because material is so variable, and conditions are so heterogeneous (both in time and in space), that attempts at replication are often futile. On the other hand, replication within blocks does allow us to estimate treatment–block interactions, and to obtain estimates of pure sampling errors. As always, the best solution will depend on the nature of the problem in hand.

Randomisation

Randomisation is something that everybody says they do, but hardly anybody does properly. Take a simple example. How do I select one tree from a forest of trees, on which to measure photosynthetic rates? I want to select the tree at random in order to avoid bias. For instance, I might be tempted to work on a tree that had accessible foliage near to the ground, or a tree that was close to the lab. Or a tree that looked healthy. Or a tree that had nice insect-free leaves. And so on. I leave it to you to list the biases that would be involved in estimating photosynthesis on any of those trees.

One common way of selecting a 'random' tree is to take a map of the forest and select a random pair of coordinates (say 157 m east of the reference point, and 72 m north). Then pace out these coordinates and, having arrived at that particular spot in the forest, select the nearest tree to those coordinates. But is this really a randomly selected tree?

If it *was* randomly selected, then it would have *exactly the same chance of being selected as every other* tree in the forest. Let us think about this. Figure 3.1 shows a plan of the distribution of trees on the ground. Even if they were originally planted out in regular rows, accidents, tree-falls, and heterogeneity in the substrate would soon lead to an aggregated spatial distribution of trees. Now ask yourself how many different random points would lead to the selection of a given tree. Start with tree (a). This will be selected by any points falling in the large shaded area. Now consider tree (b). It will only be selected if the random point falls within the tiny area surrounding that tree. Tree (a)

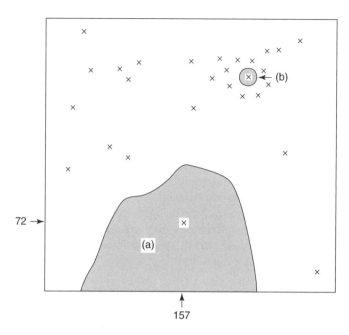

Figure 3.1 The fallacy of the random neighbour. Some people will tell you that the nearest individual to a randomly selected point is a randomly selected individual. Not in practice it isn't. Most real populations show aggregated spatial distributions, in which isolated individuals (like (a), selected by the random x and y coordinates 157 and 72) have a much greater probability of being selected than individuals, like (b), in the centre of clusters. In this example, individual (a) is about 30 times more likely to be selected than individual (b). For an individual to be selected at random, every individual must have exactly the same chance of being selected as every other. The only way to select a random individual is to number every single individual (all n of them), then draw a number at random between 1 and n

has a much greater chance of being selected than tree (b), and so *the nearest tree to a random point is not a randomly selected tree*. In a spatially heterogeneous woodland, isolated trees and trees on the edges of clumps will always have a higher probability of being picked than trees in the centre of clumps.

The answer is that to select a tree at random, every single tree in the forest must be numbered (all 24 683 of them), and then a random number between 1 and 24 683 must be drawn out of a hat. There is no alternative. Anything less than that is not randomisation.

Now ask yourself how often this is done in practice, and you will see what I mean when I say that randomisation is a classic example of 'Do as I say, and not do as I do'. As an example of how important proper randomisation can be, consider the following experiment that was designed to test the toxicity of five contact insecticides by exposing batches of flour beetles to the chemical on filter papers in petri dishes. The animals walk about and pick up the poison on their feet. The *Tribolium* culture jar was inverted, flour and all, into a large tray, and beetles were collected as they emerged from the flour. The animals were allocated to the five chemicals in sequence; 4 replicate petri dishes were treated with the first chemical, and 10 beetles were placed in each petri dish. Do you see the source of bias in this procedure?

It is entirely plausible that flour beetles differ in their activity levels (sex differences, differences in body weight, age, etc.). The most active beetles might emerge first from the pile of flour. These beetles all end up in the treatment with the first insecticide. By the time we come to finding beetles for the last replicate of the fifth pesticide, we may be grubbing round in the centre of the pile, looking for the last remaining *Tribolium*. This matters, because the amount of pesticide picked up by the beetles will depend upon their activity levels. The more active the beetles, the more chemical they pick up, and the more likely they are to die. Thus, the failure to randomise will bias the result in favour of the first insecticide because this treatment received the most active beetles.

What we should have done is this. Fill $5 \times 4 = 20$ petri dishes with 10 beetles each, adding one beetle to each petri dish in turn. Then allocate a treatment (one of the five pesticides) to each petri dish at random, and place the beetles on top of the pretreated filter paper. We allocate petri dishes to treatments most simply by writing a treatment number of a slip of paper, and placing all 20 pieces of paper in a bag. Then draw one piece of paper from the bag. This gives the treatment number to be allocated to the petri dish in question. All of this may sound absurdly long-winded, but believe me, it is vital.

The recent trend towards 'haphazard' sampling is a cop-out. What it means is that 'I admit that I didn't randomise, but you have to take my word for it that this did not introduce any important bias'. You can draw your own conclusions.

Power

The power of a test is the probability of rejecting the null hypothesis when it is false. It has to do with Type II errors: β is the probability of accepting the null hypothesis when it is false. In an ideal world, we would obviously make β as small as possible. But there is a snag. The smaller we make the probability of committing a Type II error, the greater we make the probability of committing a Type I error, and rejecting the null hypothesis when, in fact, it is correct. A compromise is called for. Most statisticians work with $\alpha = 0.05$ and $\beta = 0.2$. Now the power of a test is defined as $1 - \beta = 0.8$ under the standard assumptions. This is used to calculate the sample sizes necessary to detect a specified difference when the error variance is known or can be guessed at (see Chapter 8).

Strong inference

One of the most powerful means available to demonstrate the accuracy of an idea is an experimental confirmation of a prediction made by a carefully formulated hypothesis. There are two essential steps to the protocol of *strong inference* (Platt 1964):

- formulate a clear hypothesis

- devise an acceptable test

Neither one is much good without the other. For example, the hypothesis should not lead to predictions that are likely to occur by other extrinsic means. Similarly, the test should demonstrate unequivocally whether the hypothesis is true or false.

A great many scientific experiments appear to be carried out with no particular hypothesis in mind at all, but simply to see what happens. While this approach may be commendable in the early stages of a study, such experiments tend to be weak as an end in themselves, because there will be such a large number of equally plausible explanations for the results. Without contemplation there will be no testable predictions; without testable predictions there will be no experimental ingenuity; without experimental ingenuity there is likely to be inadequate control; in short, equivocal interpretation. The results could be due to myriad plausible causes. Nature has no stake in being understood by scientists. We need to work at it. Without replication, randomisation and good controls, we shall make little progress.

Weak inference

The phrase 'weak inference' is used (often disparagingly) to describe the interpretation of observational studies and the analysis of so-called natural experiments. It is silly to be disparaging about these data, because they are often the only data that we have. The aim of good statistical analysis is to obtain the maximum information from a given set of data, *bearing the limitations of the data firmly in mind*.

Natural experiments arise when an event (often assumed to be an unusual event, but frequently without much justification of what constitutes unusualness) occurs that is like an experimental treatment (a hurricane blows down half a forest block; a landslide creates a bare substrate; a stock market crash produces lots of suddenly poor people, etc.). According to Hairston (1989),

> The requirement of adequate knowledge of initial conditions has important implications for the validity of many natural experiments. Inasmuch as the 'experiments' are recognised only when they are completed, or in progress at the earliest, it is impossible to be certain of the conditions that existed before such an 'experiment' began. It then becomes necessary to make assumptions about these conditions, and any conclusions reached on the basis of natural experiments are thereby weakened to the point of being hypotheses, and they should be stated as such.

How long to go on?

Ideally, the duration of an experiment should be determined in advance, lest one falls prey to one of the twin temptations:

- to stop the experiment as soon as a pleasing result is obtained;

- to keep going with the experiment until the 'right' result is achieved (the Gregor Mendel effect)

In practice, most experiments probably run for too short a period, because of the idiosyncrasies of scientific funding. This short-term work is particularly dangerous in medicine and the environmental sciences, because the kind of short-term dynamics exhibited after pulse experiments may be entirely different from the long-term dynamics of the same system. Only by long-term experiments of both the pulse and the press kind, will the full range of dynamics be understood. The other great advantage of long-term experiments is that a wide range of patterns (kinds of years) is experienced.

Controls

No controls, no conclusions.

Taylor's power law

One of the most robust empirical generalisations in sampling theory is known as Taylor's power law (Taylor 1961). It states that as the sample mean increases, so the variance increases, such that a graph of log variance against log mean is a straight line whose slope is somewhere between 1.0 and 2.0. Statisticians refer to this as the *power-of-the-mean model*.

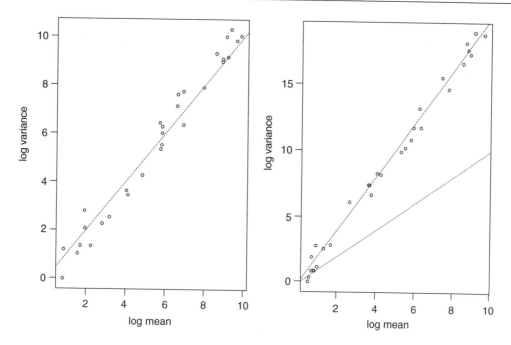

The left graph shows Taylor's power law for a Poisson process, where the variance is equal to the mean (the dotted line shows log variance = log mean). The right-hand graph is the plot for a Gamma-distributed process where the variance is proportional to the square of the mean (the dotted lines have slopes of 1 and 2 respectively).

Since one of the most important assumptions underlying standard parametric methods like regression and Anova is that sample variance is constant, you will see at once that Taylor's power law poses a major problem. There are two ways out of this dilemma. The traditional way is to take the log of the response variable (or keep taking logs repeatedly) until the variance stabilises. S-Plus offers a more attractive alternative, because it can deal directly with data that have a slope of 2 on a Taylor plot (i.e. data with a constant coefficient of variation) by means of Gamma errors (see p. 510).

Pseudoreplication

Pseudoreplication occurs when you analyse the data as if you had more degrees of freedom than you really have. There are two kinds of pseudoreplication:

- temporal pseudoreplication, involving repeated measurements from the same individual
- spatial pseudoreplication, involving several measurements taken from the same vicinity

Pseudoreplication is a problem because one of the most important assumptions of standard statistical analysis is *independence of errors*. Repeated measures through time on the same individual will have non-independent errors because peculiarities of the individual will be reflected in all of the measurements made on it (the repeated measures will be temporally correlated with one another). Samples taken from the same vicinity will have non-independent errors because peculiarities of the location will be common to all the samples (e.g. yields will all be high in a good patch and all will be low in a bad patch).

Pseudoreplication is generally quite easy to spot. The question to ask is, How many degrees of freedom for error does the experiment really have? If a field experiment appears to have lots of degrees of freedom, it is probably pseudoreplicated. Take an example from pest control of insects on plants. There are 20 plots, 10 sprayed and 10 unsprayed. Within each plot there are 50 plants. Each plant is measured five times during the growing season. Now this experiment generates $20 \times 50 \times 5 = 5000$ numbers. There are two spraying treatments, so there must be 1 degree of freedom for spraying and 4998 degrees of freedom for error. Or must there? Count up the replicates in this experiment. Repeated measurements on the same plants (the five sampling occasions) are certainly not replicates. The 50 individual plants within each quadrat are not replicates either. The reason for this is that conditions within each quadrat are quite likely to be unique, and so all 50 plants will experience more or less the same unique set of conditions, irrespective of the spraying treatment they receive. In fact, there are 10 replicates in this experiment. There are 10 sprayed plots and 10 unsprayed plots, and each plot will yield only one independent datum to the response variable (the proportion of leaf area consumed by insects, for example). Thus, there are 9 degrees of freedom within each treatment, and $2 \times 9 = 18$ degrees of freedom for error in the experiment as a whole. It is not difficult to find examples of pseudoreplication on this scale in the literature (Hurlbert 1984). The problem is that it leads to the reporting of masses of spuriously significant results (with 4998 degrees of freedom for error, it is almost impossible *not* to have significant differences). The first skill to be acquired by the budding experimenter is the ability to plan an experiment that is properly replicated.

There are various things that you can do when your data are pseudoreplicated:

- average away the pseudoreplication and carry out your statistical analysis on the means
- carry out separate analyses for each time period
- use proper time series analysis or mixed effects models (Chapters 34 and 35)

Sample covariance

Positive covariance between experimental units would occur if some samples were grouped together in good habitat and others grouped in poor habitat. Under these conditions, the sample means would differ for reasons that had nothing to do with the experimental treatments that were applied to them. Negative covariance between individuals might occur in a competition experiment where, if one of the individuals grew large, then its neighbours would necessarily be small. When there is positive covariance between samples, the measurements should be added together and the statistics carried out on the average values. In the case of negative covariance, then it may be legitimate to analyse the individual responses.

Initial conditions

Many otherwise excellent scientific experiments are spoiled by a lack of information about initial conditions. How can we know if something has changed if we don't know what it was like to begin with? It is often implicitly assumed that all the experimental units were alike at the beginning of the experiment, but this needs to be demonstrated rather than taken on faith. One of the most important uses of data on initial conditions is as a check on the efficiency of randomisation. For example, you should be able to run your statistical analysis to demonstrate that the individual organisms were not significantly different in mean size at the beginning of a growth experiment. Without measurements of initial size, it is always possible to attribute the end result to differences in initial conditions. Another reason for measuring initial conditions is that the information can often be used to improve the resolution of the final analysis through analysis of covariance (Chapter 16).

Pulse and press experiments

We need to distinguish between experiments that consist of a single kick to the system (pulse experiments) and those that involve a constant push in the back (press experiments); see Bender *et al.* (1984). We should recognise, however, that the short-term response (say over 1 to 5 years) to *either* kind of experiment may exhibit *transient dynamics that are quite atypical of equilibrium or long-term behaviour*. In general, the long-term consequence of the experiment may be impossible to predict from the short-term (transient) dynamics. For example, in a study of ecological succession, a species may increase in abundance in the first year following soil disturbance, but that same species may disappear completely in the long run.

Orthogonal designs and non-orthogonal observational data

The data in this book fall into two distinct categories. In the case of planned experiments, all of the treatment combinations are equally represented and, barring accidents, there are no missing values. Such experiments are said to be *orthogonal*. In the case of observational studies, however, we have no control over the number of individuals for which we have data, or over the combinations of circumstances that are observed. Many of the explanatory variables are likely to be correlated with one another, as well as with the response variable. Missing treatment combinations are commonplace, and the data are said to be non-orthogonal. This makes an important difference to our statistical modelling because, in orthogonal designs, the deviance that is attributed to a given factor is constant, and does not depend upon the order in which that factor is removed from the model. In contrast, with non-orthogonal data, we find that the deviance attributable to a given factor *does* depend upon the order in which the factor is removed from the model. We must be careful, therefore, to judge the significance of factors in non-orthogonal studies, when they are *removed from the maximal model* (i.e. from the model including all the other factors and interactions with which they might be confounded).

Missing values may arise in any kind of study, and the more missing values there are, the more the value of the experiment is diluted. As we shall see, S-Plus is very good at dealing with missing treatments and unequal replication, but only at a cost. Degrees of freedom are lost, and standard errors are inflated, thus reducing the likelihood of detecting significant differences.

Fixed effects and random effects

Traditional statistical analyses distinguish between experimental designs where you, the experimenter, impose treatments upon subjects (fixed effects) and experiments where you go looking for different places in which to repeat the same experiment (random effects). The statistical methods appropriate to the first case are called Model I Anova and to the second, Model II (the reasons for the Roman numerals are lost in the mists of time). In many scientific experiments, the two may be mixed together. It is commonplace in field experiments, for example, to select initially different plant communities to serve as blocks, and then to apply all the treatments at random within each block. Thus the blocks are sometimes said to be random effects and the treatments fixed effects. The important difference is that with fixed effects, we think we know that the cause of variation can be attributed to our experimentally imposed treatments, whereas with random effects, we know things differ but we have no idea precisely *why* they differ. In other cases we may well know the cause of random effects (e.g. soil fertility trends between blocks) but we want to treat them as instances from a distribution, rather then specific fixed effects. Random effects are dealt with in variance components analysis (Chapter 20) and mixed effects models (Chapter 35).

Experimental designs

Let us assume that we have a single explanatory variable and that it is a factor with 4 levels (a categorical variable). Assume further that we intend to have 8 replicates of each level, giving a total sample size of $n = 4 \times 8 = 32$. Our experiment passes the 'big sample test' because $n > 30$. There are many different ways in which we could carry out such an experiment. To make things concrete, let us assume that it is a field experiment to be carried out in 32 plots. It could equally well be a trial to be carried out on 32 patients, or in 32 factories, or on 32 batches of an industrial product. The question that immediately arises is, How do we decide what treatment to apply to each plot?

Fully randomised designs

The treatments are assigned to the 32 plots in the field completely at random. For example, 8 bits of paper with 'Treatment A' written on them, 8 with 'Treatment B', 8 with 'Treatment C' and 8 with 'Treatment D' are put into a black bag and shaken up. Then we go systematically through the 32 plots in the field. For each, we pull one piece of paper out of the bag. That is the treatment allocated to that plot. The strength of this method is that it is true randomisation, hence it reduces bias.

There are several weaknesses, however: some of them practical, some of them theoretical. Practical difficulties are that repeats of the same treatment are scattered about; it may be difficult to apply the treatment to many small plots (e.g. if the treatments are applied by machine), and the risks of misidentifying a plot, and hence applying the wrong treatment are increased. If the background variation in the field is clumped, there is a risk that groups of the same treatment end up in a 'good' patch and groups of another treatment end up in a 'bad' patch. This means that treatment effects are *confounded* with background effects, which is unfortunate.

The experiment consists of 8 replicates of 4 treatments. We make the map like this:

```
map<-rep(c("A","B","C","D"),8)
```

Then the randomisation is simple. In this case we get a 'good' one:

```
matrix(sample(map),nrow=4)
```

```
     [,1] [,2] [,3] [,4] [,5] [,6] [,7] [,8]
[1,] "C"  "A"  "D"  "B"  "C"  "A"  "D"  "A"
[2,] "C"  "D"  "C"  "D"  "A"  "B"  "D"  "B"
[3,] "B"  "A"  "A"  "C"  "C"  "C"  "B"  "D"
[4,] "A"  "A"  "B"  "D"  "B"  "D"  "B"  "C"
```

But here is a 'bad' one:

```
matrix(sample(map),nrow=4)
```

```
     [,1] [,2] [,3] [,4] [,5] [,6] [,7] [,8]
[1,] "B"  "D"  "A"  "A"  "A"  "A"  "D"  "C"
[2,] "C"  "A"  "B"  "D"  "C"  "C"  "B"  "D"
[3,] "B"  "C"  "D"  "D"  "C"  "C"  "B"  "B"
[4,] "A"  "D"  "A"  "D"  "C"  "A"  "B"  "B"
```

There are big groups of treatments (e.g. the 5 Bs in the bottom right corner). There is nothing wrong with this if the environment is homogeneous. But what if the experiment were to be carried out in a field, and the bottom right-hand corner of the field were wet? In that case the experiment would be biased because treatment B is confounded with wetter than average soils.

Stratified random designs

Here the field is divided up into strata (hence the name). The strata are of a size such that each stratum has one plot for every treatment (4 plots in this case). The number of strata is the total experiment size (32 in this case) divided by the number of treatments: $32/4 = 8$. Treatments are allocated to plots with a stratum at random.

The strengths of this design is that it is genuinely random (every treatment has exactly the same chance of appearing in any given plot), but if there are systematic differences across the field, these can be taken into account during the analysis.

The weakness of the design is that the plots are arranged differently in each block and so it may be tricky to apply the treatments mechanically, and there is a risk that treatments may be applied to plots in error.

Suppose that there was variation across the field in one direction: say the field was dry at the top and wet at the bottom of a slope (Figure 3.2). How would you arrange the strata? Would you put the strata parallel with the gradient in soil moisture (left-hand), or at right angles to the gradient (right hand diagram). In the first case (case A) all the strata have the same mean soil moisture, but each stratum is internally heterogeneous. In the second case (case B) the strata are internally homogeneous, but each stratum has a different mean soil moisture.

This is a very important point to appreciate, and lots of people pick the wrong answer. If you put the strata parallel with the gradient (case A) your 4 plots will consist of one wet, one dry, one damp and one dryish. So when you randomly allocate your 4 treatments, you *know in advance* that treatment response is going to be confounded with soil moisture effects. And this confounding is going to occur in every stratum. You have to hope that the randomisation averages out this bias for you. Murphy's law doesn't usually work like that.

Suppose you put the strata at right angles to the gradient (case B). Now all 4 plots have the same moisture content, and the treatment effects will be unbiased in each stratum. You don't expect the *mean* response to be the same in each stratum (because soil moisture matters), but you do hope that you get a consistent effect of treatment across different soil moistures.

The weakness is that *the response to treatment may depend on soil moisture*. This is called an *interaction* effect, and you need replication of each treatment in each stratum in order to deal convincingly with this (Chapter 15).

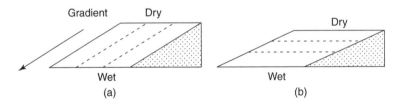

Figure 3.2 Stratified experimental designs. In cases where there is an obvious gradient in the substrate (e.g. from the dry top, to the damp bottom, of a slope), then variation in soil moisture will introduce extra variation in yield over and above the variation introduced by our experimental treatments. One sensible solution is to apply the experimental treatments in a stratified design. The question then arises as to whether the strata should be laid down parallel with the gradient, as in case (a), or at right angles to the gradient, as in case (b). The correct answer is (b), as explained in the text. In case (a) the mean yield will be the same in all the strata, but the variance within each stratum will be increased by differences in soil moisture. In case (b) the mean yield will differ from stratum to stratum, but this is an advantage, because the block differences can be factored out during the analysis of variance. The big benefit of case (b) is that the within-stratum variance is smallest, because it has not been inflated by differences in soil moisture

In the present example with 4 treatments, it would make sense to divide the field into 8 strata. Here is a map which gives the stratum number for each of the 4 plots in a column:

```
matrix(rep(1:8,rep(4,8)),nrow=4)
```

```
      [,1] [,2] [,3] [,4] [,5] [,6] [,7] [,8]
[1,]    1    2    3    4    5    6    7    8
[2,]    1    2    3    4    5    6    7    8
[3,]    1    2    3    4    5    6    7    8
[4,]    1    2    3    4    5    6    7    8
```

Now we want to apply each of the 4 treatments at random independently for each stratum (i.e. in each column). There are two steps to this: we need to produce a short vector (called *trt*) containing the 4 treatments codes, then we need to go round a loop, for each of the 8 strata, and independently randomise the 4 treatment codes using **sample**. The results are concatenated in a vector called *r* (this is null to begin with, but ends up of length 32):

```
trt<-c("A","B","C","D")
r<-NULL
for (i in 1:8) r<-c(r,sample(trt))
```

Now we can produce the randomised treatments as a map of the field:

```
matrix(r,nrow=4)
```

```
      [,1] [,2] [,3] [,4] [,5] [,6] [,7] [,8]
[1,]  "C"  "C"  "D"  "B"  "D"  "B"  "A"  "A"
[2,]  "B"  "A"  "A"  "A"  "A"  "A"  "B"  "B"
[3,]  "D"  "B"  "B"  "C"  "C"  "D"  "D"  "C"
[4,]  "A"  "D"  "C"  "D"  "B"  "C"  "C"  "D"
```

As you see, each treatment appears once in each stratum, and each stratum has been independently randomised. As it turns out in this case, no two strata are identical but in principle, of course, they could be.

Latin squares

In practice the underlying gradients may not be obvious at the outset. The field may be flat, for example, but there could be gradients in soil nutrients that you cannot see at the time you design the experimental layout. There could be multiple gradients, running in different directions.

A pragmatic response to this is to assume that the field is heterogeneous in two dimensions and to have strata running both up and down and left to right across the field. In the present example we have 4 treatments, so a sensible plan is to have $4 \times 4 = 16$ plots in a square, and to replicate this whole design twice to obtain our total experiment size of 32. For the first block of 16 plots we allocate the 4 treatments at random to the first row of the square. Perhaps it looks like this:

D, A, C, B

Now, for the second row of the square, we apply what is called *constrained randomisation*: we apply the treatments at random subject to the constraint that the same treatment cannot occur in the same column as it did in row number 1. What this means is that treatment D cannot appear in column 1, so we select from A, B and C at random. Likewise, A cannot appear in column 2, so we select from D and whichever of B or C was *not* allocated to column 1. And so on. It might turn out like this:

D, A, C, B
C, D, B, A

Now in row 3 we randomise as far as we can: the first column cannot be D or C, so we select from A and B. Suppose it was B. The experimental plan now looks like this:

D, A, C, B
C, D, B, A
B

We have A, C and D left. Both A and D are already in that column, so we have to choose C (randomisation does not come into it):

D, A, C, B
C, D, B, A
B, C

Now we have A and D left to allocate. D could go into either position, but A could only go in column 3. So that is the end of it: the third row has to be

D, A, C, B
C, D, B, A
B, C, A, D

What about the fourth row? A moment's thought will convince you that there is no more randomisation left to do. We have no degrees of freedom in choosing the treatments for row 4. Column 1 has to be A, column 2 has to be B, and so on. The final treatment allocation to plots in this case is

D, A, C, B
C, D, B, A
B, C, A, D
A, B, D, C

The advantage of the Latin square is that we can factor out any row effects and column effects at the analysis stage. The disadvantage is that plot location is difficult, so mistakes could be made in treatment application. In our present experiment we would carry out an independent randomisation for the second block of 16 plots to obtain the full experiment.

The thing about a Latin square is that we are completely free to randomise the first row, but randomisation of the second row is constrained:

```
sample(trt)
```

```
[1] "C" "B" "D" "A"
```

The second row cannot contain any of these treatments in the same columns as they appeared in the first row. So the first column of row 2 can be anything other than "C":

```
sample(trt[trt!="C"])
```

```
[1] "D" "B" "A"
```

Note the use of subscripts [ ] with the not equal operator !=. We take the first code "D". Now repeat for the second column: it can't be "B" because that appears in row 1 or "D" because we have already used that in row 2:

```
sample(trt[trt!="B"& trt!="D"])
```

```
[1] "A" "C"
```

So the second column of row 2 is "A". The third column of row 2 can only be drawn from "B" or "C":

```
sample(trt[trt!="A"& trt!="D"])
```

```
[1] "B" "C"
```

so it is a "B". Now we have no freedom in choosing the last treatment: it has to be "C". So now the first two rows are complete:

```
[1] "C" "B" "D" "A"
[2] "D" "A" "B" "C"
```

In the third row, the first column cannot be "C" or "D":

```
sample(trt[trt!="C"& trt!="D"])
```

```
[1] "A" "B"
```

so it is "A". The second column already contains "B" and "A", so

```
sample(trt[trt!="A"& trt!="B"])
```

```
[1] "D" "C"
```

it gets a "D". This means that column 3 must be "C" and column 4 must be "B":

```
[1] "C" "B" "D" "A"
[2] "D" "A" "B" "C"
[3] "A" "D" "C" "B"
```

For the last row we have no degrees of freedom. Column 1 has to be "B", and so on:

```
[1] "C" "B" "D" "A"
[2] "D" "A" "B" "C"
[3] "A" "D" "C" "B"
[4] "B" "C" "A" "D"
```

This is the complete Latin square. A general function for designing Latin squares is developed in the web pages.

Split plots

It is often the case that practical considerations rule out fully randomised, stratified or Latin square designs. Some treatments may be possible only if they are applied to much larger areas than one of our plots. Mechanised irrigation is a good example. It is much easier to irrigate $1000\,m^2$ than it is to irrigate 100 separate plots, each of $10\,m^2$. Treatments like fertilisers that are spread by tractor-driven machinery are much easier to apply in long strips than they are to apply to randomly spaced small square plots. Some expensive equipment (like growth chambers) cannot be replicated: several different treatments need to be packed into the same growth chamber. Such circumstances call for split plots.

In a split-plot experiment, *different treatments are applied to plots of different sizes.* Suppose our 4 treatments involved 2 levels of irrigation and 2 levels of fertiliser input. This is called a factorial experiment (see below). The nature of the irrigation equipment means that it has to be applied to much larger plots than the fertiliser. We must have some replication for irrigation, so the minimum acceptable design has 4 big plots: 2 irrigated and 2 not irrigated. Given our design, each of these would contain 8 smaller plots. We might divide each large plot up into 4 strata and apply fertiliser (or not) at random independently in each of the 4 strata. In such a case we would say that fertiliser treatment was nested within stratum, nested within irrigation treatment.

The drawback of split-plot designs is that the small plot treatment effects are assessed with greater power than the large plot treatments because their replication is so much higher. Also, the statistical analysis is more complicated, so there is more scope for doing the analysis wrong. There is always a premium on keeping things as simple as possible.

Factorial designs

Factorial experiments are carried out to investigate interactions between different explanatory variables. Interactions occur when *the response to one factor depends on the level of another factor.* For example, irrigation might increase yield at high nutrient input but reduce yield at low nutrient input.

The problem with factorial experiments is that they rapidly grow out of control when the experiment involves several explanatory variables. Suppose we want to carry out a drug trial on people. We know that the following factors are likely to be important: sex, age, race, place of work, educational background, income and smoking habits. Sex obviously has two levels (male and female). Age is more difficult; young adult, middle-aged and old would be a sensible three-level compromise. Race is difficult too: Caucasian, Asian and African would be one way of doing it. This is okay unless you have lots of Polynesians to consider. Place of work is really difficult; you could say office, factory, outdoors or unemployed, but this will be unsatisfactory if you want to be able to discriminate between different kinds of factories. Educational background might be classified as pre-high-school, high school and university. Income is another difficult one. What you are trying to get at is whether poverty or great wealth are important issues, so perhaps a three-level factor is best: below the official poverty level, income above the high tax threshold, or in between. Smoking is relatively easy: smoker or non-smoker is the simplest option.

Let's take stock of the number of treatment combinations involved in this experiment: there are 2 sexes, 3 ages, 3 races, 4 places of work, 3 educational backgrounds, 3 incomes, 2 smoking levels. Because this is intended to be a factorial experiment, we get the number of treatment combinations by multiplying together all the factor levels:

$$\text{experiment} = 2 \times 3 \times 3 \times 4 \times 3 \times 3 \times 2 = 1296$$

Because this is a factorial experiment we must have replication of each treatment combination, so we need at least $2 \times 1296 = 2592$ subjects in our experiment. And they need to be selected at random from a larger population. I think you get the point. Factorial experiments quickly become unmanageably large. We need to be able to cut down on the treatment combinations.

Nested designs   It is tempting to assume that any experiment which has two or more factors can be analysed as a factorial design. There are several pitfalls for the unwary:

- a factorial design must have replication for each of the interaction terms that needs to be estimated

- the treatment combinations must be independent

- the treatment combinations must be assigned at random

The commonest mistake is to assume that the treatment combinations are independent when in fact they are not. Take a simple example involving an experiment on the effects of diet on insect growth at different temperatures. We have 4 controlled temperature (CT) rooms at 10, 15, 20 and 25°C and in each CT room insects are cultured on 5 different diets, with each diet repeated 3 times at each temperature. Now ask yourself the question, How many CT rooms would be needed if this were a factorial design? There are 5 diets and 3 replicates, so we would need 15 CT rooms at each temperature, so with 4 temperatures this is 60 CT rooms in all. Since we have only 4 CT rooms, it is pretty clear that the present experiment is *not* a factorial design. But the structure of the data does not tell us this, because we have 60 numbers, and we could easily make the mistake of analysing the results as if they had come from a factorial design. Failure to recognise split-plot and other nested designs is one of the commonest mistakes in scientific statistics. It is important because *it leads to the use of the wrong error term* in hypothesis testing, and causes the standard errors to be wrongly calculated if the correct design is not recognised.

Recognising nested designs   The simple rule of thumb for recognising a factorial experiment is this. Multiply together all your treatment combinations and replicates (4 temperatures × 5 diets × 3 replicates = 60). Then ask, do I have this many *independent* repeats of the experiment. If you do, then you have a full factorial. If you don't, then you have some kind of nested or split-plot design, and you will need to think carefully about how you specify the error term in the model.

In the present example, the repeats are not independent of one another because we only have four CT rooms, one at each temperature. Thus, if one of the CT rooms was odd in some way (perhaps someone sharing the CT room was using a volatile growth regulator), then *all* the results from that CT room would be affected (all the diet treatments and all the replicates of each diet treatment). In the present case, there are no degrees of freedom for temperature, because temperature and growth room are completely confounded. The way to analyse this experiment is to treat it as a split-plot design. The CT rooms are blocks, and we compare the diets by one-way Anova nested within blocks, using the block–diet interaction term (with its 12 degrees of freedom) to test the significance of differences between the diet means (i.e. we do not use the overall error degrees of freedom with its $4 \times 5 \times (3 - 1) = 40$ degrees of freedom). Differences due to temperature effects will emerge as the block sum of squares, but we shall not be able to estimate temperature–diet interactions, because these are confounded with differences between the CT rooms. Because they all occur in the same CT room, the 'replicates' are actually pseudoreplicates when it comes to comparing diets at different temperatures. The repeats in each CT room are not completely useless, of course, because they allow 4 independent comparisons of the 5 diets (one in each CT room, with $5 \times (3 - 1) = 10$ degrees of freedom for error in each room).

The skills involved in spotting pseudoreplication only come with practice. This, again, is not something that it is easy to pick up from reading about it (but see p. 249).

Efficient regression designs

Too little thought tends to be given to the precise nature of the question being asked in regression studies (see Chapter 14 for a full introduction). Two contrasting examples should make this point.

Suppose that we have the financial and manpower resources to carry out 14 experimental measurements. Should we gather 7 replicates at just 2 extreme levels of the x axis, or 1 replicate at each of 14 different levels of the explanatory variable? Should the levels of x be equally spaced or unequally spaced along the x axis (see Figure 3.3)? These are design questions that can only be resolved by thinking carefully exactly what it is about the relationship between y and x that we wish to establish.

In the first case, let us assume that we suspect there is a threshold level of x below which y shows no response (this might be a temperature threshold, below which there is no development in an insect). The question in this case is where, precisely, is the threshold located?

In the second case, let's suppose that we know there is a smooth relationship between y and x, but we suspect the relationship is not a straight line, and we wish to test whether there is evidence for a quadratic term in the model (i.e. do the data provide support for a model that contains an x^2 term in addition to a term for x?). This might occur if the value of y increased more slowly at higher levels of x than at low (as in a study of investment, where profits might show diminishing returns).

We begin by considering some of the options. Figure 3.3 shows six different ways that our 14 measurements might be distributed along the x axis. Which of these is best? The answer, of course, depends upon the question. For the first case, we wish to locate the position of a threshold. Design

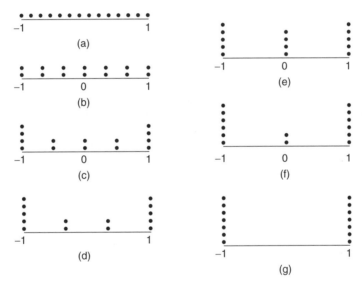

Figure 3.3 Efficient regression designs. Suppose costs dictate that 14 repeats of the experiment can be carried out. (a) The simplest option is to distribute all the measurements at equally spaced intervals along the x axis. The benefits of this design are that it is best at locating threshold effects and detecting other forms of non-linearity. The cost of this design is that it does not give the lowest possible standard error of the slope, because SSX is not as large as it could be; this is option (g). (b) The advantage of this design is that it has replication at each level of x, and so it allows an estimate of the sampling error to be made independently of the regression line. This in turn allows a test for the significance of any departure of the regression from linearity. (c) This is a compromise in which there is replication (hence an independent estimate of sampling error), but extra replication is placed at the extremes of the x axis (to reduce the standard error of the slope by making SSX larger). (d) Another compromise design with lower standard error but less ability to detect non-linearities. (e) This design would be reasonable if non-linearity were expected to be of a power law form (p. 377). It would not be sensible with S-shaped non-linearities, however. (f) Like case (e) but more extreme; it gives a good estimate of the standard error of the slope when the relationship turns out to be linear. (g) If the relationship is linear, this design gives the lowest estimate of the standard error of the slope. The design is incapable of detecting non-linearity

(a) may be the best bet if we have no idea where the threshold lies, while design (g) is completely hopeless. But design (g) is not always hopeless, because it is the design that gives the lowest standard error for the slope. The standard error of a regression slope is $\sqrt{s^2/SSX}$, and so we always want to make SSX as large as possible. This is achieved by having lots of measurements at the extreme left-hand right-hand ends of the x axis, and fewer in the middle (Chapter 14).

The solution in the second case is a little more subtle. The best design cannot be (g) because it can provide no evidence of non-linearity, since a straight line will always fit perfectly between two points. The number of degrees of freedom for non-linearity is given by $k - 2$, where k is number of levels of x. In order to be able to test for non-linearity, therefore, we need at least three levels of x. The best design is a compromise between degrees of freedom allocated to non-linearity, and degrees of freedom allocated to estimating pure sampling error. On balance, designs (d) and (f) look best as a compromise between error estimation and detection of non-linearity. If we are reasonably sure that a quadratic term is the most complex model we need to fit, then perhaps the choice (f) would be preferred, with choice (d) if we were less certain about the nature of the non-linearity; for example, the curve might be sigmoid, in which case the mean of a treatment midway between the extremes of the x axis could fall on the straight line joining the two extreme points, thus providing no evidence of non-linearity; see Draper and Smith (1981).

Further reading

Box, G.E.P., Hunter W.G., *et al.* (1978) *Statistics for Experimenters: An Introduction to Design, Data Analysis and Model Building.* New York, John Wiley.

Cochran, W.G. and Cox G.M. (1957) *Experimental Designs.* New York, John Wiley.

Fisher, R.A. (1954) *Design of Experiments.* Edinburgh, Oliver and Boyd.

Hairston, N.G. (1989) *Ecological Experiments: Purpose, Design and Execution.* Cambridge, Cambridge University Press.

Hicks, C.R. (1973) *Fundamental Concepts in the Design of Experiments.* New York, Holt, Rinehart and Winston.

Hurlbert, S.H. (1984) 'Pseudoreplication and the design of ecological field experiments.' *Ecological Monographs* **54**: 187–211.

Keppel, G. (1991) *Design and Analysis: A Researcher's Handbook.* Upper Saddle River, New Jersey, Prentice Hall.

Mead, R. (1989) *The Design of Experiments.* Cambridge, Cambridge University Press.

Platt, J.R. (1964) 'Strong inference.' *Science* **146**: 347–353.

Winer, B.J., Brown D.R. *et al.* (1991) *Statistical Principles in Experimental Design.* New York, McGraw-Hill.

4

Central tendency

Most data show some propensity to cluster around a central value. No matter what the distribution of the original data, so long as it has finite variance, averages from repeated bouts of sampling show a remarkable tendency to cluster around the arithmetic mean; this is the central limit theorem (see below). There are several ways of estimating the central tendency of a set of data and they often differ from one another. The differences between the estimates can be highly informative about the nature of the data set.

Mode

Perhaps the simplest measure of central tendency is the mode: *the most frequently represented class of data*. Technically, the mode is the interval with the greatest probability density. One of the drawbacks of the mode is that its value depends upon where we choose to put the class boundaries, so there can be an element of subjectivity in it. It is more meaningful, therefore, to use the mode as a measure of central tendency for count data, where each integer value forms its own class. Some distributions have several modes, and these can give a poor estimate of central tendency.

```
attach(distribution)
names(distribution)
[1] "fx" "fy" "fz" "x"
```

To draw two histograms side by side we set up two adjacent plotting panels:

```
par(mfrow=c(1,2))
```

Now we use **barplot** to plot the histograms. Note the use of **as.character** to coerce the x numbers into a form suitable for using as labels for the bars on the x axis:

```
barplot(fx,names=as.character(x))
barplot(fy,names=as.character(x))
```

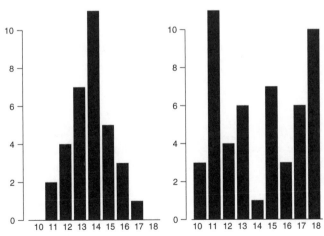

The mode of 14 is an excellent descriptor of the central tendency of the data set on the left, but the two largest modes of the data set on the right (11 and 18) are both very poor descriptors of its central tendency.

Median

The median is an extremely appealing measure of central tendency. It takes the value of the response variable that lies in the middle of the ranked set of *y* values. That is to say, 50% of the *y* values are smaller than the median, and 50% are larger. It has the great advantage of not being sensitive to outliers. Suppose we have a vector *y* that contains all of the *x* values in our left-hand histogram. That is to say, each *x* value is **rep**eated *fx* times:

```
y<-rep(x,fx)
```

Now to find the median from first principles we **sort** the *y* values (in this case they are already in ascending order, but we overlook this for the sake of generality):

```
y<-sort(y)
```

Now we need to know how many *y* values (*n*) there are; use the **length** directive for this:

```
length(y)
```

```
[1] 33
```

Because this is an odd number, the median is uniquely determined as the *y* value in position 17 of the ranked values of *y*:

```
ceiling(length(y)/2)
```

```
[1] 17
```

We find this value of *y* using subscripts [ ]:

```
y[17]
```

or, more generally, using the Up Arrow to edit in y[ ] to the **ceiling** directive:

```
y[ceiling(length(y)/2)]
```

```
[1] 14
```

Had the vector been of even-numbered length, we could have taken the *y* values on either side of the midpoint (y[length/2] and y[length/2+1]) and averaged them. A function to do this is described later. You will not be surprised to learn that there is a built-in **median** function, used directly like this:

```
median(y)
```

```
[1] 14
```

Arithmetic mean

The most familiar measure of central tendency is the arithmetic mean. It is *the sum of the y values divided by the number of y values* (*n*). Throughout the book we use $\sum$, capital Greek sigma, to mean 'add up all the values of'. So the mean of *y*, called 'y bar', can be written as

$$\bar{y} = \frac{\sum y}{n}$$

We can compute this using the **sum** and **length** functions:

    sum(y)/length(y)

    [1] 13.78788

but you will be not be surprised to learn that there is a built-in function called **mean**:

    mean(y)

    [1] 13.78788

Notice that in this case the arithmetic mean is very close to the median (14.0).

Geometric mean

The geometric mean is much less familiar than the arithmetic mean, but it has important uses in the calculation of average rates of change in economics and average population sizes in ecology. Its definition appears rather curious at first. It is the nth root of the product of the y values. Just as sigma means 'add up', so capital Greek pi, $\prod$, means multiply together. The geometric mean 'y curl' is

$$\tilde{y} = \sqrt[n]{\prod y}$$

Recall that roots are fractional powers; the square root of x is $x^{1/2} = x^{0.5}$. So to find the geometric mean of y we need to take the product of all the y values, prod(y), then raise it to the power of 1/(length of y), like this:

    prod(y)^(1/length(y))

    [1] 13.71531

As you see, in this case the geometric mean is close to both the arithmetic mean (13.788) and the median (14.0). A different way of calculating the geometric mean is to think about the problem in terms of logarithms. The 'log of times is plus', so the log of the product of the y values is the sum of the logs of the y values. All the logs used in this book are natural logs (base e = 2.781 828) unless stated otherwise (Chapter 2). So we could work out the mean of log(y) then calculate its antilog:

    meanlogy<-sum(log(y))/length(y)
    meanlogy

    [1] 2.618513

Now you need to remember (or learn) that the natural antilog function is **exp.** So the antilog of meanlogy is

    exp(meanlogy)

    [1] 13.71531

as we obtained earlier by a different route. There is no built-in function for geometric mean, but we can easily write one like this:

    geometric<-function(x) exp(sum(log(x))/length(x))

Then try the function out using our vector called y:

    geometric(y)

    [1] 13.71531

The reason that the geometric mean is so important can be seen by using a more extreme data set. Suppose we had the following annual incomes (in thousands) of five different people:

    income<-c(10,1,1,10,1000)

Now the arithmetic mean is hopeless as a descriptor of central tendency:

    mean(income)

    [1] 204.4

This measure isn't even close to *any* of the five data points. The reason for this, of course, is that the arithmetic mean is so sensitive to outliers, and the single income of 1000 has an enormous influence on the mean.

Just this once we can break the rules, and use logs to base 10 to look at this example. The $\log_{10}$ values of the 5 incomes are 1, 0, 0, 1 and 3. So the sum of the logs is 5 and the average of the logs is $5/5 = 1$. Antilog base 10 of 1 is $10^1 = 10$. This is the geometric mean of these data, and is a much better measure of central tendency (two of the five values are exactly like this). We can check using our own function:

    geometric(income)

    [1] 10

Needless to say, we get the same answer, whatever base of logarithms we use. We always use geometric mean to average variables that change multiplicatively, like ecological populations or bank accounts accruing compound interest.

Harmonic mean

Suppose you are working on elephant behaviour. Your focal animal has a square home range with sides of length 2 km. He walks the edge of the home range as follows. The first leg of the journey is carried out at a stately 1 km per hour. He accelerates over the second side to 2 km per hour. He is really into his stride by the third leg of the journey, walking at a dizzying 4 km per hour. Unfortunately, this takes so much out of him that he has to return home at a sluggish 1 km per hour.

The question is this, What is his average speed over the ground? He ended up where he started, so he has no net displacement. But our concern is with his mean velocity. What happens if we work out the average of his four speeds?

    mean(c(1,2,4,1))

    [1] 2

This is wrong, as we can see if we work out the average speed from first principles:

$$\text{velocity} = \frac{\text{distance travelled}}{\text{time taken}}$$

The total distance travelled is four sides of the square, each of length 2 km, giving a total of 8 km. Total time taken is a bit more tricky. The first leg took 2 hours (2 km at 1 km per hour), the second look 1 hour, the third took half an hour, and the last leg took another 2 hours. Total time taken was $2 + 1 + 0.5 + 2 = 5.5$ hours. So the correct average velocity is

$$v = \frac{8}{5.5} = 1.455$$

What is the trick? The trick is that this question calls for a different sort of mean. The harmonic mean has a very long definition in words. It is *the reciprocal of the average of the reciprocals*. The reciprocal of x is $1/x$, which can also be written as x^{-1}. So the formula for harmonic mean 'y hat' looks like this:

$$\hat{y} = \frac{1}{\dfrac{\sum \dfrac{1}{y}}{n}} = \frac{n}{\sum \dfrac{1}{y}}$$

Let's try this out with the elephant data:

```
4/sum(1/c(1,2,4,1))
```

```
[1]  1.454545
```

So that works okay. Let's write a general function to compute the harmonic mean of any vector of numbers:

```
harmonic<-function(x) 1/mean(1/x)
```

We can try it out on our vector y:

```
harmonic(y)
```

```
[1]  13.64209
```

To summarise, our measures of central tendency all have different values, but because the y values were well clustered, and there were no serious outliers, all the different values were quite close to one another.

Mode	Median	Arithmetic mean	Geometric mean	Harmonic mean
14	14	13.7879	13.7153	13.6421

Notice that if you try typing mode(y) you don't get 14, you get the message

```
[1]  "numeric"
```

because the function **mode** tells you the *type* of an S-Plus object, and y is of type numeric. A different mode is 'character'.

The distribution of y was quite symmetric, but many data sets are skew to one side or the other. A long tail to the right is called positive skew, and this is much commoner in practice that data sets which have a long tail to the left (negative skew, see p. 103). The frequency distribution fz is an example of positive skew:

```
par(mfrow=c(1,1))
barplot(fz,names=as.character(x))
```

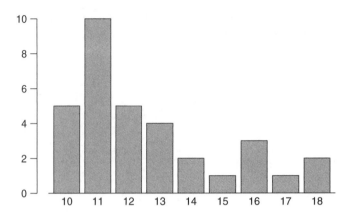

It is useful to know how the different measures of central tendency compare when the distribution is skew like this. First, we expand the frequency distribution into a vector of measurements, w, using **rep** to repeat each of the x values fz times:

```
w<-rep(x,fz)
```

We want to produce a summary of all of the information we have gathered on central tendency for the data in the vector called w. Of course, we could go through and type each function in turn, but it would be much more efficient to bundle them all together into a utility function called, say, **central** like this (the code for the user-defined function called **central** is described on p. 32). Don't try to execute this unless you have loaded the function:

```
central(w)

Arithmetic mean  = 12.606
Geometric mean   = 12.404
Harmonic mean    = 12.22
Median           = 12
Range      from 10  to 18
```

When there is positive skew, the *mode is the lowest* estimate of central tendency (11 in this case) and the *arithmetic mean is the largest* (12.606). The others are intermediate, with geometric mean > harmonic mean > median in this case.

Further reading

Casella, G. and Berger, R.L. (1990) *Statistical Inference*. Pacific Grove, California, Wadsworth and Brooks Cole.
Cox, D.R. and Hinkley, D.V. (1974) *Theoretical Statistics*. London, Chapman & Hall.
Snedecor, G.W. and Cochran, W.G. (1980) *Statistical Methods*. Ames, Iowa, Iowa State University Press.
Sokal, R.R. and Rohlf, F.J. (1995) *Biometry: The Principles and Practice of Statistics in Biological Research*. San Francisco, W.H. Freeman.

5

Probability

Most people have a reasonably good intuitive understanding of the notion of probability. Some things are certain: we are all going to die. Some things are impossible: gravity will stop working tomorrow. These extremes are given numerical values: certainty is $p = 1$ and impossibility is $p = 0$. It follows that probabilities can't have values bigger than 1 or less than 0. In between, we classify events as more or less likely by giving them values of p closer to 1 or closer to 0.

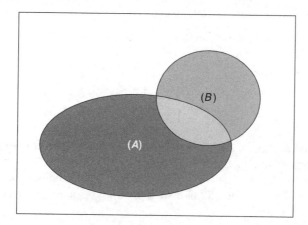

Consider the diagram above. There is the rectangular space. Then there are the partially overlapping shapes A and B. The idea is to think of the areas of these objects as probabilities. The rectangle has an area of 1. The probability of A is written $P(A)$ and this is the dark shaded area. This is about one-third of the rectangular area so we say 'the probability of A is about one-third', or we could write $P(A) = 0.333$ The area of B is a bit smaller; we say 'the probability of B is about one-fifth', or write $P(B) = 0.2$.

The key concept has to do with *the overlap between the objects*. If there *were* no overlap then the probability of A *or* B would just be the sum of the two separate probabilities, $P(A) + P(B)$.

If we are interested in the probability of A *and* B together, then the area of interest is the overlap itself: $P(A \& B) =$ the overlap. This is the hard part. If there is a *positive* correlation between the two, then getting A greatly increases the chance of getting B and hence of getting the two together. If there is a *negative* correlation between the two, then getting A greatly reduces the chance of getting B, so it is unlikely that we shall get the two together. If they are completely *independent* events, then getting A has no influence on the probability of getting B, and *the probability of getting both is just the product of their two probabilities*. So if $P(A) = 0.33$ and $P(B) = 0.2$ then $P(A \& B) = 0.33 \times 0.2 = 0.066$. Notice than we can make no such calculations about $P(A \& B)$ when the two probabilities are *not* independent, unless we know the value of the correlation coefficient (see p. 112).

If our concern is with getting *A or B* then the computation is a little bit more complicated, because *we must not count the area of overlap twice*. This means that the probability of *A* or *B* is *the probability of A plus the probability of B minus the area of overlap.* $P(A \text{ or } B) = P(A) + P(B)$—the overlap. In the extreme case of complete overlap, the probability of *A* or *B* is just the same as the probability of *A*.

Probability in symbols

$\cup$ means 'union', so $A \cup B$ is read 'either *A* or *B*'

$\cap$ means 'intersection', so $A \cap B$ is read '*A* and *B* together'

| means 'given', so $A|B$ is read '*A* given *B*' (conditional probability)

\ the backslash represents 'not', so $A \backslash B$ is read '*A* not *B*'

^{c} means 'complement', so A^c is the event that contains no *A*s

$\emptyset$ the 'empty set' representing an impossible event

Ω the 'sample space' comprising the collection of all possible objects

Some fundamental ideas are

$$P(\Omega) = 1 \quad \text{and} \quad P(\emptyset) = 0$$

$$P(A^c) = 1 - P(A)$$

For more complicated notions, the trick is to *express an event in terms of disjoint unions* and then apply probability to each disjoint element separately.

For instance, we want to know $P(A \cup B)$. Expressed as disjoint unions the events in question are $A \cup (B \backslash A)$, which is read as '*A* plus the part of *B* that is not in *A*'. Now because the regions are disjoint, we can apply *P* to get

$$P(A \cup B) = P(A) + P(B \backslash A)$$

This step is useful because we know that $P(B \backslash A)$ is the same as $P(B \backslash (A \cap B))$. These parts are disjoint, so we can apply probabilities to get $P(B) - P(A \cap B)$. Thus

$$P(A \cup B) = P(A) + P(B) - P(A \cap B)$$

This is one of the most frequently used rules in probability calculations, and it is really useful to remember the formula. The **intersection** of *A* and *B* and is written $A \cap B$ and means 'both'. The **union** of *A* and *B* is written $A \cup B$ and means 'either' (i.e. *A* or *B* or both *A* and *B*). The probability of the union of *A* and *B* is the probability of *A*, plus the probability of *B*, minus the probability of the intersection of *A* and *B*. You may find it hard at first to remember which symbol is which: union ($\cup$) looks like a letter *U* and intersection ($\cap$) doesn't ! You can always draw the figure and think about the area of overlap as the intersection. It is no panacea, of course, because you have to know the value of $P(A \cap B)$ and this in turn depends on the extent to which the events *A* and *B* are correlated (see below).

Now suppose you toss a coin. In the unlikely event that the coin lands on its edge, you are allowed to bang the table. This means that there are only two possible outcomes: heads or tails. Our intuition tells us that if we are willing to assume the coin is fair (i.e. unbiased) then heads and tails are equally likely. There are two outcomes {H,T} and when we toss the coin we get only one of these: {H} or {T}. This leads us to the assumption that if we toss the coin lots of times, the proportion of tosses that land on heads will be $\frac{1}{2} = p = 0.5$.

But what exactly do we mean by this? What happens if we toss the coin 7 times? How many possible outcomes are there? It is possible (but extremely unlikely if the coin really is fair) that they all land up heads. It is much more likely that we get 3 or 4 heads. But we might, in principle, get none at all. A moment's thought will convince you that this experiment has 8 possible outcomes, equivalent to 0, 1, 2, 3, 4, 5, 6 and 7 heads out of 7 tosses. Getting 3 or 4 is more likely than 1 or 6 but all the outcomes are possible.

There are four kinds of questions that we might want to be able to answer:

- What is the probability of getting exactly 4 heads?

- What is the probability of getting more than 2 heads?

- What is the probability of getting less than 5 heads?

- What is the probability of getting between 3 heads and 5 heads inclusive?

Let's think about each of these in turn. To work out the probability of getting exactly 4 heads in 7 tosses, we need to do two things. The first is to work out the probability of any one particular trial that gave 4 heads, e.g. {H,H,H,H,T,T,T}. The next is to work out how many different ways there are of getting exactly 4 heads out of 7 trials. We have seen one of them, but there are lots of others, e.g. {T,T,T,H,H,H,H} or {H,T,H,T,H,T,H}.

First things first. What is the probability of getting {H,H,H,H,T,T,T}? We need to consider each of the 7 tosses in turn. What is the probability that the first toss produces heads? Heads is one of two possible outcomes, so (assuming as always that the coin is fair) we say that the probability of heads $p(H) = \frac{1}{2} = 0.5$. Exactly the same reasoning applies to the second, third and fourth heads. What about the first of the tails? The same reasoning shows that $p(T)$ also $= \frac{1}{2} = 0.5$. So we have a series of 7 events, each with probability $= 0.5$. What, then, is the probability of {H,H,H,H,T,T,T}?

At the moment, we are not in a position to work this out. We need a rule for working out the probability of combinations of events (e.g. 4 consecutive heads followed by 3 consecutive tails). The rule is very important, and apparently very simple:

The probability of two independent events is the product of their separate probabilities

This means that the probability of getting 2 consecutive heads is $0.5 \times 0.5 = 0.25$. This is also the probability of getting 2 consecutive tails. So what is the probability of getting one head and one tail in 2 throws. We could work it out longhand, or we could use the information we have already worked out. If we don't get *either* 2 heads or 2 tails, then we *must* get one head and one tail. Remember that certainty is $p = 1$. So the probability of one head and one tail is $1 - 0.25 - 0.25 = 0.5$.

So let's work out the probability of getting {H,H,H,H,T,T,T}. It is

$$\frac{1}{2} \times \frac{1}{2} \times \frac{1}{2} \times \frac{1}{2} \times \frac{1}{2} \times \frac{1}{2} \times \frac{1}{2} = \left(\frac{1}{2}\right)^7 = \frac{1}{128} = 0.007\,8125$$

This is an unlikely event (we expect to get it only about 8 times in 1000 trials). A moment's thought will show you that, since heads and tails are equally likely (both have the same probability $p = 0.5$), every single possible outcome of 7 tosses has this same probability of 1 in 128.

In order to compete the calculation, we need to know the number of ways of getting 4 heads out of 7 trials. We could get a pencil and write them all down, then count them up:

{H,H,H,H,T,T,T}, {H,H,H,T,H,T,T}, {H,H,H,T,T,H,T}, {H,H,H,T,T,T,H}, etc.

But this is extremely tedious and very prone to error; for example, it would be really easy to forget to write down one of the combinations. Fortunately, there is a simple formula that will work out the answer for us. It is called the combinatorial formula, and it makes use of the *factorial* function. The mathematical notation for factorial is the exclamation mark ! Factorials apply to whole numbers (integers). The factorial of x is the product of all the integers from 1 up to and including x. Thus 4 factorial, written 4!, is $1 \times 2 \times 3 \times 4 = 24$. In S-Plus the function is **factorial:**

```
factorial(4)
```

```
[1]  24
```

The combinatorial formula works out *the number of ways of combining x items out of n*. It looks like this:

$$\frac{n!}{x!(n-x)!}$$

So what is n and what is x? In our example of 4 heads out of 7 tosses, $n = 7$ (the sample size) and $x = 4$ (the number of heads). We work out the number of ways of getting 4 out of 7 by replacing n by 7 and x by 4 in the combinatorial formula:

$$\frac{7!}{4!(7-4)!}$$

We need to do some cancelling out here. Let's expand the factorials to begin with:

$$\frac{7 \times 6 \times 5 \times 4 \times 3 \times 2}{4 \times 3 \times 2 \times 3 \times 2}$$

Note that there is no point in writing $\times 1$ at the end of each factorial (so we leave it off). Note also that we subtract 4 from 7 before we work out the factorial (i.e. 3!). The 2s and 3s cancel out. And the 4s. Also, 2×3 in the denominator makes 6, which cancels with the 6 in the numerator. So we are left with $7 \times 5 = 35$ as our answer. There are 35 ways of getting exactly 4 heads out of 7 tosses of a coin, whether it is fair or not. S-Plus speeds up this calculation with its built-in function called **choose:**

```
choose(7,4)
```

```
[1]   35
```

We can now get the answer we want. The probability of getting exactly 4 heads out of 7 tosses is:

the probability of one case times the numbers ways of getting it

Here this is

```
1/128*35
```

```
[1]   0.2734375
```

We will get exactly 4 heads out of 7 tosses just over one-quarter of the time (27% of the time to be exact). We could go through exactly the same process to work out the probability of getting 1 head, 2 heads, and so on. This is very tedious, however. S-Plus allows us to specify lists in many of its functions. So instead of applying **choose** separately for 0, 1, 2, 3, 4, 5, 6 and 7 heads, we invoke the function just once. Instead of providing a single value for the number of heads (as we wrote 4 in the last example) we provide a list of 8 numbers, 0 to 7, like this:

```
choose(7,0:7)

[1]   1   7   21   35   35   21   7   1
```

This says that there is only 1 way of getting no heads out of 7 tosses (they must all be tails). There are 7 ways of getting 1 head: it could be on the first toss, or on the second, and so on. Likewise there is only 1 way of getting 7 heads (they must all be heads). There are 7 ways of getting 6 heads, because the inevitable 1 tail could be the first, or the second, or the third, and so on. The combinatorial formula comes into its own for computing the intermediate numbers of ways of getting things. We worked out the long way that there are 35 ways of getting 4 heads. You should be able to work out that there must be 35 ways of getting 3 heads as well (because we could have done exactly the same calculation, but for 4 *tails* instead of 4 heads.). The output tells us that there are 21 ways of getting 2 heads in 7 tosses and there are 21 ways of getting 5 heads in 7 tosses.

Because in this particular case, all of the outcomes of the 7 tosses have the same probability of 1/128 (because $p(H) = p(T) = 0.5$) we can easily automate the calculation of the 8 probabilities of obtaining 0, 1, 2, 3, 4, 5, 6 or 7 heads in 7 tosses by multiplying the output of **choose(7,0:7)** by 1/128 like this:

```
choose(7,0:7)/128

[1]   0.0078125   0.0546875   0.1640625   0.2734375   0.2734375   0.1640625
[7]   0.0546875   0.0078125
```

What should these numbers add up to? Since the number of heads has to be a number between 0 and 7, the probabilities should add up to exactly 1. Let's see how good a job S-Plus has done:

```
sum(choose(7,0:7)/128)

[1]   1
```

A pretty good job is the answer. One of the things we often want to do, is to draw barplots to show the probability of different outcomes. S-Plus makes this very simple:

```
barplot(choose(7,0:7)/128)
```

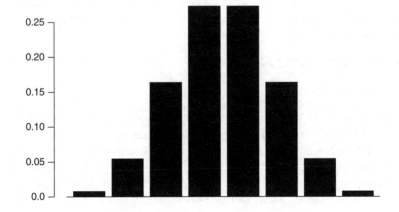

This is fine but it would be nice to have labels for the bars to show which number of heads they relate to, and a label for the *y* axis showing that the numbers are probabilities. We need to embellish the **barplot** function by providing three extra arguments:

barplot(choose(7,0:7)/128,ylab="Probability",xlab="Heads",names = as.character(0:7))

Note that we need to turn the numbers 0 to 8 into characters before we can used them as **names** for the bars.

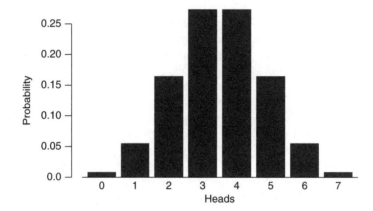

Combinatorial calculations

The National Lottery in Britain consists of selecting six different numbers between 1 and 49. Each selection costs you £1, and it is worth working out your chances of winning the jackpot when 6 balls out of 49 are drawn on live TV. This is a simple application of the combinatorial formula: *n* is 49 and *x* is 6. So the sample space is

choose(49,6)

[1]   13983816

So you have about a 1 in 14 million chance of getting the jackpot. Your bank manager in not likely to be impressed by your application for a loan of £14 million to enable you to buy one of every ticket, because the expected payout on a jackpot is much less than £14 million. Also, suppose that it takes 20 seconds to fill out and process each lottery ticket. Then the act of purchasing your tickets will take you nearly 9 years:

choose(49,6)*20/(60*60*24*365)

[1]   8.868478

even if you didn't sleep at all! This is clearly not a sensible plan.

Let us take a slightly more involved example from the card game of bridge. A pack of 52 cards is dealt at random into 4 hands of 13 cards each. The pack consists of 4 suits, each of 13 different-valued cards: the top card in each suit is the ace. What is the probability that in 7 deals there is at least one hand in which each player gets one ace? There are several steps to this. First, how many different games are there? The 52 cards can be permed in 52! ways. Each hand can be permed in

13! ways. There are 4 hands, so the combination of 4 hands can be permed in $(13!)^4$ ways. So the bridge hands are

$$\frac{52!}{(13!)^4}$$

factorial(52)/(factorial(13)^4)

[1]   5.364474e+028

which means that you are not likely to get bored by seeing the same hands over and over again. Now the 4 aces can be distributed, one to each hand, in 4! ways. When this is done, there are

$$\frac{48!}{(12!)^4}$$

ways of distributing the other cards. With these quantities we can work out the probability that the 4 aces end up, one in each hand, on any one deal:

$$P(A) = \frac{4! \times 48!/(12!)^4}{52!/(13!)^4}$$

factorial(4)*(factorial(48)/(factorial(12)^4)) /(factorial(52)/(factorial(13)^4))

[1]   0.1054982

So the answer to the first part of the question is about 1 in 10. Now we need to work out the probability of this happening at least once in 7 hands. The probability of it happening for the first time on the seventh deal is the probability of it *not* happening 6 times and then happening once. The probability of it not happening is $1-0.1 = 0.9$, so the probability of this happening 6 times is 0.9^6. Then the success, with probability 0.1, happened on the seventh hand so the probability of this particular way of getting the event is

0.9^6*0.1

[1]   0.0531441

But the success could have occurred on the very first deal, with probability 0.1. What we need to do is to add up the probabilities over all the ways that success could come about (on the first hand, the second, the third, and so on):

$$P(A \text{ occurs once in 7 deals}) = \sum_{i=1}^{7} P(B_i) = \sum_{i=1}^{7} (0.9^{i-1}\, 0.1)$$

sum(0.1*0.9^(0:6))

[1]   0.5217031

A very important piece of computing is demonstrated by this example. We do *not* use a loop from 1 to 7, make the 7 calculations, then add them up. We could do that, but it would be very inefficient. We generate the series of $i - 1$ values using 0:6 to create a vector of probabilities, then we apply the vector function **sum** to obtain the answer we need. The chance of getting one ace in each of the 4 hands once in 7 deals is about 50%.

Buffon's needle

The famous French naturalist devised a clever experiment to determine the numerical value of π. He ruled a sheet of paper with vertical lines exactly 1 unit length apart. Then he repeatedly dropped a needle, again of length exactly 1, onto the ruled paper at random. Often the needle fell at an angle more or less parallel with the ruled lines and so neither end of the needle intersected a line. Sometimes the needle fell more or less at right angles to the ruled line and one end or the other intersected a line. What Buffon showed was that the proportion of throws that gives an intersection, p, was related to π in the following remarkably simple way:

$$p = \frac{2}{\pi}$$

Here we use S-Plus to carry out a simulation experiment to see how good the rule is in practice. Instead of throwing the needle anywhere on a sheet of paper, let's randomise its position by locating it somewhere along a line from $x = 0$ to $x = 1$. Centred on this random position, the needle is then spun and comes to rest at an angle (θ) selected from a uniform random distribution somewhere between 0 and 2π (the full circle). It will help to draw a picture in order to see what is involved.

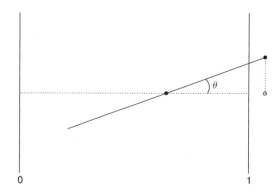

A throw of the needle results in an intersection if (as here) the x coordinate of the right-hand end of the needle ($\circ$) is >1. An intersection on the left line occurs if the x coordinate of the left end of the needle is <0. Obviously, both outcomes cannot occur on the same throw of the needle. An intersection on the left can occur only if the random location ($\bullet$) is less than 0.5, and on the right only if (as here) the location is ≥ 0.5. For an intersection of the left-hand end of the needle, the x coordinate of the end of the needle is given by $x - 0.5|\cos\theta|$. For the right end, the x coordinate is $x + 0.5|\cos\theta|$. The absolute values are necessary to take account of the two lower quadrants in which $\cos\theta$ is negative. All we need to do is count up the number of negative coordinates of the left-hand end, and the number of right-hand x coordinates that are greater than 1.

You might be tempted to write your simulation model in a loop. Go round 10 000 times and generate a random throw of the needle by generating two random numbers: a location between 0 and 1, and an angle between 0 and 2π. This would work but it is poor programming. S-Plus does things most efficiently as vectors. All we need to do is generate two vectors of random numbers using the uniform random number generator **runif**. The model looks like this. First we generate the 10 000 random locations and angles:

```
x<-runif(100000)
theta<-2*pi*runif(100000)
```

Then, for each throw, we work out the x coordinate of the left- and right-hand ends of the needle using **ifelse**:

```
left<- ifelse (x<0.5,x-0.5*abs(cos(theta)),0)
right<- ifelse (x>=0.5,x+0.5*abs(cos(theta)),0)
```

We can see how the simulation performed by calculating the proportion of throws that led to intersections, then using this to approximate the value of π. The proportion is just the sum of the TRUE cases (left < 0) and (right > 1) divided by the number of throws (10 000):

```
p<-(sum(left<0)+sum(right>1))/100000
p
```

```
[1]   0.63749
```

The value of π is supposed to be approximated by $2/p$; let's see:

```
2/p
```

```
[1]   3.137304
```

and the right answer is

```
pi
```

```
[1]   3.141593
```

Conditional probability

What is the probability of A given that B has occurred ? This is called the conditional probability of A given B and is written $P(A|B)$. A quick look at the Venn diagram (p. 75) should convince you that since A can only occur in the region of overlap $(A \cap B)$, because it is given that B has occurred, then the probability of A given B must be

$$P(A|B) = \frac{P(A \cap B)}{P(B)}$$

A numerical example might help. Suppose we want to know the probability that the total of two fair dice throws is greater than 6, given that the first throw showed a 3? It is very easy to work out the answer longhand: the second dice would need to show either 4, 5 or 6 for the total to exceed 6. This is 3 out of a total of 6 possibilities for the second throw of the dice, so the probability must be $\frac{1}{2}$. Using the formula gives the same answer but in a somewhat more roundabout way. The numerator is the number of successful cases for the two throws (3,4), (3,5) and (3, 6) $= 3$ cases. The denominator is the number of cases for the first throw (any one of the 6 faces). So

$$P(A|B) = \frac{P(A \cap B)}{P(B)} = \frac{|A \cap B|}{|B|} = \frac{3}{6} = \frac{1}{2}$$

This notion underlies many of the party game probability questions. For instance, what is the probability that in a family of two children both are boys, given that at least one is a boy. Most people immediately respond $\frac{1}{2}$. Wrong! This is the answer to a different question: For a family of two children, what is the probability that both are boys given that the younger child is a boy?

Our original question is treated like this. The sample space of all possible two-child families in terms of boys (B) and girls (G) is $\{BB, GG, GB, BG\}$. We assume that $P(BB) = P(GG) = P(GB) = P(BG) = \frac{1}{4}$. What we want to discover is

$$P(BB|\text{at least one boy})$$

The conditional sample space is therefore $\{BB, BG, GB\}$ from which only one outcome (BB) is a success. So the answer to our question is $\frac{1}{3}$ not $\frac{1}{2}$.

More formally, $P(BB|\text{at least one boy})$ is written

$$P(BB|GB \cup BG \cup BB) = \frac{P(BB \cap (GB \cup BG \cup BB))}{P(GB \cup BG \cup BB)}$$

This rather trivial example leads into one of the most important lemmas (demonstrated propositions) in all of probability theory. A **partition** is a family of events $B_1, B_2, \ldots, B_n$ such that the events are disjoint:

$$B_i \cap B_j = \emptyset \text{ when } i \neq j \quad \text{and} \quad \bigcup_{i=1}^{n} B_i = \Omega$$

When the events $B_1, B_2, \ldots, B_n$ are a partition of Ω then

$$P(A) = \sum_{i=1}^{n} P(A|B_i) P(B_i)$$

For any events A and B

$$P(A) = P(A|B)P(B) + P(A|B^c)P(B^c)$$

which, in words, says the probability of A is equal to the probability of A given B multiplied by the probability of B, plus the probability of the disjoint probability of A given 'not B' (the complement of B, which is B^c) multiplied by the probability of 'not B'. It is not immediately obvious why this is so important. However, it is useful for solving 'balls from urns' problems like the following classic. Suppose there are two urns. The first contains 2 white balls and 3 blue balls. The second urn contains 3 white balls and 4 blue balls. Here comes the twist. One ball is drawn at random from the first urn and put into the second urn. Now a ball is drawn at random from the second urn. What is the probability that it is blue?

Obviously, the result depends upon whether it was a white ball or a blue ball that was moved from urn I to urn II during the first stage in the proceedings. Therefore we need to work out two things: the probability of selecting a blue ball from urn II when it was a *blue* ball that was added (so urn II contains 3 white balls and 5 blue balls) = 5/8; and the probability of selecting a blue ball when it was a *white* ball that was added (so urn II contains 4 white balls and 4 blue balls) = 1/2. Now the probability that the ball taken from urn I was blue is 3/5 (we define this as $P(B)$) and not blue is 2/5 (this is $P(B^c)$). Now we have all the probabilities needed to work out the probability that the ball taken from urn II is blue (call this $P(A)$). In terms of symbols, our probability of 5/8 is $P(A|B)$ and our probability of 1/2 is $P(A|B^c)$. Thus the equation becomes

$$P(A) = \frac{5}{8} \cdot \frac{3}{5} + \frac{1}{2} \cdot \frac{2}{5} = \frac{23}{40}$$

which is much more complicated than most people would expect the answer to be. You can answer other kinds of questions as well. If the chosen ball is blue, what is the probability that it was the one taken from urn I? This is slightly more tricky.

After the ball was moved from urn I, there are 4 plus 3/5 blue balls and 3 plus 2/5 white balls in urn II (8 balls in all). Only 3/5 of the blue balls in urn II are the moved blue ball. Now, we know that the ball removed from urn II was blue, so the population of balls was not 8 but 4 plus 3/5. So the answer to our question is

$$P(\text{from urn I|blue}) = \frac{\frac{3}{5}}{4\frac{3}{5}} = \frac{3}{23}$$

We can generalise these examples as follows. Think about events E and F, and hypothesis H. The probability of the event is just $P(E)$. The **conditional probability** is the probability of the event *given the hypothesis*; it is written $P(E|H)$. There are 4 important axioms of probability:

- $P(E|H) \geq 0$ for all E and H (there is no such thing as negative probability)

- $P(H|H) = 1$ for all H (what we know to be true *is* true)

- $P(E \cup F|H) = P(E|H) + P(F|H)$ only when there is no overlap (when $EFH = \emptyset$)

- $P(E|FH)P(F|H) = P(EF|H)$

Emerging from these axioms are some other straightforward points:

- $P(E|H) \leq 1$ for all E and H (there's no such thing as a probability greater than 1)

- $P(E|FH) = P(EF|H)/P(F|H)$ so long as $P(F|H)$ is not equal to zero

Bayes' theorem

Bayes' theorem is an important general rule which says that

$$P(H|E)P(E) = P(EH) = P(H)P(E|H)$$

and, provided that $P(E)$ is not 0, then

$$P(H|E) = \frac{P(H)P(E|H)}{P(E)}$$

Let's examine some examples of the use of this rule. There are two kinds of twins: if the zygote split in two *after* fertilisation, the twins are called monozygotic (identical twins) and a single sperm was involved. If two separate eggs were fertilised, then two sperm were involved and the twins are called dizygotic. Usually, but not always, the monozygotic twins look very like one another, but dizygotic twins can also be very similar. The point is that you cannot necessarily tell dizygotic twins from monozygotic twins just by looking. There is a useful extra bit of information, however. Since it is the father (with his XY chromosomes) who determines the sex of the offspring (mother is XX), and since monozygotic twins come from the same sperm, they *must* be the same sex. Dizygotic twins are sometimes of the same sex, but not necessarily. One of the sperm that fertilised them might have been carrying an X chromosome and the other a Y chromosome. Let M be monozygotic and D be dizygotic and denote boys by B and girls by G.

Twins can be of three kinds: 2 boys, 2 girls or a girl and a boy (BB, GG, or GB). Now what else do we know? If the twins *are* monozygotic, then only BB or GG can occur, and both are equally likely. So

$$P(GG|M) = P(BB|M) = \tfrac{1}{2}$$

which we read as 'the probability of girl:girl, given that they are monozygotic, is one-half'. We also know that $P(GB|M) = 0$ (if they are monozygotic, then the two cannot be of different sexes). For dizygotic twins, GG and BB are less likely than GB. There are four possibilities: GG, BB, GB and BG, so GG has probability 1/4, as does BB, but one of each sex has probability of $2/4 = 1/2$. We write this down as

$$P(GG|D) = P(BB|D) = \tfrac{1}{4}$$

with the extra information that $P(GB|D) = 1/2$ (not zero as with monozygotic twins).

The information from the two kinds of twins is combined like this. The overall probability of GG is the sum of the two conditional probabilities:

$$P(GG) = P(GG|M)P(M) + P(GG|D)P(D)$$

and we know that $P(D) = 1 - P(M)$. Replacing the conditional probabilities with the values we have just calculated, gives us

$$P(GG) = \tfrac{1}{2}P(M) + \tfrac{1}{4}P(D) = \tfrac{1}{2}P(M) + \tfrac{1}{4}[1 - P(M)]$$

We can rearrange this to obtain an expression for the probability of monozygotic:

$$P(M) = 4P(GG) - 1$$

So, although you cannot be certain about the origin of a particular pair of BB or GG twins, you can get a good estimate of the proportion of monozygotic twins in the population as a whole just by looking at the proportion of all twins that are GG. So if you observe 30% of all twins to be GG, then the proportion of twins that are monozygotic is given by $4 \times 0.3 - 1 = 1.2 - 1 = 0.2$ And if the twins are BG or GB, you know that they are definitely *not* monozygotic.

Next consider the celebrated example from R.A. Fisher's famous book (R.A. Fisher 1925). The data refer to criminal convictions of twins: monozygotic twins (M) and dizygotic twins (D). If they have a criminal record they are C and if not they are N. This defines a 2×2 contingency table as follows.

Twins	Criminal	Not criminal	Total
Monozygotic	10	3	13
Dizygotic	2	15	17
Total	12	18	30

Suppose our hypothesis H is that an individual was drawn at random from this table. The probability that it is monozygotic is therefore 13/30, and the probability that it has a criminal record is 12/30 (check that you know where these numbers come from). If there was no association between criminality and the cellular origin of the twins, then the probability of being monozygotic *and* having a criminal record should be the product of the two probabilities, $12/30 \times 13/30$.

Given that they *had* a criminal record, however, the observed probability that they were monozygotic was 10/12. Now we write these simple observations in the form of probability statements:

$$P(C|H) = \frac{12}{30}$$

$$P(M|CH) = \frac{10}{12}$$

Finally, we saw that the probability that they were *both* monozygotic and had a criminal record was

$$P(MC|H) = \frac{10}{30}$$

We can combine these statements to demonstrate the value of the fourth axiom:

$$P(M|CH)P(C|H) = P(MC|H)$$

$$\frac{10}{12} \times \frac{12}{30} = \frac{10}{30}$$

and this will be true whatever the particular numerical values in the contingency table.

Beware the transposed conditional

'Blatant sexism,' screams the headline, 'only 2% of women promoted.' On the face of it, this looks like a pretty clear-cut case of sex discrimination, especially if the candidates for promotion came from a population with a 50:50 sex ratio. But we are not told this. We need to delve more deeply. Reading the small print we discover that 4 women were promoted out of a total of 200 people promoted. So at least the arithmetic is correct! Reading even further, it turns out that the number of female candidates was 40, so the success rate was 4 out of 40, or 10%. The 196 promoted males (200−4) came from of pool of male aspirants of a whopping 3270. So male success rate was about 6% (5.99% to be pedantic). So, if anything, the data show positive discrimination in favour of female candidates (10% success rather than 6%).

It is useful to learn how to label these percentages properly in terms of conditional probabilities. The probability of promotion for female candidates was 10%. This is $P(S|W)$: the probability of success (S) given that the candidate was a woman (W). Likewise, $P(S|M)$ is the probability of success (S) given that the candidate was a man (M); this is 6%.

So what exactly was the 2% in the headline? It is the proportion of promoted candidates that were women; that is to say, the probability of being a woman, given that you were promoted $P(W|S)$. This is a classic mistake in the presentation of probability information. It is known, rather grandly, as the *fallacy of the transposed conditional*. We are given $P(W|S)$ when what we really want to know is $P(S|W)$.

The extra information we need is the probability that the candidate was a woman $P(W)$ or a man $P(M)$. The tacit assumption in the headline was that both these probabilities were 50%, but we now know that this was false. The total number of candidates was 40 women plus 3270 men = 3310. So $P(W) = 40/3310 = 0.0121$ and $P(M) = 3270/3310 = 0.9879$. Obviously these add up to 100% (none of the applicants was hermaphrodite in this case).

Now, the useful bit is this; it gives us the correct average probability of promotion independent of gender:

$$P(S) = P(S|W)P(W) + P(S|M)P(M)$$

```
0.1*0.0121+0.0599*0.9879
```

```
[1]   0.06038521
```

So it really does look as if the women with their 10% success rate did *better* than the true average success rate of 6.04% that we have just worked out. We investigate whether the difference between the two proportions is statistically significant or not in Chapter 11.

There is an important extra lesson to be learned at this point: *the average proportion is not the same as the average of the proportions.* The average success rate in our sexism example is 6.04% but the average of the two separate success rates was $(10\% + 6\%)/2$ which is 8% (a very different answer). The correct way to work out the average success rate is to calculate the total number of successes $(4 + 196 = 200)$ and the total number of applicants $(40 + 3270 = 3310)$ then divide the one by the other. To get a percentage, we multiply the probability by 100 to get $200 \times 100/3310 = 6.04\%$

This kind of abuse of probability theory is rife in all walks of life. In the legal profession, the two arguments even have their own names: they are known as the *prosecutor's fallacy* and the *defender's fallacy*, respectively. Let us define I as innocent, G as guilty and E as the evidence. The prosecutor's fallacy is the fallacy of the transposed conditional (see above). It consists of quoting $P(E|I)$, the probability of the evidence given that the defendant is innocent, instead of $P(I|E)$, the probability of innocence, given the evidence. These two are equal only if the probability of innocence $P(I)$ is taken as $\frac{1}{2}$, which is hardly in accord with the notion of innocent until proven guilty!

Likelihood

The concept of likelihood may well be unfamiliar to you. We introduce it here with a simple example, then develop the notion in subsequent chapters. But the central idea is very straightforward. Some models are much better than others, and we want to find the best ones. One way of defining 'best' is to look for *the model that makes our data most likely.*

Suppose you threw three dice and added up the total. What would you expect the value to be? If you can't work it out, simulate it. Get S-Plus to do the experiment lots of times and see what happens. The way to simulate one throw of the dice is

```
ceiling(6*runif(1))
```

```
[1]   4
```

meaning we got a 4 in this case. We use **ceiling** rather than **floor** to get the integer value because we want the numbers to go from 1 to 6 rather than from 0 to 5. We want to do this three times and add up the total:

```
sum(ceiling(6*runif(3)))
```

```
[1]   17
```

This random example was almost as big as it possibly could be (the maximum value is obviously $3 \times 6 = 18$). Can you see what the expectation of this experiment is, if we repeat it lots of times? The total score of all six faces on one die is $1 + 2 + 3 + 4 + 5 + 6 = 21$. So a given throw of one die has an expectation of $21 \times 1/6 = 3.5$ With three dice, the total score is 63, so the expectation is 10.5. Lets see what S-Plus gets if we repeat the experiment 1000 times:

```
total<-numeric(1000)
for (i in 1:1000) total[i]<-sum(ceiling(6*runif(3)))
hist(total)
```

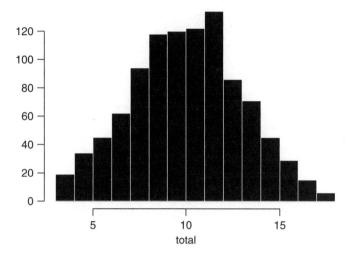

So the distribution of totals in different trials goes all the way from 3 to 18 as we would guess, but the expectation is 10 or 11. Now think about this experiment in slightly different terms. We could do the simulation because we *knew* that the probability of each of the six numbers was 1/6. Suppose now that we *didn't* know this, and we wanted to use the data from the experiment to deduce the value of this parameter. Let's call it p.

In symbols, the experiment looks like this:

$$T = 3 \times 21 = 63$$

$$p = \text{unknown}$$

$$E(x) = pT = p \times 63$$

The maximum likelihood estimate of p is defined like this. Given the data from our experiment and given the model that we intend to fit to these data ($E(x) = 63p$), what value of p makes these data most likely?

We shall try three different values of p to make the point: 1/5, 1/6 and 1/8. Suppose we run the experiment four times and get the data $\{10, 12, 5, 14\}$. If p was 1/4, then how likely would these data be? Our expectation would be $63 \times 1/4 = 15.75$; all of the data are less than this, and the differences between observed and expected are $\{-5.75, -3.75, -10.75, -1.75\}$. The sum of the absolute values of the differences is 22. If p was 1/8, our expectation would be $63 \times 1/8 = 7.875$: one of the values is smaller than this and three are bigger. The differences between what we observe and what we expect if $p = 1/8$ are $\{2.125, 4.125, -2.875, 6.125\}$. The sum of the absolute values of the differences is 15.25. Finally, what if $p = 1/6$. Then, as we have already seen, our expectation is $63 \times 1/6 = 10.5$ and the differences between observed and expected are $\{-0.5, 1.5, -5.5, 3.5\}$. The sum of the absolute values of the differences is 11.

The question is, Which value of p (1/4, 1/6 or 1/8) makes the data $\{10, 12, 5, 14\}$ most likely? The answer is clearly 1/6. We judge this because the predictions of our model ($E(x) = 63p$) are *closest to the data* for this value of p. We haven't yet said how we judged this closeness, but it obviously has something to do with the vector of differences between observed and predicted. One very simple measure might be the sum of the absolute values of the differences. Let's plot these against the different values of p:

```
p<-c(1/4,1/6,1/8)
d<-c(22,11,15.25)
plot(p,d,type="l")
```

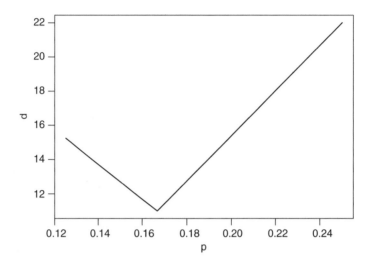

We get a very crude graph, but it makes the point: different values of p make the data more or less likely. The value of p that made the data most likely (and hence made the difference between the data and the model, d, smallest) was $p = 1/6$. That is our maximum likelihood estimate of p in the model $E(x) = 61p$.

The statistical techniques of maximum likelihood aim to find the values of the parameters of models that made the data most likely. These ideas are developed and made more precise by looking at likelihood in the context of the Normal distribution in Chapter 7.

Further reading

Grimmett, G.R. and Stirzaker, D.R. (1992) *Probability and Random Processes*. Oxford, Clarendon Press.
Kendall, M.G. and Stewart, A. (1979) *The Advanced Theory of Statistics*. Oxford, Oxford University Press.
Lee, P.M. (1997) *Bayesian Statistics: An Introduction*. London, Edward Arnold.
O'Hagen, A. (1988) *Probability: Methods and Measurement*. London, Chapman & Hall.
Riordan, J. (1978) *An Introduction to Combinatorial Analysis*. Princeton, New Jersey, Princeton University Press.

6

Variance

A measure of variability is perhaps the most important quantity in statistical analysis. The greater the variability in the data, the greater will be our uncertainty in the values of parameters estimated from the data, and the lower will be our ability to distinguish between competing hypotheses about the data.

Consider the following data, y, which are simply plotted in the order in which they were measured:

```
y<-c(13,7,5,12,9,15,6,11,9,7,12)
plot(y,ylim=c(0,20),pch=16)
```

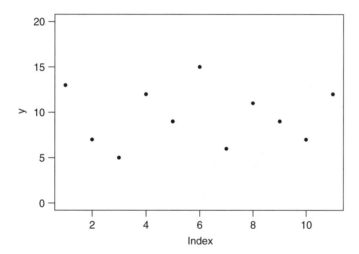

Visual inspection indicates substantial variation in y. But how to measure it? One way would be to specify the range of y values:

```
range(y)
```

```
[1]    5   15
```

The minimum value of y is 5 and the maximum is 15. The variability is contained within the range, and to that extent it is a very useful measure. But it is not ideal for general purposes. For one thing, it is totally determined by outliers, and gives us no indication of more typical levels of variation. Nor is it obvious how to use range in other kinds of calculations (e.g. in uncertainty measures). Finally, the range increases with the sample size, because if you add more numbers then eventually you will add one larger than the current maximum, and if you keep going you will find one smaller than the current minimum. This is a fact, but it is not a property of our unreliability estimate that we particularly want, We would like uncertainty to go *down* as the sample size went up.

How about fitting the average value of *y* to the data and measuring how far each individual *y* value departs from the mean?

abline(mean(y),0)

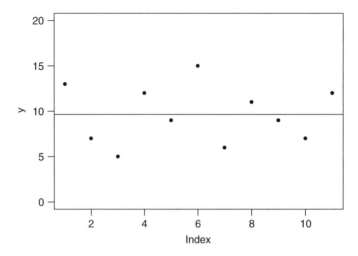

This divides the data into 5 points that are larger than the mean and 6 points that are smaller than the mean. The distance, *d*, of any point *y* from the mean $\bar{y}$ is

$$d = y - \bar{y}$$

for (i in 1:11) lines(c(i,i),c(mean(y),y[i]))

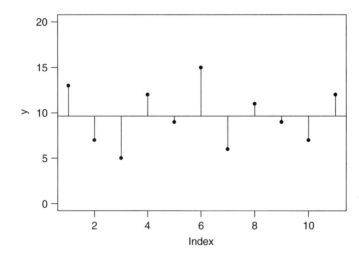

Now a variety of possibilities emerge for measuring variability. How about adding together the different values of *d*? This turns out to be completely hopeless, because the sum of the *d* values is always zero, no matter what the level of variation in the data! The proof is simple (Box 6.1).

Box 6.1 The sum of the differences $\sum (y - \bar{y})$ is zero

Start by writing down the differences explicitly

$$\sum d = \sum (y - \bar{y})$$

Take $\sum$ through the brackets. The important point is that $\sum \bar{y}$ is the same as $n\bar{y}$, so

$$\sum d = \sum y - n\bar{y}$$

and we know that $\bar{y} = \sum y / n$, so

$$\sum d = \sum y - \frac{n \sum y}{n}$$

Cancel n, leaving

$$\sum d = \sum y - \sum y = 0$$

But isn't Box 6.1 just a typical mathematician's trick, because the plus and minus values simply cancel out? Why not ignore the signs and look at the sum of the absolute values of d?

$$\sum |d| = \sum |y - \bar{y}|$$

This is actually a very appealing measure of variability, because it does not give undue weight to outliers. It was spurned in the past because it made the sums much more difficult, and nobody wants that. It has had a new lease of life since computationally intensive statistics have become much more popular.

The other simple way to get rid of the minus signs is to square each d before adding them up:

$$\sum d^2 = \sum (y - \bar{y})^2$$

This quantity has an absolutely central role in statistics. Given its importance, you might have thought they would have given it a really classy name. But no, it is called the 'sum of squares'. The sum of squares is the basis of all the measures of variability used in linear statistical analysis.

Is this all we need in our variability measure? Well not quite, because every time we add a new data point to our graph the sum of squares will get bigger. Sometimes by only a small amount when the new value is close to $\bar{y}$, but sometimes by a lot, when the new value is much bigger or smaller than $\bar{y}$. The solution is straightforward. We calculate the *average* of the squared deviations (known as the 'mean square'). But there is a small hitch. We could not calculate $\sum d^2$ before we knew the value of $\bar{y}$. And how did we know the value of $\bar{y}$? Well, we didn't know the value. We had to estimate it from the data. This leads us into a very important, but simple concept.

Degrees of freedom

Suppose we had a sample of five numbers and their average was 4. What was the sum of the five numbers? It must have been 20, otherwise the mean would not have been 4. So now we think about each of the five numbers in turn.

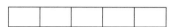

We are going to put numbers in each of the five boxes. If we allow that the numbers could be positive or negative real numbers, we ask how many values the first number could take. Once you see what I'm doing, you will realise it could take any value. Suppose it was a 2.

2				

How many values could the next number take? It could be anything. Say it was a 7.

2	7			

And the third number? Anything. Suppose it was a 4.

2	7	4		

The fourth number could be anything at all. Say it was 0.

2	7	4	0	

Now then. How many values could the last number take? Just one. It *has* to be another 7 because the numbers have to add up to 20.

2	7	4	0	7

To recap. We have total freedom in selecting the first number. And the second, third and fourth numbers. But we have no choice at all in selecting the fifth number. We have 4 degrees of freedom when we have 5 numbers. In general, we have $n - 1$ degrees of freedom if we estimated the mean from a sample of size n.

More generally still, we can propose a formal definition of degrees of freedom:

Degrees of freedom is the sample size, n, minus the number of parameters, p, estimated from the data

You should memorise this. In the example we just went through we had $n = 5$ and we had estimated just one parameter from the data—the sample mean, $\bar{y}$. So we had $n - 1 = 4$ d.f.

In a linear regression we estimate two parameters from the data when we fit the model

$$y = a + bx$$

We estimate the intercept, a, and the slope b. Because we have estimated two parameters, we have $n - 2$ d.f. In a one-way analysis of variance with 5 genotypes, we estimate 5 means from the data (one for each genotype) so we have $n - 5$ d.f. And so on. The ability to work out degrees of freedom is an incredibly useful skill. It enables you to spot mistakes in experimental designs and in statistical analyses. It helps you to spot pseudoreplication in the work of others, and to avoid it in work of your own.

Variance

We are now in a position to define the most important measure of variability in all of statistics: a variance, s^2, is always

$$\text{variance} = \frac{\text{sum of squares}}{\text{degrees of freedom}}$$

In our case, working out the variance of a single sample, the variance is

$$s^2 = \frac{\sum (y - \overline{y})^2}{n - 1}$$

This is so important, we need to take stock at this stage. The data in the following table come from three market gardens. The data show the ozone concentrations in parts per hundred million (pphm) on ten summer days.

Garden A	Garden B	Garden C
3	5	3
4	5	3
4	6	2
3	7	1
2	4	10
3	4	4
1	3	3
3	5	11
5	6	3
2	5	10

We want to calculate the variance in ozone concentration for each garden. There are four steps to this

- Determine the sample mean for each garden

- Subtract the sample mean from each value

- Square the differences and add them up to get the sum of squares

- Divide the sum of squares by degrees of freedom to obtain the variance

We begin by typing the data for each garden into vectors called A, B and C:

```
A<-c(3,4,4,3,2,3,1,3,5,2)
B<-c(5,5,6,7,4,4,3,5,6,5)
C<-c(3,3,2,1,10,4,3,11,3,10)
```

and calculating the three sample means:

```
mean(A)
[1]  3
mean(B)
[1]  5
mean(C)
[1]  5
```

Now we calculate vectors of length 10 containing the differences $y - \overline{y}$:

```
dA <- A-3
dB <- B-5
dC <- C-5
```

then determine the **sum** of the squares of these differences:

```
SSA<-sum(dA^2)
SSB<-sum(dB^2)
SSC<-sum(dC^2)
```

We find that $SSA = 12$, $SSB = 12$ and $SSC = 128$. The three variances are obtained by dividing the sum of squares by the degrees of freedom:

```
s2A<-SSA/9
s2B<-SSB/9
s2C<-SSC/9
```

To see the values of the three variances, we type their names separated by semicolons:

```
s2A;s2B;s2C
```

```
[1]  1.333333
[1]  1.333333
[1]  14.22222
```

Of course there is a short-cut formula to obtain the sample variance directly:

```
s2A<-var(A)
```

There are three important points to be made from this example:

- Two populations can have different means but the same variance (gardens A and B)

- Two populations can have the same mean but different variances (gardens B and C)

- Comparing means when the variances are different is an extremely bad idea.

The first two points are straightforward. The third point is profoundly important. Let's look again at the data. Ozone is only damaging to lettuce crops at concentrations in excess of a threshold of 8 pphm. Now looking at the average ozone concentrations we would conclude that both gardens are the same in terms of pollution damage. Wrong! Look at the data, and count the number of days of damaging air pollution in garden B. None at all. Now count garden C. This is completely different, with damaging air pollution on 3 days out of 10.

The moral is clear. *If you compare means from populations with different variances, you run the risk of making fundamental scientific mistakes.* In this case we might have concluded there was no pollution damage, when in fact there were damaging levels of pollution 30% of the time in garden C.

This begs the question of how you know whether or not two variances are significantly different. There is a very simple rule of thumb (the details are explained in Chapter 11): if the larger variance is more than 4 times the smaller variance, then the two variances are significantly different. In our example the variance ratio is

```
14.2222 / 1.3333
```

```
[1]  10.66692
```

This is much greater than 4, which means the variance in garden C is significantly higher than in the other two gardens. Note that because the variance is the same in gardens A and B, it is legitimate to carry out a significance test on the difference between their two means. This is Student's t test, explained in full below.

Short-cut formulas for calculating variance

This really was a fairly roundabout sort of process. The main problem with the formula defining variance is that it involves all those subtractions, $y - \overline{y}$. It would be good to find a way of calculating the sum of squares that didn't involve all these subtractions.

Let's expand the bracketed term $(y - \overline{y})^2$ to see if we can make any progress towards a subtraction-free solution:

$$(y - \overline{y})^2 = (y - \overline{y})(y - \overline{y}) = y^2 - 2y\overline{y} + \overline{y}^2$$

So far so good. Now we put the summation through each of the three terms separately:

$$\sum y^2 - 2\overline{y}\sum y + n\overline{y}^2 = \sum y^2 - 2\frac{\sum y}{n}\sum y + n\left[\frac{\sum y}{n}\right]^2$$

where only the ys take the summation sign, because we can replace $\sum \overline{y}$ by $n\overline{y}$. We replace $\overline{y}$ with $\sum y/n$ on the right-hand side. Now cancel n and collect the terms:

$$\sum y^2 - 2\frac{\left[\sum y\right]^2}{n} + n\frac{\left[\sum y\right]^2}{n^2} = \sum y^2 - \frac{\left[\sum y\right]^2}{n}$$

This gives us a formula for computing the sum of squares that avoids all the tedious subtractions.

To use the short-cut formula we need only the sum of the y and the sum of the y^2. It is very important to understand the difference between $\sum y^2$ and $\left[\sum y\right]^2$. It is worth doing a simple numerical example to make plain the distinction between these important quantities.

Let y be $\{1, 2, 3\}$. This means that $\sum y^2$ is $\{1 + 4 + 9\} = 14$. It also means that $\sum y$ is $\{1 + 2 + 3\} = 6$, so $\left[\sum y\right]^2 = 6^2 = 36$.

The square of the sum is always much larger than the sum of the squares. To recap:

$$\text{sum of squares} = \sum y^2 - \frac{\left[\sum y\right]^2}{n}$$

Note that in terms of computer programming technique, it is much better practice to use the *longhand* method $\sum(y - \overline{y})^2$ for computing the sum of squares, because it is much less subject to rounding errors. Our short-cut formula could be wildly inaccurate if we had to subtract one very large number from another. Box 6.2 explains how to calculate variance in terms of expectations.

Using variance

Variance is used in two main ways

- for establishing measures of unreliability

- for testing hypotheses

Standard error

Consider the properties that you would like a measure of unreliability to possess. As the variance of the data increases, what happens to the unreliability of estimated parameters? Does it go up or down? Unreliability goes up as variance increases, so we would want to have the variance on the top of any divisions in our formula for unreliability (i.e. in the numerator):

$$\text{unreliability} \propto s^2$$

Box 6.2 Variance in terms of expectations

The calculation of variance can be both simplified and generalised by use of expectations. Let the density function of X be $f(x)$. For example, let the probabilities associated with $x = 0, 1, 2$ be $f(0) = 1/4$, $f(1) = 1/2$ and $f(2) = 1/4$. The expectation of X (or the expected value of X) is the mean, which is defined like this:

$$\mathbf{E}(X) = \sum_{xf(x)>0} xf(x)$$

In our example, the mean is therefore $0 \times \frac{1}{4} + 1 \times \frac{1}{2} + 2 \times \frac{1}{4} = 0 + \frac{1}{2} + \frac{1}{2} = 1$

More generally, for any real-numbered function $g(X)$, then

$$\mathbf{E}(g(X)) = \sum_x g(x)f(x)$$

This is a really useful formula that you should try to memorise.

Continuing our example, suppose that $g(x) = x^2$. Now we can calculate the expected value of x^2 as

$$0^2 \times \tfrac{1}{4} + 1^2 \times \tfrac{1}{2} + 2^2 \times \tfrac{1}{4} = 0 + \tfrac{1}{2} + 1 = 1.5$$

This expectation has important applications in calculating variances and covariances. The variance of X is the expected value of $(X - \bar{x})^2$, which is written $\mathbf{E}((X - \mathbf{E}(X))^2)$ because $\bar{x}$ is $\mathbf{E}(X)$ the expected value of X. This means that we can write the variance as

$$\sigma^2 = \sum_x (x - \bar{x})^2 f(x)$$

Expanding the brackets gives

$$\sigma^2 = \sum_x x^2 f(x) - 2\bar{x} \sum_x xf(x) + \bar{x}^2 \sum_x f(x)$$

The middle term on the right-hand side is $-2\bar{x}\bar{x} = -2\bar{x}^2$ and in the right-hand term $\sum_x f(x) = 1$, so we have one positive and two negative squares of the means, and the equation simplifies to

$$\sigma^2 = \sum_x x^2 f(x) - \bar{x}^2$$

Returning to our example, we can use the new formula to compute the variance:

$$0^2 \times \tfrac{1}{4} + 1^2 \times \tfrac{1}{2} + 2^2 \times \tfrac{1}{4}(-1)^2 = 0 + \tfrac{1}{2} + 1 - 1 = 0.5$$

In summary, we can write the variance as

$$\text{var}(X) = \mathbf{E}\left((X - \mathbf{E}(X))^2\right)$$

which has the same value as

$$\text{var}(X) = \mathbf{E}\left(X^2\right) - (\mathbf{E}(X))^2$$

What about sample size? Would you want your estimate of unreliability to go up or down as sample size, n, increased? You would want unreliability to go down as sample size went up, so you would put sample size on the bottom of the formula for unreliability (i.e. in the denominator):

$$\text{unreliability} \propto \frac{s^2}{n}$$

Now consider the units in which unreliability is measured. What are the units in which our current measure are expressed. Sample size is dimensionless, but variance is based on the sum of squared differences, so it has dimensions of mean squared. So if the mean was a length in cm the variance would be an area in cm^2. This is an unfortunate state of affairs. It would make good sense to have the dimensions of the unreliability measure and the parameter whose unreliability it is measuring to be the same. That is why all unreliability measures are enclosed inside a big square root term.

Unreliability measures are called *standard errors*. What we have just calculated is the standard error of the mean:

$$SE_{\bar{y}} = \sqrt{\frac{s^2}{n}}$$

This is a very important equation and should be memorised. Let's calculate the standard errors of each of our market garden means:

sqrt(s2A/10)

[1] 0.3651484

sqrt(s2B/10)

[1] 0.3651484

sqrt(s2C/10)

[1] 1.19257

In written work, one shows the unreliability of any estimated parameter in a formal, structured way like this: 'The mean ozone concentration in garden A was 3.0 ± 0.365 (1 s.e., $n = 10$)'.

You write plus or minus, then the unreliability measure then, in brackets, tell the reader what the unreliability measure is (one standard error in this case) and the size of the sample on which the parameter estimate was based (10 in this case). This may seem rather stilted, unnecessary even. But the problem is that unless you do this, the reader will not know what kind of unreliability measure you have used. For example, you might have used a 95% confidence interval or a 99% confidence interval instead of one standard error (1 s.e.).

Confidence interval

A confidence interval shows the likely range in which the mean would fall if the sampling exercise were to be repeated. It is a very important concept that people often find difficult to grasp at first. It is pretty clear that the confidence interval will get wider as the unreliability goes up, so

$$\text{confidence interval} \propto \text{unreliability measure} \propto \sqrt{\frac{s^2}{n}}$$

But what do we mean by 'confidence'? This is the hard thing to understand. Ask yourself this question: Would the interval be wider or narrower if we wanted to be *more* confident that out repeat sample mean will fall inside the interval? It may take some thought, but you should be able to convince yourself that the more confident you want to be, the *wider* the interval will need to be. You can see this clearly by considering the limiting case of complete and absolute certainty. Nothing is certain in statistical science, so the interval would have to be infinitely wide.

We can produce confidence intervals of different widths by specifying different levels of confidence. *The higher the confidence, the wider the interval.* How exactly does this work? How do we turn the proportionality in the equation above into equality? The answer is by resorting to an appropriate theoretical distribution (see below). Suppose our sample size is too small to use the Normal distribution ($n < 30$, as here), then we traditionally use Student's t distribution (Chapter 11). The values of Student's t associated with different levels of confidence are tabulated but also available in the function **qt**, which gives the *quantiles of the t distribution*. Confidence intervals are always two-tailed; the parameter may be larger or smaller than our estimate of it. Thus, if we want to establish a 95% confidence interval, we need to calculate (or look up) the value of Student's t associated with $\alpha = 0.025$ (i.e. with $0.01(100\%-95\%)/2$). The value is found like this for the left-hand (0.025) and right-hand (0.975) tails:

```
qt(0.025,9)

[1]  -2.262157

qt(0.975,9)

[1]  2.262157
```

The first argument is the probability and the second is the degrees of freedom. This says that values as small as -2.262 standard errors below the mean are to be expected in 2.5% of cases ($p = 0.025$), and values as large as $+2.262$ standard errors above the mean with similar probability ($p = 0.975$). *Values of Student's t are numbers of standard errors to be expected with specified probability and for a given number of degrees of freedom.* The values of t for 99% are bigger than these (0.005 in each tail):

```
qt(0.995,9)

[1]  3.249836
```

and the value for 99.5% is bigger still (0.0025 in each tail):

```
qt(0.9975,9)

[1]  3.689662
```

Values of Student's t like these appear in the formula for calculating the width of the confidence interval, and their inclusion is the reason why the width of the confidence interval goes up as our degree of confidence is increased. The other component of the formula, the standard error, is not affected by our choice of confidence level. So, finally, we can write down the formula for the confidence interval (CI) of a mean based on a small sample ($n < 30$):

$$CI = \text{Student's } t \text{ from tables} \times \text{standard error}$$

which in our case is

$$CI_{95\%} = t_{(\alpha=0.025, \text{d.f.}=9)}\sqrt{\frac{s^2}{n}}$$

```
qt(0.975,9)*sqrt(1.33333/10)

[1]  0.826022
```

For garden B, therefore, we could write the confidence interval like this: 'The mean ozone concentration in garden B was 5.0 ± 0.826 (95% CI, $n = 10$)'.

Quantiles

Quantiles are important summary statistics. They show the values of y that are associated with specified percentage points of the distribution of the y values. For instance, the 50% quantile is the same thing as the median. The most commonly used quantiles are aimed at specifying the location of the middle of a data set and the tails of a data set. By the middle of a data set we mean the values of y between which the middle 50% of the numbers lie. That is to say, the values of y that lie between the 25% and 75% quantiles. By the tails of a distribution we mean the extreme values of y; for example, we might define the tails of a distribution as the values that are smaller than the 2.5% quantile or larger than the 97.5% quantile. To see this, we can generate a vector called z containing 1000 random numbers drawn from a Normal distribution, with a mean of 0 and a standard deviation of 1 (known as the standard Normal distribution). This is very straightforward:

```
z<-rnorm(1000,0,1)
```

We can see how close the mean really is to 0.0000:

```
mean(z)
```

```
[1] -0.01325934
```

Not bad. It is out by just over 1.3%. But what about the tails of the distribution. We know that for an infinitely large sample, the standard Normal should have 2.5% of its z values less than -1.96 and 97.5% of its values less than $+1.96$ (see below). So what is this sample of 1000 points like? We concatenate the two fractions 0.025 and 0.975 to make the second argument of **quantile**:

```
quantile(z,c(0.025,0.975))
```

```
    2.5%      97.5%
-1.913038  2.013036
```

Out by more than 2.5%. It could be just a sample size thing. What if we try with 10 000 numbers?

```
z<-rnorm(10000,0,1)
```

```
quantile(z,c(0.025,0.975))
```

```
    2.5%      97.5%
-1.985679  1.956595
```

Much better. Clearly the random number generator is a good one, but equally clearly, we should not expect samples of order 1000 to be able to produce exact estimates of the tails of a distribution. This is an important lesson to learn if you intend to use simulation (Monte Carlo methods) in testing statistical models. It says that 10 000 tries is much better than 1000. Computing time is cheap, so go for 10 000 tries in each run.

Robust estimators

One of the big problems with our standard estimators of central tendency and variability, the mean and the variance, is that they are extremely sensitive to the presence of outliers. We already know how to obtain a robust estimate of central tendency that is not affected by outliers—the median.

There are several modern methods of obtaining robust estimators of standard deviation that are less sensitive to outliers. The first is called **mad**, the median absolute deviation. Its value is scaled to be a consistent estimator of the standard deviation from the Normal distribution:

```
y<-c(3,4,6,4,5,2,4,5,1,5,4,6)
```

For this data set, it looks like this:

```
mad(y)

[1]  1.4826
```

which is close to, but different from, the standard deviation of y:

```
sqrt(var(y))

[1]  1.505042
```

Let's see how the different measures perform in the presence of outliers. We can add an extreme outlier (say $y = 100$) to our existing data set, using concatenation, and call the new vector $y1$:

```
y1<-c(y,100)
```

How has the outlier affected the mean (it used to be 4.083 33)?

```
mean(y1)

[1]  11.46154
```

It has nearly tripled it! And the standard deviation (it used to be 1.505 042)?

```
sqrt(var(y1))

[1]  26.64149
```

A more than 17-fold increase. Let's see how the outlier has affected the **mad** estimate of standard deviation:

```
mad(y1)

[1]  1.4826
```

Hardly at all (it was 1.505 before the outlier was added).

These robust estimators suggest a simple function to test for the presence of outliers in a data set. We need to make a decision about the size of the difference between the regular standard deviation and the **mad** value in order for us to be circumspect about the influence of outliers. In our extreme example comparing the deviations of $y1$ the ratio was 26.64 compared to 1.486 (a nearly 18-fold difference). What if we pick a 4-fold difference as our threshold? It will highlight extreme cases like y vs. $y1$ but it will allow a certain leeway for realistic kinds of data. Let's call the function **outlier**, and write it like this:

```
outlier<-function(x) {
    if(sqrt(var(x))>4*mad(x))   print("Outliers present")
    else   print("Deviation reasonable") }
```

We can test it by seeing what it makes of the original data set, y:

outlier(y)

[1] "Deviation reasonable"

What about the data set including $y = 100$?

outlier(y1)

[1] "Outliers present"

It is good practice to compare robust and standard estimators. I am not suggesting that we do away with means and variances; just that we be aware as to how sensitive they are to extreme values.

Skew

Mean is the first moment of a distribution and variance is the second moment. Skew (or skewness) is the dimensionless version of the third moment about the mean:

$$m_3 = \frac{\sum (y - \bar{y})^3}{n}$$

which is rendered dimensionless by dividing by the cube of the standard deviation of y (because this is also measured in units of y^3):

$$s_3 = \text{s.d. } (y)^3 = \left(\sqrt{s^2} \right)^3$$

The skew is then given by

$$\text{skew} = \frac{m_3}{s_3}$$

It measures the extent to which a distribution has long, drawn-out *tails* on one side or the other. The full formula is the expectation of the cube of the difference between y and mean of y divided by the cube of the standard deviation:

$$\gamma_1(y) = \frac{E(y - \mu)^3}{\sigma^3}$$

and is calculated like this:

$$\gamma_1 = \frac{\frac{1}{n} \sum (y - \bar{y})^3}{\left[\frac{1}{n} \sum (y - \bar{y})^2 \right]^{3/2}}$$

A Normal distribution has skew $= 0$. Negative values of γ_1 means skew to the left (negative skew) and positive values mean skew to the right. To test whether a particular value of skew is significantly different from 0 (hence the distribution from which it was calculated is significantly non-normal) we divide the estimate by its approximate standard error:

$$SE_{\gamma_1} = \sqrt{\frac{6}{n}}$$

It is straightforward to write an S-Plus function to calculate the degree of skew for any vector of numbers, x, like this:

```
skew<-function(x){

    m3<-sum((x-mean(x))^3)/length(x)

    s3<-sqrt(var(x))^3

    m3/s3 }
```

Note the use of length(x) to work out the sample size, n, whatever the size of the vector x. The last expression inside a function is not assigned to a variable name, and is returned as the value of skew(x) when this is executed from the command line.

Kurtosis

Kurtosis is a measure of non-normality that has to do with the peakiness, or flat-toppedness, of a distribution. The Normal distribution is bell-shaped, a kurtotic distribution is other than bell-shaped. In particular, a more flat-topped distribution is said to be platykurtotic, and a more pointy distribution is said to be leptokurtotic. Kurtosis is the dimensionless version of the fourth moment about the mean:

$$m_4 = \frac{\sum (y - \bar{y})^4}{n}$$

which is rendered dimensionless by dividing by the square of the variance of y (because this is also measured in units of y^4):

$$s_4 = \text{var}(y)^2 = \left(s^2\right)^2$$

Kurtosis is then given by

$$\text{kurtosis} = \frac{m_4}{s_4} - 3$$

The term -3 is included because a Normal distribution has $m_4/s_4 = 3$. This formulation therefore has the desirable property of giving zero kurtosis for a Normal distribution, while a flat-topped (platykurtic) distribution has a negative value of kurtosis, and a pointy (leptokurtic) distribution has a positive value of kurtosis.

Kurtsosis is the expectation of the fourth power of the difference between y and the mean of y, divided by the fourth power of the standard deviation:

$$\gamma_2(y) = \frac{E(y - \mu)^4}{\sigma^4} - 3$$

and is calculated like this:

$$\gamma_2 = \frac{\frac{1}{n}\sum (y - \bar{y})^4}{\left[\frac{1}{n}\sum (y - \bar{y})^2\right]^2} - 3$$

The approximate standard error of kurtosis is

$$SE_{\gamma_2} = \sqrt{\frac{24}{n}}$$

An S-Plus function to calculate kurtosis might look like this:

```
kurtosis<-function(x) {
m4<-sum((x-mean(x))^4)/length(x)
s4<-var(x)^2
m4/s4 - 3 }
```

An example of Skew and Kurtosis

We can test the two functions on the data frame called skewdata:

```
attach(skewdata)
names(skewdata)
```

```
[1] "y1" "y2"
```

```
par(mfrow=c(1,2))
hist(y1)
hist(y2)
```

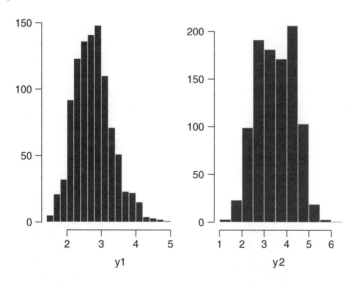

Evidently the two distributions are quite different shapes: it looks as if $y1$ is skew to the right (positively skew) and that $y2$ is rather flat-topped (platykurtic). We start by looking at $y1$:

```
skew(y1)
```

```
[1] 0.5057302
```

```
kurtosis(y1)
```

```
[1] 0.3449858
```

Now $y2$:

```
skew(y2)
```

```
[1] -0.00459501
```

```
kurtosis(y2)
```

```
[1] -0.716341
```

The question now arises as to how we are to judge whether these values of skew and kurtosis represent significant departures from normality, or whether they fall within the range of values that might be expected by chance alone? For example, the skew of $y2$ is very small (-0.0046) but how small is small enough?

We need to be able to test whether or not our calculated values are significantly different from zero. To do this we need to know the standard errors of skew and kurtosis (see above). For the present case we have $n = 1000$, so they are

```
sqrt(6/1000)
```

```
[1] 0.07745967
```

```
sqrt(24/1000)
```

```
[1] 0.1549193
```

As usual, we say that an estimate is significantly different from zero when it is 2 or more standard errors above or below zero. Thus, for $y2$ we see that it is not significantly skew ($t = 0.004\,595\,01/0.077\,459\,67 = 0.059$), but it is very significantly platykurtic ($t = -0.716\,341/0.154\,9193 = -4.62$). The distribution of $y1$ is significantly positively skew ($t = 0.505\,7302/0.077\,45\,967 = 6.529$), and is also somewhat leptokurtic ($t = 0.344\,9858/0.154\,9193 = 2.227$).

In response to these findings, we might consider log transformation of the y values prior to any statistical modelling. A histogram is a quick and effective way of seeing whether log transformation is an improvement.

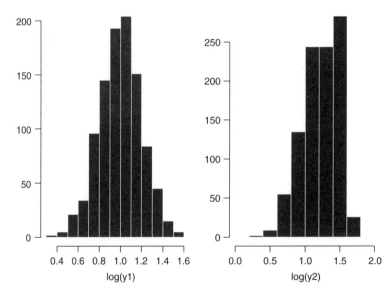

We see that log transformation has been highly effective at improving the normality of $y1$, but it has been much less effective for $y2$

```
skew(log(y1))
```

```
[1]  -0.07788302
```

The transformation has actually introduced significant negative skew in the second case:

```
skew(log(y2))
```

```
[1]  -0.5368897
```

```
kurtosis(log(y2))
```

```
[1]  -0.1314477
```

The best course of action might be to carry out the analyses on $\log y$ but to check carefully after modelling that the residuals of $\log y$ are reasonably well behaved.

Weighted means and variances

In most cases we give equal weight to all the data points in an analysis. Sometimes, however, we want to give more weight to some points than to other. For instance, we might want to give more weight to data points that we judge to be more reliable than others. Common weights are:

- sample size (more replicates, more weight)

- the reciprocal of variance (higher variance, lower weight)

This is how it works. The **weighted mean** is based on a vector of values for the response variable, y, and a vector of weights, w. Both vectors must be the same length. The sum of the weights is $\sum_{i=1}^{n} w_i$. The weighted sum of the response variable is $\sum_{i=1}^{n} w_i y_i$. The weighted mean is then simply

$$\tilde{y} = \frac{\sum wy}{\sum w}$$

Notice that the sample size, n, does not appear in the formula (when all the weights are 1 then $\sum w = n$). The **weighted variance** is based on the weighted sum of squares $\sum_{i=1}^{n} w_i (y_i - \tilde{y})^2$. Note that we use the weighted mean in place of the usual arithmetic mean in calculating the individual residuals. The weights are applied to the squares of the residuals.

As the following example shows, using weights can have a substantial effect on the significance of differences in simple tests of hypotheses. Suppose that we have data suitable for a two-sample t test with $n = 7$ replicates in each of two treatments, *xvar* and *yvar*

```
attach(pair)
names(pair)
```

```
[1] "xvar" "yvar" "wt1" "wt2"
```

The t test on the raw data shows no significant difference:

```
t.test(xvar,yvar)
```

```
           Standard Two-Sample t-Test

data:  xvar and yvar
t = -1.7321, df = 12, p-value = 0.1089
alternative hypothesis: true difference in means is not
equal to 0
```

```
95 percent confidence interval:
 -4.5158763   0.5158763
sample estimates:
 mean of x mean of y
          5         7
```

The built-in **t.test** function has no facility for carrying out tests with weights, so we need to use **lm**, like this. First we need to combine the two samples into a single vector:

vari<-c(xvar,yvar)

Now we need to create a factor to describe the treatment to which each row belongs:

ev<-factor(rep(c(1,2),c(7,7)))

Finally, the two vectors of weights need to be combined into a single vector:

wts<-c(wt1,wt2)

Now we can carry out the modelling. First, we repeat the t test using **lm** to check that the result is exactly the same:

```
model1<-lm(vari~ev)
summary.aov(model1)
```

```
          Df Sum of Sq  Mean Sq F Value     Pr(F)
      ev   1        14 14.00000       3 0.1088643
Residuals 12        56  4.66667
```

The p values are identical, as we expected. Now we do the weighted analysis like this:

```
model2<-lm(vari~ev,weights=wts)
summary.aov(model2)
```

```
          Df Sum of Sq Mean Sq  F Value      Pr(F)
      ev   1     26.45  26.450 5.111111 0.04315039
Residuals 12     62.10   5.175
```

The difference between the two weighted means is significant. Notice that the sums of squares and mean squares are different in the two models (see the formulas above). The degrees of freedom are unaltered. Because the weights make such a difference, you need to be absolutely clear about your justification for using one set of weights over another. It is very bad practice to repeat the analysis using different weights until you discover a pleasing result.

Correlation

Everyone is familiar with the maxim *Correlation does not imply causation*, but we often forget to take any notice, especially when there might be a publication in it! Most of the science we do is looking for *direct effects*. These are causal relationships of the kind where change in the explanatory variable causes change in the response variable via a mechanism that we hope to understand (e.g. we can recreate the response under controlled experimental conditions, or we have a mechanistic model that produces matching predictions).

But life is often not like this. A *common cause* may totally determine both y and x, inducing a correlation between them in the complete absence of any causal mechanism. Alternatively, *another variable*, z, may influence y in addition to the effect of x, reducing the magnitude of the correlation between y and x. In real systems there may be direct effects, shared common causes, interaction effects and a multitude of other variables involved. Many, complex and interacting are the difficulties involved in the correct interpretation of correlation analysis.

If we take 30 random samples from a Normal distribution (x) and then take a further sample of 30 independent random samples (y) we would expect to find no relationship between the two sets of numbers:

```
x<-rnorm(30)
y<-rnorm(30)
plot(x,y)
```

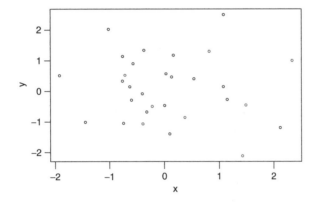

This is what statisticians call 'the sky at night': a scatterplot showing no relationship between the two independent variables x and y. The key word here is 'independent'. If we took a random sample x, but then allowed that y was dependent on x, the situation would be changed entirely. Take the extreme case of dependency: y equals x. Then the procedure looks like this:

```
x<-rnorm(30)
y<-x
plot(x,y)
```

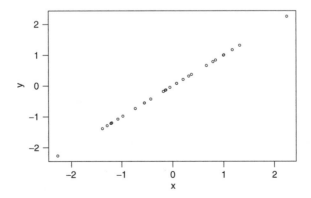

In this case of complete dependency, the only variation is in the normally distributed x values. *The y values are perfectly correlated with the x values.* In most practical cases any relationship between two variables is likely to be intermediate between the extreme cases of complete dependence, and total independence.

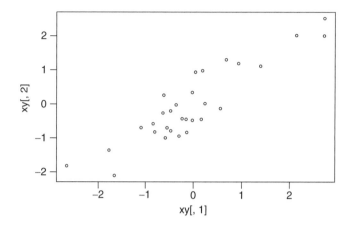

This shows the relationship between two variables selected at random from the **multivariate normal distribution**. Both variables have mean $= 0$ and standard deviation $= 1$, but they are related by a correlation coefficient $\rho = 0.9$. The figure was generated using the following two commands:

```
xy<-rmvnorm(30,rho=0.9)
```

The default number of variables generated my **rmvnorm** is 2, which is what we want here. The first argument is the number of numbers in the sample ($n = 30$) and the second argument is the correlation that we want to generate between the two variables ($\rho = 0.9$):

```
plot(xy[,1],xy[,2])
```

The object xy has two columns because we asked (implicitly) to generate two vectors of random numbers (the default of **rmvnorm**). We could give these separate names, but here we refer to them by their subscripts: the x variable is xy[,1] and the y variable is xy[,2]. The blank before the comma means 'all of the rows in the following column', and the parameter after the comma is the column number.

You recall that the variance in x is

$$\text{var}(x) = \frac{\sum (x - \bar{x})^2}{n - 1}$$

and the variance in y is

$$\text{var}(y) = \frac{\sum (y - \bar{y})^2}{n - 1}$$

What we need to do now is work out the **covariance** of x and y (Box 6.3).

Box 6.3 Covariance

We need a quantitative measure of the extent to which variation in one variable is associated with variation in another. The covariance of x and y is defined as the **expectation** of the vector product $(x - \bar{x})(y - \bar{y})$:

$$\text{cov}(x, y) = \mathbf{E}[(x - \bar{x})(y - \bar{y})]$$

Positive correlation means that this expectation is positive. This is because negative residuals of x will tend to occur together with negative residuals of y, so the product is positive. With negative correlation, positive x residuals will tend to occur with negative residuals of y (and vice versa), so the expectation is negative. We start by multiplying out the brackets:

$$(x - \bar{x})(y - \bar{y}) = xy - \bar{x}y - x\bar{y} + \bar{x}\bar{y}$$

Now applying expectations, and remembering that the expectation of x is $\bar{x}$ and the expectation of y is $\bar{y}$, (see Box 6.2) we get

$$\text{cov}(x, y) = \mathbf{E}(xy) - \bar{x}\mathbf{E}(y) - \mathbf{E}(x)\bar{y} + \bar{x}\bar{y} = \mathbf{E}(xy) - \bar{x}\bar{y} - \bar{x}\bar{y} + \bar{x}\bar{y}$$

so $-\bar{x}\bar{y} + \bar{x}\bar{y}$ cancel out, leaving

$$\text{cov}(x, y) = \mathbf{E}(xy) - \mathbf{E}(x)\mathbf{E}(y)$$

Notice that when x and y are uncorrelated, $\mathbf{E}(xy) = \mathbf{E}(x)\mathbf{E}(y)$ so the covariance is 0 in this case (see p. 190). The next step is to work out the expectation of the product xy:

$$\mathbf{E}(xy) = \sum_{x,y} xyf(x, y)$$

where the sum of products $\sum_{x,y} xy$ for every combination of x and y is weighted by the **joint mass function** $f(x, y)$. In the joint mass function, the row and column totals both sum to 1. For instance, suppose that x takes one of three values $\{1, 2, 3\}$ and y takes one of three values $\{-1, 0, 2\}$. There are 9 contingencies (combinations of x and y) and there are 18 data points in total. The joint mass function is given by the probabilities in the body of the table; these are empirically determined in this case (they are the result of a series of 18 trails) and they sum to 1.

	$y = -1$	$y = 0$	$y = 2$	$f(x)$
$x = 1$	1/18	3/18	2/18	6/18
$x = 2$	2/18	0	3/18	5/18
$x = 3$	0	4/18	3/18	7/18
$f(y)$	3/18	7/18	8/18	1

We now work out all of the terms for this example:

$$\mathbf{E}(xy) = \sum_{x,y} xyf(x, y) = \frac{29}{18}$$

Summing over the row totals, we have

$$\mathbf{E}(y) = \sum_{y} yf(y) = \frac{13}{18}$$

Summing over the column totals, we have

$$\mathbf{E}(x) = \sum_{x} xf(x) = \frac{37}{18}$$

Now we can compute the two variances (Box 6.2):

$$\text{var}(x) = \mathbf{E}\left(x^2\right) - [\mathbf{E}(x)]^2 = \frac{233}{324}$$

$$\text{var}(y) = \mathbf{E}\left(y^2\right) - [\mathbf{E}(y)]^2 = \frac{461}{324}$$

And the covariance:

$$\text{cov}(x, y) = \mathbf{E}(xy) - \mathbf{E}(x)\mathbf{E}(y) = \frac{41}{324}$$

The short-cut formula, involving the corrected sum of products, *SSXY*, is explained in Chapter 14.

Covariance itself is not a very useful measure of the extent of any correlation between x and y because its value depends so strongly on the two scales of measurement. A useful measure that removes this scale dependence is the **correlation coefficient**, ρ:

$$\rho(x, y) = \frac{\text{cov}(x, y)}{\sqrt{\text{var}(x) \times \text{var}(y)}}$$

which is dimensionless and lies in the range $-1 \leq \rho \leq +1$. The correlation coefficient for the numerical example in Box 6.3 is therefore

$$\rho(x, y) = \frac{41}{\sqrt{233 \times 461}} = 0.125$$

because the 324 in the denominators cancels out.

For sample data, we calculate the correlation coefficient r from the corrected sum of products, *SSXY*, and the corrected sums of squares of x and y (*SSX* and *SSY*)

$$r = \frac{SSXY}{\sqrt{SSX \times SSY}}$$

as explained in detail in Chapter 14.

Variance of a sum of two variables

We are interested to know the value of the variance of x plus y, $\text{var}(x + y)$. We begin with some numerical work on two Normal distributions with means of 5 and 8, and standard deviations of 1 and 2 respectively. Take 100 random numbers from each:

```
x<-rnorm(100,5,1)
y<-rnorm(100,8,2)
```

The true variances of these variables are 1 and 4 respectively. Let's see what the random number generator has produced:

```
var(x)
```

```
[1] 1.09237
```

var(y)

```
[1] 4.185538
```

Not bad at all. But our question relates to the value of var($x + y$):

var(x+y)

```
[1] 5.507257
```

In this case it looks as if var($x + y$) could be var(x) + var(y) since $5 = 1 + 4$. Could it be that simple? Perhaps the result just depends on our choice of parameter values. Let's take variances of 9 and 16. If the rule is true, then we predict that the variance of the sum will be 25. Let's see:

```
x<-rnorm(100,15,3)
y<-rnorm(100,18,4)
var(x+y)
```

```
[1] 20.14884
```

Not especially convincing. We should try to find an analytical solution.

We start by writing the variance of the sum in terms of expectations (see Box 6.2).

$$\text{var}(x + y) = \mathbf{E}\left[x + y - \mathbf{E}(x + y)\right]^2$$

Now, because $\mathbf{E}(x + y)$ can be written as $\mathbf{E}(x) + \mathbf{E}(y)$, we can write

$$\text{var}(x + y) = \mathbf{E}[(x - \mathbf{E}(x)) + (y - \mathbf{E}(y))]^2$$

Now multiply out the square term:

$$\text{var}(x + y) = \mathbf{E}\left[(x - \mathbf{E}(x))^2 + (y - \mathbf{E}(y))^2 + 2(x - \mathbf{E}(x))(y - \mathbf{E}(y))\right]$$

Take expectations of each term in sequence:

$$\text{var}(x + y) = \mathbf{E}(x - \mathbf{E}(x))^2 + \mathbf{E}(y - \mathbf{E}(y))^2 + 2\mathbf{E}(x - \mathbf{E}(x))(y - \mathbf{E}(y))$$

At this stage, note that $\mathbf{E}(x - \mathbf{E}(x))^2$ is the variance of x, and $\mathbf{E}(y - \mathbf{E}(y))^2$ is the variance of y (see Box 6.2). But what is the term $2\mathbf{E}(x - \mathbf{E}(x))(y - \mathbf{E}(y))$? This turns out to be 2 times the covariance of x and y (see Box 6.3). So, finally, we can write

$$\text{var}(x + y) = \text{var}(x) + \text{var}(y) + 2\,\text{cov}(x, y)$$

So we can see that the simple rule for the variance of a sum works if, and only if, the covariance of x and y is exactly zero. Put another way, the variance of the sum of x and y is var(x) + var(y) when x and y are independent (i.e. if they are not correlated). When they are positively correlated, the variance of their sum will be *bigger* than the sum of their variances. When they are negatively correlated, the variance of their sum will be *less* than the sum of their variances.

Let's return to our numerical example. We found var($x + y$) to be 20.14884 rather than 25, suggesting that there was a slight negative correlation between x and y:

```
var(x)+var(y)+2*var(x,y)
[1]  20.14884
```

That's more like it. Notice that in S-Plus, the variance function applied to two vectors var(x,y) returns the covariance as we would hope intuitively that it would.

Covariance is related to the correlation coefficient $\rho(x, y)$ see p. 190) by

$$\mathrm{cov}(x, y) = \rho(x, y)\sqrt{\mathrm{var}(x)\,\mathrm{var}(y)}$$

We should check to see that this is what S-Plus gives:

```
var(x,y)
```

```
[1]  -1.945145
```

then check the answer, longhand, using the formula for covariance:

```
cor(x,y) * sqrt(var(x)*var(y))
```

```
[1]  -1.945145
```

So that's okay then.

What do we mean when we say that two variables, x and y, are positively correlated? This means that when x is big then y will be big as well (and vice versa). Variance is about squared differences from the mean. So when two variables are positively correlated, the variance of their sum should be bigger than when they are independent. What if the two variables are negatively correlated? Then big values of x will tend to be associated with small vales of y. Their sum will therefore be closer to the average than if they were independent. The variance of the sum will therefore be lower when the two variables are negatively correlated than when they are independent.

Let's see this in action. We can generate *correlated random numbers* using the multivariate Normal distribution function **rmvnorm**. As you would expect, this has more arguments than the **rnorm** function because we need to specify not just two means and two variances, but the correlation between x and y as well. Let's start with the case when x and y are positively correlated with correlation coefficient $r = +0.6$:

```
xy<-rmvnorm(1000, mean=c(50,60),cov=matrix(c(1,0.6,0.6,1),2),sd= c(2,3))
```

It is worth checking the variances of the two samples:

```
var(xy[,1])
```

```
[1]  4.155267
```

```
var(xy[,2])
```

```
[1]  9.03546
```

First we get a visual impression of the correlation between x and y:

```
plot(xy[,1], xy[,2])
```

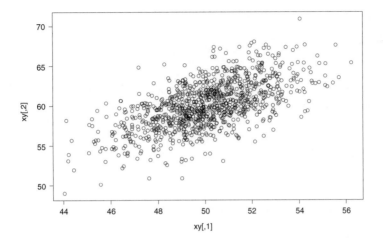

What about the variance of the sums and differences of x and y? Remember, because they are positively correlated (rather than independent), we expect the variance of their sums to be bigger than the sum of their variances (i.e. >13.20, which is $4.16 + 9.04$).

> var(xy[,1] + xy[,2])

    [1] 20.70859

That's okay. And their differences?

> var(xy[,1] - xy[,2])

    [1] 5.672859

This is much smaller than 13.20 because the two variables are positively correlated. Big values of x will tend to occur together with big values of y, so the difference between x and y is expected to be small. We can conclude that *the variance of a sum is only equal to the variance of a difference when the two variables are independent.*

What about the numeric value of the difference between the two variance estimates? Is that equal to 2 times the covariance as predicted? The **variance–covariance matrix** is easy to obtain, since xy is already a matrix:

> var(xy)

            [,1]   [,2]
    [1,] 4.155267 3.758934
    [2,] 3.758934 9.035460

Notice that the diagonal has the estimated variance of x (≈ 4) and the estimated variance of y (≈ 9) and the off-diagonals are both the covariance of x and y. We expect the variance of the sum to be roughly $2 \times 3.76 = 7.52$ bigger than the variance of $(x + y)$: $13 + 7.52 = 20.52$. That's pretty close. Similarly, the variance of the difference was measured as 5.67 and predicted to be $13 - 7.52 = 5.48$.

How is the covariance of 3.759 calculated? It is the correlation coefficient ρ multiplied by the square root of the product of the two separate variances:

$$\operatorname{cov}(x, y) = \rho \sqrt{s_x^2 s_y^2}$$

```
ssw<-sqrt(4.155267*9.03546)

0.6*ssw

[1] 3.676426
```

Why is this not exactly equal to the observed covariance of 3.758 934? If you think about it, we don't know that the correlation between x and y really is 0.6. We *wanted* it to be, and we asked the random number generating function to do its best, but this is only a set of 100 random numbers after all, and anything could have happened. We can work out what the actual correlation coefficient was:

```
cor(xy[,1],xy[,2])

[1] 0.6134655
```

So there we have it. The correlation was 0.6134 not 0.6. Now the numbers do add up:

```
cor(xy[,1],xy[,2])*ssw

[1] 3.758934
```

Further reading

Casella, G. and Berger, R.L. (1990) *Statistical Inference*. Pacific Grove, California, Wadsworth and Brooks Cole.
Cox, D.R. and Hinkley, D.V. (1974) *Theoretical Statistics*. London, Chapman & Hall.
Snedecor, G.W. and Cochran, W.G. (1980) *Statistical Methods*. Ames, Iowa, Iowa State University Press.
Sokal, R.R. and Rohlf, F.J. (1995) *Biometry: The Principles and Practice of Statistics in Biological Research*. San Francisco, W.H. Freeman.

7

The Normal Distribution

Consider the following simple exponential function:

$$y = \exp(-|x|^m)$$

As the power m in the exponent increases, the function becomes more and more like a step function. The following panels show the relationship between y and x for $m = 1, 2, 3$ and 8 respectively.

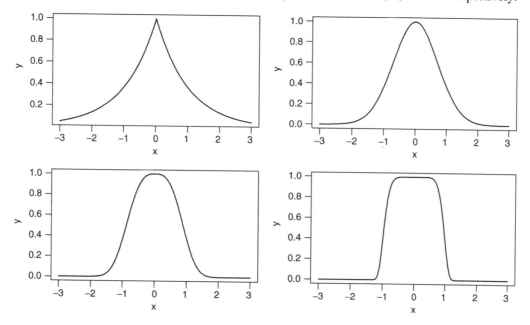

The second of these panels (top right), where $y = \exp(-x^2)$, is the basis of an extremely important and famous probability density function. Once it has been scaled, so that the integral (the area under the curve from minus infinity to plus infinity) is unity, this is the Normal distribution. Unfortunately, the scaling constants are rather cumbersome. When the distribution has mean $= 0$ and standard deviation $= 1$ (the standard Normal distribution) the equation becomes

$$p(z) = \frac{1}{\sqrt{2\pi}} \, e^{-z^2/2}$$

Suppose we have measured the heights of 100 people. The mean height was 170 cm and the standard deviation was 8 cm. The Normal distribution looks like this:

```
ht<-seq(150,190,0.01)
plot(ht,dnorm(ht,170,8),type="l")
```

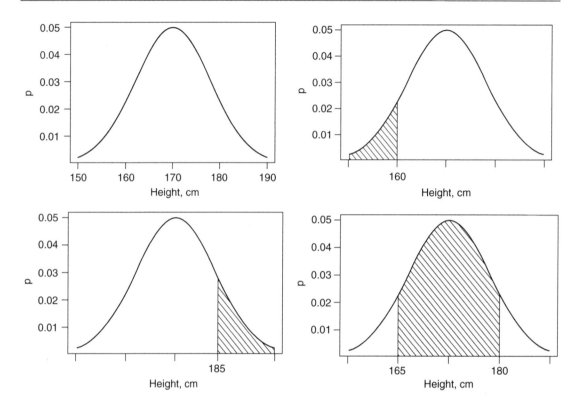

We can ask three sorts of questions about data like these. We can ask the probability that a randomly selected individual will be:

- shorter than a particular height
- taller than a particular height
- between one specified height and another

The area under the whole curve is exactly 1; everybody has a height between minus infinity and plus infinity. True, but not particularly helpful. Suppose we want to know the probability that one of our people, selected at random from the group, will be less than 160 cm tall. We need to convert this height into a value of z; that is to say, we need to convert 160 cm into *a number of standard deviations from the mean*. What do we know about the standard Normal distribution? It has a mean of zero and a standard deviation of one. So we can convert any value y, from a distribution with mean $\bar{y}$ and standard deviation s, very simply by calculating

$$z = \frac{(y - \bar{y})}{s}$$

So we convert 160 cm into a number of standard deviations. It is less than the mean height (170 cm) so its value will be negative:

$$z = \frac{(160 - 170)}{8} = -1.25$$

Now we need to find the probability of a value of the standard Normal distribution taking a value of -1.25 or smaller. This is the area under the left-hand tail of the distribution. The function we need for this is **pnorm**: we provide it with a value of z (or, more generally, with a quantile) and it provides us with the probability we want:

pnorm(-1.25)

[1] 0.1056498

So the answer to our first question is just over 10%. The second question is, What is the probability of selecting one of our people and finding that they are taller than 185 cm? The first two parts of the exercise are exactly the same as before; first we convert our value of 185 cm into a number of standard deviations:

$$z = \frac{(185 - 170)}{8} = 1.875$$

then we ask what probability is associated with this, using **pnorm**:

pnorm(1.875)

[1] 0.9696036

But this is the answer to a different question. This is the probability that someone will be *less* than 185 cm tall (that is what the function **pnorm** has been written to provide). All we need to do is to work out the complement of this:

1-pnorm(1.875)

[1] 0.03039636

So the answer to the second question is about 3%. Finally, we might want to know the probability of selecting a person between 165 cm and 180 cm? We have a bit more work to do here, because we need to calculate two z values:

$$z_1 = \frac{(165 - 170)}{8} = -0.625 \quad \text{and} \quad z_2 = \frac{(180 - 170)}{8} = 1.25$$

The important point is this: we want the probability of selecting a person between these two z values, so we *subtract the smaller probability from the larger probability*. This is the shaded area in the bottom right panel on the previous page.

pnorm(1.25)-pnorm(-0.625)

[1] 0.6283647

Thus we have a 63% chance of selecting a medium-sized person (taller than 165 cm and shorter than 180 cm) from this sample with a mean height of 170 cm and a standard deviation of 8 cm.

Maximum likelihood with the Normal distribution

The probability density of the Normal distribution is

$$f(y|\mu, \sigma) = \frac{1}{\sigma\sqrt{2\pi}} \exp\left[-\frac{(y - \mu)^2}{2\sigma^2}\right]$$

which is read as saying the probability of getting a data value y, given (|) a mean of μ and a variance of σ^2, is calculated from this rather complicated-looking two-parameter exponential function. For any given combination of μ and σ^2, it gives a value between 0 and 1. Recall that likelihood is the product of the probability densities for each of the values of the response variable, y. So if we have n values of y in our experiment, the likelihood function is

$$L(\mu, \sigma) = \prod_{i=1}^{n} \left(\frac{1}{\sigma \sqrt{2\pi}} \exp\left[-\frac{(y_i - \mu)^2}{2\sigma^2} \right] \right)$$

where the only change is that y has been replaced by y_i and we multiply together $\left(\prod_{i=1}^{n} \right)$ the probabilities for each of the n data points. There is a little bit of algebra we can do to simplify this: we can get rid of the product operator, $\prod$, in two steps. First, for the constant term: that, multiplied by itself n times, can just be written as $1/(\sigma \sqrt{2\pi})^n$. Second, remember that the product of a set of antilogs (exp) can be written as the antilog of a sum of the values of x_i like this: $\prod \exp(x_i) = \exp(\sum x_i)$. This means that the product of the right-hand part of the expression can be written as

$$\exp\left[-\frac{1}{2\sigma^2} \sum_{i=1}^{n} (y_i - \mu)^2 \right]$$

so we can rewrite the likelihood of the Normal distribution as

$$L(\mu, \sigma) = \frac{1}{\left(\sigma \sqrt{2\pi} \right)^n} \exp\left[-\frac{1}{2\sigma^2} \sum_{i=1}^{n} (y_i - \mu)^2 \right]$$

The two parameters μ and σ are unknown, and the purpose of the exercise is to use statistical modelling to determine their maximum likelihood values from the data (the n different values of y). So how do we find the values of μ and σ that maximise this likelihood? The answer involves calculus: first we find the derivative of the function with respect to the parameters, then set it to zero, and solve.

It turns out that because of the exponential function in the equation, it is easier to work out the log of the likelihood and maximise this instead. Obviously, the values of the parameters that maximise the log likelihood $l(\mu, \sigma) = \log(L(\mu, \sigma))$ will be the same as those that maximise the likelihood:

$$l(\mu, \sigma) = -\frac{n}{2} \log(2\pi) - n \log \sigma - \frac{1}{2\sigma^2} \sum (y_i - \mu)^2$$

From now on, we shall assume that summation is over the index i from 1 to n. Now for the calculus. We start with the mean, μ. The derivative of the log likelihood with respect to μ is

$$\frac{dl}{d\mu} = \sum (y_i - \mu)/\sigma^2$$

Set the derivative to zero and solve for μ:

$$\sum (y_i - \mu)/\sigma^2 = 0 \quad \text{so} \quad \sum (y_i - \mu) = 0$$

Taking the summation through the bracket and noting that $\sum \mu = n\mu$, we have

$$\sum y_i - n\mu = 0 \quad \text{so} \quad \sum y_i = n\mu \quad \text{and} \quad \mu = \frac{\sum y_i}{n}$$

The maximum likelihood estimate of μ is the arithmetic mean. Next we find the derivative of the log likelihood with respect to σ:

$$\frac{dl}{d\sigma} = -\frac{n}{\sigma} + \frac{\sum (y_i - \mu)^2}{\sigma^3}$$

recalling that the derivative of $\log x$ is $1/x$ and the derivative of $-1/x^2$ is $2/x^3$. Solving, we get

$$-\frac{n}{\sigma} + \frac{\sum (y_i - \mu)^2}{\sigma^3} = 0 \quad \text{so} \quad \sum (y_i - \mu)^2 = \sigma^3 \left(\frac{n}{\sigma}\right) = \sigma^2 n$$

$$\sigma^2 = \frac{\sum (y_i - \mu)^2}{n}$$

The maximum likelihood estimate of the variance σ^2 is the mean squared deviation of the y values from the mean. This is a biased estimate of the variance, however, because it does not take account of the fact that we estimated the value of μ from the data. To unbias the estimate, we need to lose one degree of freedom to reflect this fact, and divide the sum of squares by $n - 1$ rather than by n (see p. 94 and REML estimators in Chapter 35).

A simple example might demystify this. The data are nine values of growth which we shall analyse as a regression in Chapter 14. Here we are concerned only with calculating the log likelihood:

```
attach(regression)
names(regression)

[1] "growth"   "tannin"

growth
```

```
 1    2    3    4    5    6    7    8    9
12   10    8   11    6    7    2    3    3
```

For the calculations we need the variance in growth σ^2 and the sum of the squares of the differences between growth y_i and mean growth μ. The sum of squares is

```
sum((growth-mean(growth))^2)
```

```
[1] 108.8889
```

and the maximum likelihood estimate of the variance is the sum of squares divided by the sample size, $n = 9$:

```
108.88889 / 9
```

```
[1] 12.09877
```

Now we can calculate the log likelihood for the growth data:

```
-9/2*log(2 * pi)-9 * log(12.09877^0.5)-108.8889/(2*12.09877)
```

```
[1] -23.98941
```

This is the log likelihood for these data. The calculations were tedious, but they did not involve anything complicated. Should you need the value of the log likelihood, there is a built-in function **logLik** to calculate it for you. However, it only works after you have fitted a model to the data. We did that implicitly here, by fitting only the mean value of y. The way to make a linear statistical model **lm** of that process is as follows:

```
model<-lm(growth~1)
```

The details will be explained later. The point is that we now have a model object called model, and we can extract the log likelihood very simply:

```
logLik(model)
```

```
[1]  -23.98941
```

Note that, like all S-Plus functions, the first letter is lower case, but the L in the middle is upper case.

Normal distributions in S-Plus

The Normal distribution is fundamental to most of parametric statistics, and there are three ways of presenting it:

- the density function (the probability density associated with z)

- the probability function (gives p, the integral of the density function up to z)

- quantiles (inverse of the probability function, gives z when provided with p)

Normal density function

Probability density is obtained by specifying the values of z, the mean and the standard deviation. If we omit the mean and standard deviation, they are assumed to be 0 and 1 respectively (i.e. the standard Normal): Suppose we want to know the probability of getting $z = 1$ from a standard Normal distribution, we type

```
dnorm(1)
```

```
[1]  0.2419707
```

Perhaps more usefully, we could use this function to produce a line plot of the Normal for a specified range of z values (say from -2.5 to 2.5 in steps of 0.05):

```
z<-seq(-2.5,2.5,0.05)
y<-dnorm(z)
plot(z,y,type="l")
```

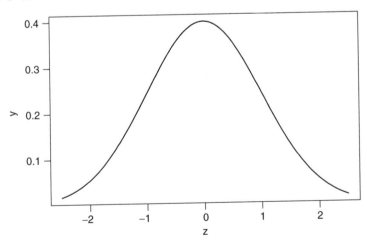

This is the famous bell-shaped curve that underlies so much of the theory in classical parametric statistics. Our interest is often in the area under the two tails of the curve. That is to say, we are concerned with the probability of obtaining unusually large values of y (right-hand tail) or unusually small values of y (left-hand tail) . You can see at once that values of z bigger than 2 and less than -2 are unlikely. The number 2, it will emerge, is a very useful rule of thumb for saying whether a particular deviation is unusually large or unusually small.

Normal probability function

In practice, therefore, we tend to be more interested in the probability function than in the density function, because we want to know about probabilities, especially about probabilities in the tails of the distribution.

Cumulative probability uses the **pnorm** function. Suppose we wanted to produce the cumulative probability plot of the last graph with mean $= 0$ and standard deviation $= 1$:

```
y<-pnorm(z)
plot(z,y,type="l")
```

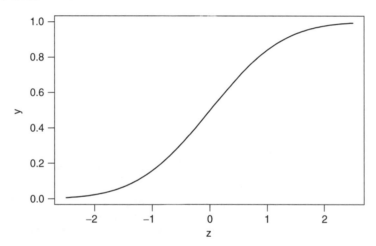

The y axis is now the probability that a randomly selected individual from this Normal distribution with mean $= 0$ and standard deviation $= 1$ will have a value of z *or smaller*. The probability that it will have a value smaller than -2.5 is close to zero. The probability that it will have a value smaller than $+2.5$ is close to 1. Between about $z = -1$ and $z = +1$ the probability increases roughly linearly with increasing z. It turns out there is a 2.5% chance that we will get a value of z less than -1.96 and a 2.5% chance that we will get a value of z greater than 1.96. This is where our rule of 2 comes from; it says there is only a 5% chance that a randomly selected individual will be more extreme than 2 standard deviations (one way or the other, larger or smaller) if the distribution has a standard deviation of 1 (as here).

We use the **pnorm** function in hypothesis testing when we want to know whether a particular value of z is extreme or not. For probabilities in the right-hand tail we need 1 minus the value returned by **pnorm** (p. 118).

Normal quantiles

Quantiles of the Normal distribution are the inverse of the probability density. With quantiles, you provide a value of the probability, p, and the function **qnorm** returns the value of z associated with this probability. This is useful, because it saves you the need to look up tables of z values to find *critical values*.

To plot the quantiles of the Normal distribution against the cumulative probabilities associated with them, we need to generate a series of 100 or so p values between 0 and 1:

```
p<-seq(0,1,0.01)
plot(p,qnorm(p),type="l")
```

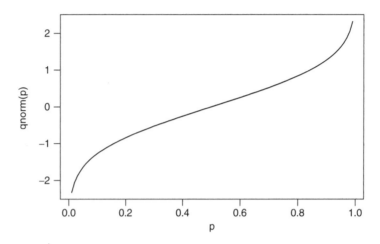

The plot looks like an S on its side. You need to recall that quantiles are measured in departures from the mean. Since the Normal distribution is symmetrical (it has zero skew; see p. 103), the probability is exactly 1/2 that a random sample will be smaller than the mean. Thus, when p is 0.5, the quantile is 0 because the sample is exactly equal to the mean. We use **qnorm** most often for asking about the tails of the distribution. Suppose we wanted to know the lower and upper critical values of a Normal distribution for computing a 99% confidence interval. Confidence intervals are always two-tailed, so for a 99% CI we put one-half of 1% ($p = 0.005$) in each tail of the distribution. The probability values whose z values we want are therefore 0.005 (the left-hand tail) and 0.995 (the right-hand tail). We make these into a vector, using concatenation, **c**, like this c(0.005,0.995) then use quantiles of the Normal distribution to get our two answers like this:

```
qnorm(c(0.005,0.995))
```

```
[1]  -2.575829 2.575829
```

This tells us that a 99% Normal confidence interval goes from -2.576 standard errors below the estimated parameter value, to $+2.576$ standard errors above it.

Random number generation

The final Normal probability function is the random number generator, **rnorm**. This is very useful in testing experimental designs where you might want to generate artificial data, in order to test the kind of analysis you propose to do. You could then check that the replication was adequate to detect the kinds of differences that you hope to be able to say are significant. The function has three arguments: first, the number of random numbers you want, second their mean, and third the standard deviation. Here is a histogram of 1000 random numbers with mean $= 10$ and standard deviation $= 2$:

```
hist(rnorm(1000,10,2))
```

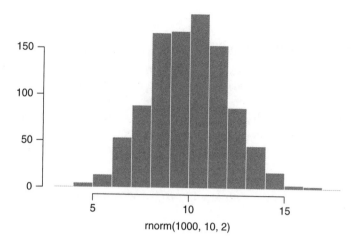

rnorm(1000, 10, 2)

The y axis is unlabelled (it is the frequency with which different random numbers occurred in the classes (bins) chosen by the histogram function). The x axis is labelled with the first argument provided to the **hist** function; this is often a variable name (which makes a good axis label) but here it is the recipe used to generate the numbers). The histogram demonstrates that when a Normal distribution has a mean of 10 and a standard deviation of 2, we are most unlikely to obtain values smaller than 5 or larger than 15 by chance alone.

Comparing data with a Normal distribution

Various tests for normality are described on p. 103. Here we are concerned with the task of comparing a histogram of real data with a smooth Normal distribution with the same mean and standard deviation, in order to look for evidence of non-normality (e.g. skew or kurtosis).

The histogram of our data is an empirical estimate of the probability density associated with different values of x, so it is the **dnorm** function we shall use to draw the smooth line (rather than **pnorm** or **qnorm**). When drawing curves over histograms there are three things to remember. The height of a histogram depends upon

- the total sample size, n (the more samples there are, the taller the bars)

- the standard deviation, s (the greater the standard deviation, the lower the bars)

- the break points of the bars(more inclusive bars—bigger bins are taller)

The key to understanding this exercise is knowing how to scale the vertical axis of the smooth Normal distribution. An example should make this plain:

```
dist<-function(x){
    x[x<0]<-0
    xm<-mean(x)
    xs<-sqrt(var(x))
    xl<-length(x)
    lowx<-min(-3,(min(x)-mean(x))/xs)
    highx<-max(3,(max(x)-mean(x))/xs)
    xv<-seq(lowx*xs,highx*xs,(highx-lowx)/100)
```

```
xv<-xv+xm
hist(x,breaks=-0.5:25.5)
lines(xv,xl*dnorm(xv,xm,xs)) }
```

The function tests for negative values of x and replaces them with zeros. It will draw the curve over whichever is greater: either 3 standard errors or the observed range of data values. 100 x values are generated in a sequence to produce the smooth line of the Normal distribution with mean xm, and standard deviation xs. The histogram is plotted using integer bins from 0 up to 25 (this range obviously depends on the particular application). The smooth lines are drawn using the calculated range of x values (xv), to a height specified by **dnorm** with the appropriate values of mean (xm) and standard deviation (xs) multiplied by the total frequency (xl).

To test the function we plot a frame of four histograms. They are all based on 1000 random numbers from a Normal distribution with mean = 10 but with different standard deviations $s = 1, 2, 3$ and 4 for the different plots:

```
par(mfrow=c(2,2)
dist(rnorm(1000,10,1))
dist(rnorm(1000,10,2))
dist(rnorm(1000,10,3))
dist(rnorm(1000,10,4))
```

Notice how the scale on the y axis changes as the standard deviation increases. This function is not a polished product. For example, it does not rescale the y axis to make sure it can accommodate the largest values generated by the **lines** function. You might like to rectify this as an exercise.

Let's compare a real data set with a Normal distribution to look for skew and kurtotis:

```
attach(fishes)
names(fishes)
```

```
[1] "mass"
```

mean(mass)

[1] 4.194275

max(mass)

[1] 15.53216

We need to return to the default of having a single graph per page:

par(mfrow=c(1,1))

Now the histogram of the mass of the fish is produced, specifying integer bins that are 1 g in width, up to a maximum of 16.5 g:

hist(mass,breaks=-0.5:16.5)

For the purposes of demonstration, we generate everything we need *inside* the **lines** function. We generate the sequence of x values for plotting (0 to 16), and the height of the density function. The height of the density function is the number of fish (length(mass)) times the probability density for each x value in the sequence, for a Normal distribution with mean(mass) and standard deviation sqrt(var(mass)) as its parameters. Here is the code we use:

lines(seq(0,16,0.1),length(mass)*dnorm(seq(0,16,0.1),mean(mass),sqrt(var(mass))))

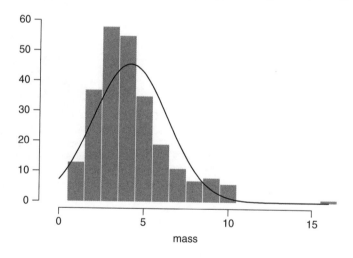

The distribution of fish sizes is clearly *not* normal. There are far too many fishes of 3 and 4 g, too few of 6 or 7 g, and too many really big fish (more than 8 g). This kind of skewed distribution is probably better described by a Gamma distribution than a Normal distribution (see p. 489).

Central limit theorem

One of the reasons that the Normal distribution is so important and so all-pervasive is as follows. If a random variable can be expressed as the sum of a large number of approximately independent components, and none of these components is much bigger than the others, then the sum will be approximately normally distributed.

This means that the average of a sample of independent and identically distributed random variables approaches the Normal distribution as the number of random variables in the average goes to infinity.

So more or less *any* realistic distribution of raw data will produce a Normal distribution of means estimated from repeated samples from that distribution. The central limit theorem states:

> For any distribution with a *finite variance*, the mean of a *random sample* from that distribution tends to be normally distributed

The central limit theorem works remarkably well, even for really badly behaved data. Take this negative binomial distribution that has a mean of 3.88, a variance/mean ratio of 4.87, and $k = 1.08$ (on the left). On the right, you see that the means of repeated samples of size $n = 30$ taken from this highly skew distribution are close to Normal in their distributions.

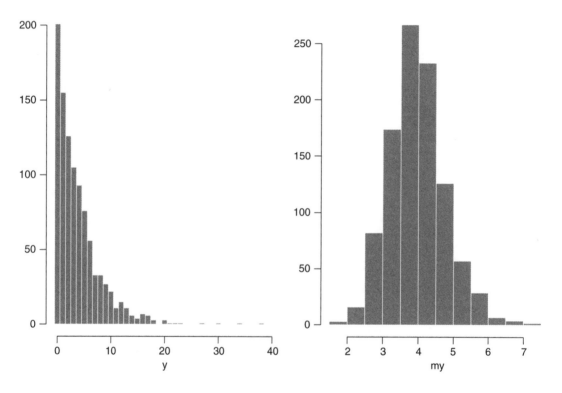

The panels were produced like this. The left-hand histogram is a frequency distribution of 1000 negative binomial random numbers (with parameters size = 1, probability = 0.2). The frequency distribution is viewed using **table**. There were 201 zeros in this particular realisation, and a single value of 38. Note the use of the sequence to produce the required **break points** in the histogram

```
par(mfrow=c(1,2))
y<-rnbinom(1000,1,0.2)
mean(y)
```

```
[1] 3.879
```

```
var(y)
```

```
[1] 18.90326
```

```
table(y)
```

0	1	2	3	4	5	6	7	8	9	10	11	12	13	14	15	16
201	155	126	105	93	76	56	33	33	27	22	11	15	11	6	4	7

17	18	20	21	22	23	27	30	34	38
6	3	3	1	1	1	1	1	1	1

```
hist(y,breaks=-0.5:38.5)
```

The right-hand panel shows the distributions of means of samples of $n = 30$ from this same negative binomial distribution. One thousand different samples were taken and their average recorded in the vector called *my* from which the histogram is produced:

```
my <- numeric(1000)
for (i in 1:1000) {
        y <- rnbinom(30, 1, 0.2)
        my[i] <- mean(y) }
hist(my)
```

In fact, the central limit theorem works remarkably well for sample sizes much smaller than 30, as we shall see later.

Further reading

Ferguson, T.S. (1996) *A Course in Large Sample Theory*. London, Chapman & Hall.
Rosner, B. (1990) *Fundamentals of Biostatistics*. Boston, PWS-Kent.
Silvey, S.D. (1970) *Statistical Inference*. London, Chapman & Hall.

8

Power calculations

The two Rs, randomisation and replication, are the most important issues in experimental design. A fundamental question that you need to address at the planning stage concerns *the number of samples you need to test the hypothesis* of interest. The answer depends on several things:

- the variance of the response variable: you will need larger samples when the variance is large than when it is small
- the size of the difference in the response variable that you want to be able to detect as significant: the smaller the difference, the larger the sample you will need to do it
- the risk of a Type I error (rejecting a true null hypothesis)
- the risk of a Type II error (accepting a false null hypothesis)

In practice the sample size often depends on entirely pragmatic considerations: the resources at your disposal, the number of people to do the work, the amount of space available, the number of experimental animals you can afford, and so on. These considerations often mean that a smaller sample is taken than a statistician would advise. This is extremely unfortunate, because money is often wasted on experiments that could not possibly reject the hypothesis you want to test. You would be amazed at the number of experiments that fail simply because issues of statistical power were not confronted at the planning stage. *Remember, if an experiment is not worth doing, it is not worth doing well.*

Power of the test

Power is the probability that a study rejects the null hypothesis when it is false. Statistical power therefore has to do with Type II errors. The probability of accepting the null hypothesis when it is true is β. The power of the test is $1 - \beta$. In practice, people often work with $\beta = 0.2$ and hence power $= 0.8$. As we have said already, there is always a trade-off between Type I and Type II errors: the harder you make it to commit a Type I error, the more likely it becomes that you will commit a Type II error. A reasonable working compromise is to take α (the probability of a Type I error) as 0.05 and power as 0.8.

The higher the variance, the greater the uncertainty in the values of any parameters estimated from the data. This means that to obtain a particular level of certainty we need more samples in cases where the variance is larger. You will need to have some idea of the variance in the response variable (e.g. from the literature or from previous experiments). If you are stuck, ask a colleague who has carried out similar work in the past. Alternatively, it might be worth carrying out an inexpensive pilot trial aimed solely at estimating the variance. It is essential that you get some idea, however rough, of the value of the variance in your response variable.

The size of the difference you want to detect may be an unfamiliar concept. It is not something that many working scientists are used to asking themselves. Many people appear to work as if their job is to detect differences, not to say what size of differences they want to detect. But as soon as you think about it, it becomes clear that you need a larger sample if you want to be able to say that a 10% difference is significant, than if you want to say that a 20% difference is significant. In practice, many practitioners plan their work in order to be able to detect 25% or 50% differences in the mean of some environmental response. This is for pragmatic reasons: smaller differences will often be the result of background variation (e.g. environmental heterogeneity) and it would be prohibitively expensive to attain the levels of replication necessary to detect differences much smaller than this. In work with opinion polls, however, the pollsters would often like to be able to detect significant differences as small as 1% or 2%, and this is why they need such enormous sample sizes (see below). A good way to think of it is as 'the smallest meaningful difference'.

Calculating sample sizes

The object is to determine the sample size, n, that will determine that a difference of a specified size will be detected as significant when the sample variance is s^2. It is assumed that α and the power of the test have been specified in advance, e.g. $\alpha = 0.05$, $(1 - \beta) = 0.8$. We can cast the problem in three different ways:

- what is the sample size necessary to detect a particular difference?

- what is the detectable difference, given a particular sample size and power?

- what is the power, given a specified sample size and a particular target difference?

The theory is straightforward. In the single-sample case we have a variance s^2, and we want to be able to say that a difference δ is significant if it occurs, with $\alpha = 0.05$ and $\beta = 0.2$. The sample size required, n, is given by

$$n = \left(\frac{s(z_\alpha + z_{1-\beta})}{\delta} \right)^2$$

This says that the smaller the difference δ we want to detect, the larger the sample size we shall need. The sample size goes up with the power $1 - \beta$ and with the sample standard deviation s. The values of z come from tables of the standard Normal distribution with parameters α and $(1 - \beta)$. The two-tailed z values for the traditional $\alpha = 0.05$ and $\beta = 0.2$ are

```
qnorm(1-0.025)

[1] 1.959964

qnorm(1-0.2)

[1] 0.8416212
```

Rule of thumb for one-sample replication

Because the z values from tables do not vary from case to case (so long as we stick with $\alpha = 0.05$ and $p = 0.8$, giving $(z_{0.05} + z_{0.8})^2 \approx 8$) all we need are the variance s^2 and the difference δ we want to be able to detect. Then the sample size required is simply

$$n = 8 \times \frac{s^2}{\delta^2}$$

For example, if we want to be able to detect a change of 2.0 in a population with variance $s^2 = 10.0$ then we shall need about $8 \times 10/2^2 = 20$ replicates. If we want to be able to detect a change of 0.1 in a population with variance $s^2 = 0.04$ then we shall need about $8 \times 0.04/0.1^2 = 32$ replicates.

Replication in the two-sample case

The two-sample case is slightly more complicated, because the two samples could have different variances, and we might want to have different replication in the two samples (n_1 and n_2). Calculate n as above for sample 1, then

$$n_1 = n\left[1 + \frac{s_2^2}{w_2 s_1^2}\right] \quad \text{and} \quad n_2 = w_2.n_1$$

where w_2 is the proportional replication in sample 2 compared with sample 1.

Power calculations in S-Plus

The S-Plus function **normal.sample.size** computes power and sample size requirements for one and two-sample t tests (there is a handy dialog box under Statistics > Power and Sample Size). Suppose that the mean of the control group is likely to be 16.0 and that we want to be able to detect a 20% increase or decrease in the mean (i.e. the treatment mean is $16.0 \times 1.2 = 19.2$). Previous work suggests that the variance is roughly 4 (so the standard deviation is 2 and we assume it to be the same in both treatments). Taking the conventional α and power values of 0.05 and 0.8 respectively, we use the function like this:

```
normal.sample.size(mean=16, mean2=19.2,sd1=2,sd2=2)

      mean1 sd1 mean2 sd2 delta alpha power n1 n2 prop.n2
  1      16   2  19.2   2   3.2  0.05   0.8  7  7        1
```

This says that we should take 7 samples from each treatment. A pragmatic approach, therefore, would be to take $n_1 = n_2 = 10$ to leave a margin for error in our estimate of the variance.

Effect size

Another way you can use the function is to determine the size of the effect you could determine with a given level of replication. In our example we might ask how much better we would do (compared with our 20% change) if we took 10 replicates instead of 7 from each treatment. Now $n_1 = n_2 = 10$ and we write

```
normal.sample.size(mean=16, n1=10, n2=10,sd1=2,sd2=2)

      mean1 sd1    mean2 sd2    delta alpha power n1 n2 prop.n2
  1      16   2 18.50581   2 2.505814  0.05   0.8 10 10        1
```

They key figure here is 18.505 81. This is the second mean with replication of 10, so the detectable effect size is $18.505\,81 - 16 = 2.505\,81$ (i.e. a 16% change in the mean). This is substantially better than the 20% difference we aimed at originally.

Power

Finally, we could use the function to calculate the power of the test. For this, we supply the two means, the two standard deviations and the two sample sizes. What power would we have to detect a difference in means as small as 2.0 with equal replication of 10?

```
normal.sample.size(mean=16, mean2=18,n1=10, n2=10,sd1=2,sd2=2)

  mean1 sd1 mean2 sd2 delta alpha      power n1 n2 prop.n2
1    16   2    18   2     2  0.05 0.6087795 10 10       1
```

This is pretty hopeless. Our power of 61% is not much better than tossing a coin. A 50:50 outcome has power $= 50\%$.

Power and sample size for proportion data

Another kind of question that often arises in survey work is how many individuals to sample in order to detect a given change in opinion. Suppose that 55% had expressed a particular opinion in the last opinion poll. How many people do I need to canvas in order to be reasonably certain that the percentage has increased to 60%? The S-Plus function to work this out is called **binomial.sample.size**. The function determines what to compute on the basis of the arguments provided. If p_{alt} or p_2 is given, but n_1 is not, then sample size is computed. If p_{alt} or p_2 is given along with n_1, then the power is computed. If only n_1 is provided, the minimum detectable difference is computed using the default power of 0.80. In this example we know that $p = 0.55$ and $p_{alt} = 0.6$. The function is used like this:

```
binomial.sample.size(p=.55,p.alt=.6)

  p.null p.alt delta alpha power  n1
1   0.55   0.6  0.05  0.05   0.8 810
```

So the answer is more than 800 people. Perhaps I had better think again? What could I achieve if I could only afford to interview 500 people?

```
binomial.sample.size(p=.55,n1=500)

  p.null      p.alt      delta alpha power  n1
1   0.55 0.6123313 0.06233134  0.05   0.8 500
```

Actually, that's not too bad. I could still expect to discover a change from 55% to 61.2%. Finally, I could ask what power my original specified difference in proportions would have at this reduced replication by specifying both the alternative proportion ($p_{alt} = 0.6$) and the reduced sample size ($n_1 = 500$):

```
binomial.sample.size(p=.55,p.alt=.6,n1=500)

  p.null p.alt delta alpha     power  n1
1   0.55   0.6  0.05  0.05 0.5787404 500
```

This shows that my power has declined from 80% to 58%, which doesn't sound anywhere like as promising. As ever, you get what you pay for.

The moral is that there is no point spending money on sampling when there is no realistic prospect of being able to detect a significant difference. Power testing should be a prerequisite for any experimental or observational study. It will almost certainly save you from wasting time or money (or both).

Power calculations for non-normal and other 'difficult' data

These power calculations based on Normal or binomial distribution theory are reasonably robust in practice. In some applications, however, you may want to work with samples where the variance is a function of the mean (e.g. count data), or where the errors are expected to be highly skew (e.g. count data with a low mean and many zeros in the data). In such a case you might want to resort to simulation to answer your power and sample size questions. In the following example we want to compare the means of two negative binomially distributed samples.

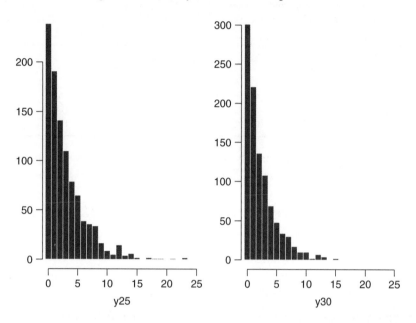

Negative binomial data with parameters (1 and 0.25) and (1 and 0.3) were generated by the **rnbinom** function, giving means of 3.011 (variance = 11.76) and 2.275 (variance = 7.036) respectively. We are interested in the power of tests of the null hypothesis that the two means are the same. The code below generates 1000 repeats of the following experiment. Samples of size n are taken from negative binomial distributions with parameters 1 and 0.3 and 1 and 0.25 respectively:

```
d<-numeric(1000)
n<-350
for (i in 1:1000) {

y1<-rnbinom(n, 1, .3)
y2<-rnbinom(n, 1, .25)
y<-c(y1,y2)
fa<-factor(c(rep(1,n),rep(2,n)))

model<-glm(y~fa,poisson)

d[i]<-as.vector(summary.aov(model)$"Pr(F)")[1] }
```

The two samples are compared using **glm** with Poisson errors corrected from overdispersion (Chapter 29). The result of each simulated comparison is extracted from the summary of the model (this is saved as a vector of p values in the vector called d). When the simulation is complete, we ask what proportion of the tests gave significant p values. We want this proportion (the power) to be 80% or more. We repeat the whole exercise for different values of the sample size, n. The results of trials using seven different sample sizes are shown in Table 8.1.

Table 8.1 The body of the table shows power (%) for detecting the difference between p_1 and p_2 as significant in two samples each of size n (top row), compared using a log-linear model (with Poisson errors), corrected for overdispersion using an empirical scale parameter. The data, pictured above, have a negative binomial distribution with variance/mean ratios $\gg 1$. For 80% power, the simulations suggest that sample sizes of 350, 100 and about 25 are required as p_2 increases from 0.3 to 0.5.

	n	20	50	100	200	350	400	500
p_1	p_2							
0.25	0.3	11	20	32	55	81	85	92
0.25	0.35	19	51	81	98	100	100	100
0.25	0.5	75	100	100	100	100	100	100

It is instructive to see what Normal theory power calculations would have suggested for the three cases covered here. In the first case (means 3.011 and 2.275 and 3.43 and standard deviations 2.65) we have

```
normal.sample.size(mean2=2.275, mean=3.011, sd1=3.43, sd2=2.65)

    mean1  sd1   mean2   sd2   delta  alpha power   n1  n2 prop.n2
1   3.011 3.43  2.275  2.65  -0.736   0.05   0.8  273 273       1
```

asking for 273 samples rather than about 350 suggested by the simulations. In the second case (means 3.011 and 1.71 and 3.43 and 2.11) standard deviations we have

```
normal.sample.size(mean2=1.71, mean=3.011, sd1=3.43, sd2=2.11)

    mean1  sd1 mean2   sd2  delta  alpha power n1 n2 prop.n2
1 3.011 3.43  1.71 2.11  -1.301   0.05   0.8 76 76       1
```

asking for 76 rather than about 100. In the last case there was a threefold difference between the means:

```
normal.sample.size(mean2=0.9, mean=3.011, sd1=3.43, sd2=1.41)

    mean1  sd1 mean2 sd2  delta  alpha power n1 n2 prop.n2
1 3.011 3.43   0.9 1.41 -2.111   0.05   0.8 25 25       1
```

which is almost exactly the same sample size (25) as suggested by the simulation.

The moral is that if you don't want to do the simulations, then taking the sample sizes predicted from Normal theory and multiplying them by about 1.3 is a reasonable rule of thumb when the variance/mean ratios are bigger than about 2.

Further reading

Fisher, L.D. and Van Belle, G. (1993) *Biostatistics*. New York, John Wiley.
Rosner, B. (1990) *Fundamentals of Biostatistics*. Boston, PWS-Kent.

9

Understanding data:graphical analysis

The importance of knowing your data

It is a common failing to rush into statistical analysis without first obtaining a thorough understanding of the data. S-Plus encourages good habits of data exploration by providing a superb range of graphical facilities. The object of the exercise is to:

- understand the distribution of the response variable

- look for trends with the explanatory variables

- consider the need for transformation

- look for potentially influential observations

- find errors that have occurred during data entry

- test the assumptions of the statistical models that you intend to employ

There are two fundamentally different kinds of explanatory variables, continuous and categorical, and each leads to a completely different sort of graph. In cases where the explanatory variable is continuous, like length or weight or altitude, the appropriate plot is a **scatterplot**. In cases where the explanatory variable is categorical, like colour or gender, then the appropriate plot is a **boxplot**.

Plotting with continuous explanatory variables: scatterplots

The **response variable** y is plotted on the vertical axis of a graph; the **explanatory variable** x is plotted on the horizontal axis of a graph. It is extremely simple to obtain a scatterplot of y against x in S-Plus:

```
attach(plotdata)
names(plotdata)
```
```
[1] "x" "y"
```

The **plot** function needs only two arguments (although it can have many more). First it requires the name of the explanatory variable (x in this case), and second it requires the name of the response variable (y in this case):

```
plot(x,y)
```

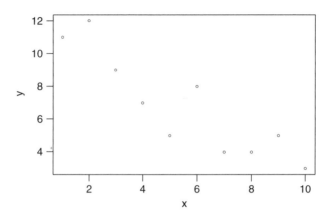

For the purposes of data exploration, this may be all you need. But for publication it is useful to be able to change the appearance of the plot. It is often a good idea to have a longer, more explicit label for the axes than is provided by the variable names that are used as default options (*x* and *y* in this case). Suppose we want to change the label *x* into the longer label 'Explanatory variable'. To do this we use the **xlab** function with the text of the label enclosed within double quotes. Use the Up Arrow to get back the last command line, and insert a comma after *y*, then type xlab= then the new label in double quotes, like this:

```
plot(x,y,xlab="Explanatory variable")
```

You might want to alter the label on the *y* axis as well. The function you need for that is **ylab**. Use the Up Arrow key again, and insert a *y* label, like this:

```
plot(x,y,ylab="Response variable",xlab="Explanatory variable")
```

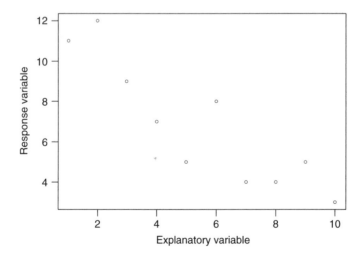

It is easy to change the plotting symbols. At the moment, you are using the default plotting character (pch=1) which is an open circle. If you want + (plus signs), use plotting character 3. Use Up Arrow, then insert a comma after y, then type pch=3

```
plot(x,y,pch=3,ylab="Response variable",xlab="Explanatory variable")
```

Open triangles are obtained using pch=2, and so on:

plot(x,y,pch=2,ylab="Response variable",xlab="Explanatory variable")

Adding lines to a scatterplot

There are two kinds of lines that you might want to add to a scatterplot: lines that the computer estimates for you (e.g. regression lines) and lines that you specify yourself (e.g. lines indicating some theoretical values for y and x).

Regression lines

The simplest line to fit to some data is the linear regression of y on x:

$$y = a + bx$$

This has two variables (y and x) and two parameters (intercept a and slope b). The model is interpreted as saying that y is a function of x, so that changing x might be expected to cause changes in y. The opposite is not true, and changing y by some other means would not be expected to bring about a change in x.

The basic statistical function to derive the two parameters of the linear regression (intercept a and slope b) is the linear model **lm** (Chapter 14). This uses the standard model form as its argument:

$$y \sim x$$

This is read a 'y tilde x' and means y is to be estimated as a linear function of x. We learn all about this in Chapter 14. To draw the regression line through the data, we employ the straight line drawing function **abline** (intercept and slope). We can combine the regression analysis and the line drawing into a single function like this:

abline(lm(y~x))

One of the nice features of **abline** is that it automatically draws the line to fit exactly within the limits set by the frame of our graph.

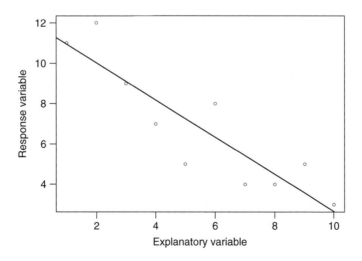

User-specified lines

Sometimes you want to draw your own line on a scatterplot. A common case is where you want to draw a line of slope = 1 that goes through the origin. Or you might what to draw a horizontal line or a vertical line for some purpose. Suppose that in this case we want to draw a straight line that reflects a theoretical model known from previous work. This line goes through the (x, y) points (0,12) and (10,4).

To draw the line, we use **c** to create two lists: a list of x points c(0,10), and a list of y points c(12,4). Because there are two numbers in each list, the computer will draw a single straight line in response to the **lines** function. The first argument contains the x points of the lines, and the second the y points of the lines. We shall use a third argument to change the line type to number 2 (lty=2) in order to contrast with the solid regression line that we have just drawn:

```
lines(c(0,10),c(12,4),lty=2)
```

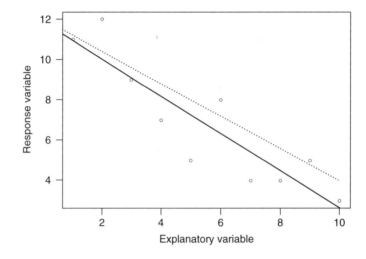

Adding extra points to a graph

The important point is that material is overlaid on your graph until another **plot** function is issued. You have seen lines added to the scatterplot using **abline** and **lines**. You can add new points to the scatterplot using **points**.

Sometimes we want several data sets on the same graph, and we would want to use different plotting symbols for each data set in order to distinguish between them. Suppose that we want to add the following points to our current graph. We have five new values of the explanatory variable; let's call them v and give them the values 2, 4, 6, 8 and 10:

```
v<-c(2,4,6,8,10)
```

We need the matching five new values of the response variable, w:

```
w<-c(8,5,6,6,2)
```

To add these points to the graph, we simply type **points** (not **plot** as this would draw fresh axes and we would lose what we had already done):

```
points(v,w,pch=3)
```

The points are added to the scatterplot using plotting character pch=3 (which is the plus sign +). A warning message is printed, because one of our points (10,2) lies outside the plotting region that was defined by the original **plot.** Note that the plotting region is not subsequently rescaled as **lines** or **points** are added. Later on we shall see how to deal with this in the context of writing general plotting routines that will always accommodate the largest and smallest values of x and y.

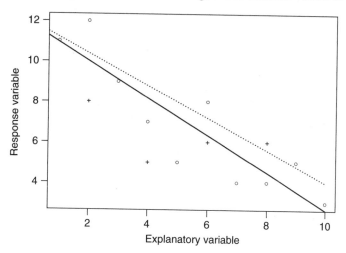

If we want to fit a separate regression line for these new data, we just type

```
abline(lm(w~v),lty=3)
```

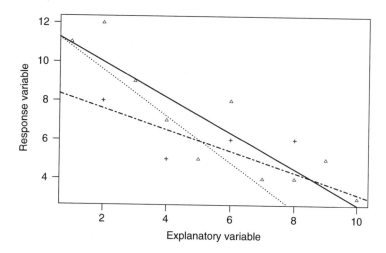

This is about as complicated as you would want to make any figure. Adding more information would begin to detract from the message.

Plotting with categorical explanatory variables: boxplots

Categorical variables are **factors** with two or more **levels**. For most kinds of animal, sex is a factor with two levels, male and female, and we could make a variable called *sex* like this:

```
sex<-c("male","female")
```

The categorical variable is the factor called *sex*, and the two levels are 'male' and 'female'. In principle, factor levels can be names or numbers. We shall often use names because they are so much easier to interpret, but some older software (like GLIM) can only handle numbers as factor levels.

The next example uses the factor called *month* to investigate weather patterns at Silwood Park in south-east England. We begin by attaching the data frame:

attach(SilwoodWeather)

To see the names of the variables and the order in which they appear in the data frame, we use **names**:

names(SilwoodWeather)

```
[1] "upper" "lower" "rain" "month" "yr"
```

The variables in the data frame are upper and lower daily temperatures in degrees Celsius, rainfall that day (in millimetres), and the names of the month and year to which the data belong. There is one more bit of housekeeping we need to do before we can plot the data. We need to declare month to be a factor. At the moment, S-Plus just thinks it is a number (1 to 12):

month<-factor(month)

You can always check on the status of a variable by asking whether it's a factor using the **is.factor** function:

is.factor(month)

```
[1] TRUE
```

Yes, it is a factor. Now we can **plot** using a categorical explanatory variable (*month*):

plot(month,upper)

The syntax is exactly the same, but because the first variable is a factor we get a boxplot rather than a scatterplot.

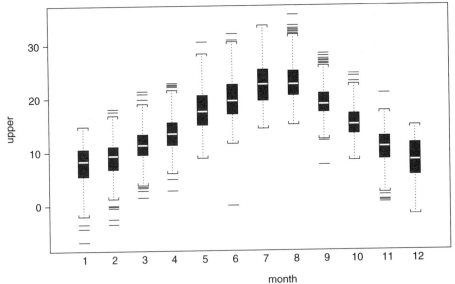

The boxplot summarises a great deal of information very clearly. First, it is very good at showing up errors. Clearly, you would never get a maximum temperature of zero in June! The horizontal line shows the **median** response for each month. The bottom and top of the box show the 25 and 75 **percentiles** respectively (i.e. the location of the middle 50% of the data). The horizontal line joined to the box by the dashed line (sometimes called the whisker) shows 1.5 times the **interquartile range** of the data. Points beyond this (outliers) are drawn individually as horizontal lines. Boxplots not only show the location and spread of data but also indicate skewness (asymmetry in the sizes of the upper and lower parts of the box). For example, in February the range of lower temperatures was much greater than the range of higher temperatures. You might like to add the axis labels 'Month' and 'Maximum daily temperature' as an exercise.

Plots with error bars

When the aim is to provide plots of mean values estimated for different levels of a categorical explanatory variable, it is important to show the level of uncertainty associated with each of the estimates. The **error.bar** function will produce pointwise error bars when provided with three arguments:

- the x axis location of each point

- the y axis location (the height of the bar) for each point

- the length of the error bar, up and down, from each point

The error bars can be all the same length, or all different lengths. The function will scale the y axis automatically to accommodate the tops of the highest bars.

The example is analysed in full in Chapter 18; the data are the biomass of individual plants in a competition experiment with a control and four levels of reduced competition (two levels of neighbour shoot removal, n_{25} and n_{50}, and two levels of root pruning, r_{10} and r_5):

```
attach(compexpt)
names(compexpt)

[1] "biomass" "clipping"
```

The idea is to plot the mean biomass in each of the five treatments along with the 95% confidence interval for the mean, based on the pooled error variance. We need to produce a vector containing the five means:

```
y<-tapply(biomass,clipping,mean)
```

and another containing the common 95% confidence intervals. To work this out, we need the pooled error variance from the one-way Anova:

```
summary(aov(biomass~clipping))
```

	Df	Sum of Sq	Mean Sq	F Value	Pr(F)
clipping	4	85356.5	21339.12	4.301536	0.008751641
Residuals	25	124020.3	4960.81		

so the pooled error variance is 4960.81. The 95% confidence interval is t from tables $\times$ the standard error. The value of Student's t (two-tailed, $\alpha = 0.05$) with 5 degrees of freedom (6 replicates per treatment) is

```
qt(0.975,5)

[1] 2.570582
```

The standard error of the mean is $\sqrt{s^2/n}$ where $s^2 = 4960.81$ and $n = 6$ is

    sqrt(4960.81/6)

    [1] 28.75416

We need to make a vector containing this confidence interval with as many values as there are error bars to be drawn in our figure, five in this case. We call this vector z:

    z<-rep(qt(0.975,5)*sqrt(4960.81/6),5)

Then the error bar plot is straightforward. We number the points 1 to 5 using 1:5 as a surrogate set of x coordinates, provide y (the mean values) and z (the lengths, up and down, of the 95% confidence intervals) like this:

    error.bar(1:5,y,z,ylab="biomass",xlab="treatment")

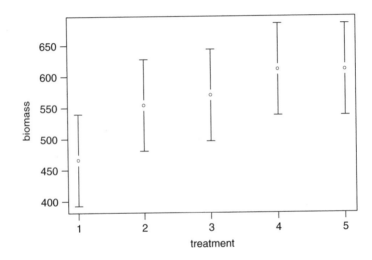

The error bars all overlap, but since the bars are 95% confidence intervals this does *not* necessarily mean there are no significant differences within the data frame. The means for the four competition treatments all lie above the top of the confidence interval for the controls (treatment 1), so it looks as if all the competition treatments increased mean biomass, but none of the competition treatments appears to be significantly different from any of the others. The example is analysed in full, using contrasts, on p. 325.

Barplots with error bars

Many people prefer to present results like these as barplots, rather than points. The basic barplot is produced like this:

    barplot(y,ylab="biomass",xlab="clipping",names=levels(clipping))

where the heights of the bars represent mean biomass in the five competition treatments (y, see above). The bars are given the names of the levels of the five different clipping treatments:

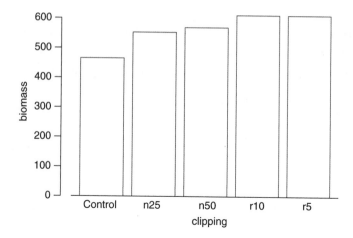

The extra issues are these:

- we want to overlay the error bars on an existing barplot, so we want to override the **error.bar** default, which produces a fresh plot; use add=T for this

- the tops of the highest error bars may fall outside the plotting area produced by the **barplot** command; use ylim in the **barplot** command to allow for this

- it is not obvious what x coordinates should be used for locating the error bars in the centres of the bars produced by **barplot** (see below for the solution)

- the default bars have a gap in their centre (see above); override this with gap=F

We start by calculating the heights of the five bars:

```
y<-tapply(biomass,clipping,mean)
```

In this case we choose to use error bars, z, that are ± 1 standard error of the mean, with the length of the bar worked out separately for each mean (i.e. not using the pooled error variance):

```
z<-tapply(biomass,clipping,stdev)/sqrt(6)
z
```

```
   control        n25        n50        r10         r5
23.51796  44.97011  19.59025  23.99815  24.47277
```

This means that we shall need to rescale the barplot because the tallest error bar is going to be almost 25 g higher than the current maximum of 600 g.

Most importantly, we need to find out where, exactly, the five bars should be centred on the x axis. The relevant x coordinates are held as a vector associated with the barplot object. To extract them, assign the barplot object a name, say *xval*, then look at *xval*:

```
xval<-barplot(y,ylab="biomass",xlab="clipping",names=levels(clipping))
xval
```

```
[1]  0.7 1.9 3.1 4.3 5.5
```

These are the x coordinates of the centres of the five bars of the barplot, and the x coordinates we need for the **error.bar** function. We can now replot the barplot leaving space for the top of the highest error bar with ylim:

```
barplot(y,ylim=c(0,700),ylab="biomass",xlab="clipping",names=levels(clipping))
```

Then we add the error bars to the plot like this:

```
error.bar(xval,y,z,gap=F,add=T)
```

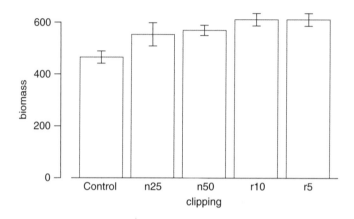

The control plants have significantly lower biomass than in the other four treatments, but there is no significant difference between the mean biomass in any of the root or neighbour pruning treatments (see p. 325 for a full analysis). Note that standard error bars (here) and confidence interval bars (above) give quite different visual impressions of the level of variability present in the data. This issue is discussed at length on p. 275.

Drawing smooth lines

The default in **plot** with a continuous explanatory variable is to produce a scatterplot using open circles as the plotting symbol. Often, however, we want to produce graphs that are smooth lines rather than scatters of points. To do this, we use the option type="line"

Suppose we want to draw the graph $y = 2x^3$ over the range of x values from 0 to 1. First we generate a suitable series of x values:

- smooth curves need about 100 segments to look good
- we need to specify the maximum and minimum values of x appropriately

```
x<-seq(0,1,0.01)
```

For the purposes of demonstration, we shall produce two plots side by side, so we alter the graphics parameter mfrow to allow this:

```
par(mfrow=c(1,2))
```

Now we plot the curve, first as a series of points and then as a smooth line:

```
plot(x, 2*x^3)
plot(x, 2*x^3,type="line")
```

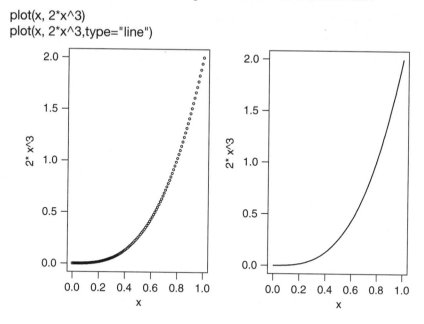

The option type="line" can be abbreviated to type="l" (but make sure you type the letter *l* and not the numeral 1 when you come across this).

Changing the scales of the axes

S-Plus does a good job of selecting appropriate scales for our axes based on the maximum and minimum values of *x* and *y*, but there are occasions when we need to overrule the default settings. A common case is where we want two adjacent graphs to have the same axes even though they are based on different sets of numbers.

The appropriate plot options for setting limits to the axes are xlim and ylim. Each requires a single argument, which is a vector of length 2 containing the minimum and maximum values required. In any **plot** command you can specify one or both of them (the default is neither). Here are the values of two data sets:

```
x1<-0:5
y1<-c(13,11,8,3,4,5)
x2<-seq(2,12,2)
y2<-c(8,4,6,6,1,2)
```

We plot them on separate graphs using par(mfrow=c(1,2)):

```
par(mfrow=c(1,2))
plot(x1,y1)
plot(x2,y2,pch=16)
```

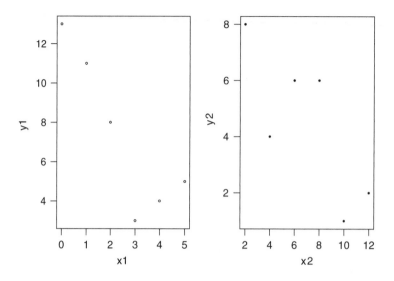

Notice that both the *x* axes and *y* axes are different on each graph. To make it easier to compare the two data sets, we want to plot them on the same scales. We need to work out the maximum and minimum values of each variable before we start plotting. The **range** function is useful for this:

```
range(c(x1,x2))
```

```
[1]  0  12
```

```
range(c(y1,y2))
```

```
[1]  1  13
```

It looks reasonable to scale the *x* axis from 0 to 12 and the *y* axis from 0 to 15 (you don't have to use the maximum and minimum values):

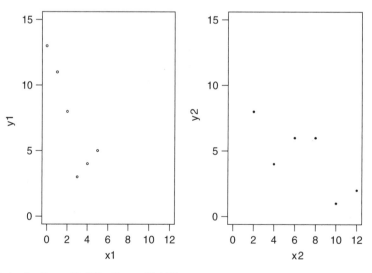

```
plot(x1,y1,xlim=c(0,12),ylim=c(0,15))
plot(x2,y2,xlim=c(0,12),ylim=c(0,15),pch=16)
```

Remember that overriding the default scaling may be useful:

- when you want two comparable graphs side by side

- when you want to overlay several lines or sets of points on the same axes

- recall that the initial **plot** function sets the axes scales; this can be a problem if subsequent lines or points are off-scale

Text in graphs

It is very straightforward to add text to graphics. All you do is specify the (x, y) coordinates of the bottom left-hand corner of the place where you want the text to be, then supply the text string enclosed in double quotes. Suppose we want to add the labels (a) and (b) to the two graphs we have just drawn. Inspection suggests that a good place to put the labels would be in position (10,14). You will need to redraw the figures: text is added (like points and lines) before the next **plot** command is issued. This means that we must add the label (a) to the left-hand plot *before* we draw the axes of the right-hand plot. We need also to decide on the size of the characters to be added. The current default can be assessed by looking at the axes labels (e.g. x1 and y1); these look too small, so let's make the additional text twice the size of the defaults by setting the option cex=2 (cex stands for 'character expansion'):

```
plot(x1,y1,xlim=c(0,12),ylim=c(0,15))
text(10,14,"(a)",cex=2)
plot(x2,y2,xlim=c(0,12),ylim=c(0,15),pch=16)
text(10,14,"(b)",cex=2)
```

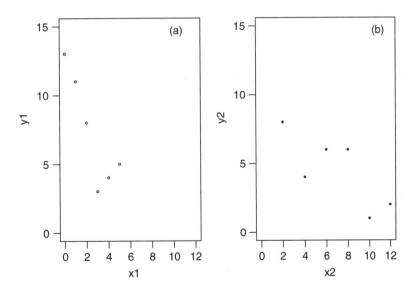

Text can be oriented at any angle. To demonstrate this, we create a vector called *labels*, containing the first 10 letters of the alphabet in lower case:

```
labels<-letters[1:10]
labels
```

```
[1]  "a" "b" "c" "d" "e" "f" "g" "h" "i" "j"
```

Plot type= "n" is useful for scaling the axes but suppressing the creation of any filler for the space within the axes:

```
plot(1:10,1:10,type="n")
text(1:10,1:10,labels,cex=2)
plot(1:10,1:10,type="n")
text(1:10,10:1,labels,cex=2,crt=180)
```

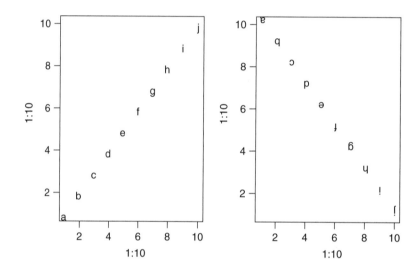

Note the use of character rotation crt (degrees counterclockwise from the horizontal) to turn the letters upside down in the right-hand plot. These letters are twice default size (cex=2). Note also the reversal of the sequence argument, so that the y values of the character locations in the right-hand plot go down from 10 to 1 as x goes up from 1 to 10.

Logarithmic axes

You will often want to plot one or both of your axes on a logarithmic scale. The options log="xy", log="x" or log="y" control the kinds of axes, producing log-log, log-x or log-y axes respectively. Here are the same data plotted four ways:

```
attach(curvedata)
names(curvedata)
```

```
[1]  "xvalues" "yvalues"
```

Now we make a frame for four adjacent plots, a 2×2 array:

```
par(mfrow=c(2,2))
```

From top left to bottom right, we plot the data using **lines** with untransformed axes, then as log y against x, as y against log x, and finally as a log-log plot:

```
plot(xvalues,yvalues,type="l")
plot(xvalues,yvalues,log="y",type="l")
plot(xvalues,yvalues,log="x",type="l")
plot(xvalues,yvalues,log="xy",type="l")
```

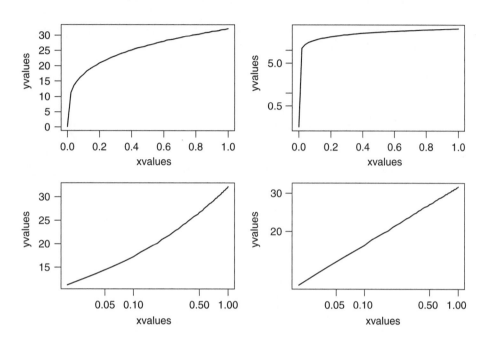

The relationship is approximately linear only for the log-log plot (bottom right), suggesting a power law relationship between y and x (Chapter 21).

Summary

- **plot**: plot(x,y) gives a scatterplot if x is continuous, or a boxplot if x is a factor
- **type**: options include lines type="l" or null (axes only) type="n"
- **lines**: lines(x,y) plots a smooth function of y against x using the x and y values provided
- **points**: points(x,y) adds another set of data points to a plot
- line types: e.g. dotted line, useful in multiple plots, lty=2 (an option in plot or lines)
- plotting characters for different data sets: pch=2 or pch="*" (an option in points or plot)
- setting limits to x and y axis scales xlim=c(0,25), ylim=c(0,1) (an option in plot)

Multivariate plots

Data sets often consist of many continuous variables, and it is important to find out if the variables are interrelated and how. In this example of air pollution (SO_2) data there are seven continuous variables:

```
attach(Pollute)
names(Pollute)

[1] "Pollution" "Temp" "Industry" "Population" "Wind" "Rain"
[7] "Wetdays"
```

The simple but powerful function **pairs** produces a matrix of plots of y against x and x against y for all of the variables in the data frame (all seven in this case):

```
pairs(Pollute)
```

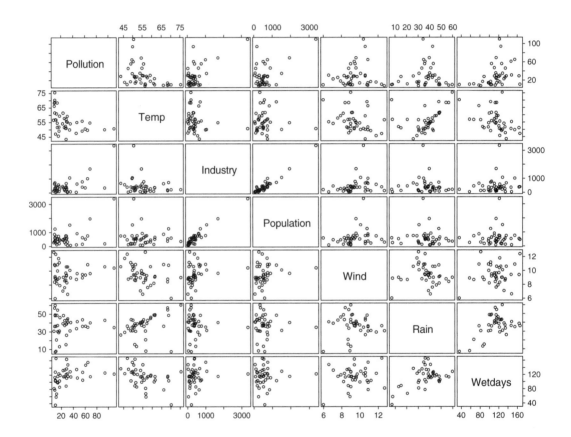

Interpretation of the matrix takes some getting used to. The rows of the matrix have the response variable (the y axis) as labelled by the variable name that appears in that row. For example, all the graphs on the middle row (the fourth row) have Population on the y axis. Likewise, the columns of the matrix all have the same explanatory variable (the x axis) as labelled by the variable name that appears in that column. For example, all the graphs in the rightmost column have Wetdays on their x axis.

This kind of multiple scatterplot is good at showing some patterns but poor at showing others. Where the data are well spaced then relationships can show up clearly. But when the data are clumped at one end of the axis it is much more difficult. Likewise, the plots are poor when variables are categorical (integer). Not unexpectedly, there is a very close correlation between Industry and

Population, but the relationship between Pollution and Wind is far from clear. We shall carry out the statistical analysis of these data later on (Chapters 24 and 31 and p. 190).

Conditioning plots

A real difficulty with multivariate data is that the relationship between two variables may be obscured by the effects of other processes.

When you carry out a two-dimensional plot of y against x, then all the effects of the other explanatory variables are squashed flat onto the plane of the paper. The function **coplot** produces a conditioning plot automatically in cases like pollution this example, where all the explanatory variables are continuous. It is extremely easy to use. Suppose we want to look at Pollution as a function of Temp but we want to condition this on different levels of Rain. This is all we do:

coplot(Pollution~Temp | Rain)

The plot is described by a model formula: Pollution is on the y axis, Temp on the x axis, with six separate plots conditioned on the values of Rain shown in the top panel.

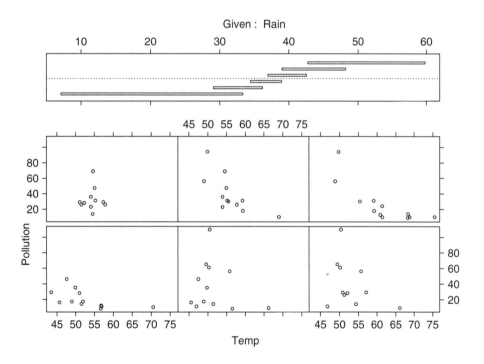

The panels are ordered from lower left, row-wise, to upper right, from lowest rainfall to highest. The upper panel shows the range of values for Rain chosen by **coplot** to form the six panels. Note that the range of rainfall values varies from panel to panel. The largest range is in the bottom left plot (range 8 to 33) and the smallest for the bottom right plot (35 to 39). Note that the ranges of Rain overlap one another at both ends; this is called a **shingle**. All the usual extra options can be employed with **coplot** (e.g. fitting regressions or smoothers separately for each panel); we cover this later.

Tree-based models

The most difficult problems in interpreting multivariate data are that:

- the explanatory variables may be correlated with one another
- there may be interactions between explanatory variables
- relationships between the response and explanatory variables may be non-linear

An excellent way of investigating interactions between multiple explanatory variables involves the use of tree-based models. Tree models are constructed on a very simple, stepwise principle. The computer works out which of the variables explains most of the variance in the response variable, then determines the threshold value of the explanatory variable that best partitions the variance in the response. Then the process is repeated for the *y* values associated with *large values* of the first explanatory variable, asking which explanatory variable explains most of this variation. The process is then repeated for the *y* values associated with *low values* of the first explanatory variable. The process is repeated until there is no residual explanatory power. The procedure is very simple, involving the function **tree**. For the pollution data set, we type

```
regtree<-tree(Pollution ~ . , data=Pollute)
```

where the model formula reads 'Pollution (our response variable) is a function of *all* of the explanatory variables in the data frame called Pollute'. The power comes from the 'dot option'. The tilde ~ means 'is a function of'. The dot . means 'everything'. So make sure you get the syntax right. The punctuation is tilde dot comma.

Now regtree contains the regression tree and we want to see it, and to label it:

```
plot(regtree)
text(regtree)
```

This is interpreted as follows. The most important explanatory variable is Industry, and the threshold value separating low and high values of Industry is 748.

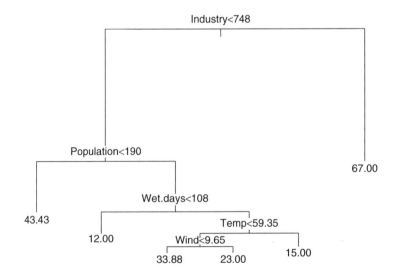

The right-hand branch of the tree indicates the mean value of air pollution for high levels of Industry (67.00). The fact that this limb is unbranched means that no other variables explain a significant amount of the variation in pollution levels for high values of Industry. The left-hand limb does not show the mean values of pollution for low values of Industry, because there are other significant explanatory variables. Mean values of pollution are only shown at the extreme ends of branches. For low values of Industry, the tree shows us that Population has a significant impact on air pollution. At low values of Population (<190) the mean level of air pollution was 43.43. For high values of Population, Wetdays is significant. Low numbers of wet days (<108) have mean pollution levels of 12.00 while Temp has a significant impact on pollution for places where the number of wet days is large. At high temperatures ($>59.35°F$) the mean pollution level was 15.00 while at lower temperatures the run of Wind is important. For still air (Wind <9.65) pollution was higher (33.88) than for higher wind speeds (23.00). The statistical analyses of regression trees, including the steps involved in model simplification, are dealt with in Chapter 31.

The virtues of tree-based models are numerous:

- they are easy to appreciated and to describe to other people

- the most important variables stand out

- interactions are clearly displayed

- non-linear effects are captured effectively

- the complexity (or lack of it) in the behaviour of the explanatory variables is plain to see

Plotting symbols

These are known as plotting characters and are controlled by pch = "c" where c is the character to be used for plotting points. If pch is a decimal point (period or fullstop) then a centred plotting dot is used. Usually, you will specify the plotting symbol by a number, using pch = n where the following values of n generate different symbols: square (0), octagon (1), triangle (2), cross (3), × (4), diamond (5) and inverted triangle (6). Superimposed symbols are (7) $4 = 0+4$, (8) $= 3+4$, (9) $= 3+5$, (10) $= 1+3$, (11) $= 2+6$, (12) $= 0+3$, (13) $= 1+4$ and (14) $= 0+2$. Solid symbols are square (15), octagon (16), triangle (17), and diamond (18). Using the numbers 32 through 126 for pch yields the 95 ASCII characters from space through tilde. The numbers between 161 and 252 yield characters, accents, ligatures or nothing, depending on the font. Here is some code which prints all the plotting symbols from 0 to 81 from bottom left to top right on a scale of (c(0,1), c(0,1)).

```
pchars<- function(){
k<- -1
plot(c(0,1),c(0,1),lab=c(0,0,0),xlab="",ylab="",type="n")
for (i in 1:9) {
for (j in 1:9){
k<-k+1
points(j/10,i/10,pch=k)  }}}

pchars()
```

H	I	J	K	L	M	N	O	P
?	@	A	B	C	D	E	F	G
6	7	8	9	:	;	<	=	>
-	.	/	0	1	2	3	4	5
$	%	&	'	(	)	*	+	,
☑	□	○	△	+		!	"	#
◆	▼	✚	✖	—	∣	※	♂	♀
✦	⊕	⊠	⊞	⊠	◩	■	●	▲
□	○	△	+	×	◇	▽	⊠	✳

Note that the plot has unlabelled axes (xlab="") with no tick marks (lab=c(0,0,0)).

Different plots on the same axes

Sometimes there may be a good reason for putting two completely different response variables on the same graph. This can be a sensible thing to do if two parallel time series need to be overlaid to look for coincidence of peaks or troughs, for instance. Alternatively, you may want to include a thumbnail plot as an insert on a blank part of the main plot.

In the first case, where two graphs share the same x axis but have different scales on their y axes, use par(new=T, xaxs="d"). The data refer to fluctuations in animal numbers counted in August, and the number of gales in February of the following year:

```
attach(gales)
names(gales)

[1] "year"  "number"  "February"
```

Start by rescaling the margins, so there is room to put a separate axis with tick marks and a label on the right-hand side of the plot:

```
par(mar=c(5,4,4,5)+0.1)
```

Now plot the first graph in the normal way:

```
plot(year,number,ylim=c(0,2000),xlim=c(1950,2002),type="l")
```

Next we use new=T to override the clearance of the first graph when the next plot directive is issued, and xaxs="d" to specify direct x axes in the next plot:

```
par(new=T,xaxs="d")
```

Now the next plot is drawn in the normal way, except that we specify axes=F (we don't want to overwrite the numbers axis) and ylab="" (we don't want to write 'February' on top of 'number'). The type of graph to be plotted is "h", which gives vertical lines rather than points:

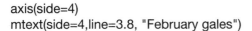

To finish off the graph, we want to put a different axis on the right-hand side (this is side = 4), to show the number of gales in February, using line=3.8 to space the text away from the right-hand axis:

```
axis(side=4)
mtext(side=4,line=3.8, "February gales")
```

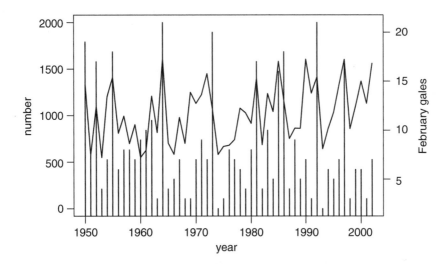

To insert thumbnail graphics, it is best to use **subplot**. This takes the plot directive of the graph to be inserted as its first argument, then a vector of two *x* coordinates (using the scale of the first graph) indicating where you want the bottom left and upper right corners of the inserted graph to be. The third argument is a vector of the matching *y* coordinates. Suppose we want to put a thumbnail graph of February gales in the bottom right-hand corner of the time series for numbers. Make the first plot in the normal way:

```
plot(year,number,ylim=c(0,2000),xlim=c(1950,2002),type="l")
```

Now work out where you want the insert to be. Remember to leave space for the axes labels in defining the lower left and upper right coordinates of the insert. The coordinates are in the units of the scale of the first graph, just drawn. They refer to corners of the graph axes, not the corners of the whole plotting area (this is a common misunderstanding). Let's put the axes between years 1980 and 2000 and between numbers 200 and 600:

```
subplot(plot(year,February,type="h"),x=c(1980,2000),y=c(200,600))
```

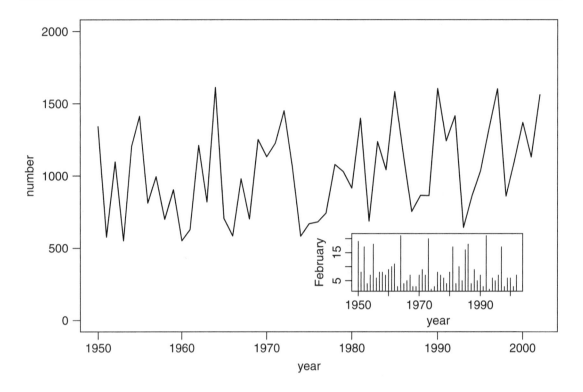

Having gone to all this trouble, I can't actually see why you would ever want to do this in earnest!

Panel plots

We have seen how to draw multiple graphs on the same page using mfrow to determine the number of rows and columns of graphs, and using subscripts to select the values of x and y that are used to produce each plot. This is fine, but there is an outstandingly powerful built-in function for producing plots like this, known as panel plots. The great advantage of panel plots is that you can condition them on combinations of other explanatory variables. This is a powerful tool in data exploration prior to multiple regression analysis, in non-linear curvefitting, and in exploring results from designed experiments where graphs might be required for different combinations of categorical explanatory variables.

The data frame nlme contains height data on plants of different genotypes grown at different concentrations of phosphorus (P). The whole data set looks like this:

```
attach(nlme)
names(nlme)

[1] "Genotype" "P" "Height"

plot(P,Height)
lines(smooth.spline(P,Height))
```

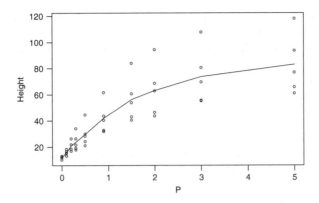

It would be good to see the graphs for each genotype separately. This is what panel plot is really good at. To see the trend, we shall fit a non-parametric smoother through the data for each genotype. The function to produce panel plots is **xyplot** and it has two parts. The first part shows the material to be plotted, but it is expressed like a model formula y~x rather than the familiar plot(x,y) sequence. After the model formula comes the conditioning operator | (vertical bar) followed by a list of the conditioning factors. In this case there is just one factor, Genotype. The second part of **xyplot** contains a function definition to specify what kinds of things are to be shown in each panel. You may want a scatterplot, plus one or more kinds of lines, for instance. In this case we want a smooth spline (**lines**) and a scatterplot (panel.xyplot). The function definition is enclosed inside curly brackets { } and the whole thing finishes with the right bracket of the **xyplot** function. In full, it looks like this:

```
xyplot(Height~P | Genotype,
    panel=function(x,y,subscripts, ...) {
    lines(smooth.spline(x,y))
    panel.xyplot(x,y)
})
```

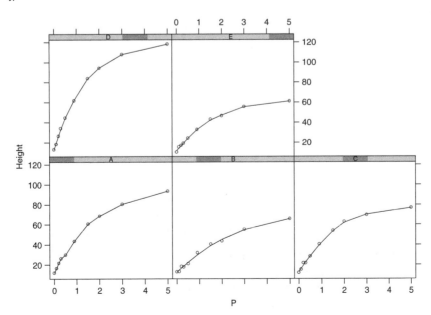

The resulting panel plot shows the dramatic differences in the phosphorus responses of the five different genotypes, whose names (A to E) appear in the bar above each panel. The default is to produce the panels from bottom left to top right, but if you don't want this, you can change it by using the as.table=T option, inserted before the panel=function line. Your panels will then begin in the upper left corner of the page. There are numerous options for panel plotting, which you can browse with online help.

Text in a scatterplot

Sometimes the identity of a data point is more important than its precise location in the scatterplot. In such cases we use **text** instead of **points**. Here are the murder rate data for different states categorised by regions of the USA:

```
attach(murders)
names(murders)
```

```
[1] "state" "population" "murder" "region"
```

```
coplot(murder~population|region)
```

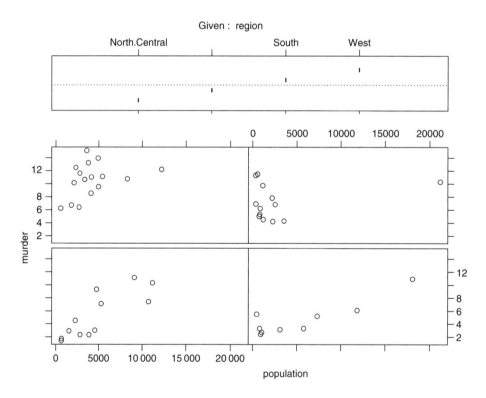

It would be much more interesting if we could see which state was which. This is how to do it:

```
xyplot(murder ~ population | region,
    groups = as.character(state),
    panel = function(x, y, subscripts, groups)
    text(x, y, groups[subscripts]))
```

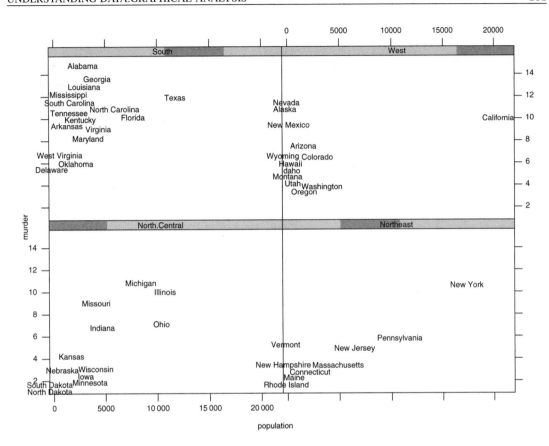

The panels are arranged in alphabetical order on regions from bottom left. There are some interesting patterns: the murder rate (per unit population) increases with total population in the regions Northeast and North Central but not in the regions South and West. The highest murder rates are in low population states in the South like Alabama and Georgia.

Further reading

Chambers, J.M., Cleveland, W.S. *et al.* (1983) *Graphical Methods for Data Analysis*. Belmont, California, Wadsworth.

10

Understanding data: tabular analysis

Graphs are excellent at conveying broad patterns, but tables are much better for conveying detail. Initial data inspection should aim to expose all the important features of the data frame, and to do this it is essential to produce summary tables of all the key features.

Summary statistics

Here are the main data summary functions that can be applied to entire vectors:

- mean(y) gives the arithmetic mean of the numbers in vector y
- max(y) shows the largest value of y
- min(y) shows the minimum value of y
- range(y) shows the smallest and largest values of y
- median(y) finds the median (50%) value of y

The variability of the numbers in y can be described by a variety of functions:

- var(y) gives the variance of y
- stdev(y) calculates the standard deviation of y (the square root of var(y))
- mad(y) is a robust measure of deviation, insensitive to outliers
- cor(x,y) calculates the correlation between x and y

We have written functions (p. 104) to calculate distributional shape parameters like:

- skew(y) to test for the presence of long tails
- kurtosis(y) to test for unusually flat or peaky distributions

Classified tables of summary statistics

Excel users will have come across Pivot Table, which is one of the most useful of all Excel's many features. It allows the summary functions listed above to be applied separately to different parts of the data frame on the basis of the levels of one or more categorical explanatory variables. The S-Plus equivalent is the function called **tapply** (GLIM users will recognise many of the features of $tabulate). To see it in action, we attach the data on weight gain of different ages, sexes and genotypes:

```
attach(Gain)
names(Gain)

[1] "Weight" "Sex" "Age" "Genotype" "Score"
```

The function **tapply** can be extremely simple. Here we calculate the mean weight for each sex:

```
tapply(Weight,Sex,mean)
```

produces the simple table

```
   female      male
9.070312  8.238705
```

Three arguments are typically supplied to **tapply**:

- the response variable to be summarised (Weight in this case)

- the explanatory variable, whose levels are to be used as a basis for the summary (Sex in this case)

- the summary function to be applied to the response variable (mean in this case); it can be any built-in or home-made function (listed earlier)

More complicated tables can be produced by specifying more than one explanatory variable. This is achieved using **list**, like this:

```
tapply(Weight,list(Genotype,Sex),mean)

          female      male
CloneA   8.849022  7.990882
CloneB   9.743349  9.032124
CloneC   7.806758  6.945931
CloneD   9.752316  8.735516
CloneE   7.918245  7.170862
CloneF  10.352179  9.556917
```

Here we have a two-dimensional summary table of mean weights, with the first variable in the list (Genotype) appearing as the **rows** of the summary table, and the second variable as the **columns** (Sex). For even more complicated cases, **tapply** produces multiple two-dimensional tables:

```
tapply(Weight,list(Genotype,Age,Sex),mean)
```

This creates a three-way classification. As before, the levels of the first named explanatory variable (Genotype) appear as the rows of each table and the second named variable (Age) as the columns. You get a separate two-dimensional table for each level of the third named explanatory variable (Sex):

```
,, female
             1         2         3         4         5
CloneA  8.128603  8.516595  8.987419  8.956231  9.656265
CloneB  9.187976  9.618807  9.816943  9.710912 10.382107
```

```
CloneC  7.685253  7.384319  7.907386   7.946400   8.110433
CloneD  9.069348  9.580191  9.568836  10.048262  10.494940
CloneE  7.521931  7.394792  7.767412   8.263021   8.644069
CloneF  9.714613  9.721850 10.799047  10.955917  10.569467

,, male
              1         2         3          4          5
CloneA  7.445630  8.000223  7.705105   8.348737   8.454716
CloneB  8.778213  8.340045  9.092006   9.105217   9.845137
CloneC  6.100194  6.513769  7.236200   7.228504   7.650989
CloneD  7.874351  8.511567  8.607625   8.994112   9.689927
CloneE  6.479922  6.828619  7.376615   7.245333   7.923820
CloneF  9.143637  9.170336  9.379750  10.037730  10.053133
```

You can produce summary tables of other statistics. Here are the standard deviations for each age:

tapply(Weight,Age,stdev)

```
       1        2        3        4        5
1.123672 1.101811 1.107623 1.156095 1.067085
```

And here are the maxima for each genotype:

tapply(Weight,Genotype,max)

```
  CloneA    CloneB    CloneC    CloneD    CloneE    CloneF
9.656265  10.38211  8.110433  10.49494  8.644069  10.95592
```

Saving summary vectors

The objects created by **tapply** can be saved and used for other purposes like producing plots or carrying out statistical modelling. Suppose we wanted to plot the mean weight at each age against age. We save the vector of mean weights like this:

mwt<-as.vector(tapply(Weight,Age,mean))

Note the use of the function **as.vector** to coerce the summary table into the form of a vector (i.e. by removing the headings). The trick here is to see that you need to produce a new vector, of the same length as mwt, to contain the ages. You cannot put plot(Age,mwt) because the two vectors are different lengths. What is needed is a new vector with a different name to contain the information on age. The simplest way to make this is to copy the last **tapply** command and edit it, so that it summarises Age rather than Weight:

age<-as.vector(tapply(Age,Age,mean))

Because S-Plus is case sensitive, age is a different variable from Age. Now we can do the plot:

plot(age,mwt,type="b")

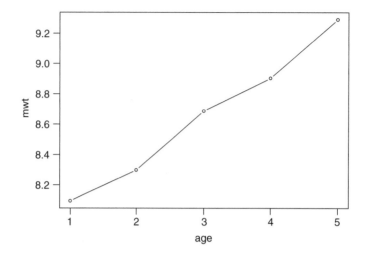

Plotting type="b" joins the dots by plotting both symbols and lines.

Shortened vectors of factor levels

We often want to condense a data frame by averaging over one or more of the explanatory variables. In the example above, for instance, we might want a table that averaged the mean weights over the two sexes:

```
meansex<-tapply(Weight,list(Genotype,Age),mean)
meansex
```

```
            1         2         3         4          5
CloneA  7.787116  8.258409  8.346262  8.652484   9.055490
CloneB  8.983094  8.979426  9.454475  9.408065  10.113622
CloneC  6.892724  6.949044  7.571793  7.587452   7.880711
CloneD  8.471850  9.045879  9.088230  9.521187  10.092434
CloneE  7.000927  7.111706  7.572013  7.754177   8.283944
CloneF  9.429125  9.446093 10.089399 10.496823  10.311300
```

This reduced table has two explanatory variables (Genotype in the rows, and Age in the columns) rather than three (Age, Genotype and Sex). If we want to keep this summary table for subsequent analysis, we will want to have extra vectors to define which age and genotype are associated with each mean weight. We cannot use Genotype and Age because they are the wrong length:

```
length(Genotype)
```

```
[1] 60
```

```
length(meansex)
```

```
[1] 30
```

We need to make new vectors of length 30 for genotype and age. As we saw earlier, this is easy for a continuous variable like Age; we just average it in exactly the same way as we averaged Weight, above:

```
ages<-as.vector(tapply(Age,list(Genotype,Age),mean))
```

Note, however, that this procedure does not work for a categorical variable like Genotype:

```
genos<-as.vector(tapply(Genotype,list(Genotype,Age),mean))
genos
```

```
 [1] NA NA NA NA NA NA NA NA NA NA NA NA NA NA NA NA NA NA NA NA NA NA
[23] NA NA NA NA NA NA NA NA
```

S-Plus does not understand what it is supposed to do if we ask it to work out the mean of a set of factor levels (recall that these are character strings). The trick is to convert the factor levels to numbers, before we carry out the averaging, using **as.numeric** like this:

```
genos<-as.vector(tapply(as.numeric(Genotype),list(Genotype,Age),mean))
genos
```

```
 [1] 1 2 3 4 5 6 1 2 3 4 5 6 1 2 3 4 5 6 1 2 3 4 5 6 1 2 3 4 5 6
```

The numbers are stored column-wise, rather than row-wise, by the **as.vector** function, so these level numbers are the correct genotype assignations for the numbers in the summary table called meansex. All we need to do now, in order to retrieve the names for the factor levels (to replace the less informative numbers), is use the vector called genos as subscripts for the **levels** of the factor Genotype:

```
levels(Genotype)[genos]
```

```
 [1] "CloneA" "CloneB" "CloneC" "CloneD" "CloneE" "CloneF" "CloneA"
 [8] "CloneB" "CloneC" "CloneD" "CloneE" "CloneF" "CloneA" "CloneB"
[15] "CloneC" "CloneD" "CloneE" "CloneF" "CloneA" "CloneB" "CloneC"
[22] "CloneD" "CloneE" "CloneF" "CloneA" "CloneB" "CloneC" "CloneD"
[29] "CloneE" "CloneF"
```

This will almost certainly seem deeply mysterious on first reading. But persevere, because it is a very useful piece of programming. Here is the full code for reducing Genotype (length $= 60$) to genotype (length $= 30$) with the factor levels in genotype in the correct sequence (as determined by the ordering of the variables in the list in the **tapply** function):

```
genos<-as.vector(tapply(as.numeric(Genotype),list(Genotype,Age),mean))
genotype<-levels(Genotype)[genos]
```

Now we can use the summary table and its associated (shortened) explanatory variables for further analysis. For example, here are the maxima of the sex-averaged Weight values (meansex) across all Age values for each genotype:

```
tapply(meansex,genotype,max)
```

```
  CloneA   CloneB   CloneC   CloneD   CloneE   CloneF
 9.05549 10.11362 7.880711 10.09243 8.283944 10.49682
```

The important thing to remember is that the length of the summary vectors is given by the product of the numbers of levels of the variables in the classifying list of the **tapply** function. In the case above, Genotype has 6 levels and Age has 5 levels, so the shorter vector is of length $6 \times 5 = 30$.

Further reading

Hampel, F.R., Ronchetti, E.M. *et al.* (1986) *Robust Statistics: The Approach Based on Influence Functions*. New York, John Wiley.

Hoaglin, D.C., Mosteller, F. *et al.* (1983) *Understanding Robust and Exploratory Data Analysis*. New York, John Wiley.

Huber, P.J. (1981) *Robust Statistics*. New York, John Wiley.

11

Classical tests

There is absolutely no point in carrying out an analysis that is more complicated than it needs to be. Occam's razor applies to the choice of statistical model just as strongly as to anything else—simplest is best.

The so-called classical tests deal with some of the most frequently used kinds of analysis (Box 11.1). In particular, they are the models of choice for:

- comparing two variances (Fisher's F test)
- comparing two sample means with normal errors (Student's t test)
- comparing two sample means with non-normal errors (Wilcoxon's rank test)
- correlating two variables (Pearson's correlation or Spearman's rank correlation)
- testing for independence in contingency tables (chi-square or Fisher's exact)
- comparing two proportions (the binomial test)

Box 11.1 Classical tests in S-Plus

- **binom.test** for comparing a binomial sample with a particular value of p
- **chisq.test** for testing for independence in a contingency table
- **cor.test** for testing whether two vectors are correlated
- **fisher.test** for an exact test of independence in a 2×2 contingency table
- **friedman.test** for performing Friedman's rank sum test (unreplicated blocked data)
- **kruskal.test** for performing a Kruskal–Wallis rank sum test (one-way layout)
- **mantelhaen.test** for performing a Mantel–Haenszel chi-square test on a three-dimensional contingency table
- **mcnemar.test** for performing a McNemar chi-square test on a two-dimensional contingency table.
- **prop.test** for comparing two or more binomial proportions
- **t.test** for comparing two sample means using Student's t test
- **var.test** for comparing two Normal sample variances using Fisher's F ratio
- **wilcox.test** for comparing two sample medians using the rank sum test

---

We cover each of these in turn, starting with one-sample tests then moving on to two-sample tests. Note, however, that the same hypotheses can be tested using other methods (e.g. simulation or generalised linear modelling); these will be described in later chapters.

Single-sample estimation

Suppose we have a single sample. The questions we might want to answer are these:

- what is the mean value?

- is the mean value significantly different from current expectation or theory?

- what is the level of uncertainty associated with our estimate of the mean value?

In order to be reasonably confident that our inferences are correct, we need to establish some facts about the distribution of the data:

- are the values normally distributed or not?

- are there outliers in the data?

- if data were collected over a period of time, is there evidence for serial correlation?

Non-normality, outliers and serial correlation can all invalidate inferences made by standard parametric tests like Student's t test. Much better in such cases to use a robust non-parametric technique like Wilcoxon's signed rank test.

We can investigate the issues involved with Michelson's famous data on estimating the speed of light collected in 1879. The actual speed is $299\,000\,\mathrm{km\,s^{-1}}$ plus the values in our data frame called light:

```
attach(light)
names(light)

[1]  "speed"

hist(speed)
```

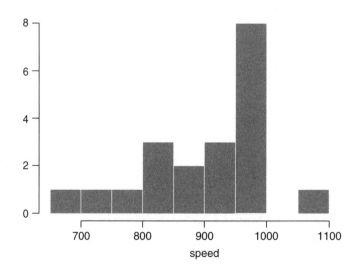

Clearly, these data are not normally distributed. There is a long tail to the left (i.e. the data are negatively skew). We get a **summary** of the non-parametric descriptors of the sample like this:

```
summary(speed)

 Min.    1st Qu.    Median    Mean    3rd Qu.    Max.
 650      850        940       909     980        1070
```

From this, you see at once that the median (940) is substantially bigger than the mean (909), as a consequence of the strong negative skew in the data we saw in the histogram. The **interquartile range** is the difference between the first and third quartiles: $980 - 850 = 130$. This is useful in the detection of outliers. Here is a good rule of thumb:

An outlier is a value more than 1.5 times the interquartile range above the third quartile, or below the first quartile.

In this case, outliers would be measurements of speed that were less than $850 - 195 = 655$ or greater than $980 + 195 = 1175$. You will see that there are no large outliers in the data set, but one small outlier (650).

Inference in the one-sample case

We want to test the hypothesis that Michelson's estimate of the speed of light is significantly different from the value of 299 990 thought to prevail at the time. The data have all had 299 000 subtracted from them, so the test value (our null hypothesis) is 990. Because of the non-normality, the use of Student's t test in this case is ill-advised. The correct test is Wilcoxon's signed rank test. The code for this is in a library (note the . between wilcox and test):

```
library(ctest)
wilcox.test(speed,mu=990)

Warning: Cannot compute exact p-value with ties
    Wilcoxon signed rank test with continuity correction
data: speed
V = 22.5, p-value = 0.00213
alternative hypothesis: true mu is not equal to 990
```

We accept the *alternative* hypothesis because $p = 0.002\,13$ (i.e. much less than 0.05). We conclude that the sample mean is significantly different from 990.

For the purpose of demonstration only, we show the equivalent Student's t test:

```
t.test(speed,mu=990)

       One-sample t-Test
data:    speed
t = -3.4524, df = 19, p-value = 0.0027
alternative hypothesis: true mean is not equal to 990
95 percent confidence interval:
 859.8931 958.1069
```

In this case the p value ($p = 0.0027$) is almost exactly the same as with the more appropriate Wilcoxon test ($p = 0.0021$). The 95% confidence interval (which stretches from 859.9 to 958.1) does not include the hypothesised value of 990.

Comparing two variances

Before we can carry out a test to compare two sample means, we need to test whether the sample variances are significantly different. The test could not be simpler. It is called Fisher's F test after the famous statistician and geneticist R. A. Fisher, who worked at Rothamsted in south-east England. To compare two variances, all you do is *divide the larger variance by the smaller variance*.

Obviously, if the variances are the same, the ratio will be 1. In order to be significantly different, the ratio will need to be significantly bigger than 1 (because the larger variance goes on top, i.e. in the numerator). How will we know a significant value of the variance ratio from a non-significant one? The answer, as always, is to look up the critical value of the variance ratio in tables. In this case we want tables of critical values of Fisher's F. The S-Plus function for this is **qf**, which stands for 'quantiles of the F distribution'. For our example of ozone levels in market gardens (see p. 95) there were 10 replicates in each garden, so there were $10 - 1 = 9$ degrees of freedom for each garden. In comparing two gardens, therefore, we have 9 d.f. in the numerator and 9 d.f. in the denominator. Suppose we work at the traditional $\alpha = 0.05$, then since F tests are always one-tailed (the bigger variance always goes on top), we find the critical value of F like this:

```
qf(0.95,9,9)
```

```
[1] 3.178893
```

This means that a calculated variance ratio will need to be greater than or equal to 3.18 in order for us to conclude that the two variances are significantly different at $\alpha = 0.05$. To see the test in action, we can compare the variances in ozone concentration for market gardens B and C (see p. 95):

```
attach(f.test.data)
names(f.test.data)
```

```
[1] "gardenB" "gardenC"
```

First, we compute the two variances:

```
var(gardenB)
```

```
[1] 1.333333
```

```
var(gardenC)
```

```
[1] 14.22222
```

The larger variance is clearly in garden C, so we compute the F ratio like this:

```
F.ratio<-var(gardenC)/var(gardenB)
F.ratio
```

```
[1] 10.66667
```

The variance in garden C is more than 10 times as big as the variance in garden B. The critical value of F for this test (with 9 d.f. in both the numerator and the denominator) is 3.81 (see **qf** above), so *since the calculated value is larger than the critical value, we reject the null hypothesis*.

The null hypothesis was that the two variances were not significantly different, so we accept the alternative hypothesis that the two variances are significantly different. In this case, therefore, it would be wrong to compare the two sample means using Student's t test.

There is a built-in function called **var.test** for speeding up the procedure. All we provide are the names of the two variables containing the raw data whose variances are to be compared (we don't need to work out the variances first):

var.test(gardenB,gardenC)

```
        F test for variance equality
data: gardenB and gardenC
F = 0.0938, num df = 9, denom df = 9, p-value = 0.0016
alternative hypothesis: true ratio of variances is not
equal to 1
95 percent confidence interval:
 0.02328617 0.37743695
sample estimates:
 variance of x variance of y
       1.333333      14.22222
```

Note that the variance ratio, F, is given as roughly 1/10 rather than roughly 10 because we did not tell **var.test** to put the bigger variance from garden C on top. But the p value of 0.0016 is correct, and we reject the null hypothesis. These two variances are highly significantly different.

Comparing two means

The question is this: Given what we know about the variation from replicate to replicate within each sample (the within-sample variance), how likely is it that our two sample means were drawn from populations with the same average. If this is highly likely, then we shall say that our two sample means are not significantly different. If it is rather unlikely, then we shall say that our sample means are significantly different. Perhaps a better way to proceed is to work out the probability that the two samples were indeed drawn from populations with the same mean. If this probability is very low (say less than 5% or less than 1%) then we can be reasonably certain (95% or 99% in these two examples) than the means really are different from one another. Note, however, that we can never be 100% certain; the apparent difference might just be due to random sampling—we happened to get a lot of low values in one sample, and a lot of high values in the other.

There are two simple tests for comparing two sample means:

- **Student's t test** when the means are independent, the variances constant, and the errors are normally distributed

- **Wilcoxon rank sum test** when the means are independent but errors are *not* normally distributed

What to do when these assumptions are violated (e.g. when the variances are different) is discussed later on.

Student's t test

Student was the pseudonym of W.S. Gosset, who published his influential paper in *Biometrika* in 1908. He was prevented from publishing under his own name by the archaic employment laws at the time, which allowed his employer, the Guinness Brewing Company, to prevent him publishing independent work. Student's t distribution, later perfected by R. A. Fisher, revolutionised the study of small-sample statistics where inferences need to be made on the basis of the sample variance s^2 with the population variance σ^2 unknown (indeed, usually unknowable). The test statistic is the number of standard errors by which the two sample means are separated:

$$t = \frac{\text{difference between the two means}}{SE \text{ of the difference}} = \frac{\overline{y}_A - \overline{y}_B}{SE_{\text{diff}}}$$

Now we know the standard error of the mean (see p. 97) but we have not yet met the standard error of the difference between two means. For two independent (non-correlated) variables, *the variance of a difference is the sum of the separate variances* (Box 11.2).

Box 11.2 The variance of a difference between two samples

Think about the sum of squares of a difference:

$$\sum \left[(y_A - y_B) - (\mu_A - \mu_B) \right]^2$$

If we average this, we get the variance of the difference (forget about degrees of freedom for a minute). Let's call this average $\sigma^2_{\bar{y}_A - \bar{y}_B}$ and rewrite the sum of squares

$$\sigma^2_{\bar{y}_A - \bar{y}_B} = \text{average of } [(y_A - \mu_A) - (y_B - \mu_B)]^2$$

Multiply out the square then expand the brackets to give

$$(y_A - \mu_A)^2 + (y_B - \mu_B)^2 - 2(y_A - \mu_A)(y_B - \mu_B)$$

We already know that the average of $(y_A - \mu_A)^2$ is the variance of population A and the average of $(y_B - \mu_B)^2$ is the variance of population B. So the variance of the *difference* between the two sample means is the *sum* of the variances of the two samples, plus this term $2(y_A - \mu_A)(y_B - \mu_B)$. But we also know that $\sum d = 0$ (see p. 93) so because the samples from A and B are independently drawn they are uncorrelated, which means that taking any point from sample A and B we have

$$\text{average of } (y_A - \mu_A)(y_B - \mu_B) = (y_A - \mu_A)\{\text{average of } (y_B - \mu_B)\} = 0$$

This important result needs to be stated separately:

$$\sigma^2_{\bar{y}_A - \bar{y}_B} = \sigma^2_A + \sigma^2_B$$

So if two samples are drawn independently from populations with different variances, then *the variance of the difference is the sum of the two sample variances.*

This important result allows us to write down the formula for the standard error of the difference between two sample means:

$$SE_{\text{diff}} = \sqrt{\frac{s_A^2}{n_A} + \frac{s_B^2}{n_B}}$$

At this stage we have everything we need to carry out Student's t test. Our null hypothesis is that the two sample means are the same, and we shall accept this unless the value of Student's t is so large that it is unlikely that such a difference could have arisen by chance alone. For the ozone example introduced on p. 95, each sample has 9 degrees of freedom, so we have 18 d.f. in total. Another way of thinking of this is to reason that the complete sample size as 20, and we have estimated two parameters from the data, $\bar{y}_A$ and $\bar{y}_B$, so we have $20 - 2 = 18$ d.f. We typically use 5% as the chance of rejecting the null hypothesis when it is true (this is the Type I error rate). Because we didn't know in advance which of the two gardens was going to have the higher mean ozone concentration (we usually don't), this is a two-tailed test, so the *critical value* of Student's t is

qt(0.975,18)

[1] 2.100922

This means that our test statistic needs to be bigger than 2.1 in order to reject the null hypothesis, and hence to conclude that the two means are significantly different at $\alpha = 0.05$. The data frame is attached like this:

attach(t.test.data)
names(t.test.data)

[1] "gardenA" "gardenB"

We can carry out the test longhand like this, starting by calculating the two variances s2A and s2B:

s2A<-var(gardenA)
s2B<-var(gardenB)

Now the value of Student's t is *the difference divided by the standard error of the difference*. The difference between the two means is the numerator, and the square root of the sum of the two variances is in the denominator:

(mean(gardenA)-mean(gardenB))/sqrt(s2A/10+s2B/10)

which gives the value of Student's t as

[1] -3.872983

With t tests you can ignore the minus sign; it is only the absolute value of the difference between the two sample means that concerns us. So the calculated value of the test statistic is 3.87 and the critical value from tables is 2.10 (above). This is what we say: *Since the calculated value is larger than the critical value, we reject the null hypothesis.*

Notice that the wording is exactly the same as it was for the F test (above). Indeed, the wording is always the same, and you should try to memorise it. The abbreviated form is easier to remember: *larger reject, smaller accept*. The null hypothesis was that the two means were not significantly different, so we reject this and accept the alternative hypothesis that *the two means were significantly different*.

You won't be surprised to learn that there is a built-in function to do all the work for us. It is called, helpfully, **t.test** and is used simply by providing the names of the two vectors containing the samples on which the test is to be carried out (gardenA and gardenB in our case).

t.test(gardenA,gardenB)

There is rather a lot of output. You often find this. The simpler the statistical test, the more voluminous the output:

```
        Standard Two-Sample t-Test
data    gardenA and gardenB
t = -3.873, df = 18, p-value = 0.0011
alternative hypothesis: true difference in means is not
equal to 0
95 percent confidence interval:
```

```
-3.0849115 -0.9150885
sample estimates:
 mean of x mean of y
         3         5
```

The result is exactly the same as we obtained longhand. The value of t is -3.873 and since *the sign is irrelevant in a t test* we reject the null hypothesis because the test statistic is larger than the critical value of 2.1. The mean ozone concentration is significantly higher in garden B than in garden A. The computer printout also gives a p value and a confidence interval. Note that, because the means are significantly different, *the confidence interval on the difference does not include zero* (in fact, it goes from -3.085 up to -0.915). The p value is useful in written work, like this:

Ozone concentration was significantly higher in garden B (mean $= 5.0$ pphm) than in garden A (mean $= 3.0$ pphm; $t = 3.873$, $p = 0.001$, d.f. $= 18$).

Wilcoxon rank sum test

The Wilcoxon rank sum test is a non-parametric alternative to Student's t test; we could use it if the errors looked to be non-normal. Let's look at the errors in gardens A and C:

```
attach(w.test.data)
names(w.test.data)
```

```
[1] "gardenA" "gardenC"
```

```
par(mfrow=c(1,2))
hist(gardenA,breaks=c(0.5:11.5))
hist(gardenC,breaks=c(0.5:11.5))
```

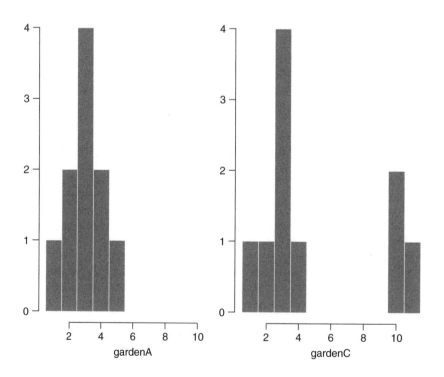

The errors in garden A look to be reasonably Normal, but the errors in garden C are definitely not Normal.

The Wilcoxon rank sum test statistic, W, is defined like this. Both samples are put into a single array with their sample names (A and C in this case) clearly attached. Then the aggregate list is sorted, taking care to keep the sample labels with their respective values. Then a rank is assigned to each value, with ties getting appropriate average rank: (two-way ties get (rank i + (rank i + 1))/2, three-way ties get (rank i + (rank$i$ + 1) + (rank i + 3))/3, and so on. Finally the ranks are added up for each of the two samples, and significance is assessed on size of the smaller sum of ranks.

This involves some interesting computing. First we make a combined vector of the samples:

```
combined<-c(gardenA,gardenC)
combined
```

```
 1  2  3  4  5  6  7  8  9 10  1  2  3  4    5  6  7    8  9 10
 3  4  4  3  2  3  1  3  5    2  3  3  2  1  10  4  3  11  3 10
```

Then we make a list of the sample names, A and C:

```
sample<-c(rep("A",10),rep("C",10))
sample
```

```
[1] "A" "A" "A" "A" "A" "A" "A" "A" "A" "A" "C" "C" "C" "C" "C" "C" "C" "C" "C" "C"
```

Now the trick is to use the built-in function **rank** to get a vector containing the ranks, smallest to largest, within the combined vector:

```
rank.combi<-rank(combined)
rank.combi
```

```
 [1]  9.5  15.0  15.0   9.5   4.0   9.5   1.5   9.5  17.0   4.0   9.5   9.5   4.0
[14]  1.5  18.5  15.0   9.5  20.0   9.5  18.5
```

Notice that the ties have been dealt with by averaging the appropriate ranks. Now all we need to do is calculate the sum of the ranks for each garden. We could use **sum** with conditional subscripts for this:

```
sum(rank.combi[sample=="A"])
```

```
[1]  94.5
```

```
sum(rank.combi[sample=="C"])
```

```
[1]  115.5
```

Alternatively, we could use **tapply** with **sum** as the required operation:

```
tapply(rank.combi,sample,sum)
```

```
    A      C
 94.5  115.5
```

In either case we compare the smaller of the two values (94.5) with values in tables (e.g. Snedecor and Cochran 1980, p. 555), and reject the null hypothesis if our value of 94.5 is *smaller* than the value in tables. For samples of sizes 10 and 10 like ours, the value in tables is 78. Our value is much bigger than this, so we accept the null hypothesis. The two sample means are not significantly different.

We can carry out the whole procedure automatically, and avoid the need to use tables of critical values of Wilcoxon's rank sum test, by using the built-in function **wilcox.test**:

wilcox.test(gardenA,gardenC)

which produces the following output:

```
      Wilcoxon rank-sum test

data:  gardenA and gardenC
rank-sum normal statistic with correction Z = -0.7841,p-value = 0.433
alternative hypothesis: true mu is not equal to 0

Warning messages:
   cannot compute exact p-value with ties
```

This is interpreted as follows. The function uses a Normal approximation algorithm to work out a z value, and from this a p value for the assumption that the means are the same. This p value of 0.433 is much bigger than 0.05, so we accept the null hypothesis. Unhelpfully, it then prints the alternative hypothesis in full, which a careless reader could take as meaning that 'true mu is not equal to 0' ('mu' is the difference between the two means). However, we have just demonstrated, rather convincingly, that the true mu *is* equal to 0. The idea of putting the message in the output is to show that we were doing (the default) two-tailed test, but this is a silly way of saying it, with a serious risk of misinterpretation. The warning message at the end draws attention to the fact that there are ties in the data, hence the p value cannot be calculated exactly (this is seldom a real worry).

The non-parametric test is much more appropriate than the t test when the errors are not Normal, and the non-parametric test is about 95% as powerful with Normal errors, and can be *more* powerful than the t test if the distribution is strongly skewed by the presence of outliers. But the Wilcoxon test does not make the really important point for this case, because like the t test, it says that ozone pollution in the two gardens is not significantly different. Yet we know, because we have looked at the data, that gardens A and C are definitely different in terms of their ozone pollution, but it is the *variance* that differs, not the mean. Neither the t test nor the rank sum test can cope properly with situations where the variances are different, but the means are the same. This draws attention to a very general point: *scientific importance and statistical significance are not the same thing*.

Lots of results can be highly significant but of no scientific importance at all (e.g. because the effects are cancelled out by later events). Likewise, lots of scientifically important processes may not be statistically significant. For instance, just because there is no damage from ozone in garden A, and the mean pollution levels in gardens A and C are not significantly different, it is *not* correct to conclude that there is no air pollution damage in garden C. In fact, we know from looking at the data (p. 95) that damaging air pollution levels occur in garden C on 30% of the days that were sampled.

Tests on paired samples

Sometimes, two-sample data come from paired observations. In this case we might expect a correlation between the two measurements, either because they were made on the same individual, or were taken from the same location. You might recall that earlier we found that the variance of a *difference* was the average of

$$(y_A - \mu_A)^2 + (y_B - \mu_B)^2 - 2(y_A - \mu_A)(y_B - \mu_B)$$

which is the variance of sample A plus the variance of sample B minus 2 times the covariance of A and B (see p. 174). When the covariance of A and B is *positive*, this is a great help because pairing *reduces* the variance of the difference, and should make it easier to detect significant differences between the means. Pairing is not always effective, because the correlation between y_A and y_B may be weak. It would be disastrous if the correlation were to turn out to be negative!

One way to proceed is to reduce the estimate of the standard error of the difference by taking account of the measured correlation between the two variables. A simpler alternative is to calculate the difference between the two samples from each pair, then do a one-sample test comparing the mean of the differences to zero. This halves the number of degrees of freedom, but it reduces the error variance substantially in cases where there is a strong positive correlation between y_A and y_B.

The following data are a composite biodiversity score based on a kick sample of aquatic invertebrates:

```
x<-c(20,15,10,5,20,15,10,5,20,15,10,5,20,15,10,5)
y<-c(23,16,10,4,22,15,12,7,21,16,11,5,22,14,10,6)
```

The elements of x and y are paired because the two samples were taken on the same river, upstream (y) or downstream (x) of the same sewage outfall. If we ignore the fact that the samples are paired, it appears that the sewage outfall has no impact on biodiversity score ($p = 0.6856$):

```
t.test(x,y)
```

```
        Standard Two-Sample t-Test
data:    x and y
t = -0.4088, df = 30, p-value = 0.6856
alternative hypothesis: true difference in means is not
equal to 0
95 percent confidence interval:
 -5.246747  3.496747
sample estimates:
 mean of x mean of y
      12.5     13.375
```

However, if we allow that the samples are paired (simply by specifying the option paired=T), the picture is completely different:

```
t.test(x,y,paired=T)
```

```
        Paired t-Test
data:    x and y
t = -3.0502, df = 15, p-value = 0.0081
alternative hypothesis: true mean of differences is not
equal to 0
95 percent confidence interval:
 -1.4864388 -0.2635612
sample estimates:
 mean of x - y
        -0.875
```

The difference between the means is highly significant ($p = 0.0081$). The moral is clear. If you do have information on blocking (the fact that the two samples came from the same river in this case) then you should use it in the analysis. It can never do any harm, and sometimes (as here) it can do a huge amount of good.

The sign test

The sign test is one of the simplest of all statistical tests. Suppose that you cannot *measure* a difference, but you can *see* it (e.g in judging a diving contest, or in carrying out an analysis of evolutionary phylogeny). For example, in a study with 9 fossils, 8 were more 'conical', and 1 was less 'conical' than the ancestral species. How strong is the evidence that our 9 fossils are significantly more conical in shape than the ancestor? The answer comes from a two-tailed binomial test. How likely is a response of 1/9 or 8/9 if the populations are actually the same (i.e. $p = 0.5$)? We use a binomial test for this, specifying the number of 'successes' (1) and the total sample size (9):

```
binom.test(1,9)
```

This produces the following output:

```
              Exact binomial test
data:   1 out of 9
number of successes = 1, n = 9, p-value = 0.0391
alternative hypothesis: true p is not equal to 0.5
```

We would conclude that the shape is significantly different from the ancestor because $p < 0.05$. It is easy to write a function to carry out a sign test to compare two samples, x and y

```
sign.test <- function(x, y)
{
    if(length(x) != length(y))
        stop("The two variables must be the same length")
    d <- x - y
    binom.test(sum(d > 0), length(d))
}
```

The function starts by checking that the two vectors are the same length, then works out the vector of the differences, d. The binomial test is then applied to the number of positive differences (sum(d>0)) and the total number of numbers (length(d)). Here is the sign test used to compare the ozone levels in gardens A and B (see above):

```
sign.test(gardenA,gardenB)
```

```
            Exact binomial test
data:   sum(d > 0) out of length(d)
number of successes = 0, n = 10, p-value = 0.002
alternative hypothesis: true p is not equal to 0.5
```

Note that the p value (0.002) from the sign test is larger than in the equivalent t test ($p = 0.0011$) that we carried out earlier. This will generally be the case; other things being equal, the parametric test will be more powerful than the non-parametric equivalent.

Binomial tests to compare two proportions

On page 87 we observed that 4 females were promoted out of 40 candidates, and 196 men were promoted out of 3270 candidates. The question naturally arises as to whether the apparent positive discrimination in favour of women is statistically significant, or whether this sort of difference (10% success for women versus 6% success for men) could arise through chance alone. This is easy in S-Plus using the built-in binomial proportions test:

```
prop.test(c(4,196),c(40,3270))

        2-sample test for equality of proportions with continuity
correction
data:  c(4, 196) out of c(40, 3270)
X-square = 0.5229, df = 1, p-value = 0.4696
alternative hypothesis: two.sided
95 percent confidence interval:
 -0.06591631 0.14603864
sample estimates:
 prop'n in Group 1 prop'n in Group 2
        0.1             0.05993884
```

There is no evidence in favour of positive discrimination ($p = 0.4696$). A result like this will occur more than 45% of the time by chance alone. Just think what would have happened if one of the successful female candidates had not applied. Then the same promotion system would have produced a female success rate of 3/39 instead of 4/40 (7.7% instead of 10%). In small samples, small changes have big effects.

Chi-square contingency tables

A great deal of information comes in the form of *counts* (whole numbers or integers); the number of animals that died, the number of branches on a tree, the number of days of frost, the number of companies that failed, the number of patients that died. With count data, the number 0 is often the value of a response variable (consider, for example, what a 0 would mean in the context of the examples just listed).

According to the Oxford Dictionary, 'contingency' means a thing dependent on an uncertain event. In statistics the contingencies are all the events that could possibly happen. The contingency table shows the counts of how many times each of the contingencies actually happened in a particular sample. Consider the following example that has to do with the relationship between hair colour and eye colour in white people. For simplicity, we just chose two contingencies for hair colour: 'fair' and 'dark'. Likewise we just chose two contingencies for eye colour: 'blue' and 'brown'. These two **categorical variables,** eye colour and hair colour, both have two **levels** ('blue' and 'brown', 'fair' and 'dark' respectively). Between them, they define four possible outcomes: fair hair and blue eyes, fair hair and brown eyes, dark hair and blue eyes, and dark hair and brown eyes. We take a sample of people and count how many of them fall into each of these four categories. Then we fill in the 2×2 contingency table like this.

	Blue eyes	Brown eyes
Fair hair	38	11
Dark hair	14	51

These are our observed frequencies (or counts). The next step is very important. In order to make any progress in the analysis of these data, we need a **model** which predicts the expected frequencies. What would be a sensible model in a case like this? There are all sorts of complicated models that you might erect, but the simplest model (Occam's razor, or the principle of parsimony) is that hair colour and eye colour are **independent.** We may not believe this is actually true, but the hypothesis has the great virtue of being falsifiable. It is also a very sensible model to choose because it makes it easy to predict the expected frequencies based on the assumption that the model is true. We need to do some simple probability work. What is the probability of getting a random individual from this sample whose hair was fair? A total of 49 people (38 + 11) had fair hair out of a total sample of 114 people. So the probability of fair hair is 49/114 and the probability of dark hair is 65/114. Notice that because we have only two contingencies, these two probabilities add up to 1, (49 + 65)/114. What about eye colour? What is the probability of selecting someone at random from this sample with blue eyes? A total of 52 people had blue eyes (38 + 14) out of the sample of 114, so the probability of blue eyes is 52/114 and the probability of brown eyes is 62/114. As before, these sum to 1, (52 + 62)/114. It helps to add the subtotals to the margins of the contingency table like this.

	Blue eyes	Brown eyes	Row totals
Fair hair	38	11	49
Dark hair	14	51	65
Column totals	52	62	114

Now comes the important bit. We want to know the expected frequency of people with fair hair and blue eyes, to compare with our observed frequency of 38. Our model says that the two are independent. This is essential information, because it allows us to calculate the expected probability of fair hair and blue eyes. *If, and only if, the two traits are independent, the probability of having fair hair and blue eyes is the product of the two probabilities.* So, following our earlier calculations, the probability of fair hair and blue eyes is 49/114 × 52/114. We can do exactly equivalent things for the other three cells of the contingency table.

	Blue eyes	Brown eyes	Row totals
Fair hair	$\dfrac{49}{114} \times \dfrac{52}{114}$	$\dfrac{49}{114} \times \dfrac{62}{114}$	49
Dark hair	$\dfrac{65}{114} \times \dfrac{52}{114}$	$\dfrac{65}{114} \times \dfrac{62}{114}$	65
Column totals	52	62	114

Now we need to know how to calculate the expected frequency. It couldn't be simpler. It is just the probability multiplied by the total sample ($n = 114$). So the expected frequency of blue eyes and fair hair is

$$\frac{49}{114} \times \frac{52}{114} \times 114 = 22.35$$

which is much less than our observed frequency of 38. It is starting to look as if our hypothesis of independence of hair and eye colour is false.

You might have noticed something useful in the last calculation: two of the sample sizes cancel out. Therefore, the expected frequency in each cell is just the row total (R) times the column total (C) divided by the grand total (G) like this:

$$E = \frac{R \times C}{G}$$

We can now work out the four expected frequencies.

	Blue eyes	Brown eyes	Rowtotals
Fair hair	22.35	26.65	49
Dark hair	29.65	35.35	65
Column totals	52	62	114

Notice that the row and column totals (the marginal totals) are retained under the model. It is clear that the observed frequencies and the expected frequencies are different. But in sampling, everything always varies, so this is no surprise. The important question is whether the expected frequencies are *significantly* different from the observed frequencies?

We assess the significance of the difference by comparing the observed and expected frequencies using a chi-square test. We calculate a test statistic χ^2 (Pearson's chi-square) as follows:

$$\chi^2 = \sum \frac{(O - E)^2}{E}$$

where capital Greek sigma $\sum$ just means 'add up all the values of'. It makes the calculations easier if we write the observed and expected frequencies in parallel columns, so that we can work out the corrected squared differences most easily.

	O	E	$(O - E)^2$	$\dfrac{(O - E)^2}{E}$
Fair hair and blue eyes	38	22.35	244.92	10.96
Fair hair and brown eyes	11	26.65	244.92	9.19
Dark hair and blue eyes	14	29.65	244.92	8.26
Dark hair and brown eyes	51	35.35	244.92	6.93

All we need to do now is to add up the four components of chi-square to get $\chi^2 = 35.34$

The question now arises, Is this a big value of chi-square or not? This is important, because if it *is* a bigger value of chi-square than we would expect by chance, then we should reject the null hypothesis. But if it is within the range of values that we would expect by chance alone, then we should accept the null hypothesis.

We always proceed in the same way at this stage. We have a calculated value of the test statistic, $\chi^2 = 35.34$. We *compare this value with the relevant value in statistical tables*. We have one degree of freedom for a 2×2 contingency table. In general, there are

$$\text{d.f.} = (r - 1) \times (c - 1)$$

degrees of freedom (d.f.) in a contingency table, where $r =$ the number of rows in the table, and $c =$ the number of columns. So we have d.f. $= 1$ in this case.

The next thing we need to do is say how certain we want to be about the falseness of the null hypothesis. The more certain we want to be, the larger the value of chi-square we would need to reject the null hypothesis. It is conventional to work at the 95% level. That is our certainty level, so our uncertainty level is $100 - 95 = 5\%$. Expressed as a fraction, this is called alpha ($\alpha = 0.05$). Technically, alpha is the probability of *rejecting* the null hypothesis when it is *true*. This is called a Type I error. A Type II error is *accepting* the null hypothesis when it is *false*.

In chi-square tables we now look up the **critical value** of chi-square; that is to say, the largest value that would arise by chance alone when the null hypothesis is true. The columns of the table are defined by different alpha values, and the rows by different degrees of freedom. We have $\alpha = 0.05$ and d.f. $= 1$, so the critical value is 3.841.

The logic goes like this. Since the calculated value is *greater* than the critical value from tables, we *reject* the null hypothesis. You should memorise this sentence; put the emphasis on 'greater' and 'reject'.

What have we learned so far? We have rejected the null hypothesis that eye colour and hair colour are independent. But that's not quite the end of the story, because we have not established the *way* in which they are related. For example, is the correlation between them positive or negative? To do this we simply look at the data and compare the observed and expected frequencies. If fair hair and blue eyes were positively correlated, would the observed frequency of getting them together be greater or less than the expected frequency? A moment's thought should convince you that the observed frequency will be greater than the expected frequency when the traits are positively correlated (and less when they are negatively correlated). In our case we expected only 22.35 but observed 38 people with both fair hair and blue eyes. So it is clear that fair hair and blue eyes are highly significantly positively associated.

In S-Plus the procedure is very straightforward. We start by defining the counts as a 2×2 matrix like this:

```
count<-matrix(c(38,14,11,51),nrow=2)
count

      [,1] [,2]
[1,]    38   11
[2,]    14   51
```

Notice that you *enter the data column-wise* (not row-wise). Then the test uses the **chisq.test** function, with the matrix of counts as its only argument:

```
chisq.test(count)

Pearson's chi-square test with Yates' continuity correction

data: count
X-square = 33.112, df = 1, p-value = 0
```

The calculated value of chi-square is slightly different from ours, because Yates' correction has been applied as the default (Sokal and Rohlf 1995, p. 736). If you switch the correction off (**correct=F**), you get the value we calculated by hand:

```
chisq.test(count,correct=F)

Pearson's chi-square test without Yates' continuity correction

data: count
X-square = 35.3338, df = 1, p-value = 0
```

It makes no difference at all to the interpretation; there is a highly significant positive association between fair hair and blue eyes in this group of people.

Fisher's exact test

Fisher's exact test is used for the analysis of contingency tables in which *one or more expected frequencies is less than 5*. The individual counts are *a*, *b*, *c* and *d* like this.

2×2 Table	Col. 1	Col. 2	Row totals
Row 1	a	b	$a+b$
Row 2	c	d	$c+d$
Column totals	$a+c$	$b+d$	n

The probability of any one outcome is given by

$$p = \frac{(a+b)!(c+d)!(a+c)!(b+d)!}{a!b!c!d!n!}$$

where *n* is the grand total.

Our data concern the distribution of 8 ants' nests over 20 trees of two species (A and B). There are 2 categorical explanatory variables (ants and trees), and 4 contingencies, ants (present or absent) and trees (A or B). The response variable is the vector of 4 counts of trees {6, 4, 2 and 8}.

	Tree A	Tree B	Row totals
With ants	6	2	8
Without ants	4	8	12
Column totals	10	10	20

It is easy to calculate the probability for this particular outcome:

```
factorial(8)*factorial(12)*factorial(10)*factorial(10)/
        (factorial(6)*factorial(2)*factorial(4)*factorial(8)*factorial(20))
```

```
[1] 0.07501786
```

but that is only part of the story. We need to compute the probability of outcomes that are more extreme than this. There are two of them. Suppose only one ant colony was found on tree B. Then the table values would be 7, 1, 3, 9 but the row and column totals would be exactly the same—*the marginal totals are constrained*. The numerator always stays the same, so this case has probability

```
factorial(8)*factorial(12)*factorial(10)*factorial(10)/
        (factorial(7)*factorial(3)*factorial(1)*factorial(9)*factorial(20))
```

```
[1] 0.009526078
```

There is an even more extreme case if no ant colonies at all were found on tree B. Now the table elements become 8, 0, 2, 10 with probability

```
factorial(8)*factorial(12)*factorial(10)*factorial(10)/
        (factorial(8)*factorial(2)*factorial(0)*factorial(10)*factorial(20))
```

```
[1] 0.0003572279
```

and we need to add these three probabilities together:

0.07501786+0.009526078+0.000352279

```
[1]  0.08489622
```

But there was no *a priori* reason for expecting the result to be in this direction. It might have been tree A that had relatively few ant colonies. We need to allow for extreme counts in the opposite direction by doubling this probability (all Fisher's exact tests are two-tailed):

2*(0.07501786+0.009526078+0.000352279)

```
[1]  0.1697924
```

This shows that there is no evidence of a correlation between tree species and ant colonies. The observed pattern, or a more extreme one, could have arisen by chance alone with probability $= 0.17$.

There is a built-in function called **fisher.test,** which saves us all this computation. It takes as its argument a 2×2 matrix containing the counts of the four contingencies. We make the matrix like this (compare with the alternative method of making a matrix, above):

```
x<-as.matrix(c(6,4,2,8))
dim(x)<-c(2,2)
x

           [,1]   [,2]
  [1,]      6      2
  [2,]      4      8
```

and run the test like this:

```
fisher.test(x)

         Fisher's exact test

data:  x
p-value = 0.1698
alternative hypothesis: two.sided
```

Correlation between two variables

The following paired data show mean biodiversity scores from samples of invertebrates taken upstream and downstream from sewage outfalls on nine rivers (labelled A to I in the data frame called paired):

```
attach(paired)
names(paired)

[1]  "Location"  "Upstream"  "Downstream"
```

We begin by asking whether there is a correlation between upstream and downstream biodiversities across rivers:

```
cor(Upstream,Downstream)

[1]  0.8820102
```

There is a strong positive correlation. Not surprisingly, species-rich rivers are rich in invertebrates both above and below the sewage outfall. If you want the significance of a correlation (i.e. the p value associated with the calculated value of r) then use **cor.test** rather than **cor**. This test has non-parametric options for **Kendall's tau** or **Spearman's rank** depending on the method you specify (method="k" or method="s"). The default method is Pearson's product-moment correlation (method="p"):

```
cor.test(Upstream,Downstream)
```

```
        Pearson's product-moment correlation
data:  Upstream and Downstream
t = 4.9521, df = 7, p-value = 0.0017
alternative hypothesis: true coef is not equal to 0
sample estimates:
        cor
0.8820102
```

The correlation is highly significant ($p = 0.0017$). Just for practice, let's try to extract the correlation coefficient from the three variances: s12 the upstream variance, s22 the downstream variance and sD2 the variance of the differences (Upstream–Downstream):

```
s12<-var(Upstream)
s22<-var(Downstream)
sD2<-var(Upstream-Downstream)
```

Applying the formula we obtained earlier (p. 114), the correlation should be given by

$$\rho = \frac{\sigma_y^2 + \sigma_z^2 - \sigma_{y-z}^2}{2\sigma_y\sigma_z}$$

So, using the values we have just calculated, we get

```
(s12+s22-sD2)/(2*sqrt(s12)*sqrt(s22))
```

```
[1] 0.8820102
```

which checks out. We can also see whether the variance of the difference is equal to the sum of the component variances:

```
sD2
```

```
[1] 0.01015
```

```
s12+s22
```

```
[1] 0.07821389
```

No, it is not. They would only be equal if the two samples were independent. In fact, we know they are positively correlated, so the variance of the difference should be *less* than the sum of the variances by $2rs_1s_2$:

```
s12+s22-2*0.882*sqrt(s12)*sqrt(s22)
```

```
[1] 0.01015079
```

That's more like it.

Comparing two variances when the samples are correlated

Suppose we wanted to assess the significance of the differences between the upstream and downstream variances. We cannot do a simple F test (see above) because this assumes that the samples are independent (and we know that they are highly correlated).

We have already seen that sums and differences of correlated variables have different variances; they differ by $4\rho\sigma_1\sigma_2$ (see above). The covariance of the sums and the differences of the same data looks like this:

$$\mathrm{cov}(DS) = \mathrm{cov}(y - z)(y + z) = \sigma_y^2 - \sigma_z^2$$

because the two terms in cov(yz) cancel out. So the correlation between the differences and the sums, ρ_{DS}, is given by

$$\rho_{DS} = \frac{\sigma_y^2 - \sigma_z^2}{\sqrt{(\sigma_y^2 + \sigma_z^2)^2 - 4\rho^2\sigma_y^2\sigma_z^2}}$$

It is not at all obvious why this is useful in testing the significance of the difference between the two variances until we express it in terms of the variance ratio $\phi = \sigma_y^2/\sigma_z^2$:

$$\rho_{DS} = \frac{\phi - 1}{\sqrt{(\phi + 1)^2 - 4\rho^2\phi}}$$

If the variances are equal (so that the null hypothesis is true) then $\phi = 1$, which means that $\rho_{DS} = 0$. Alternatively, if $\sigma_y^2 > \sigma_z^2$ then $\rho_{DS} > 0$ or if $\sigma_y^2 < \sigma_z^2$ then $\rho_{DS} < 0$. So the variance test for correlated variables involves *a correlation test on the sums and the differences of the two correlated variables*:

```
sum<-Upstream+Downstream
dif<-Upstream-Downstream
cor.test(sum,dif)
```

```
        Pearson's product-moment correlation

data:   sum and dif
t = -0.927, df = 7, p-value = 0.3848
alternative hypothesis: true coef is not equal to 0
sample estimates:
        cor
 -0.330677
```

The variances are not significantly different ($p = 0.38$). The negative sign of the correlation indicates that the second (downstream) variance is larger than the first (upstream) variance:

```
s12
```

```
[1]  0.03273611
```

```
s22
```

```
[1]  0.04547778
```

Scale-dependent correlations

Another major difficulty with correlations is that scatterplots can give a highly misleading impression of what is going on. The moral of this exercise is very important: things are not always as they seem. The data show the number of species of mammals in forests of differing productivity:

```
attach(productivity)
names(productivity)
```

```
[1]  "x"  "y"  "f"
```

```
plot(x,y)
```

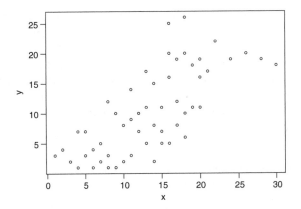

There is a very clear positive correlation: increasing productivity (x) is associated with increasing species richness (y). The correlation is highly significant:

```
cor.test(x,y,method="spearman")
```

```
        Spearman's rank correlation
data:  x and y
normal-z = 5.4717, p-value = 0
alternative hypothesis: true rho is not equal to 0
sample estimates:
        rho
 0.7516389
```

But what if we look at the relationship for each region (f) separately, using **xyplot**?

```
xyplot(y~x| f)
```

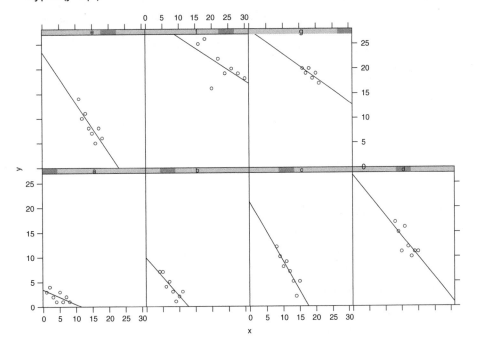

I've added the regression lines for emphasis (see p. 312) but the pattern is obvious. For every single case, increasing productivity is associated with *reduced* mammal species richness within each region.

The moral of this example is clear: you need to be extremely careful when looking at *correlations across different scales*. Things that are positively correlated over short timescales may turn out to be negatively correlated in the long term. Things that appear to be positively correlated at large spatial scales may turn out (as in this example) to be negatively correlated at small scales.

Correlation and covariance

The function **cor** returns the correlation matrix of a data frame, or a single value showing the correlation between one vector and another:

```
attach(Pollute)
names(Pollute)
[1] "Pollution" "Temp" "Industry" "Population" "Wind" "Rain" "Wet.days"

cor(Pollute)
```

	Pollution	Temp	Industry	Population	Wind	Rain	Wet.days
Pollution	1.00000000	-0.43360020	0.64516550	0.49377958	0.09509921	0.05428389	0.36956363
Temp	-0.43360020	1.00000000	-0.18788200	-0.06267813	-0.35112340	0.38628047	-0.43024212
Industry	0.64516550	-0.18788200	1.00000000	0.95545769	0.23650590	-0.03121727	0.13073780
Population	0.49377958	-0.06267813	0.95545769	1.00000000	0.21177156	-0.02606884	0.04208319
Wind	0.09509921	-0.35112340	0.23650590	0.21177156	1.00000000	-0.01246601	0.16694974
Rain	0.05428389	0.38628047	-0.03121727	-0.02606884	-0.01246601	1.00000000	0.49605834
Wet.days	0.36956363	-0.43024212	0.13073780	0.04208319	0.16694974	0.49605834	1.00000000

The phrase 'data dredging' is used disparagingly to describe the act of trawling through a table like this, desperately looking for big values which might suggest relationships that you can publish. This behaviour is not to be encouraged. The correct approach is model simplification (see p. 442). Note that the correlations are identical in opposite halves of the matrix (in contrast to regression, where regression of *y* on *x* would be different from a regression of *x* on *y*).

The correlation between two vectors produces a single value:

```
cor(Pollution,Wet.days)

[1]  0.3695636
```

We now use **cor.test** to test whether the correlation between Pollution and Wet.days of 0.3696 just obtained is significant or not. First, **Pearson's product-moment** correlation:

```
cor.test(Pollution,Wet.days)

        Pearson's product-moment correlation

data:  Pollution and Wet.days
t = 2.4838, df = 39, p-value = 0.0174
```

This says that the correlation is significant at $p = 0.0174$.

Kendall's tau is simply an option method="k" of the same **cor.test**:

```
cor.test(Pollution,Wet.days,method="k")

        Kendall's rank correlation tau
data:  Pollution and Wet.days
```

```
normal-z = 3.2975, p-value = 0.001
alternative hypothesis: true tau is not equal to 0
sample estimates:
        tau
    0.3573171
```

This tests leads to the same conclusion, but the non-parametric Kendall test gives much higher significance in this case ($p = 0.001$ vs. 0.0174). You should compare this result with the full multiple regression analysis on p. 442, the generalised additive model on p. 602 and the tree model on p. 583. Correlations with single explanatory variables can be highly misleading if (as is typical) there is substantial correlation among the explanatory variables.

Partial correlation

With more than two variables, you often want to know the correlation between x and y when, say, z is held constant. The *partial correlation coefficient* measures this. It enables correlation due to a shared common cause to be distinguished from direct correlation:

$$r_{xy.z} = \frac{r_{xy} - r_{xz}r_{yz}}{\sqrt{(1 - r_{xz}^2)(1 - r_{yz}^2)}}$$

cor(Pollution,Wet.days)

```
[1]  0.3695636
```

cor(Pollution,Wind)

```
[1]  0.09509921
```

cor(Wet.days,Wind)

```
[1]  0.1669497
```

So the partial correlation between Pollution and Wet.days controlling for Wind is

(0.3695636-(0.0951*0.16695))/sqrt((1-0.0951^2)*(1-0.16695^2))

```
[1]  0.3603544
```

Suppose we had four variables and we wanted to look at the correlation between x and y holding the other two, z and w, constant:

$$r_{xy.zw} = \frac{r_{xy.z} - r_{xw.z}r_{yw.z}}{\sqrt{(1 - r_{xw.z}^2)(1 - r_{yw.z}^2)}}$$

You will need partial correlation coefficients (PCCs) if you want to do path analysis. In this book we prefer to use tree models and various kinds of model simplification following multiple regression. Nevertheless, if you need them, you can use built-in functions to get the values of partial correlation coefficients as follows.

The sum of squares attributable to a given variable can be determined by deleting it from a model containing the other variables, using update with **anova**. Divide this by *SST* and you get what you might call a partial r^2. Take the square root of this to get a partial correlation coefficient. Suppose we want to work with Pollution as a function of Wet.days and Wind. We fit a fullish model like this:

```
model<-lm(Pollution~Wet.days+Wind)
```

and we get *SST* by fitting the intercept on its own, like this:

```
model0<-lm(Pollution~1)
```

Now we delete Wet.days from the full model using update:

```
model1<-update(model,~.-Wet.days)
```

and use **anova** to obtain the necessary sums of squares:

```
anova(model,model1,model0)
```

```
Analysis of Variance Table
Response: Pollution
```

	Terms	Resid. Df	RSS	Test	Df	Sum of Sq	F Value	Pr(F)
1	Wet.days + Wind	38	19002.74					
2	Wind	39	21838.59	-Wet.days	-1	-2835.859	5.670902	0.0223624
3	1	40	22037.90		-1	-199.308	0.398558	0.5316143

Thus, *SST* was 22 037.9 (see the column headed RSS), and partial *SSR* for Wet.days was 2835.859 (see the column headed Sum of Sq). So the partial *r* is

```
sqrt (2835.859 / 22037.9)
```

```
[1]  0.3587213
```

which differs form our earlier calculation only through the rounding errors incurred there.

Missing values and correlation matrices

Missing values are a big problem in studies of correlation (Little and Rubin 1987; Schafer 1996). The 'complete' and 'available' options of **cor** are consistent if missing values are *uninformative*. This means they are 'missing completely at random', so that whether or not a value is missing does not depend the values of any of the other variables, nor on whether or not they are observed or missing. There are two options. If na.method="available" then means and variances are computed for each variable using all non-missing values, and covariances for each pair of variables are computed using observations with no missing data for that pair. If na.method="include" then a missing value in the jth column of x causes row j of the output to contain all NAs, and a missing value in the kth column of y (or x if y is not supplied) causes column k of the output to contain all NAs. This amounts to a lot of lost information. The best plan is often to think carefully about whether to include a vector that contains lots of missing values in the data frame in the fist place, because its inclusion can knock out so many potentially valuable data points from other variables.

Kolmogorov–Smirnov tests

People know this test for its fine name, rather than for what it actually does. It is an extremely simple test for asking one of two different questions:

- are two sample distributions the same, or are they significantly different from one another?

- does a particular sample distribution arise from a particular hypothesised distribution?

The two-sample problem is the one most often used. The apparently simple question is actually very broad. It is obvious that two distributions could be different because their means were different. But two distributions with exactly the same mean could be significantly different if they differed in variance, or in skew or kurtosis (see p. 104). The Kolmogorov–Smirnov test works on *cumulative distribution functions* (CDFs). These give the probability that a randomly selected value of $X \leq x$:

$$F(x) = P[X \leq x]$$

This sounds somewhat abstract. Suppose we had counts of insect wing sizes for two geographically separated populations and we wanted to test whether the distribution of wing lengths was the same in the two places:

```
attach(wings)
names(wings)

[1] "Size"       "LocationA" "LocationB"
```

There is a useful function **cdf.compare** for testing whether an empirical distribution is likely to have come from a particular specified distribution. To compare our two distributions with appropriate Normal distributions, and make two plots side by side, we write

```
par(mfrow=c(1,2))
cdf.compare(LocationA,distribution="normal")
cdf.compare(LocationB,distribution="normal")
```

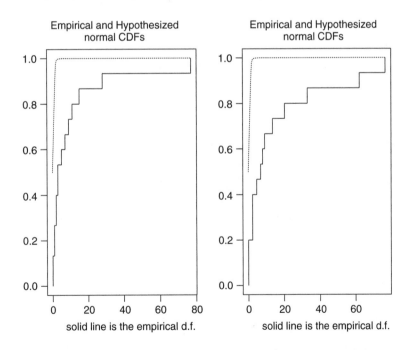

Clearly, neither empirical distribution (the solid lines for locations A and B) is anything like a Normal distribution (the smooth dotted lines). The question, however, is whether the two distributions are significantly different from one another? We use the same function to inspect the CDFs from the two locations:

```
cdf.compare(LocationA,LocationB)
```

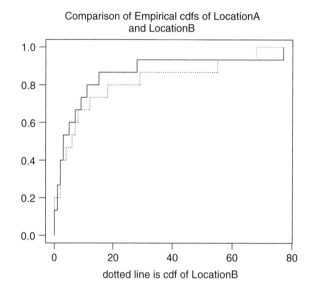

Comparison of Empirical cdfs of LocationA
and LocationB

dotted line is cdf of LocationB

The solid line (location A) rises more rapidly, while the dotted line (location B) has more larger wing lengths. The two distributions are different, but are they significantly different? The Kolmogorov–Smirnov goodness of fit test, **ks.gof,** will tell us:

```
ks.gof(LocationA,LocationB)

        Two-Sample Kolmogorov-Smirnov Test

data:  LocationA and LocationB
ks = 0.1333, p-value = 0.9998

alternative hypothesis:
   cdf of LocationA does not equal the cdf of LocationB
for at least one sample point.
```

Not only are they not significantly different, the two distributions are almost exactly the same! The probability that they come from the same distribution is $p = 0.9998$. The null hypothesis is accepted and the alternative (unhelpfully printed after the non-significant answer) is rejected. When two distributions are significantly different, the value of $p < 0.05$. Alternative analyses of the same data might involve a log-linear model of the counts (see p. 537) or a chi-square test of goodness of fit (see p. 468).

Further reading

Conover, W.J. (1980) *Practical Nonparametric Statistics.* New York, John Wiley.
Lehmann, E.L. (1986) *Testing Statistical Hypotheses.* New York, John Wiley.
Snedecor, G.W. and Cochran, W.G. (1980) *Statistical Methods.* Ames, Iowa, Iowa State University Press.
Sokal, R.R. and Rohlf F.J. (1995) *Biometry: The Principles and Practice of Statistics in Biological Research.* San Francisco, W.H. Freeman.
Sprent, P. (1989) *Applied Nonparametric Statistical Methods.* London, Chapman & Hall.

12

Bootstrap and jackknife

The development of computationally intensive statistics offers an alternative to classical Normal theory that is both simple and extremely robust. The same model can be fitted to resampled data 10 000 times in just a few seconds. This chapter deals with three kinds of tests

- bootstrap (where the data are repeatedly sampled with replacement in order to estimate confidence intervals for various parameters)

- jackknife (where each data point is omitted in turn for influence testing)

- permutation tests (where the treatment codes are repeatedly reshuffled in order to test hypotheses about treatment effects)

For every parameter we estimate from data, we need to establish an **unreliability estimate**. We use this to judge the uncertainty associated with any inferences we may want to make about our point estimate, and to establish a confidence interval for the true value of the parameter. Up to now, we have used parametric measures like standard errors that are based on the assumption of normality of errors (see p. 97). If the assumption of normality is wrong, then our unreliability estimates will also be wrong, but it is hard to know *how* wrong they will be, using standard analytical methods. An alternative way of establishing unreliability estimates is to **resample** our data using what are called **bootstrap** methods (Efron and Tibshirana 1993; Shao and Tu 1995).

Resampling

Take the 10 numbers 0 to 9:

```
x<-0:9
x
```

```
[1] 0 1 2 3 4 5 6 7 8 9
```

The function **sample** takes the numbers supplied to it and shuffles them; that is to say, the numbers are put into a random sequence where every number is represented and no number appears more than once:

```
sample(x)
```

```
[1] 3 5 6 2 9 7 8 4 1 0
```

In this particular randomisation the first came last (0) and the last (9) came fifth. Because all of the numbers of the original vector are present in the sampled vector, all of the statistical properties of the sampled and original vectors are identical. Here, for example, is a comparison of the mean of *x* and the mean of the shuffled vector:

```
mean(x)
```

```
[1] 4.5
```

```
mean(sample(x))
```

```
[1]  4.5
```

The **sample** function is very useful at the design phase for randomising the allocation of individuals to experimental treatments. Suppose that we had an experiment in which 20 individuals were to be allocated to one of four treatments (i.e. replication = 5) then we could line up the individuals, and allocate them to treatments systematically from the following list:

```
sample(rep(c("A","B","C","D"),5))
```

```
[1]  "B"  "C"  "B"  "A"  "D"  "C"  "C"  "A"  "D"  "A"  "D"  "B"  "D"
     "B"  "A"  "D"  "A"  "C"  "B"  "C"
```

so the first individual was allocated to treatment 'B', the second to treatment 'C', the third to treatment 'B' and so on.

Perhaps the most useful feature of **sample**, however, is its ability to sample with replacement. Without saying it explicitly, what we have been doing up to this point is sampling without replacement. Once a number had been used, it wouldn't be used again. Sampling with replacement means:

- some numbers will appear in the sample more than once

- some numbers won't appear in the sample at all

Let's sample with replacement from the numbers 0 to 9 generated earlier:

```
sample(x,replace=T)
```

```
[1]  0  2  4  1  9  3  9  1  2  2
```

You see three 2s and two 9s in this particular realisation of the sample, and no 5, 6, 7 or 8. Another realisation will be different. We can use the **table** function to count the different numbers:

```
table(sample(x,replace=T))
```

```
1  2  3  4  6  7  8  9
1  1  1  3  1  1  1  1
```

so in this randomisation there were three 4s but no 0s or 5s. Doing it again gives a different answer:

```
table(sample(x,replace=T))
```

```
1  2  4  5  7  8
1  2  2  2  1  2
```

This time there were no 0s or 3s or 6s or 9s, but duplicates of 2, 4, 5 and 8. Since sampling with replacement produces different vectors each time, the parameters (like mean and variance) will differ from each realisation to the next. Here are the means of six different samples:

```
for (i in 1:6) print (mean(sample(x,replace=T)))
```

```
[1]  4.1
[1]  4.5
[1]  4.5
[1]  4.5
[1]  4.4
[1]  3.2
```

For these six replicate trials, the mean varied from a low of 3.2 to a high of 4.5 (which we know to be the true mean). You can see, I suspect, that if you were to repeat this process lots of times, you could establish a confidence interval for the sample mean. This is called a bootstrap confidence interval because you appear to have got something for nothing—you have 'pulled yourself up by your own bootstraps'. Suppose we did this 10 000 times. We would have 10 000 different mean values, but the distribution of these numbers might be highly informative. In particular, it might be very informative about the quantiles of the distribution of the mean (i.e. the values of the mean that characterise the tails of the distribution). We could ask what are the unusually large values of the mean (e.g. the 97.5 percentiles) or the unusually small values (e.g. the 2.5 percentiles). Let's try it and see what happens. We created a vector (say, *xmeans*) to take the 10 000 values of the mean of the samples taken from x:

```
xmeans<-numeric(10000)
```

Now we take the means of 10 000 realisations of sample(x,replace=T) and store each mean value in *xmeans*, like this:

```
for (i in 1:10000) xmeans[i]<-mean(sample(x,replace=T))
```

The overall mean is very close to the known true mean of 4.5:

```
mean(xmeans)
```

```
[1]  4.50861
```

Here we are more interested in the shape and the spread of the distribution of values in *xmean*. We can look at them using **hist**:

```
hist(xmeans)
```

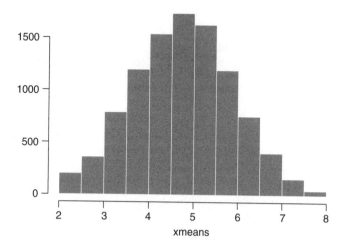

Notice that the distribution of means is bell-shaped and very well behaved, even though the data themselves were from a uniform (rectangular) distribution, and hence far from Normal (see central limit theorem, p. 127). Just looking at the histogram, it is clear that a mean smaller than 2 or larger than 7 is most unlikely. This provides us with a crude confidence interval. We can do much better by working out the quantiles of the distribution like this:

```
quantile(xmeans,c(.025,.975))
```

```
2.5%   97.5%
2.7    6.3
```

This is the 95% bootstrapped confidence interval for the mean of x. We can be 95% sure that the true mean lies between 2.7 and 6.3 (we know that the true mean is exactly 4.5). How does this compare with a parametric confidence interval based on the clearly erroneous assumption that the residuals around the mean are normally distributed? To get the Normal theory confidence interval we need to calculate the standard error of the mean:

```
se<-sqrt(var(x)/length(x))
```

and then look up the value of Student's t in tables:

```
tvalue<-qt(.975,length(x))
```

then calculate the confidence interval (t from tables times the standard error):

```
ci<-tvalue*se
ci
```

```
[1]  2.133281
```

Now we can add the confidence interval of 2.133 281 to the mean, and take it away, to obtain the 95% confidence limits:

```
mean(x)+ci
```

```
[1]  6.63328
```

```
mean(x)-ci
```

```
[1]  2.366719
```

In this case the bootstrap confidence intervals are narrower than the (inappropriate) Normal theory confidence intervals which gave a lower bound of 2.37 (rather than 2.7), and an upper bound of 6.63 (rather than 6.3).

Bootstrap estimates

Here we revisit the speed of light data that we analysed in Chapter 11. The null hypothesis is that the mean is 990. The sample mean is 909. The question is this: How likely is it that a sample of size $n = 20$ would have a mean of 909 if it came from a population with a mean of 990. We shall resample the data with replacement 10 000 times and ask where, in terms of quantiles, a value of 990 lies. We shall say that the sample mean is significantly different if 990 lies above the 97.5 percentile of the distribution of resampled means.

```
attach(light)
names(light)
```

```
[1]  "speed"
```

```
mean(speed)
```

```
[1]  909
```

Now we write a loop to go round 10 000 times, resample with replacement, and save the mean in a vector called *d*:

```
d<-numeric(10000)
for (i in 1:10000) d[i]<-mean(sample(speed,replace=T))
```

Now we plot the histogram of resampled means and ask about the quantile associated with 97.5% (two-tailed):

```
hist(d)
quantile(d,.975)

97.5%
  951
```

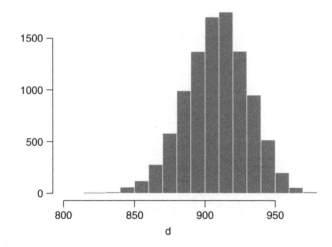

Both the histogram and the quantile (951) show that a value of 990 is most unlikely if the null hypothesis is true. We therefore reject the null hypothesis and conclude that the speed of light is significantly lower than 990.

Bootstrap function in S-Plus

The **bootstrap** function is extremely easy to use; you just specify the variable (speed), and the statistic to be bootstrapped (mean) like this:

```
bootstrap(speed,mean)
```

After the default 1000 iterations (you can easily increase it), the output looks like this:

```
Call:
bootstrap(data = speed, statistic = mean)

Number of Replications: 1000

Summary Statistics:
      Observed    Bias    Mean      SE
mean       909   0.6805   909.7   22.99
```

In bootstrapping, **bias** is simply the difference between the sample mean and the bootstrapped mean. If the bias is large, it means that the sample contains one or more influential points or the distribution is highly skew. Here the bias is very small and we can be confident in rejecting the null hypothesis that the mean is 990. The bootstrapped standard error of the mean (22.99) is very similar to the full-sample standard error:

sqrt(var(speed)/length(speed))

[1] 23.46218

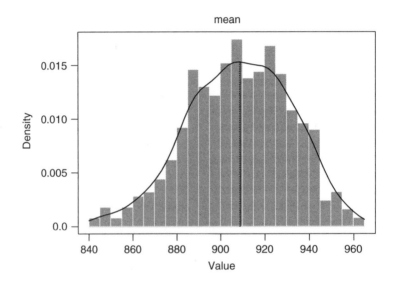

The **plot** function applied to a bootstrap object shows the sample mean as a solid vertical line (at 909) and a dashed vertical line at the mean of the replicates; the difference between the two lines is the bias. The bars of the histogram are the replicated means, and the line is a smoothed estimate of the probability density.

    bs<-bootstrap(speed,mean)
    plot(bs)
    summary(bs)

The **summary** function applied to a bootstrap object produces extra information, especially on the **BCa** percentiles. This stands for 'bias-corrected and adjusted' percentiles; these are accurate only for replication >1000. Note that the bias is different for this second bootstrap of the same data (-0.402 vs. 0.6805):

```
Call:
bootstrap(data = speed, statistic = mean)

Number of Replications: 1000

Summary Statistics:
     Observed   Bias   Mean     SE
mean      909  -0.402  908.6  23.33
```

```
Empirical Percentiles:
       2.5%    5%   95%   97.5%
mean   860    869   944   950.5

BCa Confidence Limits:
       2.5%    5%    95%   97.5%
mean  857.4  865.7 942.5  948.4
```

Jackknife

The jackknife is a test of the influence of individual values of the response variable on parameter estimates, while the bootstrap is a way of establishing confidence intervals for estimated parameter values. The jackknife is different from the bootstrap in that we only recalculate the parameter of interest n times (instead of, say, 10 000 times). For the speed of light data we have $n = 20$, so we shall do 20 calculations each based on $n - 1 = 19$ values of the response variable. We leave out each data point in turn and ask how influential it was in determining the overall sample mean:

```
jk<-numeric(20)
for (i in 1:20) jk[i]<-mean(speed[-i])
hist(jk)
```

Note the use of negative subscripts [-i] to leave out the data points one at a time.

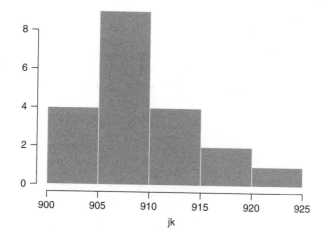

With only 20 data points the histogram is very crude, but we can see at once that leaving out the most influential point only increased the sample mean from 909 to about 925 (actually 922.6). We reject the null hypothesis and conclude that our sample mean is significantly lower than 990.

Jackknife in S-Plus

For these data, the jackknife picks up no estimate of bias at all (compare with the small bias under bootstrapping):

```
jackknife(speed,mean)

Call:
jackknife(data = speed, statistic = mean)
```

```
Number of Replications: 20

Summary Statistics:
       Observed  Bias  Mean    SE
mean        909     0   909 23.46
```

Bootstrapping in regression analysis

For more complex analyses, the arguments of the call to the bootstrap function are correspondingly more intricate, but the principle is exactly the same. Here we bootstrap the regression analysis that is worked out longhand in Chapter 14. We want to obtain a bootstrapped estimate of the confidence interval for the slope of the relationship between weight gain and dietary tannin:

```
attach(regression)
names(regression)
```

```
[1] "growth" "tannin"
```

```
bs<-bootstrap(regression,coef(lm(growth~tannin)))
bs
```

```
Call:
bootstrap(data = regression, statistic = coef(lm(growth ~ tannin)))

Number of Replications: 1000

Summary Statistics:
              Observed     Bias    Mean      SE
(Intercept)     11.756  0.09657  11.852  0.9290
     tannin     -1.217 -0.01511  -1.232  0.1938
```

You can see that there is very little bias in either the intercept or the slope.

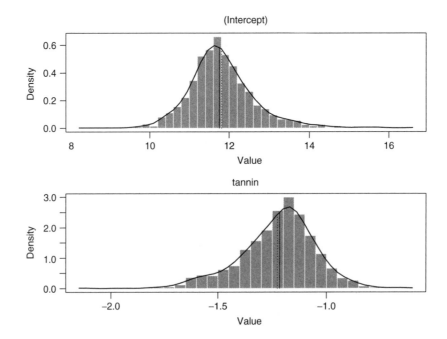

These graphs are obtained with plot(bs). Both the intercept and the slope are very well behaved; the bars show the replicates and the smooth line is an estimated probability density. The solid vertical line is the sample estimate and the dashed line is the bootstrapped estimate (the difference between the two lines is the bias).

The extra information obtained with summary(bs) consists of the empirical percentiles and BCa confidence limits:

```
       Empirical Percentiles:
                     2.5%        5%        95%       97.5%
       (Intercept)  10.349    10.556    13.3228    13.7521
            tannin  -1.628    -1.563    -0.9672    -0.9118

       BCa Confidence Limits:
                     2.5%        5%        95%       97.5%
       (Intercept)  10.497    10.750    13.6860    14.278
            tannin  -1.646    -1.587    -0.9938    -0.936
```

Because there is essentially no bias in this example, the two estimates are very similar.

Jackknife-after-bootstrap

Jackknife-after-bootstrap is a technique for estimating the standard error of some functional of the bootstrap distribution of parameter estimates. For example, it may be used to estimate standard errors for the bootstrap estimate of standard error. Jackknife-after-bootstrap also calculates relative influences reflecting the degree of influence each observation has upon the functional under consideration. Here it is applied to the bootstrapped regression (bs) we have just carried out:

jack.after.bootstrap(bs)

```
       Functional of Bootstrap Distribution of Parameters:
                       Func   SE.Func
       (Intercept)    11.811   0.8056
            tannin    -1.227   0.1715

       Observations with Large Influence on Functional:
       $"(Intercept)":
          (Intercept)
       4       2.251
```

The output is interpreted as follows. Func means parameter (intercept and slope in this case) and SE Func means the standard error of that parameter. The lower panel highlights influential data points, and the parameter estimates that they influence: here, data point number 4 has a strong influence on the estimate of the intercept (2.251 is the value of the absolute relative influence). None of the data points has strong influence on the estimate of the slope. So the jackknife after bootstrap estimate of the intercept is 11.811 with a standard error of 0.8056 (for comparison, the regression gives 11.56 with a standard error of 1.04, and the bootstrap gives 11.76 with a standard error of 0.929). The jackknife-after-bootstrap estimate of the slope is -1.227 with $SE = 0.172$ (for comparison, the regression gives -1.217 with $SE = 0.219$ and the bootstrap gives -1.217 with $SE = 0.194$). All are very similar in this case. To save on space, you might write this as follows: 'the jackknife after bootstrap estimates of the standard errors of slope and intercept were smaller than, but within 21% of, the regression estimates'. Plotting a jackknife-after-bootstrap regression object produces two plots.

The first shows the influence of individual data points on the estimates of intercept, and the second on the slope. Observation number 4 is influential on the estimate of the intercept, but there are no influential points for the slope (although point number 7 comes close to being influential).

Permutation tests

We can use **sample** in a completely different kind of way. Up to this point, we have sampled the *values of the response variable* (we have taken different values of *y* for analysis). An alternative is to *shuffle the values of the explanatory variables.* The idea behind this is simple: if there is no difference between the means of two treatments, then it will make no difference which treatment label is associated with which data point. In contrast, if there *is* a difference between the treatment means, then it matters greatly which treatment label is associated with which data point.

The example comes from a survey of slug densities in two fields; Rookery and Nursery:

```
attach(slugsurvey)
names(slugsurvey)
```

```
[1] "slugs" "field"
```

```
levels(field)
```

```
[1] "Nursery" "Rookery"
```

We start by working out the difference in mean slug density in the two fields using **tapply**:

```
tapply(slugs,field,mean)
```

```
Nursery Rookery
  1.275   2.275
```

So the observed difference in means was $2.275 - 1.275 = 1.0$ with more slugs in Rookery than in Nursery. In order to test the null hypotheses that mean slug densities are the same in the two fields, we can use a computationally intensive permutation test.

The trick here is to shuffle the vector of field labels. The unshuffled version looks like this:

```
field
```

```
 [1] Nursery Nursery Nursery Nursery Nursery Nursery Nursery Nursery Nursery Nursery
[11] Nursery Nursery Nursery Nursery Nursery Nursery Nursery Nursery Nursery Nursery
[21] Nursery Nursery Nursery Nursery Nursery Nursery Nursery Nursery Nursery Nursery
[31] Nursery Nursery Nursery Nursery Nursery Nursery Nursery Nursery Nursery Nursery
[41] Rookery Rookery Rookery Rookery Rookery Rookery Rookery Rookery Rookery Rookery
[51] Rookery Rookery Rookery Rookery Rookery Rookery Rookery Rookery Rookery Rookery
[61] Rookery Rookery Rookery Rookery Rookery Rookery Rookery Rookery Rookery Rookery
[71] Rookery Rookery Rookery Rookery Rookery Rookery Rookery Rookery Rookery Rookery
Levels: Nursery  Rookery
```

All the data from Nursery come first, then all the data from Rookery. Once we sample (without replacement), the field names look like this:

```
sample(field)
```

```
 [1] Rookery Nursery Nursery Rookery Nursery Rookery Nursery Rookery Nursery Rookery
[11] Nursery Nursery Rookery Rookery Rookery Nursery Rookery Rookery Nursery Nursery
[21] Nursery Rookery Rookery Nursery Rookery Nursery Nursery Nursery Nursery Nursery
[31] Nursery Nursery Nursery Nursery Rookery Nursery Nursery Rookery Rookery Nursery
[41] Rookery Rookery Rookery Rookery Rookery Rookery Rookery Nursery Rookery Nursery
```

```
[51] Nursery Nursery Nursery Rookery Nursery Nursery Nursery Rookery Rookery Rookery
[61] Rookery Rookery Nursery Rookery Rookery Nursery Rookery Nursery Rookery Rookery
[71] Rookery Nursery Rookery Nursery Rookery Nursery Rookery Nursery Nursery Rookery
Levels:  Nursery  Rookery
```

Later, when we analyse the data on slug densities in two fields using a range of parametric methods, we shall obtain equivocal results. Anova on the square root transformed counts says that the means were highly significantly different, but a log-linear model corrected for overdispersion says that they were not significantly different (see p. 542).

By using **sample** on the factor levels, the *labels have been randomised* relative to the values of the response variable to which they apply (slugs). Now we can calculate the difference between the means of the two subsamples and store its value. After 10 000 realisations we can see how the observed difference between field means (1.0) compares to the differences between random subsamples of the same size taken from the whole data set. This is called a **permutation test**:

```
diff<-numeric(10000)
for (i in 1:10000){
fieldname<-sample(field)
diff[i]<-tapply(slugs,fieldname,mean)[2]-tapply(slugs,fieldname,mean)[1]}
```

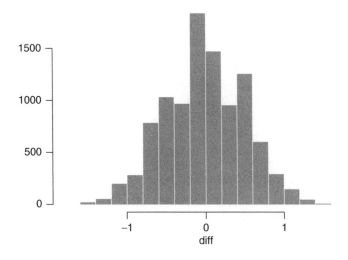

As you would have predicted, the mean difference between the randomly labelled subsamples is zero. But values as large as our observed difference between the fields of +1.0 *do* occur by chance alone (see the right-hand tail). We need to know the 97.5 percentile of diff in order to see whether 1.0 is larger or smaller than the critical (permutation) estimate.

```
quantile(diff,0.975)
```

```
97.5%
    1
```

Sod's law! Our value lies exactly on the upper 95 percentile. If we try a harsher test (say two-tailed 99%), our observed difference of 1.0 is not significant:

```
quantile(diff,0.995)
```

```
99.5%
  1.3
```

So our interpretation has to be circumspect. The permutation test points in the same direction as the log-linear models corrected for overdispersion (perhaps the field means are *not* significantly different; they are certainly not as highly significantly different as suggested by the simple parametric tests (e.g. Anova on the square root transformed counts) which assumed Normal errors which gave $p \approx 0.003$ (see p. 543).

An alternative test is to calculate bootstrapped confidence intervals for mean slug density in the two fields, then see if the confidence intervals overlap. To make things as simple as possible, we make two separate vectors, one for each field:

```
nursery<-slugs[field=="Nursery"]
rookery<-slugs[field=="Rookery"]
mean(nursery)
```

```
[1] 1.275
```

```
mean(rookery)
```

```
[1] 2.275
```

We want to know whether the mean slug density of 2.275 in Rookery is significantly greater than the mean of 1.275 observed in Nursery. First we bootstrap 10 000 estimates of the mean for Nursery:

```
nmeans<-numeric(10000)
for (i in 1:10000) nmeans[i]<-mean(sample(nursery,replace=T))
quantile(nmeans,c(.025,.975))
```

```
     2.5%       97.5%
0.649375   2.025000
```

Next we do the same for Rookery:

```
rmeans<-numeric(10000)
for (i in 1:10000) rmeans[i]<-mean(sample(rookery,replace=T))
quantile(rmeans,c(.025,.975))
```

```
2.5%  97.5%
1.65   3.00
```

The intervals *do* overlap substantially, so the difference between the means is certainly not clear-cut.

Permutation tests in more complex experiments

In a factorial or split-plot experiment you may want to constrain the randomisation of the factor levels. For instance, in a block design, you may want the treatment labels to be randomised *within each block* so that differences between the block means are retained in the analysis. If you did not do this, then it is possible that all the treatments within one block might turn out to be the same in some of the permutations.

The data refer to mean failure rates (the number of rejects per 1000 components manufactured) in 7 different factories for 5 error management strategies (labelled A to E); the standard technique is treatment D:

```
attach(factories)
names(factories)
```

```
[1] "factory" "treatment" "rate"
```

We start with data inspection, and look at the amount of variation between factories (the block effect) in terms of mean failure rate:

```
tapply(rate,factory,mean)
```

Doncaster	Gwent	Newport	Rotherham	Sheffield	Sunderland	Worksop
12.05664	8.746094	16.80297	9.70543	11.74314	15.73486	13.22627

There is almost a twofold difference in failure rate between the best factory (Gwent) and the worst (Newport). What about the treatment effects?

```
tapply(rate,treatment,mean)
```

A	B	C	D	E
11.19894	12.5437	10.26018	15.43952	13.42582

It looks as if treatment C reduces the failure rate relative to past standard practice (treatment D). We cannot test for any interaction between treatment and factory because there is no replication of treatments within blocks ($n = 5 \times 7 = 35$).

Conventional Anova suggests there is an extremely significant difference between the treatment means once differences between the factories have been taken into account:

```
model<-aov(rate~treatment+Error(factory))
summary(model)
```

```
Error: factory
            Df Sum of Sq   Mean Sq  F  Value  Pr(F)
Residuals  6   260.7013  43.45021

Error: Within
            Df  Sum of Sq  Mean Sq   F Value Pr(F)
treatment   4    113.2757 28.31893 246.9802     0
Residuals  24      2.7519  0.11466
```

We now repeat the analysis using a permutation test. The trick is that we want to maintain the identity of the factories (the block effects). This means that we need to randomise the treatment labels independently within each factory:

```
treats<-NULL
for(i in 1:7) treats<-c(treats,sample(levels(treatment)))
```

This is achieved by looping through each factory separately (we can do this with (i in 1:7) because results from the factories are grouped in sequence in the data frame; if they were not, then we could sort them so that they were in ordered groups) then concatenating a new set of randomisations of the five levels of treatment onto the existing vector called **treats**. Next we declare treats to be a factor:

```
treats<-factor(treats)
```

In this example we are going to extract the F value from the Anova table as our measure of the significance of the treatment effects. It is also a useful exercise in *subscripting model objects*. The F value we want is in the second of the two Anova tables (`Error: Within`); this is addressed by double subscripts [[2]]. Within this table, the sub-element we want is the column called F Value, so we extract this using the $ operator. Finally, within the column of F values, it is the first element that we want (the F test for treatments); this is obtained with subscript [1]. So the whole thing looks like this:

```
summary(model)[[2]]$"F Value"[1]
```

```
[1] 246.9802
```

Complicated, but very useful for writing general routines involving model summary tables. Now we fit the **aov** model 1000 times and save the F values in a vector d:

```
d<-numeric(1000)
for(j in 1:1000) {
treats<-NULL
for(i in 1:7) treats<-c(treats,sample(levels(treatment)))
treats<-factor(treats)
d[j]<-summary(aov(rate~treats+Error(factory)))[[2]]$"F Value"[1]}
```

A histogram of the 1000 F values shows what has been happening:

```
hist(d)
```

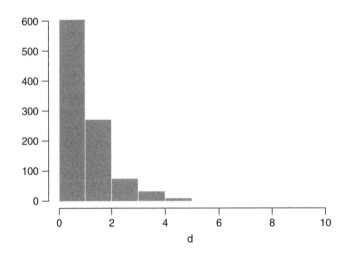

The F values from the Anova (reflecting that the differences between the means that arose by chance alone) are strongly clustered around 0 (as we would expect, having shuffled the treatment levels), but some large values do occur. We use **max** to obtain the largest F value found in 1000 permutations:

```
max(d)
```

```
[1] 9.800537
```

That is a highly significant F value with 4 and 24 d.f., but we know it is the result of chance alone, because we have randomised the treatment codes. We would expect some large values in 1000 simulations (about 50 of them if we are working at $\alpha = 0.05$). Our main interest, therefore, is in the quantile; we want the one-tailed (95%) value of F:

```
quantile(d,0.95)

  95%
2.967431
```

This is very close to the value of F you would expect by chance:

```
qf(.95,4,24)

[1] 2.776289
```

Recall that our measured F value from the Anova was 246.98. This is extremely convincing support for the view that treatment means are significantly different; they are not an artefact of the differences between the factory means. We know this because the differences between the factories were retained in our simulations as a result of the constrained randomisation (every treatment occurred just once in every factory).

Further reading

Efron, B. and Tibshirani, R.J. (1993) *An Introduction to the Bootstrap*. San Francisco, Chapman & Hall.
Robert, C.P. and Casella, G. (1999) *Monte Carlo Statistical Methods*. New York, Springer-Verlag.
Shao, J. and Tu, D. (1995) *The Jackknife and Bootstrap*. New York, Springer-Verlag.

13

Statistical models in S-Plus

Fitting models to data is the central function of S-Plus. The process is essentially one of exploration; there are no fixed rules and no absolutes. The object is to determine a minimal adequate model from the large set of potential models that might be used to describe the given set of data. In this book we discuss five types of model:

- the null model
- the minimal adequate model
- the current model
- the maximal model
- the saturated model

The stepwise progression from the saturated model (or the maximal model, whichever is appropriate) through a series of simplifications to the minimal adequate model is made on this basis of *deletion tests*; *F* tests or chi-square tests that assess the significance of the increase in deviance that results when a given term is removed from the current model.

Models are representations of reality that should be both accurate and convenient. However, it is impossible to maximise a model's realism, generality and holism simultaneously, and the principle of parsimony or Occam's razor (see p. 51) is a vital tool in helping to choose one model over another. Thus, we would only include an explanatory variable in a model if it significantly improved the fit of a model. Just because we went to the trouble of measuring something, does not mean we have to have it in our model. Parsimony says that, other things being equal, we prefer:

- a model with $n - 1$ parameters to a model with n parameters
- a model with $k - 1$ explanatory variables to a model with k explanatory variables
- a linear model to a model which is curved
- a model without a hump to a model with a hump
- a model without interactions to a model containing interactions between factors

Other considerations include a preference for models containing explanatory variables that are easy to measure over variables that are difficult or expensive to measure. Also, we prefer models that are based on a sound mechanistic understanding of the process over purely empirical functions.

The model formula

The structure of the model is specified in the model formula like this:

$$\text{response variable} \sim \text{explanatory variable(s)}$$

where the **tilde** symbol $\sim$ reads 'is modelled as a function of'. So a simple linear regression of y on x would be written y ~ x, and a one-way anova where sex is a two-level factor would be written y ~ sex. The right-hand side of the model formula shows:

- the number of explanatory variables and their identities; their attributes (e.g. continuous or categorical) are usually defined prior to the model fit

- the interactions between the explanatory variables (if any)

- non-linear terms in the explanatory variables

On the right of the tilde, one also has the option to specify offsets or error terms in some special cases. Like the response, the explanatory variables can appear as transformations, or as powers or polynomials (Table 13.1).

It is very important to note that symbols are used differently in model formulas than in arithmetic expressions. In particular:

+ indicates inclusion of an explanatory variable in the model (not addition)

- indicates deletion of an explanatory variable from the model (not subtraction)

* indicates inclusion of explanatory variables and interactions (not multiplication)

/ indicates nesting of explanatory variables in the model (not division)

| indicates conditioning (e.g. y ~ x | z is read as 'y as a function of x given z')

There are several other symbols that have special meaning in model formulas. The colon : means an interaction, so that A:B means the two-way interaction between A and B, and N:P:K:Mg means the four-way interaction between N, P, K and Mg.

Some terms can be written in an expanded form:

A*B*C is the same as A+B+C+A:B+A:C+B:C+A:B:C

A/B/C is the same as A+B%in%A+C%in%B%in%A

(A+B+C)^3 is the same as A*B*C

(A+B+C)^2 is the same as A*B*C - A:B:C

Interactions between explanatory variables

Interactions between two two-level categorical variables A*B mean that two main effect means and one interaction mean are evaluated. On the other hand, if factor A has 3 levels and factor B has 4 levels, then 7 parameters are estimated for the main effects (3 means for A and 4 means for B). The number of interaction terms is $(a-1)(b-1)$ where a and b are the numbers of levels of the factors A and B respectively. So in this case, S-Plus would estimate $(3-1)(4-1) = 6$ parameters for the interaction.

Interactions between two continuous variables are fitted differently. If x and z are two continuous explanatory variables, then x*z means fit x+z+x:z and the interaction term x:z behaves as if a new variable had been computed that was the pointwise product of the two vectors x and z. The same effect could be obtained by calculating the product explicitly:

```
x.by.z <- x * z
```

then using the model formula y ~ x + z + x.by.z. Note that the representation of the interaction by the *product* of the two continuous variables is an assumption, not a fact. The real interaction might be of an altogether different functional form (e.g. x * z^2).

Interactions between a categorical variable and a continuous variable are interpreted as an analysis of covariance; a separate slope and intercept are fitted for each level of the categorical variable. So y ~ A*x would fit three regression equations if the factor A had three levels; this would estimate six parameters from the data, three slopes and three intercepts.

The slash operator is used to denote nesting. Thus, with categorical variables A and B, the expression y ~ A/B means fit 'A plus B within A'. This could be written in two other equivalent ways:

$$y \sim A + A{:}B$$

$$y \sim A + B \% \text{ in } \% A$$

both of which alternatives emphasise that there is no point in attempting to estimate a main effect for B (it is probably just a factor label like 'tree number 1' that is of no scientific interest).

Some functions for specifying non-linear terms and higher-order interactions are useful. To fit a polynomial regression in *x* and *z*, we could write

$$y \sim poly(x,3) + poly(z,2)$$

to fit a cubic polynomial in *x* and a quadratic polynomial in *z*.

To fit interactions, but only up to a certain level, the ^ operator is useful. The formula

$$y \sim (A + B + C)^2$$

fits all the main effects and two-way interactions (i.e. it excludes the three-way interaction that A*B*C would have included).

The **I** function (capital *i*) stands for 'as is'. It overrides the interpretation of a model symbol as a formula operator when the intention is to use it as an arithmetic operator. Suppose you wanted to fit $1/x$ as an explanatory variable in a regression, you might try this:

$$y \sim 1/x$$

but this actually does something very peculiar. It fits *x* nested within the intercept! When it appears in a model formula, the slash operator is assumed to imply nesting. To obtain the effect we want, we use **I** to write

$$y \sim I(1/x)$$

We also need to use **I** when we want * to represent multiplication and ^ to mean 'to the power' rather than an interaction model expansion (see above).

Multiple error terms

When there is nesting (e.g. split plots in a designed experiment) or temporal pseudoreplication (see p. 368) you can include an **Error** function as part of the model formula. Suppose you had a three-factor factorial experiment with categorical variables A, B and C. The twist is that each treatment is applied to plots of different sizes. A is applied to replicated whole fields, B is applied at random to half fields and C is applied to smaller split-split plots within each field. This is shown in a model formula like this:

$$y \sim A*B*C + Error(A/B)$$

Note that the terms within the model formula are separated by asterisks to show that it is a full factorial with all interaction terms included, whereas the terms are separated by slashes in the **Error** statement. There is one less term in the **Error** statement than there are different sizes of plot (two in this case) and the terms are listed left to right from the largest plot to the smallest plot; see p. 354 for details and examples.

The null model

The simple command y~1 causes the null model to be fitted. This works out the grand mean (the overall average) of all the data and works out the total deviance (or the total sum of squares, *SST*, in models with Normal errors and the identity link). In some cases this may be the minimal adequate model; it is possible that none of the explanatory variables we have measured contribute anything significant to our understanding of the variation in the response variable. This is normally what you don't want to happen at the end of your three-year research project.

To remove the intercept (parameter 1) from a regression model (i.e. to force the regression line through the origin, you fit -1 like this:

$$y \sim x - 1$$

You should not do this unless you know exactly what you are doing, and exactly why you are doing it (see p. 453 for details). Removing the intercept from an Anova model where all the variables are categorical has a different effect:

$$y \sim sex - 1$$

This gives the mean for males and the mean for females in the summary table, rather than the overall mean and the difference in mean for males (see Helmert contrasts, p. 333).

In the **update** function used during model simplification, the dot . is used to specify 'what is there already' on either side of the tilde. So if your original model said

model<-lm(y~A*B)

then the **update** function to remove the interaction term A:B could be written like this:

model2<-update(model, ~ . - A:B)

Note that there is no need to repeat the name of the response variable, and the punctuation ~. means take model as it is, and remove from it ('minus') the interaction term A:B.

Model formulas for regression

Model formulas look very like equations but there are important differences. Our simplest useful equation looks like this:

$$y = a + bx$$

It is a two-parameter model with one parameter for the intercept, a, and another for the slope, b, of the graph of the continuous response variable y against a continuous explanatory variable x. The model formula for the same relationship looks like this:

$$y \sim x$$

The equals sign is replaced by a tilde, and all of the parameters are left out. It we had a multiple regression with two continuous explanatory variables x and z, the equation would look like this:

$$y = a + bx + cz$$

but the model formula would be

$$y \sim x + z$$

It is all wonderfully simple. But how does S-Plus know what parameters we want to estimate from the data? We have only told it the names of the explanatory variables. We have said nothing about how to fit them, or what sort of equation we want to fit to the data.

The key to this is to understand *what kind of explanatory variable is being fitted* to the data. If the explanatory variable x specified on the right of the tilde is a continuous variable, then S-Plus *assumes* that you want to do a regression, and hence that you want to estimate two parameters in a linear regression whose equation is $y = a + bx$.

If the equation you want to fit is more complicated than this, then you need to specify the form of the equation, and use non-linear methods (**nls** or **nlme**) to fit the model to the data (Chapter 23).

Model formulas for analysis of variance

What if you had a categorical variable (say a factor called w with k levels) and you wanted to do a one-way Anova rather than a regression? The model formula looks exactly the same as before:

$$y \sim w$$

So how does S-Plus know that you want to do an Anova rather than a regression? The answer is, it doesn't. Not unless you have told it explicitly beforehand that the variable called w is a categorical variable rather than a continuous variable: w <- factor(w). Of course, if the factor levels within w were text rather than numbers, this confusion could not arise, and S-Plus would have assumed that the variable was categorical, hence it would have known to do an Anova rather than a regression.

We have skipped over an important issue up to this point. We are familiar with representing a regression by an equation ($y = a + bx$) and the parameters in such an equation are familiar (the intercept and the slope in this case). But what does the equation for an analysis of variance look like? How does S-Plus make an equation out of the model formula y ~ w when w is a categorical variable? It is quite likely that you will not understand this on first reading, but you should persevere because the idea is really simple, and if you understand it then the interpretation of statistical models will be much more straightforward.

The equation to represent an analysis of variance where there is one factor with two levels (say w is sex with levels 'male' or 'female') looks like this:

$$y = a + bw_1 + cw_2$$

It looks exactly like a multiple regression, with *the factor levels entering the equation as if they were separate variables*. In a multiple regression the parameter a would be the intercept and the parameters b and c would be slopes. In an Anova this does not make any sense, because *all* the w_1 are 'male' and *all* the w_2 are 'female'. What you need to understand is that when S-Plus knows that w is categorical, it replaces the value for w_1 in the equation with a 1 when the data point refers to a male, and with a 0 when it does not. Likewise for w_2. This is replaced by a 1 when the data point refers to a female and by 0 otherwise. So now the equation can take one of only two forms:

$$y = a + bw_1 + c \times 0 = a + bw_1 = a + b \text{ for males}$$

because $w_1 = 1$ and $w_2 = 0$ in this case, and

$$y = a + b \times 0 + cw_2 = a + cw_2 = a + c \text{ for females}$$

because $w_1 = 0$ and $w_2 = 1$ in this case. It might now be more obvious what the parameters b and c represent. Suppose for the sake of argument that parameter a was the overall mean for males and females lumped together. Now if we want y to represent the mean for the males (as we do) then b must be *the difference between the means* of the males and the overall mean. Similarly, we want y for the females to represent the mean response of females, the parameter c must be *the difference between the means* of the females and the overall mean.

That's all there is to it. In a regression the intercept is the intercept and the other parameter is a slope. In an Anova *the intercept is a mean, and the other parameters are differences between means.*

Model formulas for analysis of covariance

Often our models contain some explanatory variables that are continuous (e.g. age) and others that are categorical (e.g. sex). Models that have mixtures of continuous and categorical explanatory variables are called analysis of covariance (Ancova; see Chapter 16). What does the equation look like in this case? The simplest way of showing the equations of an Ancova is to have a separate regression equation for each factor level. So in our example we would have one equation for the males and another equation for the females:

$$y = a_1 + b_1 x \quad \text{for the males}$$

$$y = a_2 + b_2 x \quad \text{for the females}$$

where x is age, subscript 1 refers to males and subscript 2 refers to females. The model has four parameters: two slopes (b_1 and b_2) and two intercepts (a_1 and a_2). The model formula for this is

$$y \sim w * x$$

where w is a two-level factor representing sex. When one of the variables is continuous (x) and the other is categorical (w), the asterisk $*$ is understood by S-Plus to mean 'fit a separate regression slope for each sex'. The four parameters, however, are presented in a rather different way than in our original equation. There are several options, but the simplest is to have the first parameter as an intercept, the second as a slope, the third as a difference between two intercepts, and the fourth as a difference between two slopes. This is hard to understand, but is explained in detail in Chapter 18.

A simpler model might have the same common slope for the two sexes, but two different intercepts (a three-parameter mode). The model formula for this is

$$y \sim w + x$$

The difference is that the asterisk is replaced by a plus sign. When one of two variables in a model formula is categorical (like w here) and the other one is continuous (like x here), then S-Plus interprets the plus sign as meaning 'fit a common slope, but estimate different intercepts for each level of the categorical variable'. So the equation for this model is

$$y = a_1 + bx \quad \text{for the males}$$

$$y = a_2 + bx \quad \text{for the females}$$

with three parameters (a_1, a_2 and the common slope b). There are several ways to present the parameter estimates, but the simplest is to have the first parameter as an intercept, the second as a difference between two intercepts and the third as the common slope.

Table 13.1 Examples of S-Plus model formulas

Model	Model formula	Comments
Null	y ~ 1	1 is the intercept in regression models, but here it is the overall mean y
Regression	y ~ x	x is a continuous explanatory variable
One-way Anova	y ~ sex	sex is a two-level categorical variable
Two-way Anova	y ~ sex + genotype	genotype is a four-level categorical variable
Factorial Anova	y ~ N * P * K	N, P and K are two-level factors to be fitted along with all their interactions
Three-way Anova	y ~ N*P*K - N:P:K	As above, but don't fit the three-way interaction
Ancova	y ~ x + sex	a common slope for y against x but with two intercepts, one for each sex
Ancova	y ~ x * sex	two slopes and two intercepts
Nested Anova	y ~ a/b/c	factor c nested within factor b within factor a
Split-plot Anova	y ~ a*b*c+Error(a/b)	a factorial experiment but with three plot sizes and three different error variances, one for each plot size
Multiple regression	y ~ x + z	two continuous explanatory variables, flat surface fit
Multiple regression	y ~ x * z	fit an interaction term as well (x + z + x:z)
Multiple regression	y ~ x + I(x ^ 2) + z + I(z ^ 2)	fit a quadratic term for both x and z
Multiple regression	y <- poly(x,2) + z	fit a quadratic polynomial for x and linear z
Multiple regression	y ~ (x + z + w) ^ 2	fit three variables plus all their two-way interactions
Non-parametric model	y ~ s(x) +lo(z)	y is a function of smoothed x and loess z
Transformed response and explanatory variables	log(y) ~ I(1/x) + sqrt(z)	all three variables are transformed in the model

The main kinds of model formulas are shown in Table 13.1. Formulas such as these appear in two main places:

- in model-fitting functions like **lm** (y~x), **aov** (y~x), **gam** (y~s(x)) or **tree** (y~x+z)

- in plot functions like **xyplot** (y~x | z), **coplot** (y~x | z) or **wireframe** (y~x*z)

Fitting statistical models in S-Plus

Models are fitted using one of the model-fitting functions:

- **lm** fits a linear model with normal errors and constant variance; generally this is used for regression analysis using continuous explanatory variables

- **aov** fits analysis of variance with Normal errors, constant variance and the identity link; generally used for categorical explanatory variables or Ancovas with a mix of categorical and continuous explanatory variables

- **glm** fits generalised linear models to data using categorical or continuous explanatory variables, by specifying one of a family of **error structures** (e.g. Poisson for count data or binomial for proportion data) and a particular **link** function.

- **gam** fits generalised additive models to data with one of a family of error structures (e.g. Poisson for count data or binomial for proportion data) in which the continuous explanatory variables can (optionally) be fitted as arbitrary smoothed functions using non-parametric smoothers rather than specific parametric functions

- **lme** fits linear mixed effects models with specified mixtures of fixed effects and random effects and allows for the specification of correlation structure among the explanatory variables and autocorrelation of the response variable (e.g. time series effects with repeated measures)

- **nls** fits a non-linear regression model via least squares, estimating the parameters of a specified non-linear function

- **nlme** fits a specified non-linear function in a mixed effects model where the parameters of the non-linear function are assumed to be random effects; allows for the specification of correlation structure among the explanatory variables and autocorrelation of the response variable (e.g. time series effects with repeated measures)

- **loess** fits a local regression model with one or more continuous explanatory variables using non-parametric techniques to produce a smoothed model surface

- **tree** fits a regression tree model using binary recursive partitioning whereby the data are successively split along coordinate axes of the explanatory variables so that at any node the split is chosen that maximally distinguishes the response variable in the left and the right branches. With a categorical response variable, the tree is called a classification tree, and the model used for classification assumes that the response variable follows a multinomial distribution

For most of these models, a range of **generic functions** can be used to obtain information about the model. Here are the most important and the most frequently used:

- **summary** produces parameter estimates and standard errors from **lm**, and Anova tables from **aov**; this will often determine your choice between **lm** and **aov**. For either **lm** or **aov** you can choose **summary.aov** or **summary.lm** to get the alternative form of output (an Anova table or a table of parameter estimates and standard errors; see p. 287)

- **plot** produces diagnostic plots for model checking, including residuals against fitted values, influence tests, etc.

- **anova** is a wonderfully useful function for comparing different models and producing Anova tables

- **update** is used to modify the last model fit; it saves both typing effort and computing time

Other useful generics include:

- **coef**: the coefficients (estimated parameters) from the model

- **fitted**: the fitted values, predicted by the model for the values of the explanatory variables included

- **resid**: the residuals (the differences between measured and predicted values of y)

- **predict**: uses information from the fitted model to produce smooth functions for plotting a line through the scatterplot of your data

Further reading

Bruce, A. and Gao, H.-Y. (1996) *Applied Wavelet Analysis with S-Plus*. New York, Springer-Verlag.

Chambers, J.M. and Hastie, T.J. (1992) *Statistical Models in S*. Pacific Grove, California, Wadsworth and Brooks Cole.

Everitt, B.S. (1994) *Handbook of Statistical Analyses Using S-PLUS*. New York, Chapman & Hall / CRC.

Krause, A. and Olson, M. (2000) *The Basics of S and S-Plus*. New York, Springer-Verlag.

Millard, S.P. and Krause, A. (2001) *Using S-PLUS in the Pharmaceutical Industry*. New York, Springer-Verlag.

Venables, W.N. and Ripley, B.D. (1997) *Modern Applied Statistics with S-Plus*. New York, Springer-Verlag.

14

Regression

Regression is the statistical model we use when the explanatory variable is continuous. If the explanatory variables were categorical we would use analysis of variance (Chapter 15). If you are in any doubt about whether to use regression or analysis of variance, ask yourself whether your graphical investigation of the data involved producing scatter plots or bar charts. If you produced scatterplots, then the statistics you need to use are regression. If you produced bar charts, then you need analysis of variance.

The objects of regression are twofold:

- to estimate the parameters of a model
- to estimate how good the model is at describing the data

Specifying the model

One of the most important (and often neglected) parts of the exercise is selecting which of the many possible models we should fit to the data. The principle of **parsimony** tells us that we should fit the simplest possible model. In practice this means that we should fit the model with the smallest possible number of parameters. The **null model** is that there is no relationship between y and x, i.e. that y is a constant:

$$y = a$$

If there is a relationship between y and x, then the next simplest assumption is that the relationship between y and x is linear:

$$y = a + bx$$

where a **response variable** y is hypothesised as being a linear function of the **explanatory variable** x, and the two **parameters** a and b. In the case of simple linear regression, the parameter a is called the *intercept* (the value of y when $x = 0$), and b is the *slope* of the line (or the gradient, measured as the change in y in response to unit change in x). The aims of the analysis are as follows:

- to estimate the values of the parameters a and b
- to estimate their standard errors
- to use the standard errors to assess which terms are necessary within the model (i.e. whether the parameter values are significantly different from zero)
- to determine what fraction of the variation in y is explained by the model and how much remains unexplained

Data inspection

The first step is to look carefully at the data. Is there an upward or downward trend, or could a horizontal straight line be fitted through the data? If there is a trend, does it look linear or curvilinear? Is the scatter of the data around the line more or less uniform, or does the scatter change systematically as x changes?

Consider the data in the data frame called regression. They show how weight gain (mg) of individual caterpillars declines as the tannin content of their diet (%) is experimentally increased:

```
attach(regression)
names(regression)
```

```
[1]  "growth"    "tannin"
```

```
plot(tannin,growth,pch=16)
```

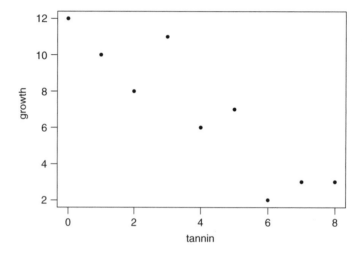

It looks as if there is a downward trend in y as x increases, and that the trend is roughly linear. There is no evidence of any systematic change in the scatter as x changes. This cursory inspection leads to several expectations:

- the intercept a is greater than zero
- the slope b is negative
- the variance in y is constant
- the scatter about the straight line is relatively slight

We can carry out a crude regression analysis by eye. The intercept a is the value of y when $x = 0$. Inspection of the graph indicates that $a \approx 12$. The slope b is the change in y per unit change in x. From the graph we can see that y declined from about 12 to about 2 (a change of -10) as x increased from 0 to 8 (a change of $+8$).

The *slope is the change in y divided by the change in x that brought it about*:

$$b = \frac{\Delta y}{\Delta x}$$

so $b \approx -10/8 \approx -1.25$. That's fine, but we need a more objective and more general means of doing this. The process is called *least squares linear regression.*

In least squares regression analysis we shall make the following assumptions:

- errors are normally distributed

- variance is constant

- the explanatory variable is measured without error

- all of the unexplained variation is confined to the response variable

Later on, we shall deal with cases that have non-normal errors and non-constant variance.

Least squares

The technique of least squares linear regression defines the *best-fit* straight line as the line which minimises *the sum of the squares of the departures of the y values from the line.* We can see what this means in graphical terms. The first step is to fit a horizontal line though the data, using **abline**, showing the average value of y:

abline(mean(growth),0)

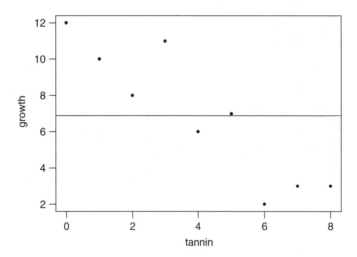

The scatter around the line defined by $\bar{y}$ is said to be the total variation in y. Each point on the graph lies a vertical distance $d = y - \bar{y}$ from the horizontal line, and we define the total variation in y as being the sum of the squares of these departures:

$$SST = \sum(y - \bar{y})^2$$

where *SST* stands for *total sum of squares.* It is the sum of the squares of the vertical distances shown here:

for (i in 1:9) lines(c(tannin[i],tannin[i]),c(growth[i],mean(growth)))

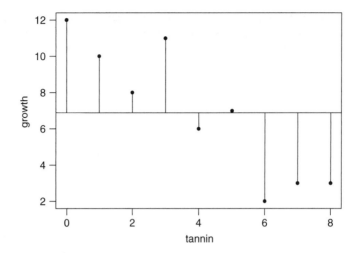

Now we want to fit a straight regression line through the data. There are two decisions to be made about such a best-fit line:

- Where should the line be located?
- What slope should it have?

```
plot(tannin,growth,pch=16)
abline(lm(growth~tannin))
```

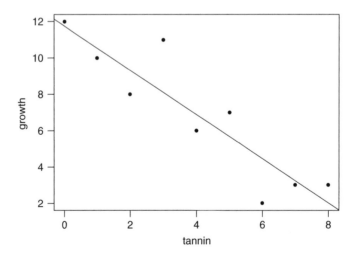

Location of the line is relatively straightforward, because a best-fit line should clearly pass through the point defined by the average values of x and y. The line can then be pivoted at the point $(\overline{x}, \overline{y})$, and rotated until the best fit is achieved. The process is formalised as follows. Each point on the graph lies a distance $e = y - \hat{y}$ from the fitted line, where the predicted value $\hat{y}$ is found by evaluating the equation of the straight line at the appropriate value of x:

$$\hat{y} = a + bx$$

```
yhat<-predict(lm(growth~tannin))
for (i in 1:9) lines(c(tannin[i],tannin[i]),c(growth[i],yhat[i]))
```

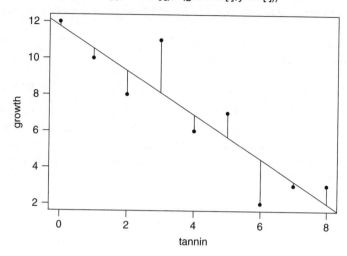

Let us define the error sum of squares as the sum of the squares of the e: this is the 'unexplained variation' in y:

$$SSE = \sum (y - \hat{y})^2$$

Now imagine rotating the line around the point $(\bar{x}, \bar{y})$. SSE will be large when the line is too steep. It will decline as the slope gets closer to the best-fit line, then it will increase again as the line becomes too shallow. We can draw a graph of SSE against the slope b, like this:

```
i<-0
sse<-numeric(101)
for (b in seq(-6,4,0.1){
    i<-i+1
    sse[i]<-sum((growth-12- b*tannin)^2)}
b<- seq(-6,4,0.1)
plot(b,sse,type="l")
```

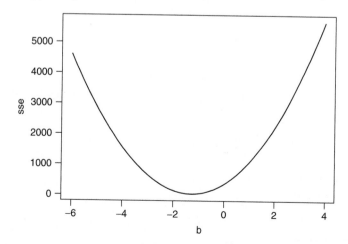

We define *the best-fit line as the one that minimises SSE*. By inspection, we can see that the maximum likelihood estimate of the slope is obviously somewhere around -1.2. To find this value analytically, we need to find the derivative of *SSE* with respect to b, set it to zero, and solve for b.

Maximum likelihood estimate of the slope

As we have said, the location of the line is fixed by assuming that it passes through the point $(\bar{x}, \bar{y})$. Thus we can rearrange the equation to obtain an estimate of the intercept a in terms of the best-fit slope b:

$$a = \bar{y} - b\bar{x}$$

We begin by replacing these average values of $\bar{y}$ and $\bar{x}$ by $\sum y/n$ and $\sum x/n$, so

$$a = \frac{\sum y}{n} - b\frac{\sum x}{n}$$

Now we change the value of the slope, b, until we minimise the error sum of squares:

$$SSE = \text{minimum} \sum (y - a - bx)^2$$

The details are presented in Box 14.1, and a pictorial impression is given on p. 225.

Box 14.1 The least squares estimate of the regression slope, b

The *best-fit* slope is found by rotating the line until the *error sum of squares, SSE*, is minimised. The error sum of squares is the sum of the squares of the individual departures, $e = y - \hat{y}$, shown above:

$$SSE = \text{minimum} \sum (y - a - bx)^2$$

noting the change in sign of bx. This is how the best fit is defined.
Next we find the derivative of *SSE* with respect to b:

$$\frac{d\,SSE}{db} = -2\sum x(y - a - bx)$$

because the derivative with respect to b of the bracketed term is $-x$, and the derivative of the squared term is 2 times the squared term. The constant -2 can be taken outside the summation. Multiplying x through the bracket gives

$$\frac{d\,SSE}{db} = -2\sum xy - ax - bx^2$$

Now apply summation to each term separately, set the derivative to zero, and divide both sides by -2 to remove the unnecessary constant:

$$\sum xy - \sum ax - \sum bx^2 = 0$$

We cannot solve the equation as it stands because there are two unknowns, a and b. However, we already know the value of a in terms of b. Also, note that $\sum ax$ can be written as $a \sum x$, so replacing a and taking both a and b outside their summations gives

$$\sum xy - \left[\frac{\sum y}{n} - b\frac{\sum x}{n} \right] \sum x - b\sum x^2 = 0$$

Now multiply out the central bracketed term by $\sum x$ to get

$$\sum xy - \frac{\sum x \sum y}{n} + b\frac{\left(\sum x\right)^2}{n} - b\sum x^2 = 0$$

Finally, take the two terms containing b to the other side, and note their change of sign:

$$\sum xy - \frac{\sum x \sum y}{n} = b\sum x^2 - b\frac{\left(\sum x\right)^2}{n}$$

and then divide both sides by $\sum x^2 - \left(\sum x\right)^2 /n$ to obtain the required estimate of b:

$$b = \frac{\sum xy - \frac{1}{n}\sum x \sum y}{\sum x^2 - \frac{1}{n}\left(\sum x\right)^2}$$

Thus the value of b that minimises the sum of squares of the departures is given simply by

$$b = \frac{SSXY}{SSX}$$

For analytical work on regression we need the values of three *corrected sums of squares*. $SSXY$ is the corrected sum of products (x times y, the measure of how x and y covary), SSX is the corrected sum of squares for x, and SST is the total sum of squares which we met earlier. For comparison, here are the three important formulas presented together:

$$SST = \sum y^2 - \frac{1}{n}\left(\sum y\right)^2$$

$$SSX = \sum x^2 - \frac{1}{n}\left(\sum x\right)^2$$

$$SSXY = \sum xy - \frac{1}{n}\left(\sum x \sum y\right)$$

Note the similarity of their structures. SST is calculated by adding up y times y then subtracting the total of the y times the total of the y divided by n (the number of points on the graph). Likewise, SSX is calculated by adding up x times x then subtracting the total of the x times the total of the x divided by n. Finally, $SSXY$ is calculated by adding up x times y then subtracting the total of the x times the total of the y divided by n. Box 14.2 gives the derivation of this short-cut formula for obtaining $SSXY$.

It is worth reiterating what these three quantities represent. *SST* measures the total variation in the *y* values about their mean (it is the sum of the squares of the *d* in the earlier plot). *SSX* represents the total variation in *x* (expressed as the sum of squares of the departures from the mean value of x), and is a measure of the range of *x* values over which the graph has been constructed. *SSXY* measures the covariation of *y* and *x* in terms of the corrected sum of products (see p. 110 for the definition of covariance). Note that *SSXY* is negative when *y* declines with increasing *x*, positive when *y* increases with *x*, and zero when *y* and *x* are uncorrelated.

Analysis of variance in regression

The idea is to partition the total variation in the response, *SST*, into two components: the component that is explained by the regression line (which we shall call *SSR*, the *regression sum of squares*) and an unexplained component (the residual variation which we shall call *SSE*, the *error sum of squares*).

Look at the relative sizes of the departures *d* and *e* in the figures we drew earlier. Now, ask yourself what would be the relative size of $\sum d^2$ and $\sum e^2$ if the slope of the fitted line were *not significantly different from zero*. A moment's thought should convince you that if the slope of the best-fit line were zero, then the two sums of squares would be exactly the same. If the slope were zero, then the two lines would lie in exactly the same place, and the sums of squares would be identical. Thus, when the slope of the line is not significantly different from zero, we would find that $SST = SSE$. Similarly, if the slope were significantly different from zero (i.e. significantly positive or negative), then *SSE* would be substantially *less* than *SST*. In the limit, if all the points fell exactly on the fitted line, then *SSE* would be zero (see Box 14.3).

Now we calculate a third quantity, *SSR*, called the *regression sum of squares*:

$$SSR = SST - SSE$$

This definition means that SSR will be large when the fitted line accounts for much of the variation in *y*, and small when there is little or no linear trend in the data. In the limit, *SSR* would be equal to *SST* if the fit were perfect (because *SSE* would then equal zero), and *SSR* would equal zero if *y* were independent of *x* (because, in this case, *SSE* would equal *SST*).

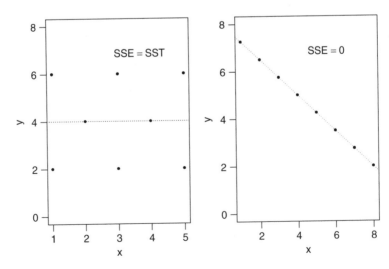

These three quantities form the basis for drawing up the Anova table.

Source	SS	d.f.	MS	F	F tables (5%)
Regression	SSR	1	SSR	$F = \dfrac{SSR}{s^2}$	qf(0.95,1,n-2)
Error	SSE	$n-2$	$s^2 = \dfrac{SSE}{n-2}$		
Total	SST	$n-1$			

The sums of squares are entered in the first column. *SST* is the corrected sum of squares of *y*, as given above. We have defined *SSE* already, but this formula is inconvenient for calculation, since it involves *n* different estimates of $\hat{y}$ and *n* subtractions. Instead, it is easier to calculate *SSR* and then to estimate *SSE* by difference. This is because the regression sum of squares is given simply by

$$SSR = b \times SSXY$$

so that

$$SSE = SST - SSR$$

So what, exactly, is the regression sum of squares? A visual impression of *SSR* is obtained as follows. It is the sum of the squares of the distances between the predicted values, $\hat{y}$, and the overall mean value, $\overline{y}$:

$$SSR = \sum (\hat{y} - \overline{y})^2$$

The regression sum of squares is the sum of the squares of the distances shown below by the vertical lines.

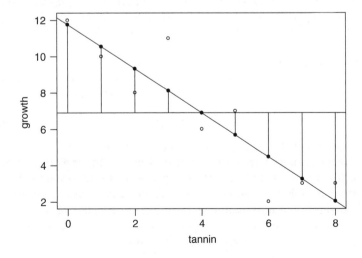

The fitted values are the solid circles and the data are the open circles. The code to produce this figure is

```
plot(tannin, growth,type="n")
abline(mean(growth),0)
model<-lm(growth~tannin)
abline(model)
for (i in 1:9) lines(c(tannin[i],tannin[i]),c(mean(growth),predict(model)[i]))
points(tannin,predict(model),pch=16)
points(tannin,growth)
```

The second column of the Anova table contains the degrees of freedom. The estimation of the total sum of squares required that one parameter $\bar{y}$, the mean value of y, be estimated from the data prior to calculation. Thus the total sum of squares has $n - 1$ degrees of freedom when there are n points on the graph. The error sum of squares could not be estimated until the regression line had been drawn through the data. This required that two parameters, the mean value of y and the slope of the line, were estimated from the data. Thus, the error sum of squares has $n - 2$ degrees of freedom. The regression degrees of freedom represents *the number of extra parameters* involved in going from the null model $y = \bar{y}$ to the full model $y = a + bx$. (i.e. $2 - 1 = 1$ degree of freedom because we have added the slope, b, to the model). As always, the component d.f. add up to the total d.f.

At this stage it is worth recalling that variance is *always* calculated as

$$\text{variance} = \frac{\text{sum of squares}}{\text{degrees of freedom}}$$

The Anova table is structured to make the calculation of variances as simple and clear as possible. For each row in an Anova table, you simply divide the sum of squares by the adjacent degrees of freedom. The third column therefore contains two variances; the first row shows the regression variance and the second row shows the all-important error variance (s^2). One of the oddities of analysis of variance is that the variances are referred to as *mean squares* (this is because a variance is defined as a mean squared deviation). The error variance s^2, also known as the *error mean square* (*MSE*), is the quantity used in calculating standard errors and confidence intervals for the parameters, and in carrying out hypothesis testing.

The **standard error of the regression slope**, b, is given by

$$SE_b = \sqrt{\frac{s^2}{SSX}}$$

Recall that standard errors are *unreliability estimates*. Unreliability increases with the error variance so it makes sense to have s^2 in the numerator (on top of the division). It is less obvious why unreliability should depend on the range of x values. Look at these two graphs that have exactly the same slopes and intercepts. The difference is that the left-hand graph has all of its x values close to the mean value of x, while the graph on the right has a broad span of x values. Which of these do you think would give the most reliable estimate of the slope?

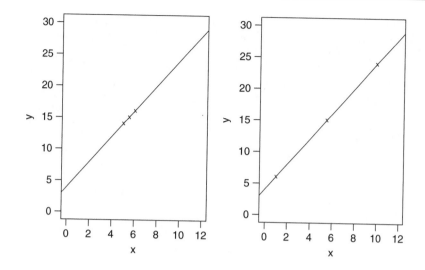

It is pretty clear that it is the graph on the right, with the wider range of x values. Increasing the spread of the x values reduces unreliability of the estimated slope and therefore appears in the denominator (on the bottom of the equation).

What is the purpose of the big square root term? It is there to make sure the units of the unreliability estimate are the same as the units of the parameter whose unreliability is being assessed. The error variance is in units of y *squared*, but the slope is in units of y *per unit change in* x.

A 95% confidence interval for b is now given in the usual way:

$$CI_b = t \text{ (from tables, } \alpha = 0.025, \text{ d.f.} = n - 2) \times SE_b$$

where t is obtained using **qt**, the quantiles of Student's t distribution (two-tailed, $\alpha = 0.025$) with $n - 2 = 7$ degrees of freedom:

qt(0.975,7)

[1] 2.364624

This is a little larger than our rule of thumb ($t \approx 2$) because the degrees of freedom are so small. An important significance test concerns the question of whether or not the slope of the best-fit line is significantly different from zero. If it is not, then the principal of parsimony suggests that the line should be assumed to be horizontal. If this were the case, then y would be independent of x (it would be a constant, $y = a$).

The **standard error of the intercept**, a, is given by

$$SE_a = \sqrt{\frac{s^2 \sum x^2}{n \times SSX}}$$

which is like the formula for the standard error of the slope, but with two additional terms. Uncertainly declines with increasing sample size n. It is less clear why uncertainty should increase with $\sum x^2$. The reason for this is that uncertainly in the estimate of the intercept increases, the further away from the intercept that the mean value of x lies. You can see this from the following graphs. On the left is a graph with a low value of $\bar{x}$ and on the right an identical graph (same slope and intercept) but estimated from a data set with a higher value of $\bar{x}$. In both cases there is a 25% variation in the slope. Compare the difference in the prediction of the intercept in the two cases.

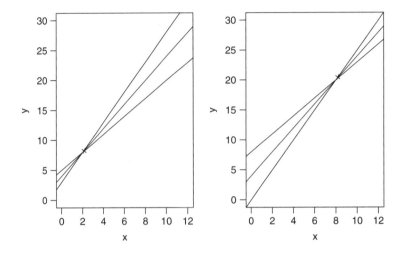

Confidence in predictions made with linear regression declines with the square of the distance between the mean value of x and the value at which the prediction is to be made, i.e. with $(x - \bar{x})^2$. Thus, when the origin of the graph is a long way from the mean value of x, the standard error of the intercept will be large, and vice versa. A 95% confidence interval for the intercept is therefore

$$CI_a = t(\text{from tables}, \alpha = 0.025, \text{ d.f.} = n - 2) \times SE_a$$

with the same two-tailed t value as before. In general, the **standard error for a predicted value** $\hat{y}$ is given by

$$SE_{\hat{y}} = \sqrt{s^2 \left[\frac{1}{n} + \frac{(x - \bar{x})^2}{SSX} \right]}$$

Note that the formula for the standard error of the intercept is just the special case of this for $x = 0$ (you should check the algebra of this result as an exercise).

Calculations involved in linear regression

The data are in a data frame called regression; the data frame contains the response variable growth and the continuous explanatory variable tannin. You see the nine values of each by typing the name of the data frame:

```
attach(regression)
names(regression)

[1] "growth" "tannin"

regression

   growth tannin
1      12      0
2      10      1
```

3	8	2
4	11	3
5	6	4
6	7	5
7	2	6
8	3	7
9	3	8

The numbers are mean weight gain (mg) of individual caterpillars and tannin content of their diet (%). You will find it useful to work through the calculations longhand as we go. The first step is to compute the 'famous five': $\sum x = 36$, $\sum x^2 = 204$, $\sum y = 62$, $\sum y^2 = 536$ and $\sum xy = 175$. Note that $\sum xy$ is $0 \times 12 + 1 \times 10 + 2 \times 8 + \cdots + 8 \times 3 = 175$. Note the use of the semicolon ; to separate two functions:

sum(tannin);sum(tannin^2)

[1] 36
[1] 204

sum(growth);sum(growth^2)

[1] 62
[1] 536

sum(tannin*growth)

[1] 175

Now calculate the three corrected sums of squares, using the formulas given earlier:

$$SST = 536 - \frac{62^2}{9} = 108.889$$

$$SSX = 204 - \frac{36^2}{9} = 60$$

$$SSXY = 175 - \frac{36 \times 62}{9} = -73$$

Then the slope of the best-fit line is simply

$$b = \frac{SSXY}{SSX} = \frac{-73}{60} = -1.21666$$

and the intercept is

$$a = \bar{y} - b\bar{x} = \frac{62}{9} + 1.21666\frac{36}{9} = 11.755$$

The regression sum of squares is

$$SSR = b \times SSXY = -1.21666 \times -73 = 88.82$$

and the error sum of squares can be found by subtraction:

$$SSE = SST - SSR = 108.89 - 88.82 = 20.07$$

Now we can complete the Anova table.

Source	SS	d.f.	MS	F	Probability
Regression	88.82	1	88.820	30.98	0.0008
Error	20.07	7	2.867		
Total	108.89	8			

The calculated F ratio of 30.98 is much larger than the 5% value in tables with 1 degree of freedom in the numerator and 7 degrees of freedom in the denominator. To find the expected F value from tables we use **qf** (quantiles of the F distribution):

qf(0.95,1,7)

[1] 5.591448

We can ask the question the other way round. What is the probability of getting an F value of 30.98 if the null hypothesis of $b = 0$ is true ? We use 1-pf for this:

1-pf(30.98,1,7)

[1] 0.0008455934

So we can unequivocally reject the null hypothesis that the slope of the line is zero. Increasing dietary tannin leads to significantly reduced weight gain for these caterpillars. We can now calculate the standard errors of the slope and intercept:

$$SE_b = \sqrt{\frac{2.867}{60}} = 0.2186$$

$$SE_a = \sqrt{\frac{2.867 \times 204}{9 \times 60}} = 1.041$$

To compute the 95% confidence intervals, we need the two-tailed value of Student's t with 7 degrees of freedom:

qt(.975,7)

[1] 2.364624

So the 95% confidence intervals for the slope and intercept are given by

$$a = 11.756 \pm 2.463$$
$$b = -1.216\,66 \pm 0.517$$

In written work we would put something like this:

'The intercept was 11.76 ± 2.46 (95% CI, $n = 9$)'.

The reader needs to know that the interval is a 95% CI (and not say a 99% interval or a single standard error) and that the sample size was 9. Then they can form their own judgement about the parameter estimate and its unreliability.

Degree of scatter

Knowing the slope and the intercept tells only part of the story. The two graphs below have exactly the same slope and intercept, but they are completely different from one another in their degree of scatter.

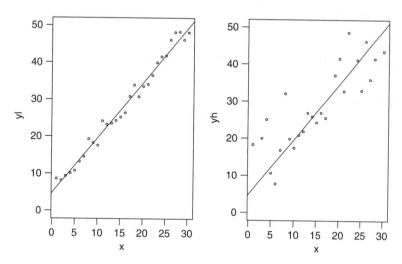

It is obvious that we need to be able to quantify the degree of scatter, so that we can distinguish cases like this. In the limit, the fit could be perfect, in which case all the points fall exactly on the regression line. SSE would be zero and $SSR = SST$. In the opposite case, there is absolutely no relationship between y and x, so $SSR = 0$, $SSE = SST$, and the slope of the regression is $b = 0$ (see p. 228). Formulated in this way, it becomes clear that we could use some of the quantities already calculated to derive an estimate of scatter. We want our measure to vary from 1.0 when the fit is perfect, to zero when there is no fit at all. Now recall that the total variation in y is measured by SST. We can rephrase our question to ask, What fraction of SST is explained by the regression line? A measure of scatter with the properties we want is given by the ratio of SSR to SST. This ratio is called the *coefficient of determination* and is denoted by r^2:

$$r^2 = \frac{SSR}{SST}$$

Its square root, r, is the familiar *correlation coefficient* (see p. 112). Note that because r^2 is always positive, it is better to calculate the correlation coefficient from

$$r = \frac{SSXY}{\sqrt{SSX \times SST}}$$

You should check that these definitions of r and r^2 are consistent.

Using S-Plus for regression

The procedure so far should have been familiar. Let us now use S-Plus to carry out the same regression analysis. At this first introduction to the use of S-Plus, each step will be explained in detail. As we progress, we shall take progressively more of the procedures for granted. There are six steps involved in the analysis of a linear model:

- get to know the data

- suggest a suitable model

- fit the model to the data

- subject the model to criticism

- analyse for influential points

- simplify the model to its bare essentials

With regression data, the first step involves the visual inspection of plots of the response variable against the explanatory variable. We need answers to the following questions:

- is there a trend in the data?

- what is the slope of the trend (positive or negative)?

- is the trend linear or curved?

- is there any pattern to the scatter around the trend?

We have already attached the data frame called **regression** (p. 232). The scatterplot is obtained like this:

    plot(tannin,growth)

It appears that a linear model would be a sensible first approximation. There are no massive outliers in the data, and there is no obvious trend in the variance of y with increasing x. Note that in the **plot** function the arguments are in the order x then y. The fitted line is added using **abline** like this:

    abline(lm(growth~tannin))

where lm means linear model, ~ is pronounced 'tilde', and the order of the variables is $y \sim x$ (in contrast to the **plot** function).

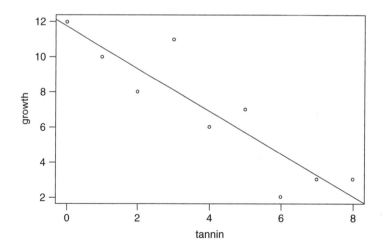

Statistical modelling of regression proceeds as follows. We pick a name, say model, for the regression object, then choose a fitting procedure. We could use any one of several but let's keep it simple and use **lm**, the linear model that we have already used in **abline**. All we do it this:

model<-lm(growth~tannin)

which is read 'model gets a linear model in which growth is modelled as a function of tannin'. Now we can do lots of different things with the object called model. The first thing we might do is summarise it:

summary(model)

```
Call: lm(formula = growth ~ tannin)
Residuals:
     Min        1Q   Median       3Q     Max
  -2.456  -0.8889  -0.2389  0.9778  2.894

Coefficients:
                  Value   Std. Error   t value  Pr(>|t|)
(Intercept)    11.7556       1.0408    11.2947    0.0000
     tannin    -1.2167       0.2186    -5.5654    0.0008

Residual standard error: 1.693 on 7 degrees of freedom
Multiple R-Squared: 0.8157
F-statistic: 30.97 on 1 and 7 degrees of freedom, the p-
value is 0.0008461

Correlation of Coefficients:
        (Intercept)
tannin -0.8402
```

The first part of the output shows the Call (the model you fitted). This is useful as a label for the output when you come back to it weeks or months later. Next is a summary of the residuals, showing the largest and smallest as well as the 25% (1Q) and 75% (3Q) quantiles in which the central 50% of the residuals lie. Next the parameter estimates (labelled Coefficients) are printed, along with their standard errors just as we calculated them by hand. The t values and p values are for tests of the null hypotheses that the intercept and slope are equal to zero (against two-tailed alternatives that they are not equal to zero). Residual standard error is 1.693 on 7 degrees of freedom; this is the square root of the error variance from our Anova table.

Multiple R-Squared 0.8157 is the ratio SSR/SST we calculated earlier. F-statistic: 30.97 on 1 and 7 degrees of freedom, and the p value is 0.000 8461; these are the last two columns of our Anova table. The final element of the printout shows the correlation between the two parameters (Intercept and tannin) to be -0.8402; this shows how change in one parameter would affect the best-fit estimate of the other parameter. Here, for instance, the bigger the intercept, the more strongly negative the slope would be.

Next we might **plot** the object called model. This produces a useful series of *diagnostic plots*:

plot(model)

The first graph shows the residuals against the fitted values. It is good news because it shows no evidence of the variance increasing for larger values of $\hat{y}$, and shows no evidence of curvature.

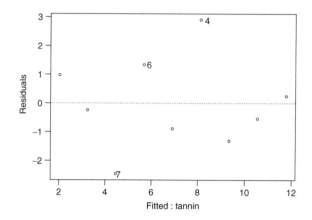

The next plot shows the ordered residuals plotted against the quantiles of the standard Normal distribution. If, as we hope, our errors really are Normal, then this plot should be linear.

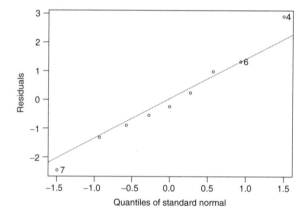

The data are reasonably well behaved, with only point number 4 having a larger residual than expected. The scale location plot is very like the first plot but shows the square root of the standardised residuals against the fitted values (useful for detecting non-constant variance).

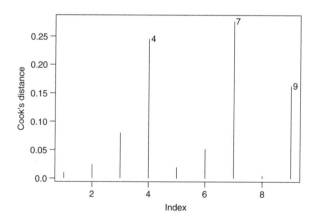

The last plot shows Cook's distance. This is an *influence measure* that shows which of the outlying values might be expected to have the largest effects on the estimated parameter values. This shows that it was point number 7 that had the greatest influence on the parameter estimates. It is often a good idea to repeat the modelling exercise with the most influential points omitted in order to assess the magnitude of their impact on the structure of the model. Note the use of negative subscripts to remove the data point in question [-7]:

```
model2<-lm(growth[-7]~tannin[-7])
summary(model2)
```

```
Coefficients:
                 Value  Std. Error   t value   Pr(>|t|)
(Intercept)    11.6892    0.8963     13.0417    0.0000
  tannin[-7]    -1.1171    0.1956     -5.7116    0.0012
Residual standard error: 1.457 on 6 degrees of freedom
Multiple R-Squared: 0.8446
F-statistic: 32.62 on 1 and 6 degrees of freedom, the p-value
is 0.001247
```

This shows that although data point number 7 was the most influential, that influence was not particularly great. Leaving it out has reduced the standard errors slightly, and altered the parameter estimates a little bit. We have no good reason to eliminate point number 7, so we revert to the original model which included all the data.

Suppose we want to know the predicted value $\hat{y}$ at tannin $= 5.5\%$. We could write out the equation in calculator mode, using the parameter estimates from the summary table, like this:

```
11.7556-1.2167*5.5
```

```
[1] 5.06375
```

Alternatively, we could use the **predict** function with the object called model. We provide the value for tannin=5.5 in a **list** in order to protect the vector of x values (called tannin, of course) from being adulterated. The name of the x values to be used for prediction (tannin in this case) must be *exactly* the same as the name of the explanatory variable in the model (see below for a full discussion of this much misunderstood point):

```
predict(model, list(tannin=5.5))
```

```
5.063889
```

The difference in the fourth decimal place is caused by the rounding error introduced by our hand calculation (we only used four decimal places).

For linear plots we use **abline** for superimposing the model on a scatterplot of the data points (see above). For curved responses it is sensible to use the **predict** function to generate the lines, by supplying it with a suitably fine-scale vector of x values; 60 or more x points produce a smooth line in most cases (we shall do this in Chapter 22).

Summary

The steps in the regression analysis were:

- data inspection
- model specification

- model fitting

- model criticism

We conclude that the present example is well described by a linear model with normally distributed errors. There are some large residuals, but no obvious patterns in the residuals that might suggest any systematic inadequacy of the model. The slope of the regression line is highly significantly different from zero, and we can be confident that, for the caterpillars in question, increasing dietary tannin reduces weight gain and that, over the range of tannin concentrations considered, the relationship is reasonably linear. Whether the model could be used accurately for predicting what would happen with much higher concentrations of tannin, would need to be tested by further experimentation. It is extremely unlikely to remain linear, however, as this would soon begin to predict substantial weight losses (negative gains), and these could not be sustained for long.

Overview

- regression involves estimating parameter values from data

- there are always lots of different possible models to describe a given data set

- model choice is a big deal

- always use diagnostic plots to check the adequacy of your model after fitting

- the two big problems are non-constant variance and non-normal errors

---

Box 14.2 The short-cut formula for the sum of products

We have seen how the short-cut formulas for the sums of squares of X and Y (SSX and SST) are calculated (p. 97). Here we are interested in the corrected sum of products, $SSXY$. This is based on the expectation of the product $(x - \bar{x})(y - \bar{y})$. Start by multiplying out the brackets:

$$(x - \bar{x})(y - \bar{y}) = xy - x\bar{y} - y\bar{x} + \bar{x}\,\bar{y}$$

Now apply the summation remembering that $\sum \bar{y} = n\bar{y}$ and $\sum x\bar{y} = \bar{y} \sum x$

$$\sum xy - \bar{y} \sum x - \bar{x} \sum y + n\bar{x}\,\bar{y} = \sum xy - n\bar{y}\,\bar{x} - n\bar{x}\,\bar{y} + n\bar{x}\,\bar{y} = \sum xy - n\bar{x}\,\bar{y}$$

because $\sum x = n\bar{x}$ and $\sum y = n\bar{y}$ (replacing the summation signs rather than the means). So replacing the two means by their summation definitions:

$$\sum xy - n \frac{\sum x}{n} \frac{\sum y}{n}$$

which, on cancelling n, gives the corrected sum of products as

$$SSXY = \sum xy - \frac{\sum x \sum y}{n}$$

You should check how $SSXY$ is related to the covariance of x and y (see p. 112).

Box 14.3 Demonstration that $SST = SSR + SSE$

We begin with the formula that relates the data y to the model $\hat{y}$ where d is the departure of the model from the data:

$$y = \hat{y} + d \quad \text{so} \quad y^2 = \hat{y}^2 + 2d\hat{y} + d^2$$

Applying summations gives

$$\sum y^2 = \sum \hat{y}^2 + \sum d^2 + 2 \sum d\hat{y}$$

The problem can now be recast in terms of a demonstration that $\sum d\hat{y} = 0$, because if it does then we have our result that $\sum y^2 = \sum \hat{y}^2 + \sum d^2$ since $SST = \sum y^2$, $SSR = \sum \hat{y}^2$ and $SSE = \sum d^2$. Start by writing out the deviation in full:

$$d = y - \hat{y}$$

We can ignore the constant, a, so

$$\hat{y} = bx$$

and the product $d\hat{y}$ can be written

$$d\hat{y} = (y - bx) \times bx$$

Now apply the summation:

$$\sum d\hat{y} = \sum bx(y - bx) = b \sum xy - b^2 \sum x^2$$

Next replace one of the b in b^2 by $\sum xy / \sum x^2$ (see p. 227) to get

$$\sum d\hat{y} = b \sum xy - b \frac{\sum xy}{\sum x^2} \sum x^2 = b \sum xy - b \sum xy = 0$$

because the $\sum x^2$ terms cancel. This proves that

$$SST = SSR + SSE$$

$$\sum (y - \overline{y})^2 = \sum (\hat{y} - \overline{y})^2 + \sum (y - \hat{y})^2$$

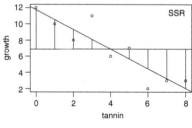

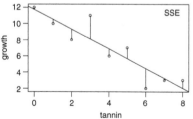

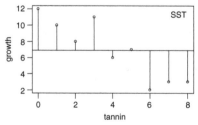

Further reading

Caroll, R.J. and Ruppert, D. (1988) *Transformation and Weighting in Regression*. New York, Chapman & Hall.

Cook, R.D. and Weisberg, S. (1982) *Residuals and Influence in Regression*. New York, Chapman & Hall.

Draper, N.R. and Smith, H. (1981) *Applied Regression Analysis*. New York, John Wiley.

Mosteller, F. and Tukey, J.W. (1977) *Data Analysis and Regression*. Reading, Mass., Addison-Wesley.

Neter, J., Wasserman, W. *et al.* (1985) *Applied Linear Regression Models*. Homewood, Illinois, Irwin.

15

Analysis of variance

Instead of fitting continuous, measured explanatory variables to data (as in regression), many experiments involve exposing experimental material to a range of discrete *levels* of one or more categorical variables known as *factors*. Thus, a factor might be a drug treatment for a particular cancer, with five levels corresponding to a placebo plus four new pharmaceuticals. Alternatively, a factor might be mineral fertiliser, where the four levels represented four different mixtures of nitrogen, phosphorus and potassium. Factors are often used in experimental designs to represent statistical *blocks*; these are internally homogeneous units in which each of the experimental treatments is repeated. Blocks may be different hospitals in a medical study, different fields in an agricultural trial, different genotypes in a plant physiology experiment, or different factories in a comparison of industrial processes.

It is important to understand that regression and Anova are identical approaches except for the nature of the explanatory variables. For example, it is a small step from having three levels of a shade factor (e.g. light, medium and heavy shade cloths) and carrying out a one-way analysis of variance, to measuring the light intensity in the three treatments and carrying out a regression with light intensity as the continuous explanatory variable. As we shall see, some experiments combine regression and analysis of variance by fitting a series of regression lines, one for each of the levels of a given factor; this is called analysis of covariance (Chapter 16).

If you want to compare the means of two treatments you do a *t* test (Chapter 11). If you want to compare three or more means, you do an Anova (analysis of variance). The fundamental question in Anova is whether or not there are any differences between these means that are worth attempting to explain. It is important, therefore, to understand exactly what the null hypothesis is when we carry out Anova. The null hypothesis typically assumes that nothing is happening; in the context of Anova, therefore, the null hypothesis, H_0, is that *all the means are the same*. What would it mean to reject such a null hypothesis? A moment's thought will show that H_0 will be false, and should be rejected, when *at least one mean is significantly different from the others*. In fact, lots of different circumstances would lead to us reject H_0, and the rejection of H_0 is more often the starting point of the analysis than the endpoint.

Statistical background

It is not at all obvious how you can analyse differences between means by looking at variances. But this is what analysis of variance is all about. An example should help to make clear how this works. To keep things simple, suppose we have just two levels of a single factor. We plot the data in the order in which they were measured: first for the first level and then for the second level. This graph shows the total sum of squares.

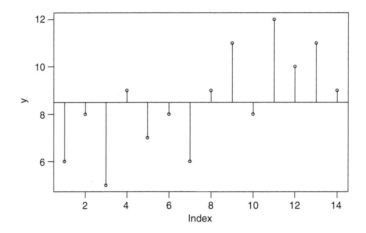

It was produced like this:

```
attach(anova.data)
plot(y)
abline(mean(y),0)
for(i in 1:14) lines(c(i,i), c(mean(y),y[i]))
```

This shows the variation in the data set as a whole. We define the total sum of squares, *SST*, as the total variation of *y* about the overall mean value $\bar{\bar{y}}$. It is calculated as the sum of the squares of the differences between each *y* value and the overall mean:

$$SST = \sum (y - \bar{\bar{y}})^2$$

Next we can fit each of the separate treatment means, $\bar{y}_A$ and $\bar{y}_B$, and calculate the sum of squares of the differences between each *y* value and its own treatment mean. We call this *SSE*, the error sum of squares, and calculate it like this:

$$SSE = \sum (y_A - \bar{y}_A)^2 + \sum (y_B - \bar{y}_B)^2$$

On the graph, the differences from which *SSE* is calculated look like this.

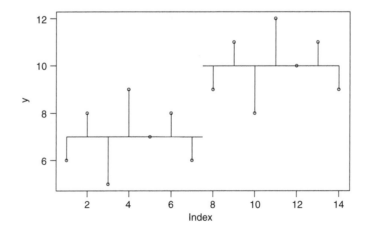

The graph was produced like this:

```
plot(y)
lines(c(1,7.5),c(means[1],means[1]))
lines(c(7.5,14),c(means[2],means[2]))
for (i in 1:7) lines(c(i,i), c(means[1],y[i]))
for (i in 8:14) lines(c(i,i), c(means[2],y[i]))
```

Now ask yourself, if the two means were the same, what would be the relationship between SSE and SST? After a moment's thought you should have been able to convince yourself that if the means *are* the same, then SSE is the same as SST, because the two horizontal lines in the last plot would be in the same position as the single line in the earlier plot. Now what if the means were significantly different from one another? What would be the relationship between SSE and SST in this case? Which would be the larger? Again, it should not take long for you to see that if the means *are* different, then SSE will be *less* than SST. Indeed, in the limit, SSE could be zero if the replicates from each treatment fell exactly on their respective means. This is how analysis of variance works. It's amazing but true. *You can make inferences about differences between means by looking at variances* (well, at sums of squares actually, but more of that later).

We can calculate the difference between SST and SSE, and use this as a measure of the difference between the treatment means; this is traditionally called the *treatment sum of squares*, and is denoted by SSA:

$$SSA = SST - SSE$$

In fact, SSA is the sum of the squares of the differences between the overall mean and the individual treatment means. That is to say, the sum of the squares of the differences between the fitted values, $\hat{y}$, and the overall mean, $\bar{\bar{y}}$, just as in regression (see p. 229).

$$SSA = \sum(\hat{y} - \bar{\bar{y}})^2$$

The distances for computing SSA look like this, with the response variable overlaid as open circles, and the fitted values as solid circles.

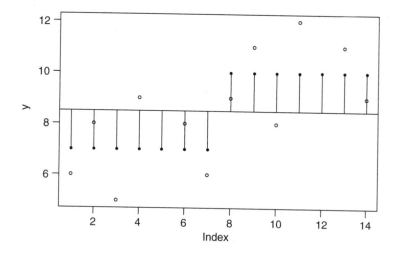

The graph was produced like this:

```
model<-lm(y~f)
plot(y)
abline(mean(y),0)
points(predict(model),pch=16)
for (i in 1:14) lines(c(i,i), c(mean(y),predict(model)[i]))
```

The technique we are interested in, however, is analysis of variance, not analysis of sums of squares. We convert the sums of squares into variances by dividing by their degrees of freedom. In our example there are two levels of the factor and so there is $2 - 1 = 1$ degree of freedom for SSA. In general, we might have k levels of any factor and hence $k - 1$ d.f. for treatments. If each factor level were replicated n times, then there would be $n - 1$ d.f. for error within each level (we lose one degree of freedom for each individual treatment mean that is estimated from the data). Since there are k levels, there would be $k(n - 1)$ d.f. for error in the whole experiment. The total number of numbers in the whole experiment is kn, so total d.f. is $kn - 1$ (the single degree is lost for our estimating the overall mean, $\overline{\overline{y}}$). As a check in more complicated designs, it is useful to make sure that the individual component degrees of freedom add up to the correct total:

$$kn - 1 = k - 1 + k(n - 1) = k - 1 + kn - k = kn - 1$$

The divisions for turning the sums of squares into variances are conveniently carried out in an Anova table.

Source	SS	d.f.	MS	F	Critical F
Treatment	SSA	$k - 1$	$MSA = \dfrac{SSA}{k - 1}$	$F = \dfrac{MSA}{s^2}$	qf(0.95, k-1, k(n-1))
Error	SSE	$k(n - 1)$	$s^2 = \dfrac{SSE}{k(n - 1)}$		
Total	SST	$kn - 1$			

Each element in the sums of squares column is divided by the number in the adjacent degrees of freedom column to give the variances in the mean square column (headed MS). The significance of the difference between the means is then assessed using an F test (a variance ratio test). The treatment variance MSA is divided by the error variance, s^2, and the value of this test statistic is compared with the critical value of F using **qf** (the quantiles of the F distribution, with $p = 0.95$, $k - 1$ degrees of freedom in the numerator, and $k(n - 1)$ degrees of freedom in the denominator). If you need to look up the critical value of F in tables, remember that you look up the numerator degrees of freedom (on top of the division) across the *top* of the table, and the denominator degrees of freedom down the rows. If the test statistic is larger than the critical value, we *reject* the null hypothesis

$$H_0 : \textit{all the means are the same}$$

and accept the alternative

$$H_1 : \textit{at least one of the means is significantly different from the others}$$

If the test statistic is less than the critical value, then the differences could have arisen by chance alone, and so we accept the null hypothesis.

Another way of visualising the process of Anova is to think of the relative amounts of sampling variation between replicates receiving the same treatment (i.e. between individual samples in the same level), and between different treatments (i.e. between-level variation). When the variation between replicates within a treatment is large compared to the variation between treatments, we are likely to conclude that the difference between the treatment means is not significant. Only if the variation between replicates within treatments is relatively small compared to the differences between treatment means, will we be justified in concluding that the treatment means are significantly different.

Calculations in Anova

The definitions of the various sums of squares can now be formalised, and ways found of calculating their values from samples. The total sum of squares, SST, is defined as

$$SST = \sum y^2 - \frac{\left(\sum y\right)^2}{kn}$$

just as in regression (Chapter 14). Note that we divide by the total number of numbers we added together to get $\sum y$ (the grand total of all the y) which is kn. It turns out that the formula that we used to define SSE is rather difficult to calculate (see above), so we calculate the treatment sums of squares SSA, which is easy, and obtain SSE by difference.

The treatment sum of squares, SSA, is defined as

$$SSA = \frac{\sum C^2}{n} - \frac{\left(\sum y\right)^2}{kn}$$

where the new term is C, the *treatment total*. This is the sum of all the n replicates within a given level. Each of the k different treatment totals is squared, added up, and then divided by n (the number of numbers added together to get the treatment total). The formula is slightly different if there is unequal replication in different treatments, as we shall see below. The meaning of C will become clear when we work through an example:

```
attach(oneway)
names(oneway)
```

```
[1] "Growth"          "Photoperiod"
```

The data come from a simple growth room experiment, in which the response variable is Growth (mm) and the categorical explanatory variable is a factor called Photoperiod with four levels: very short, short, long and very long daily exposure to light. There were six replicates of each treatment, and the data look like this:

Very short	Short	Long	Very long
2	3	3	4
3	4	5	6
1	2	1	2
1	1	2	2
2	2	2	2
1	1	2	3

The treatment totals, C_i, are simply the sums of the replicates. So

$$C_1 = 2 + 3 + 1 + 1 + 2 + 1 = 10$$

$$C_4 = 4 + 6 + 2 + 2 + 2 + 3 = 19$$

The sum of the squared treatment totals is divided by n. This is *the number of numbers that were added together to get the total* that we squared. The same principles apply to all the Anovas we shall do, no matter how complicated they get, so it is worth making absolutely sure that you understand what C_i is and why we divide $\sum C_i^2$ by n.

So, for our example, *SSA* is calculated as

$$SSA = \frac{10^2 + 13^2 + 15^2 + 19^2}{6} - \frac{57^2}{4 \times 6} = \frac{855}{6} - \frac{3249}{24} = 142.5 - 135.375 = 7.125$$

Notice that the term on the right-hand side (the correction factor as we shall call it) is also divided by the number of numbers that were added together to get the total ($\sum y$) which is squared in the numerator. Finally,

$$SSE = SST - SSA$$

to give all the elements required for completion of the Anova table.

Assumptions of Anova

You should be aware of the assumptions underlying the analysis of variance. They are all important, but some are more important than others:

- random sampling
- equal variances
- independence of errors
- Normal distribution of errors
- additivity of treatment effects

These assumptions are well and good, but what happens if they do not apply to the data you propose to analyse? We consider each case in turn.

Random sampling

If samples are not collected at random, then the experiment is seriously flawed from the outset. The process of randomisation is sometimes immensely tedious, but none the less important for that (see p. 53). Failure to randomise properly often spoils what would otherwise be well-designed experiments; see Hairston (1989) for examples. Unlike the other assumptions of Anova, there is nothing we can do to rectify non-random sampling after the event. If the data are not collected randomly, they will probably be biased. If they are biased, their interpretation is bound to be equivocal.

Equal variances

It seems odd, at first, that a technique which works by comparing variances should be based on the assumption that the variances are equal. What Anova actually assumes, however, is that the *sampling errors* do not differ significantly from one treatment to another. The comparative part of Anova works by comparing the variances that are due to differences between the treatment means with the variation between samples within treatments. The contributions towards *SSA* are allowed to vary between treatments, but the contributions towards *SSE* should not be significantly different.

Recall the folly of attempting to compare treatment means when the variances of the samples are different in the example of three commercial gardens producing lettuce (see p. 95). There were three important points to be made from these calculations:

- two treatments can have different means and the same variance

- two treatments can have the same mean but different variances

- when the variances are different, the fact that the means are identical does *not* mean that the treatments have identical effects

We often encounter data with non-constant variance, and S-Plus is ideally suited to deal with this. With Poisson-distributed data, for example, the variance is equal to the mean, and with binomial data the variance (npq) increases to a maximum and then declines with the mean (np). Many data on time to death (or time to failure of a component) exhibit the property that the coefficient of variation is roughly constant, which means that a plot of log(variance) against log(mean) increases with a slope of approximately 2 (this is known as Taylor's power law); Gamma errors can deal with this. Alternatively, the response variable can be transformed (Chapter 22) to stabilise the variance, and the analysis carried out on this new, better-behaved response variable.

Independence of errors and pseudoreplication

It is very common to find that the errors are not independent from one experimental unit to another. We know, for example, that the response of an organism is likely to be influenced by its sex, age, body size, neighbours and their sizes, history of development, genotype, and many other things. Differences between our samples in these other attributes would lead to non-independence of errors, and hence to bias in our estimation of the difference that was due to our experimental factor. For instance, if more of the irrigated plots happened to be on good soil, then increases in yield might be attributed to the additional water when, in fact, they were due to the higher level of mineral nutrient availability on those plots.

The ways to minimise the problems associated with non-independence of errors are:

- use block designs, with each treatment combination applied in every block (i.e. repeat the whole experiment in several different places)

- divide the experimental material up into homogeneous groups at the outset (e.g. large, medium and small individuals), then make these groups into statistical blocks, and apply each treatment within each of them

- have high replication

- insist on thorough randomisation

- measure initial conditions

No matter how careful the design, however, there is always a serious risk of non-independence of errors. A great deal of the skill in designing good experiments is in anticipating where such problems are likely to arise, and in finding ways of blocking the experimental material to maximise the power of the analysis of variance. This topic is considered in more detail later.

Normal distribution of errors

Data often have non-normal error distributions. The commonest form of non-normality is skewness, in which the distribution has a long 'tail' at one side or the other. A great many collections of data are skewed to the right, because most individuals in a population are small, but a few individuals are very large indeed (well out towards the right-hand end of the size axis). A second kind of non-normality that is encountered is kurtosis. This may arise because there are long tails on both sides of the mean so that the distribution is more 'pointed' than the bell-shaped Normal distribution (so-called leptokurtosis), or conversely because the distribution is more 'flat-topped' than a Normal (so-called platykurtosis). Finally, data may be bimodal or multimodal, and look nothing at all like the Normal curve. This kind of non-normality is most serious, because skewness and kurtosis are often cured by the same kinds of transformations that can be used to improve homogeneity of variance. If a set of data is strongly bimodal, then it is clear that any estimate of central tendency (mean or median) is likely to be *unrepresentative of most of the individuals* in the population (see p. 69).

Additivity of treatment effects in multifactor experiments

In two-way analysis there are two kinds of treatment (say drugs and radiation therapy) each with two or more levels. Anova is based on the assumption that the effects of the different treatments are additive. This means that if one of the drug treatments produced a response of 0.4 at low radiation, then the drug will produce a response of 0.4 at high radiation. Where this is *not* the case, and the response to one factor depends upon the level of another factor (i.e. where there is *statistical interaction*), we must use factorial experiments. Even here, however, the model is still assumed to be additive. When treatment effects are multiplicative (e.g. temperature and dose are multiplicative in some toxicity experiments), then it may be appropriate to transform the data by taking logarithms in order to make the treatment effects additive. In S-Plus, we could specify a log link function in order to achieve additivity in a case like this.

A worked example of one-way analysis of variance

To draw this background material together, we shall work through the photoperiod example by hand (see p. 247). In so doing, it will become clear what S-Plus is doing during its analysis of the same data.

To carry out a one-way Anova we aim to partition the total sum of squares *SST* into just two parts: variation attributable to differences between the treatment means, *SSA*, and the unexplained variation which we call *SSE*, the error sum of squares:

$$SST = 175 - \frac{57^2}{24} = 39.625$$

$$SSA = \frac{10^2 + 13^2 + 15^2 + 19^2}{6} - \frac{57^2}{24} = 7.125$$

$$SSE = SST - SSA = 39.625 - 7.125 = 32.5$$

Source	SS	d.f.	MS	F
Photoperiod	7.125	3	2.375	1.462
Error	32.500	20	$s^2 = 1.625$	
Total	39.625	23		

The Anova table is interpreted as follows. The calculated value of $F = 1.462$ is less than the value in tables (the value expected to occur by chance alone with probability $\alpha = 0.05$, given 3 degrees of freedom in the numerator and 20 degrees of freedom in the denominator, is 3.098) and therefore we accept H_0. Remember: calculated *less* than tables, *accept* H_0.

There is no evidence in this analysis for any difference in mean growth under the different photoperiods.

Anova in S-Plus

```
attach(oneway)
names(oneway)
```

```
[1] "Growth"      "Photoperiod"
```

We begin by calculating the mean growth at each photoperiod, using **tapply** in the usual way:

```
tapply(Growth,Photoperiod,mean)
```

```
 Long      Short      Very.long    Very.short
  2.5     2.166667    3.166667      1.666667
```

The values look fine, with mean growth increasing with the duration of illumination. But the *factor levels are printed in alphabetical order*. This usually does not matter, but here it is inconvenient because the factor Photoperiod is ordered. We can fix this very simply by declaring Photoperiod to be an **ordered** factor, and providing an ordered list of the factor levels to reflect the fact that

very short < short < long < very long

```
Photoperiod<-ordered(Photoperiod,levels=c("Very.short","Short","Long","Very.long"))
```

Now when we use **tapply**, the means are printed in the desired sequence:

```
tapply(Growth,Photoperiod,mean)
```

```
 Very.short     Short      Long     Very.long
  1.666667     2.166667    2.5      3.166667
```

The one-way analysis of variance is carried out using the aov function. The model formula is written like this:

response variable ~ explanatory variable

where it is understood that the explanatory variable is a factor. Although in this case we know that Photoperiod must be a factor because it has text as the factor levels, it is useful to know how to check that a variable is a factor (especially for variables with numbers as factor levels). We use the **is.factor** function to find this out:

is.factor(Photoperiod)

[1]   TRUE

so Photoperiod *is* a factor. All we need to do now is to give a name to the object that will contain the results of the analysis (model), and carry out the **aov**:

model<-aov(Growth~Photoperiod)

Nothing happens until we ask for the results. All of the usual generic functions can be used for Anova, including **summary** and **plot**:

summary(model)

	df	Sum of Sq	Mean Sq	F Value	Pr(F)
Photoperiod	3	7.125	2.375	1.461538	0.2550719
Residuals	20	32.500	1.625		

These are the figures we obtained by hand. It is abundantly clear *there is no significant difference between the mean growth rates with different periods of illumination.* An F value of 1.46 will arise due to chance when the means are all the same with a probability greater than 25% (for significance, you recall, we want this probability to be less than 0.05). The differences we noted at the beginning are not significant because the variance is so large (the residual mean square is $s^2 = 1.625$) and the error degrees of freedom (d.f. = 20) are rather small.

Model checking involves plot(model). You will see that while there is slight evidence of hetero-scedasticity (i.e. there is some tendency for the variance to increase as the fitted values increase).

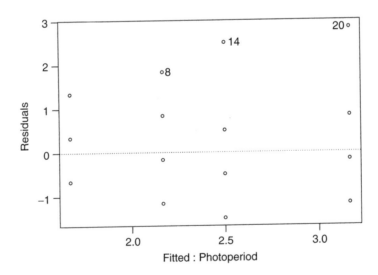

There are strong signs of non-normality in the residuals (J-shape not linear). This is perhaps not surprising given that the data are small integers. We return to these issues later on.

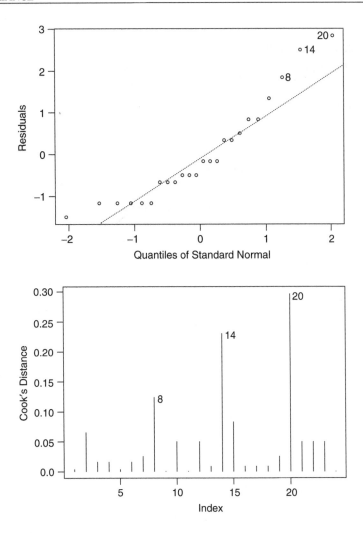

The influence plot (Cook's distance) shows that data points numbers 20, 14 and 8 have a big influence on the model. For the time being, we continue with the analysis of the same data set, but taking further information into account.

After **aov** has been used to produce a model object, you have the option of using **model.tables** to obtain tables of estimates and relevant standard errors. The function is generic, and can be used extract information about means, effects, residuals, and their standard errors. Effects are differences between the overall mean and the individual treatment means:

```
model.tables(model)

Tables of effects

Photoperiod
Very.short      Short   Long  Very.long
  -0.70833   -0.20833  0.125    0.79167
```

To see what these figures mean, we calculate the individual treatment means:

```
tapply(Growth,Photoperiod,mean)
 Very.short        Short  Long  Very.long
  1.666667      2.166667   2.5   3.166667
```

and the grand mean (the overall mean, $\bar{\bar{y}}$):

```
mean(Growth)
```

```
[1] 2.375
```

The effects are $\bar{y}_i - \bar{\bar{y}}$ for each factor level i. For very short photoperiods, the effect size is $1.666 - 2.375 = -0.7083$:

```
tapply(Growth,Photoperiod,mean)-mean(Growth)
  Very.short          Short       Long   Very.long
  -0.7083333     -0.2083333     0.125   0.7916667
```

The main option of **model.tables** is type, a character string specifying the type of tables desired. As we have seen the default is effects for tables of marginal effects for each term in the model. Choices are feffects for effects for factorial (2^k) models; means for tables of fitted means; adj.means for adjusted means; residuals for tables of residuals. The option se is a logical value indicating whether standard errors should be computed for the tables. If TRUE, a component se is returned, containing standard error information for each table. The form of standard error returned is determined by the type of table requested. If type = "effects", standard errors for individual effects are returned. If type = "means", standard errors for the *difference of two means* are returned. If type = "residuals", standard errors for residuals are returned. For effects and means the design must be balanced. Standard errors for unbalanced designs can be computed for contrasts of interest using se.contrast. If type = "adj.means", the value of se is ignored and the standard error for the adjusted means is always computed.

Two-way analysis of variance

The previous analysis had a single categorical explanatory variable. Many experiments have two or more categorical explanatory variables, and we need to know how this alters the way that Anova is carried out. We continue the analysis of the photoperiod data. We had fitted a four-level factor for Photoperiod, but the unexplained variation, SSE, was very large (32.5); so large in fact that the differences between mean growth in the different photoperiods was not significant. Fortunately, the experiment was well designed. Material from six plant genotypes was cloned and photoperiod treatments were allocated at random to each of four clones of the same genotype. Each photoperiod was allocated once to each genotype. There was no replication of genotypes within photoperiods, so we cannot carry out a factorial Anova. But we can carry out a two-way Anova with Photoperiod and Genotype as two categorical explanatory variables. The sum of squares attributable to Photoperiod is unchanged, and we can add this directly to our new, expanded Anova table. The total sum of squares will also be unchanged.

Source	SS	d.f.	MS	F
Photoperiod	7.125	3		
Genotype		5		
Error		15		
Total	39.625	23		

We know that the sums of squares for Genotype and Error must add up to 32.5, but we need to work out one of them before we can obtain the other by subtraction. Before we do that, we can fill in the d.f. column without doing any calculations. There six levels of Genotype so there must be $6 - 1 = 5$ d.f. for Genotype. We know that the d.f. column must always add up to the total degrees of freedom, so

$$\text{error d.f.} = 23 - 5 - 3 = 15$$

There is, however, an analytical way to calculate error d.f.. Technically, in an analysis like this without any replication, *the error term is the interaction between the constituent factors.* You will recall that interaction degrees of freedom are the product of the component degrees of freedom. So, in this case with $(c - 1) = 3$ d.f. for Photoperiod and $(r - 1) = 5$ d.f. for Genotype, we have

$$\text{error d.f.} = (r - 1)(c - 1) = 5 \times 3 = 15$$

The only extra calculation we need to do is to obtain the Genotype sum of squares. The procedure is exactly analogous to computing the Photoperiod sum of squares. Photoperiod differences were reflected in the column totals. Likewise Genotype differences are reflected in the row totals:

Genotype	Very short	Short	Long	Very long	Row totals
A	2	3	3	4	**12**
B	3	4	5	6	**18**
C	1	2	1	2	**6**
D	1	1	2	2	**6**
E	2	2	2	2	**8**
F	1	1	2	3	**7**

So we use the squares of the row totals in computing *SSB*. Note that we divide by the number of numbers in a row (4 in this case, one for each photoperiod). For *SSA* we divided by the number of numbers in a column (6 in that case). In Anova you generally add up the squares of subtotals and *divide by the number of numbers that were added together to get that subtotal.* The Genotype sum of squares is therefore calculated as

$$SSB = \frac{12^2 + 18^2 + 6^2 + 6^2 + 8^2 + 7^2}{4} - \frac{57^2}{24} = 27.875$$

and the new, much smaller *SSE* is obtained by difference:

$$SSE = SST - SSA - SSB = 39.625 - 7.125 - 27.875 = 4.625$$

The two-way Anova table can now be completed.

Source	SS	d.f.	MS	F
Photoperiod	7.125	3	2.375	7.703
Genotype	27.875	5	5.575	18.08
Error	4.625	15	$s^2 = 0.308$	
Total	39.625	23		

The interpretation of the experiment is now completely different. We conclude that Photoperiod has a highly significant effect on mean growth ($F = 7.703$, d.f. $= 3, 15$, $p < 0.01$). The moral is that good experimental design, aimed at reducing variation between individuals (blocking by Genotype in this case), can pay huge dividends in terms of the likelihood of detecting scientifically important differences.

We tidy up by using the function **rm** to remove the variable names from the one-way analysis:

rm(Growth, Photoperiod)

Two-way Anova in S-Plus

attach(twoway)
names(twoway)

[1] "Growth"   "Photoperiod"   "Genotype"

We begin by comparing the mean growth rates of the six genotypes, using **tapply**:

tapply(Growth,Genotype,mean)

A   B   C   D   E   F
3 4.5 1.5 1.5 2 1.75

Evidently, there are substantial differences between the mean growth rates of the different genotypes. This variation was formerly included in the error variance, so by removing the variation attributable to genotype, we will make the error variance smaller, hence we will make the significance of the difference between the photoperiod means greater. The two-way Anova is carried out simply by specifying the names of two explanatory variables (both must be factors) in the model formula, linked by a plus sign +. We call the object containing the results of the analysis model2 and proceed as follows:

model2<-aov(Growth~Genotype+Photoperiod)

Nothing happens until we specify the form of output we would like:

summary(model2)

	df	Sum of Sq	Mean Sq	F Value	Pr(F)
Genotype	5	27.875	5.575000	18.08108	0.000007092
Photoperiod	3	7.125	2.375000	7.70270	0.002404262
Residuals	15	4.625	0.308333		

The two-way Anova table shows the dramatic effect of including Genotype in the model. Far from being insignificant (as it was in the one-way Anova), the difference between the Photoperiod means is now highly significant ($p < 0.0025$). This example highlights the enormous potential benefits of introducing blocking into the experimental design. If the photoperiod treatments had been allocated to different genotypes at random, we would not have obtained a significant result. By ensuring that every photoperiod was applied to every genotype, the experimental design was greatly improved. Randomisation was carried out, but it involved deciding at random which of the four clonal plants from each genotype was allocated to a given photoperiod treatment.

Model criticism involves using plot(model2). You will see that the non-normality problems that looked so severe after the one-way analysis have been completely cured; evidently it was the differences between the genotype means that were the principal cause of the difficulty. If we had not known the identity of the genotypes, and hence been unable to carry out the two-way analysis, we would have been left with this problem. We later return to the issue of unexplained variation under the topic of *overdispersion*.

Interaction

The analysis of variance is based on the assumption of additivity. That is to say, if nitrogen causes a yield increase of 1 ton per hectare, and irrigation causes a yield increase of 0.5 ton per hectare, then the addition of nitrogen and irrigation together will cause a yield increase of 1.5 ton per hectare. If this is not the case, then we have what statisticians call interaction. In general terms, *an interaction is operating when the response to factor A depends on the level of factor B*. This should become clear from some graphical examples. In what follows, there are 3 levels of *A*, and 2 levels of *B*; the graphs were produced using **interaction.plot** (see p. 270).

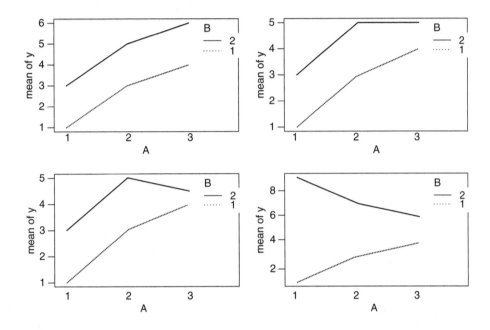

The top left panel shows the case of no interaction: the responses to *A* are *parallel* for both levels of *B*; the second level of *B* is associated with an increase in the response of 2.0 at all levels of *A*. The top right panel shows an interaction, because the response to factor *B* is lower at high levels of *A* than at low (1 unit rather than 2). The bottom left panel shows a response reversal: the response to level 3 of *A* is positive at low levels of *B* but negative at high levels. Finally, the bottom right panel shows an extreme interaction: the response to *A* is positive at low levels of *B* but negative at high levels of *B*. This case draws attention to a very important general point about Anova. Look at the bottom right-hand panel and ask yourself, What is the main effect of factor *A*? You will see that if you average the response over the two levels of factor *B*, the response is 5 at each level of *A*. In this case, *A* has no main effect. But does this mean that *A* has no effect on the response? Obviously not. At low levels of *B*, increasing *A* causes an increase in response. At high levels of *B*, increasing *A* causes a decrease in response. Changing *A* always produces a response, despite the fact that it apparently has no main effect.

Because interactions can sometimes cause main effects to cancel out, *we never investigate main effects in the Anova table until we have investigated the interaction effects*. If a factor appears in a significant interaction, then that factor needs to remain in the model, irrespective of how small its main effect may appear to be from the *F* test that appears in the Anova table.

Two-way factorial analysis of variance

Factorial experiments allow the investigation of *interactions* between categorical explanatory variables. The only requirement is that *there must be replication* at each combination of factor levels. The only new concept is the calculation of the *interaction sum of squares*. Like all components of Anova, it is based on a sum of squared totals divided by the number of numbers that were added together to get each total. For interactions the subtotals we need are often not directly available (like the row totals and column totals of the standard two-way analysis). To make calculation as straightforward as possible, it is a good idea to draw up a table of interaction totals; the levels of factor A make up the rows and the levels of factor B make up the columns. Each cell of the table contains the sum of the replicates in each factor combination. We call these totals Q. Now the interaction sum of squares, $SSAB$, is calculated like this:

$$SSAB = \frac{\sum Q^2}{n} - SSA - SSB - CF$$

because the interaction subtotals contain the main effects of factor A and factor B as well as the interaction effect we are after. That is why we subtract SSA and SSB in calculating $SSAB$.

Source	SS	d.f.	MS	F
Factor A	SSA	$a-1$	$\dfrac{SSA}{(a-1)}$	$\dfrac{MSA}{s^2}$
Factor B	SSB	$b-1$	$\dfrac{SSB}{(b-1)}$	$\dfrac{MSB}{s^2}$
Interaction $A:B$	$SSAB$	$(a-1)(b-1)$	$\dfrac{SSAB}{(a-1)(b-1)}$	$\dfrac{MSAB}{s^2}$
Error	SSE	$ab\,(n-1)$	$s^2 = \dfrac{SSE}{ab(n-1)}$	
Total	SST	$abn-1$		

The example concerns the growth of hamsters of different coat colours (different genotypes) when fed diets of different compositions:

```
attach(factorial)
names(factorial)

[1]   "growth"   "diet"   "coat"

      growth  diet   coat
1      6.6     A    light
2      7.2     A    light
3      6.9     B    light
4      8.3     B    light
5      7.9     C    light
6      9.2     C    light
7      8.3     A    dark
8      8.7     A    dark
9      8.1     B    dark
```

```
10      8.5      B     dark
11      9.1      C     dark
12      9.0      C     dark
```

There are $a = 3$ levels of diet (factor A), $b = 2$ levels of coat colour (factor B) and $n = 2$ replicates. We begin with data inspection. Because there are two factors, we shall want to produce two boxplots, so we alter the graphical parameters using mfrow=c(1,2) like this:

```
par(mfrow=c(1,2))
plot(diet,growth)
plot(coat,growth)
```

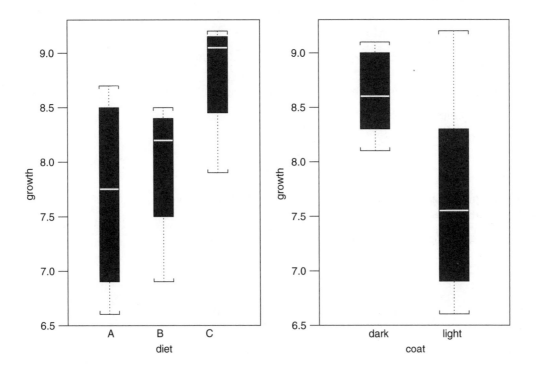

Growth is higher for diet C, and lower for animals with light-coloured coats. The required sums of squares look like this:

$$CF = \frac{\left[\sum y\right]^2}{abn}$$

$$SST = \sum y^2 - CF$$

$$SSA = \frac{\sum A^2}{bn} - CF$$

$$SSB = \frac{\sum B^2}{an} - CF$$

$$SSAB = \frac{\sum Q^2}{n} - SSA - SSB - CF$$

$$SSE = SST - SSA - SSB - SSAB$$

The correction factor (CF) is the grand total of all the numbers, squared and then divided by the total number of numbers in the experiment $(a \times b \times n = 3 \times 2 \times 2 = 12)$:

sum(growth)

[1] 97.8

sum(growth)^2

[1] 9564.84

$$CF = \frac{97.8^2}{12} = \frac{9564.84}{12} = 797.07$$

The total sum of squares requires the sum of the squares of all the individual growth responses:

sum(growth^2)

[1] 805.2

$$SST = 805.2 - 797.07 = 8.13$$

The treatment sum of squares for diet (SSA) is calculated in the usual way, using the squares of the three diet totals:

tapply(growth,diet,sum)

```
   A     B     C
30.8  31.8  35.2
```

$$SSA = \frac{30.8^2 + 31.8^2 + 35.2^2}{2 \times 2} - CF = \frac{3198.92}{4} - 797.07 = 2.66$$

Likewise, the treatment sum of squares for coat (SSB) is based on the sum of squares of the two coat totals:

tapply(growth,coat,sum)

```
dark  light
51.7   46.1
```

$$SSB = \frac{51.7^2 + 46.1^2}{3 \times 2} - CF = \frac{4798.1}{6} - 797.07 = 2.613\,333$$

In this case we divide by 6 (not 4) because each of the coat totals is the sum of six numbers (3 diets × 2 replicates).

The **interaction totals** are the sums of the two replicates in each of the six combinations of factor levels. Note the use of **list** to specify the two-dimensional summary table. Rows are the name that comes first in the list (coat), and columns are the second (diet):

tapply(growth,list(coat,diet),sum)

```
        A     B     C
dark  17.0  16.6  18.1
light 13.8  15.2  17.1
```

We need the sum of the squares of these six subtotals, which we divide by 2 (because each of the subtotals is the sum of two numbers). This quantity contains the main effects sums of squares SSA and SSB, so to obtain the interaction sum of squares, we must subtract these two, as well as the correction factor:

$$SSAB = \frac{17.0^2 + 16.6^2 + \cdots + 17.1^2}{2} - 2.66 - 2.613\,333 - 797.07 = 0.686\,67$$

Finally, we get SSE by difference:

$$SSE = 8.13 - 2.66 - 2.613\,333 - 0.686\,67 = 2.17$$

Now we can fill in the Anova table.

Source	SS	d.f.	MS	F
diet	2.66	2	1.33	3.6774
coat	2.613\,333	1	2.613\,333	7.2258
Interaction				
diet:coat	0.686\,67	2	0.343\,33	0.9493
Error	2.17	6	0.361\,67	
Total	8.13	11		

Factorial Anova in S-Plus

The analysis proceeds very simply by using the * operator. This means 'fit all the main effects and all their interactions'. Thus the model formula

growth ~ diet * coat

is shorthand for

growth ~ diet + coat + diet : coat

where the : operator means 'the interaction between'. We type

```
model<-aov(growth~diet*coat)
summary(model)
           Df  Sum Sq  Mean Sq  F value   Pr(>F)
diet        2  2.66000  1.33000   3.6774  0.09069  .
coat        1  2.61333  2.61333   7.2258  0.03614  *
diet:coat   2  0.68667  0.34333   0.9493  0.43833
Residuals   6  2.17000  0.36167
```

The output of the Anova table is interpreted like this. We always start by looking at the interaction rather than the main effects. The F value of 0.9493 falls well short of significance, so there is no evidence for an interaction between diet and coat colour. Now for the main effects. There is a significant effect of coat colour ($p < 0.04$) but not of diet ($p > 0.05$). Now we use plot(model) to carry out model criticism. The variance is constant and the errors are Normal, so that is good.

It is a good idea at this stage to do model simplification to remove any non-significant parameters. We use **update** to remove the interaction term. It works like this:

model2<-update(model, ~ . - diet:coat)

which is read as follows. The new model (called model2 here) gets an **update** of the earlier model (called model in the first argument in update). The syntax is now very important: it is , ~ . -. This says take the whole of model (the ~. bit, where . means 'all of'), and remove from it (hence the -) the interaction diet:coat (which is read 'diet by coat'). Now we have two models: a complicated one (called model) and a simpler one (called model2). S-Plus is superb at comparing models. We use the **anova** function to compare two or more models, like this:

```
anova(model,model2)

Analysis of Variance Table

Model 1: growth ~ diet + coat + diet:coat
Model 2: growth ~ diet + coat

   Res.Df Res.Sum Sq Df   Sum Sq F value Pr(>F)
1       6    2.17000
2       8    2.85667 -2  -0.68667 0.9493 0.4383
```

This says that the simpler model2 is not significantly worse in its explanatory power than the more complicated model ($p = 0.4383$). In other words, the model simplification is justified. Let's look at the output of model2:

```
summary(model2)

           Df  Sum Sq Mean Sq F value   Pr(>F)
diet        2 2.66000 1.33000  3.7246  0.07190 .
coat        1 2.61333 2.61333  7.3186  0.02685 *
Residuals   8 2.85667 0.35708
```

There is a hint of an effect of diet ($p = 0.0719$) but we are ruthless in the elimination of non-significant terms, so we shall try leaving that out as well:

```
model3<-update(model2, ~ . -diet)
anova(model2,model3)

Analysis of Variance Table

Model 1: growth ~ diet + coat
Model 2: growth ~ coat

   Res.Df  Res.Sum Sq  Df   Sum Sq  F value  Pr(>F)
1       8      2.8567
2      10      5.5167  -2  -2.6600   3.7246  0.0719 .
```

This gives us exactly the same p value as in model2, but when we come onto more complicated statistical models this will not always be the case, and we prefer to compare models by deletion rather than by t tests on their parameter values. The simplified model looks like this:

summary(model3)

```
             Df  Sum Sq  Mean Sq  F value  Pr(>F)
coat          1  2.6133   2.6133   4.7372  0.05457  .
Residuals    10  5.5167   0.5517
```

This is an interesting example of what can happen in model simplification. The effect of coat was highly significant in model2 but just fails to reach significance when diet is left out. It looks as if both factors are important, or neither factor is unequivocally important. Only a more detailed analysis will tell. Let's look at the means for the three diets:

tapply(growth,diet,mean)

```
   A     B     C
 7.70  7.95  8.80
```

It looks as if diets A and B produce a rather similar response, but perhaps diet C produces faster growth than the other two? To investigate this, we calculate a new two-level factor called diet2, which has the value 1 for diets A and B, and 2 for diet C:

diet2<-factor(1+(diet=="C"))
diet2

```
 [1] 1 1 1 1 2 2 1 1 1 1 2 2
Levels:  1 2
```

Make sure you understand what we've done here. Now we try adding this revised dietary factor to model3:

model4<-update(model3, ~ . +diet2)

and see whether it makes a significant difference using **anova**:

anova(model3,model4)

```
Analysis of Variance Table

Model 1: growth ~ coat
Model 2: growth ~ coat + diet2

  Res.Df Res.Sum Sq Df Sum Sq F value  Pr(>F)
1     10     5.5167
2      9     2.9817  1 2.5350  7.6518 0.02189 *
```

Yes, it does. Saving that extra degree of freedom, by reducing the factor levels of diet from 3 to 2 has turned a non-significant effect into a significant one. Let's see if there is an interaction with this new, simpler factor:

model5<-update(model4, ~ . +diet2:coat)
anova(model4,model5)

```
Analysis of Variance Table

Model 1: growth ~ coat + diet2
Model 2: growth ~ coat + diet2 + coat:diet2
     Res.Df  Res.Sum Sq  Df    Sum Sq  F value  Pr(>F)
1         9     2.98167
2         8     2.70000   1   0.28167   0.8346  0.3877
```

No, there isn't. This means that model4 is minimal adequate.

summary(model4)

```
            Df  Sum Sq  Mean Sq  F value    Pr(>F)
coat         1  2.6133   2.6133   7.8882   0.02042  *
diet2        1  2.5350   2.5350   7.6518   0.02189  *
Residuals    9  2.9817   0.3313
```

This example shows the benefits of model simplification. Our interpretation is now much simpler and much clearer than before. In brief, there is a significant effect of coat colour phenotype on mean growth, and there is a significant difference between diet C and the other two diets in mean growth rate. There is no evidence, however, that different coat colours respond to diet C in different ways (i.e. there is no interaction effect). Diagnostic plots on model4 show that the assumption of the Anova in regard to normality of errors seems reasonable, but there is a hint of variance declining with the fitted values. Given the low replication, this is not serious.

Three-way factorial analysis of variance

The only new element in a three-way analysis is the extra weight of calculation involved. The principles are exactly the same as in two-way factorial Anova. We have a levels of factor A, b levels of factor B and c levels of factor C. There are n replicates of each treatment combination (unequal replication can be handled, but it makes the formulas unnecessarily complicated, so we consider only the equal replication case for the hand calculations). The main effect totals are denoted by A, B and C respectively. The main effect sums of squares are calculated as in one-way Anova, subtracting CF (the correction factor) from the sums of squares of the treatments totals divided by the relevant replication:

$$CF = \frac{\left[\sum y\right]^2}{abcn}$$

$$SSA = \frac{\sum A^2}{bcn} - CF$$

$$SSB = \frac{\sum B^2}{acn} - CF$$

$$SSC = \frac{\sum C^2}{abn} - CF$$

The three two-way interactions are calculated in exactly the same way as they were in two-way Anova, based on two-way interaction tables of subtotals (see above). Only the three-way formula is given here; this is based on a new set of tables of subtotals: *there is a separate AB table for every level of C.* The sums of squares of each of the elements, T, of these tables are added together and divided by n, the number of numbers added together to obtain each of the subtotals. The three-way interaction sum of squares is then calculated like this:

$$SSABC = \frac{\sum T^2}{n} - SSA - SSB - SSC - SSAB - SSAC - SSBC - CF$$

The term $\sum T^2/n$ contains all three two-way interaction sums of squares as well as all three main effects sums of squares, so we need to subtract these in order to obtain the three-way interaction sum of squares that we are after. Now the error sum of squares is obtained by subtracting all the other component sums of squares from the total sum of squares in the usual way:

$$SSE = SST - SSA - SSB - SSC - SSB - SAC - SSBC - SSABC$$

and the Anova table can be completed.

Source	SS	d.f.	MS	F
Factor A	SSA	$a - 1$	$\dfrac{SSA}{(a-1)}$	$\dfrac{MSA}{s^2}$
Factor B	SSB	$b - 1$	$\dfrac{SSB}{(b-1)}$	$\dfrac{MSB}{s^2}$
Factor C	SSC	$c - 1$	$\dfrac{SSC}{(c-1)}$	$\dfrac{MSC}{s^2}$
Interaction $A:B$	SSAB	$(a-1)(b-1)$	$\dfrac{SSAB}{(a-1)(b-1)}$	$\dfrac{MSAB}{s^2}$
Interaction $A:C$	SSAC	$(a-1)(c-1)$	$\dfrac{SSAC}{(a-1)(c-1)}$	$\dfrac{MSAC}{s^2}$
Interaction $B:C$	SSBC	$(b-1)(c-1)$	$\dfrac{SSBC}{(b-1)(c-1)}$	$\dfrac{MSBC}{s^2}$
Interaction $A:B:C$	SSABC	$(a-1)(b-1)(c-1)$	$\dfrac{SSABC}{(a-1)(b-1)(c-1)}$	$\dfrac{MSABC}{s^2}$
Error	SSE	$abc\,(n-1)$	$s^2 = \dfrac{SSE}{abc(n-1)}$	
Total	SST	$abcn - 1$		

Now, bracing ourselves, we start the calculations by hand. The response variable is the population growth rate of *Daphnia*. There are three categorical response variables: Water (with 2 levels from the rivers Tyne and Wear), Detergent (with 4 levels, imaginatively named BrandA, BrandB, BrandC and BrandD) and *Daphnia* clone (with 3 levels, Clone1, Clone2 and Clone3). Each treatment combination was replicated three times. So our factor level codings from the equations above are $a = 2$, $b = 4$, $c = 3$ and $n = 3$.

```
    Growth.rate   Water  Detergent  Daphnia
1    2.919086     Tyne    BrandA    Clone1
2    2.492904     Tyne    BrandA    Clone1
3    3.021804     Tyne    BrandA    Clone1
```

4	2.350874	Tyne	BrandA	Clone2
5	3.148174	Tyne	BrandA	Clone2
6	4.423853	Tyne	BrandA	Clone2
7	4.870959	Tyne	BrandA	Clone3
8	3.897731	Tyne	BrandA	Clone3
9	5.830882	Tyne	BrandA	Clone3
10	2.302717	Tyne	BrandB	Clone1
11	3.195970	Tyne	BrandB	Clone1
12	2.829021	Tyne	BrandB	Clone1
13	3.027500	Tyne	BrandB	Clone2
14	5.285108	Tyne	BrandB	Clone2
15	4.260955	Tyne	BrandB	Clone2
16	4.049528	Tyne	BrandB	Clone3
17	4.623400	Tyne	BrandB	Clone3
18	5.625845	Tyne	BrandB	Clone3
19	2.634182	Tyne	BrandC	Clone1
20	3.513746	Tyne	BrandC	Clone1
21	3.714657	Tyne	BrandC	Clone1
22	3.862106	Tyne	BrandC	Clone2
23	3.955181	Tyne	BrandC	Clone2
24	3.044308	Tyne	BrandC	Clone2
25	3.358051	Tyne	BrandC	Clone3
26	5.633479	Tyne	BrandC	Clone3
27	4.613175	Tyne	BrandC	Clone3
28	2.200593	Tyne	BrandD	Clone1
29	2.233800	Tyne	BrandD	Clone1
30	3.357184	Tyne	BrandD	Clone1
31	3.111764	Tyne	BrandD	Clone2
32	4.842598	Tyne	BrandD	Clone2
33	4.362592	Tyne	BrandD	Clone2
34	2.230210	Tyne	BrandD	Clone3
35	3.104323	Tyne	BrandD	Clone3
36	4.762764	Tyne	BrandD	Clone3
37	2.319406	Wear	BrandA	Clone1
38	2.098191	Wear	BrandA	Clone1
39	3.541969	Wear	BrandA	Clone1
40	3.888784	Wear	BrandA	Clone2
41	3.960038	Wear	BrandA	Clone2
42	5.742290	Wear	BrandA	Clone2
43	5.244230	Wear	BrandA	Clone3
44	4.795048	Wear	BrandA	Clone3
45	5.380755	Wear	BrandA	Clone3
46	3.610532	Wear	BrandB	Clone1
47	2.247651	Wear	BrandB	Clone1
48	3.388947	Wear	BrandB	Clone1
49	4.061630	Wear	BrandB	Clone2
50	5.478983	Wear	BrandB	Clone2

```
51   4.303407    Wear    BrandB    Clone2
52   3.973060    Wear    BrandB    Clone3
53   4.632224    Wear    BrandB    Clone3
54   5.284316    Wear    BrandB    Clone3
55   2.480984    Wear    BrandC    Clone1
56   3.950710    Wear    BrandC    Clone1
57   2.133731    Wear    BrandC    Clone1
58   5.586468    Wear    BrandC    Clone2
59   6.918344    Wear    BrandC    Clone2
60   5.270421    Wear    BrandC    Clone2
61   2.015707    Wear    BrandC    Clone3
62   3.467042    Wear    BrandC    Clone3
63   5.028927    Wear    BrandC    Clone3
64   2.307174    Wear    BrandD    Clone1
65   2.962274    Wear    BrandD    Clone1
66   2.699762    Wear    BrandD    Clone1
67   6.569412    Wear    BrandD    Clone2
68   6.405130    Wear    BrandD    Clone2
69   5.990973    Wear    BrandD    Clone2
70   2.553241    Wear    BrandD    Clone3
71   2.592766    Wear    BrandD    Clone3
72   1.761603    Wear    BrandD    Clone3
```

We begin by calculating *SST* because it is easy, and this will make us feel better:

$$CF = \frac{277.3372^2}{2 \times 4 \times 3 \times 3} = \frac{76\,915.9}{72} = 1068.276$$

$$SST = \sum y^2 - CF = 1185.434 - 1068.276 = 117.1572$$

Now we can work out the three main effect sums of squares, using the equations above:

$$SSA = \frac{132.691^2 + 144.6461^2}{4 \times 3 \times 3} - CF = 1.98506$$

$$SSB = \frac{69.927^2 + 72.1808^2 + 71.1812^2 + 64.0482^2}{2 \times 3 \times 3} - CF = 2.21157$$

$$SSC = \frac{68.157^2 + 109.8509^2 + 99.3293^2}{2 \times 4 \times 3} - CF = 39.1777$$

Calculation of the three two-way interactions requires that we draw up **three tables of subtotals**: an *AB* table summing over *Daphnia* clones and replicates, an *AC* table summing over Detergents and replicates, and a *BC* table summing over Water and replicates. The subtotals in each of the three tables are squared, added up and divided by the number of numbers that were added together to get each subtotal ($3 \times 3 = 9$, $4 \times 3 = 12$ and $2 \times 3 = 6$ respectively). Here are the three tables, and the sums of their squares:

	BrandA	BrandB	BrandC	BrandD
Tyne	32.95627	35.20004	34.32889	30.20583
Wear	36.97071	36.98075	36.85233	33.84233

$\sum T^2 = 9653.83$

	Clone1	Clone2	Clone3
Tyne	34.41566	45.67501	52.60035
Wear	33.74133	64.17588	46.72892

$\sum T^2 = 13\,478.05$

	Clone1	Clone2	Clone3
BrandA	16.39336	23.51401	30.01961
BrandB	17.57484	26.41758	28.18837
BrandC	18.42801	28.63683	24.11638
BrandD	15.76078	31.28247	17.00491

$\sum T^2 = 6781.597$

The two-way interaction sums of squares now look like this:

$$SSAB = \frac{9653.83}{3 \times 3} - SSA - SSB - CF = 0.17486$$

$$SSAC = \frac{13\,478.05}{4 \times 3} - SSA - SSC - CF = 13.732$$

$$SSBC = \frac{6781.597}{2 \times 3} - SSB - SSC - CF = 20.6006$$

Calculation of the three-way interaction sum of squares involves working out all of the individual subtotals for the $2 \times 4 \times 3 = 24$ treatment combinations, each of which is the sum of $n = 3$ replicates:

```
 8.433794   7.959566   8.327708   9.247130   9.862586
 8.565425   7.791576   7.969209   9.922901  13.591112
12.573563  13.844020  10.861595  17.775234  12.316954
18.965514  14.599572  15.420034  14.298773  13.889600
13.604706  10.511676  10.097297   6.907610
```

Now we need to square each of these, add them up and divide by 3:

$$SSABC = \frac{34.56.017}{3} - SSA - SSB - SSC - SSAB - SSAC - SSBC - CF = 5.8476$$

The error sum of squares is the last thing we need to calculate:

$$SSE = 117.152 - 1.98506 - 2.21157 - 39.1777 - 0.17486 - 13.732 - 20.6006 - 5.8476$$

$$= 33.42776$$

Here is the complete Anova table.

Source	SS	d.f.	MS	F
Water	1.985	1	1.985	2.852
Detergent	2.116	3	0.705	1.013
Daphnia	39.178	2	19.589	28.145
Water:Detergent	0.1749	3	0.058	0.083
Water:*Daphnia*	13.732	2	6.866	9.865
Detergent:*Daphnia*	20.6006	6	3.433	4.932
Three-way interaction	5.8476	6	0.975	1.401
Error	33.4278	48	$s^2 = 0.696$	
Total	117.1572	71		

The three-way interaction is not significant, nor is the Water:Detergent interaction, but both other two-way interactions Water:*Daphnia* and Detergent:*Daphnia* are highly significant. All three factors therefore appear in at least one significant interactions, so model simplification ends here. The apparently insignificant main effects of Water and Detergent should *not* be interpreted as meaning that these factors have no significant effect on population growth rate. Always check the highest-order interactions first, and stop model simplification as soon as all of the terms have appeared in at least one significant interaction.

We can compare our answers with those produced by the computer:

```
attach(daphnia)
names(daphnia)
```

```
[1] "Growth.rate"    "Water"    "Detergent"    "Daphnia"
```

There are three factors, so initial data inspection requires three plots:

```
par(mfrow=c(1,3))
plot(Water,Growth.rate)
plot(Detergent,Growth.rate)
plot(Daphnia,Growth.rate)
```

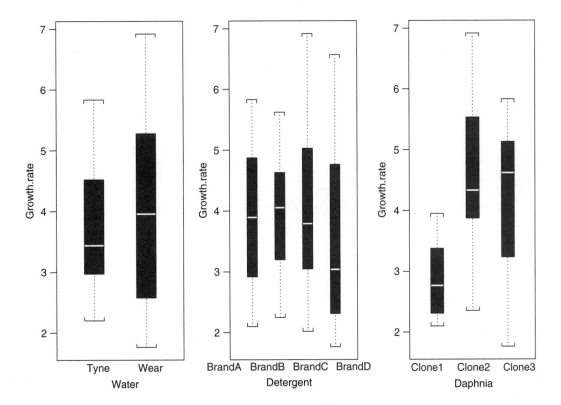

The variance is greater in the Wear than in the Tyne but there is no obvious effect on the mean growth rate. The first three Detergents are all rather similar, but the fourth seems to cause a lower growth rate. The first *Daphnia* clone has a lower growth rate than the other two. We shall test these assertions and look for interactions between the effects of Water, Detergent and *Daphnia* clone in the factorial Anova.

The model formula uses the * notation to specify the fit of all interaction terms. In this case the model calls for the estimation of 23 parameters, using up 6 degrees of freedom for the three-way interaction, 11 for the three two-way interactions and 6 for the 3 main effects:

```
model<-aov(Growth.rate~Water*Detergent*Daphnia)
summary(model)
```

	Df	Sum of Sq	Mean Sq	F Value	Pr(F)
Water	1	1.98506	1.98506	2.85042	0.0978380
Detergent	3	2.21157	0.73719	1.05855	0.3754783
Daphnia	2	39.17773	19.58887	28.12828	0.0000000
Water:Detergent	3	0.17486	0.05829	0.08370	0.9686075
Water:Daphnia	2	13.73204	6.86602	9.85914	0.0002587
Detergent:Daphnia	6	20.60056	3.43343	4.93017	0.0005323
Water:Detergent:Daphnia	6	5.84762	0.97460	1.39946	0.2343235
Residuals	48	33.42776	0.69641		

As always, the interpretation begins with the highest-order interactions. The data provide no evidence to support the inclusion of three-way interaction Water:Detergent:*Daphnia* ($p = 0.23$). There is no significant two-way interaction between Water and Detergent ($p = 0.97$), but there are highly significant two-way interactions between Water and *Daphnia* and Detergent and *Daphnia*. All three factors appear in one or more significant interactions, which means that all factors are important. The non-significant main effect for Detergent should *not*, therefore be interpreted as meaning that Detergent has no significant effect on *Daphnia* growth rate. The correct interpretation is that the effect of Detergent on growth rate varies from one *Daphnia* clone to another. Model criticism is carried out using plot(model).

In order to see how an interaction works, we can use **tapply** to calculate a two-dimensional table of mean growth rates for each combination of Detergent and *Daphnia* clone (note the use of **list** to get the two-dimensions we want):

```
tapply(Growth.rate,list(Detergent,Daphnia),mean)
```

	Clone1	Clone2	Clone3
BrandA	2.732227	3.919002	5.003268
BrandB	2.929140	4.402931	4.698062
BrandC	3.071335	4.772805	4.019397
BrandD	2.626797	5.213745	2.834151

The interaction between Detergent and Daphnia is interpreted as follows. Consider BrandA. Mean growth rate rises from Clone1 to Clone2 to Clone3. BrandB and BrandC show a different pattern, with growth rate not increasing from Clone2 to Clone3. BrandD is different again, with growth rate declining steeply from Clone2 to Clone3. In general, it is a good idea to produce plots to show these interaction effects.

This is achieved by using the **interaction.plot** function. Note that the response variable is last in the list, and that the factor to make the x axis goes first in the list:

```
interaction.plot(Detergent,Daphnia,Growth.rate)
```

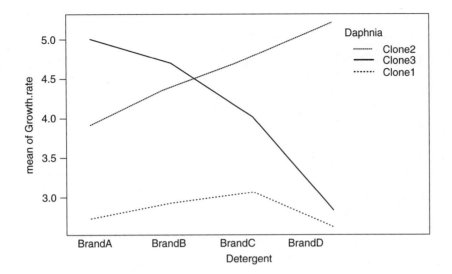

In general, interactions show up as *non-parallelness* in the lines of the interaction plot (see p. 257).

The **model.tables** function comes into its own with more complicated models. Here it is used with the options 'adjusted means' and 'standard errors' selected:

```
model.tables(model,type="adj.means",se=T)
```

```
Tables of adjusted means
Grand mean

    3.851905
se 0.098348

 Water
     Tyne   Wear
    3.6859 4.0179
se 0.1391 0.1391

 Detergent
     BrandA BrandB BrandC BrandD
     3.8848 4.0100 3.9545 3.5582
se   0.1967 0.1967 0.1967 0.1967

 Daphnia
     Clone1 Clone2 Clone3
     2.8399 4.5771 4.1387
se 0.1703 0.1703 0.1703

 Water:Detergent
Dim 1 : Water
Dim 2 : Detergent
     BrandA BrandB BrandC BrandD
Tyne 3.6618 3.9111 3.8143 3.3562
  se 0.2782 0.2782 0.2782 0.2782
```

```
Wear 4.1079 4.1090 4.0947 3.7603
  se 0.2782 0.2782 0.2782 0.2782

 Water:Daphnia
Dim 1 : Water
Dim 2 : Daphnia
      Clone1 Clone2 Clone3
Tyne 2.8680 3.8063 4.3834
  se 0.2409 0.2409 0.2409
Wear 2.8118 5.3480 3.8941
  se 0.2409 0.2409 0.2409

 Detergent:Daphnia
Dim 1 : Detergent
Dim 2 : Daphnia
        Clone1 Clone2 Clone3
BrandA 2.7322 3.9190 5.0033
    se 0.3407 0.3407 0.3407
BrandB 2.9291 4.4029 4.6981
    se 0.3407 0.3407 0.3407
BrandC 3.0713 4.7728 4.0194
    se 0.3407 0.3407 0.3407
BrandD 2.6268 5.2137 2.8342
    se 0.3407 0.3407 0.3407

Water:Detergent:Daphnia
Dim 1 : Water
Dim 2 : Detergent
Dim 3 : Daphnia

, , Clone1
    BrandA BrandB BrandC BrandD
Tyne 2.8113 2.7759 3.2875 2.5972
  se 0.4818 0.4818 0.4818 0.4818
Wear 2.6532 3.0824 2.8551 2.6564
  se 0.4818 0.4818 0.4818 0.4818

, , Clone2
    BrandA BrandB BrandC BrandD
Tyne 3.3076 4.1912 3.6205 4.1057
  se 0.4818 0.4818 0.4818 0.4818
Wear 4.5304 4.6147 5.9251 6.3218
  se 0.4818 0.4818 0.4818 0.4818

, , Clone3
    BrandA BrandB BrandC BrandD
Tyne 4.8665 4.7663 4.5349 3.3658
  se 0.4818 0.4818 0.4818 0.4818
Wear 5.1400 4.6299 3.5039 2.3025
  se 0.4818 0.4818 0.4818 0.4818
```

Because the option is adj.means, the standard errors (above) are standard errors of differences rather than standard errors of means. This makes it easier to carry out pairwise comparisons (but see p. 274).

Fractional factorials

As we have seen, factorial experiments with many factors quickly suffer from runaway expansion. In most cases, all of the combinations of factor levels will be of interest because there is no way of knowing in advance which are going to turn out to be interesting and which are not. Sometimes, however, there are circumstances that allow us to say in advance that certain high-order interactions are not interesting. In other cases, we may know that an interaction is unimportant from previous experimental work. In such circumstances, we can save space and money by employing fractional factorial designs. Suppose we have 5 factors, each with 2 levels. This is called a 2^5 experiment because there are $2 \times 2 \times 2 \times 2 \times 2 = 32$ treatment combinations. It is a field experiment on plant growth with nutrients and pests: plus and minus the nutrients N, P and K and plus and minus pesticides against insects and molluscs. With minimal replication the full factorial would require 64 plots, and this may be well beyond our resources to manage. But if we are willing to forego information on the highest-order interactions, then massive savings in resource requirements can be made. The built-in function called **fac.design** is useful here.

First create an object (called factornames here) to contain the names of the factors and their respective factor levels; two levels each is the maximum allowed for this function:

```
factornames <- list(insects = c("plus", "minus"),
molluscs = c("plus", "minus"),
N = c("plus", "minus"),
P = c("plus", "minus"),
K = c("plus", "minus"))
```

Now use **fac.design** to produce the experimental design. The first argument is the only compulsory one: it gives the factor levels (2, 2, 2, 2, and 2 in this case) generated by rep (2,5). The fraction of the factorial required (1/4 in this case) is specified by the fraction option, here expressed as a model formula, showing the interactions we are prepared to sacrifice:

```
experiment <- fac.design(rep(2,5), factor.names = factornames,
          fraction = ~ - insects:molluscs:N + N:P:K)
experiment
```

```
  insects molluscs   N       P       K
1 minus      plus  plus    plus    plus
2  plus     minus  plus    plus    plus
3  plus      plus minus   minus    plus
4 minus     minus minus   minus    plus
5  plus      plus minus    plus   minus
6 minus     minus minus    plus   minus
7 minus      plus  plus   minus   minus
8  plus     minus  plus   minus   minus

Fraction: ~ - insects:molluscs:N + N:P:K
```

The message at the bottom of the design table shows the interactions that have been sacrificed in reducing the design from 32 treatment combinations to 8 (a one-quarter fractional design). Note that each factor is balanced: four plots receive each addition, four plots do not. The main effect of nitrogen, for example, comes from a comparison of plots $\{1, 2, 7, 8\}$ with plots $\{3, 4, 5, 6\}$. The interaction between insects and molluscs comes from comparing $\{3, 5\}$ with $\{1, 7\}$ and $\{2, 8\}$. And so on.

Multiple comparisons

Multiple comparisons are a problem because they influence the reliability of α, the probability of a Type I error. If we do one test at $\alpha = 0.05$, then we can be reasonably confident that the probability of a false positive is 5%. But if we do 40 comparisons, we expect at least 2 of them to appear to be significant by chance alone. How are we to know which are real and which are spurious differences? The answer, of course, is that we can't.

If we *must* do multiple comparisons, then we had better make α smaller, so that we can remain reasonably sure that the null hypotheses we reject really are false. But there is an obvious trade-off here. The harder we contrive to make it to commit a Type I error, the easier it becomes to commit a Type II error. If we make α too small, we shall end up accepting a lot of false null hypotheses.

There are many ways of dealing with multiple comparisons. One of the simplest and best known is **Bonferroni's correction**, which reduces α in proportion to the number of comparisons we make (m). Just replace α by α/m.

But we can often do much better than this, using the built-in function **multicomp**. The method it uses has a wonderful name; it is called **Tukey's honestly significant difference** (HSD). The central idea is 'family-wise error rate protection', which means that the probability that *all* bounds hold is at least one α. The alternative specifies 'comparison-wise error rate protection', which means that the probability that any one preselected bound holds is one α.

The key point is that there is more information in these simultaneous confidence intervals than there would be in a collection of typical pairwise tests. There are three broad kinds of multiple comparisons:

- multiple comparisons with a control

- all pairwise comparisons

- user defined (with **lmat** and **adjust**)

You can specify the kinds of comparisons to be carried out by providing a keyword. Available keywords are the default mca for all pairwise differences of adjusted means; mcc for pairwise differences between a single adjusted mean of the focus factor or **lmat** column (the control) and the others; or none, if the adjusted means or **lmat** columns themselves are of interest without further differencing. One α is the desired joint confidence level (error.type="fwe") or comparison-wise confidence level (error.type="cwe"). The default is alpha=.05.

A character string can be provided to specify the desired method for critical point calculation. The default is "best.fast": other methods include:

- "lsd" Fisher's unprotected *least significant difference* method (see below)

- "tukey" requests the critical point to be the Tukey studentised-range quantile scaled by $\sqrt{2}$

- "dunnett" for comparisons-with-control intervals

- "bon" for the Bonferroni method (see above)

Least significant difference

The structure of a simple t test for comparing two means is this:

$$t = \frac{\text{difference}}{\text{standard error of that difference}}$$

Now, if instead of calculating the value of t from a given difference, we reverse the question and ask, What is the smallest difference that would come out as significant if we did a t test? The answer, of course, is the difference that, when divided by the standard error of the difference, gives exactly the value of Student's t that you would find in tables. So if we write LSD for least significant difference, then

$$t_{\text{from two-tailed tables}} = \frac{LSD}{\sqrt{2\frac{s^2}{n}}}$$

and rearranging gives

$$LSD = t_{tables} \times \sqrt{2\frac{s^2}{n}}$$

One virtue of LSD is for drawing error bars on barplots. If you draw a whisker up and down from the top of the bar for a distance of $LSD/2$, then

- non-overlapping error bars indicate that the means are significantly different
- overlapping error bars indicate that the means are *not* significantly different

The other traditional recipes for error bars (± 1 standard error, or $\pm 95\%$ confidence interval) fail on one or other of these counts. Non-overlapping 1 s.e. bars do *not* necessarily mean that the means are significantly different, because they are based on standard errors, not on the (larger) standard error of the difference between two means. Overlapping 95% confidence intervals do not necessarily mean that the means are not significantly different (the difference between the means could be almost as much as 4 s.e and yet the confidence intervals would still overlap). Journal editors are not sympathetic to this argument. We have always used ± 1 s.e., and we are not going to change now!

Multiple comparisons in S-Plus

It is unusual for a study to end with the initial Anova table, unless of course this indicates that there are no significant differences between any of the means (see p. 251). More typically we want to know whether there are significant differences between subsets of the treatment means, and how large these differences are. The aim of the function called **multicomp** is to provide more information in a set of simultaneous confidence intervals than would be provided by a collection of significance tests for differences. The example is a one-way Anova where the categorical variable (woodland habitat type) has 16 levels, and we are interested to know which kinds of habitats produce significantly different average yields of fungi in autumn:

```
attach(Fungi)
names(Fungi)

[1]  "Habitat"    "Fungus.yield"
```

First we do a straightforward one-way Anova to establish whether there are any significant differences in fungus yield between the 16 habitat types:

```
model<-aov(Fungus.yield~Habitat)
summary(model)

            Df Sum of Sq  Mean Sq F Value Pr(F)
   Habitat  15  6556.529 437.1019 54.1156     0
 Residuals 144  1163.115   8.0772
```

Yes. There are highly significant differences between the means. We can look at the data using **plot** to see what these differences might be:

```
plot(Habitat,Fungus.yield)
```

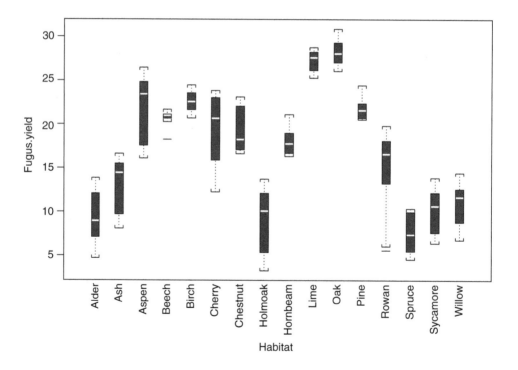

It is clear that both mean fungus yield and variability differ substantially from one habitat to another, with the highest yields in lime and oak, and the lowest in spruce.

To assess all pairwise comparisons we can use **multicomp** as follows:

```
allcomp<-multicomp(model)
plot(allcomp)
```

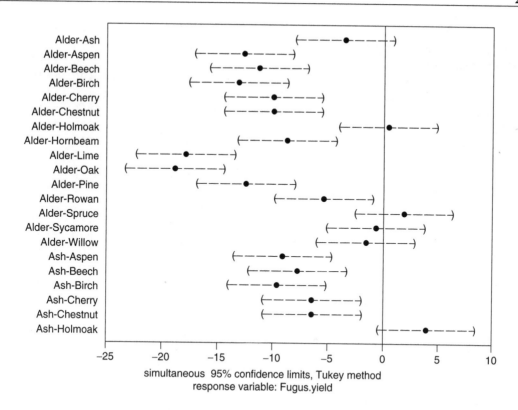

simultaneous 95% confidence limits, Tukey method
response variable: Fugus.yield

This is the first of six pages of plotted output! The significant comparisons are those that *do not intersect* the zero line (i.e. most of them). With a design as large as this, it might be best to look at the printed output, by typing **allcomp**:

```
allcomp

95 % simultaneous confidence intervals for specified
linear combinations, by the Tukey method

critical point: 3.4877
```

Note that if you were constructing a 95% confidence interval for any one of the means, you would have taken the two-tailed value of Student's t with d.f. $= 9$ of 2.262, or if you were comparing two means using least significant difference (see below) with d.f. $= 18$, $t = 2.101$. Note that the critical point used by Tukey's method of honestly significant differences is much larger than this (3.4877).

```
response variable: Fungus.yield

intervals excluding 0 are flagged by '****'
```

	Estimate	Std.Error	Lower Bound	Upper Bound	
Alder-Ash	-3.5100	1.27	-7.9500	0.920	
Alder-Aspen	-12.6000	1.27	-17.0000	-8.180	****
Alder-Beech	-11.3000	1.27	-15.7000	-6.820	****
Alder-Birch	-13.1000	1.27	-17.6000	-8.690	****
Alder-Cherry	-9.9600	1.27	-14.4000	-5.530	****

Alder-Chestnut	-9.9400	1.27	-14.4000	-5.510	****
Alder-Holmoak	0.4580	1.27	-3.9800	4.890	
Alder-Hornbeam	-8.7100	1.27	-13.1000	-4.270	****
Alder-Lime	-17.9000	1.27	-22.3000	-13.400	****
Alder-Oak	-18.8000	1.27	-23.3000	-14.400	****
Alder-Pine	-12.4000	1.27	-16.9000	-7.990	****
Alder-Rowan	-5.3700	1.27	-9.8000	-0.937	****
Alder-Spruce	1.9100	1.27	-2.5300	6.340	
Alder-Sycamore	-0.6340	1.27	-5.0700	3.800	
Alder-Willow	-1.5600	1.27	-5.9900	2.880	
Ash-Aspen	-9.1000	1.27	-13.5000	-4.670	****
Ash-Beech	-7.7400	1.27	-12.2000	-3.310	****
Ash-Birch	-9.6100	1.27	-14.0000	-5.180	****
Ash-Cherry	-6.4500	1.27	-10.9000	-2.010	****
Ash-Chestnut	-6.4300	1.27	-10.9000	-2.000	****
Ash-Holmoak	3.9700	1.27	-0.4630	8.400	
Ash-Hornbeam	-5.1900	1.27	-9.6300	-0.761	****
Ash-Lime	-14.3000	1.27	-18.8000	-9.910	****
Ash-Oak	-15.3000	1.27	-19.8000	-10.900	****
Ash-Pine	-8.9100	1.27	-13.3000	-4.480	****
Ash-Rowan	-1.8600	1.27	-6.2900	2.580	
Ash-Spruce	5.4200	1.27	0.9870	9.850	****
Ash-Sycamore	2.8800	1.27	-1.5500	7.310	
Ash-Willow	1.9600	1.27	-2.4800	6.390	
Aspen-Beech	1.3600	1.27	-3.0700	5.790	
Aspen-Birch	-0.5060	1.27	-4.9400	3.930	
Aspen-Cherry	2.6600	1.27	-1.7800	7.090	
Aspen-Chestnut	2.6700	1.27	-1.7600	7.100	
Aspen-Holmoak	13.1000	1.27	8.6400	17.500	****
Aspen-Hornbeam	3.9100	1.27	-0.5250	8.340	
Aspen-Lime	-5.2400	1.27	-9.6800	-0.811	****
Aspen-Oak	-6.2300	1.27	-10.7000	-1.790	****

And so on for several pages. Aspen and Holmoak are significantly different, but Aspen and Hornbeam are not significantly different. Note that for significance, the upper and lower bounds must not include zero. Either the upper bound must be negative (e.g. Aspen-Oak) or the lower bound must be positive (e.g. Aspen-Holmoak). In the plot, the significant comparisons are *the ones that do not cross the vertical line at zero.*

In other circumstances we may be less interested in comparing all the treatments with one another as in comparing the treatments to a single control. In a medical experiment, for example, the response to each drug treatment might be compared with the response shown by the placebo. Such all-to-one comparisons are known as MCC (multiple comparisons with a control). We need to provide an extra argument for the **multicomp** function specifying the column number of the factor level which is to act as the control. For the purposes of demonstration, let's compare all the yields with the yield of birch (column 5):

```
controlcomp<-multicomp(model,comparisons="mcc",control=5)
plot(controlcomp)
```

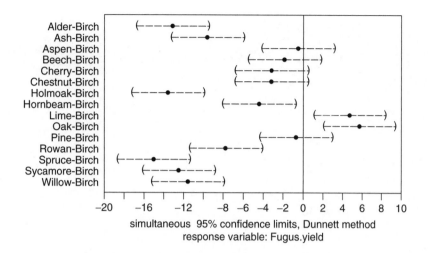

simultaneous 95% confidence limits, Dunnett method
response variable: Fugus.yield

This is a much more manageable figure. Oak and Lime yield significantly more fungus than Birch while Alder, Ash, Holmoak, Hornbeam, Rowan, Spruce, Sycamore and Willow yield significantly less.

Further reading

Hochberg, Y. and Tamhane, A.C. (1987) *Multiple Comparison Procedures*. New York, John Wiley.
Hsu, J.C. (1996) *Multiple Comparisons: Theory and Methods*. London, Chapman & Hall.
Miller, R.G. (1997) *Beyond ANOVA: Basics of Applied Statistics*. London, Chapman & Hall.
Neter, J., Wasserman, W. *et al.* (1985) *Applied Linear Regression Models*. Homewood, Illinois, Irwin.

16

Analysis of covariance

Analysis of covariance combines elements from regression and analysis of variance. There is at least one continuous explanatory variable and at least one categorical explanatory variable. The procedure works like this:

- fit two or more linear regressions of y against x (one for each level of the factor)

- estimate different slopes and intercepts for each level

- use model simplification (deletion tests) to eliminate unnecessary parameters

For example, we could use Ancova in a medical experiment where the response variable was 'days to recovery' and the explanatory variables were 'smoker or not' (categorical) and 'blood cell count' (continuous). In economics, local unemployment rate might be modelled as a function of country (categorical) and local population size (continuous).

Suppose we are modelling weight (the response variable) as a function of sex and age. Sex is a factor with two levels (male and female) and age is a continuous variable. The maximal model therefore has four parameters: two slopes (a slope for males and a slope for females) and two intercepts (one for males and one for females) like this:

$$weight_{male} = a_{male} + b_{male} \times age$$

$$weight_{female} = a_{female} + b_{female} \times age$$

This maximal model is shown in the top left panel.

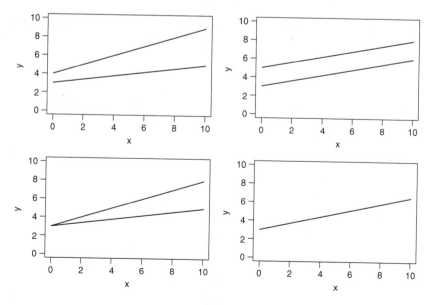

Model simplification is an essential part of analysis of covariance, because the principle of parsimony requires that we keep as few parameters in the model as possible. In this case the process of model simplification begins by asking whether we need all four parameters. Perhaps we could make do with two intercepts and a common slope (top right). Or a common intercept and two different slopes (bottom left). Alternatively, there may be no effect of sex at all, in which case we only need two parameters (one slope and one intercept) to describe the effect of age on weight (bottom right). There again, the continuous variable may have no significant effect on the response, so we only need two parameters to describe the main effects of sex on weight. This would show up as two separated, horizontal lines in the plot (one mean weight for each sex). In the limit, neither the continuous nor the categorical explanatory variables might have any significant effect on the response, in which case, model simplification will lead to the one-parameter null model $\hat{y} = \bar{y}$ (a single horizontal line).

Calculations involved in Ancova

We start with the simple example that we analysed as an Anova in Chapter 15, in which six different plant genotypes were exposed to four different light regimes. When we analysed the data as a one-way Anova, we found no significant effect of photoperiod, because the variation in growth rate between the different genotypes was so great. When we took differences between the genotype means into account by doing a two-way Anova, we found a highly significant effect of photoperiod. Here we change the explanatory variable from a categorical variable (day length with four levels) into a continuous variable (the number of hours of illumination). There are two benefits in doing this

- it saves us 2 degrees of freedom in describing the effect of photoperiod (1 slope and 1 intercept instead of 4 photoperiod means)

- it allows us to test for an interaction between genotype and photoperiod on growth response (recall that we couldn't do this using Anova because there was no replication, so the interaction term had to be used as the error term)

- other things being equal, a regression approach will generally be more powerful than an Anova-based approach, because Anova takes no account of the *ordering* of the factor levels, only of the differences between their means

The calculations are as follows. We need to do six separate regressions of growth against photoperiod (one for each genotype), keeping account of the corrected sums of products, *SSXY*, for each one separately. Here are the data again, this time showing the number of hours of illumination as the continuous explanatory variable:

Genotype	8 h	12 h	16 h	24 h
A	2	3	3	4
B	3	4	5	6
C	1	2	1	2
D	1	1	2	2
E	2	2	2	2
F	1	1	2	3

Here is the Anova table that we calculated in Chapter 15.

Source	SS	d.f.	MS	F
Photoperiod	7.125	3	2.375	7.703
Genotype	27.875	5	5.575	18.08
Error	4.625	15	$s^2 = 0.308$	
Total	39.625	23		

Genotype SS (SSG) and Total SS will remain as they were. What we plan to do is to reduce Error SS and increase the explained Photoperiod SS by taking account of the fact that the different day lengths can be ordered.

We shall do an overall regression to begin with, just estimating a single overall slope for the relationship between Growth and Photoperiod. The Anova table will therefore look like this.

Source	SS	d.f.	MS	F
Photoperiod		1		
Genotype	27.875	5		
Error		17	s^2	
Total	39.625	23		

We have reduced the degrees of freedom for Photoperiod from 3 to 1 (a useful model simplification in itself) and increased the error degrees of freedom accordingly (this is good for reducing the error variance). So we start by calculating an overall regression sum of squares using all 24 data points. First we need the famous five:

$$\sum x, \sum x^2, \sum y, \sum y^2, \sum xy$$

```
attach(photoperiod)
names(photoperiod)
```
```
[1] "Genotype"    "Growth"    "Photoperiod"
```

So x is Photoperiod and y is Growth. We get the famous five like this:

```
sum(Photoperiod);sum(Photoperiod^2)
```
```
[1] 360
[1] 6240
```
```
sum(Growth);sum(Growth^2)
```
```
[1] 57
[1] 175
```
```
sum(Growth*Photoperiod)
```
```
[1] 932
```

So the corrected sums of squares (Chapter 14) are calculated like this:

$$SST = 175 - \frac{57^2}{24} = 39.625$$

$$SSXY = 932 - \frac{57 \times 360}{24} = 77$$

$$SSX = 6240 - \frac{360^2}{24} = 840$$

Now we have everything we need to calculate the slope of the overall straight line relating Growth to Photoperiod (remember that $b = SSXY/SSX$):

$$b = \frac{77}{840} = 0.0917$$

The next step is to work out the sums of squares for the Anova table. First the explained sum of squares (the regression sum of squares, SSR):

$$SSR = b \times SSXY = 0.0917 \times 77 = 7.0583$$

We know $SST = 39.625$ already from our Anova calculations (and see above), so we obtain SSE by subtracting SSR and SSG from SST:

$$SSE = SST - SSR - SSG = 39.625 - 7.0583 - 27.875 = 4.4917$$

The Anova table can now be completed.

Source	SS	d.f.	MS	F
Photoperiod	7.0583	1	7.0583	25.575
Genotype	27.875	5	5.575	20.199
Error	4.4917	17	$s^2 = 0.276$	
Total	39.625	23		

This is already a considerable improvement over the two-way Anova: the F ratio for the effect of photoperiod has gone up from 7.703 to 25.575, and the error variance has gone down from 0.308 to 0.276 (Chapter 15). But we can do better than this. We can ask whether there is any evidence that *different genotypes showed different responses* to photoperiod. This is an interaction term, which involves fitting six different slopes to the model instead of one common slope (as we just did here).

The calculations are not hard, just tedious. We need to find SSR separately for each of the six genotypes. For this, we need six values of b and six values of $SSXY$ (Box 16.1).

Box 16.1 Calculations for analysis of covariance

First, we need the sums and sums of squares:

tapply(Growth,Genotype,sum)

A	B	C	D	E	F
12	18	6	6	8	7

tapply(Growth^2,Genotype,sum)

A	B	C	D	E	F
38	86	10	10	16	15

tapply(Photoperiod,Genotype,sum)

A	B	C	D	E	F
60	60	60	60	60	60

tapply(Photoperiod^2,Genotype,sum)

A	B	C	D	E	F
1040	1040	1040	1040	1040	1040

Finally, we calculate the sums of products:

tapply(Photoperiod*Growth,Genotype,sum)

A	B	C	D	E	F
196	296	96	100	120	124

All the SSX are the same:

$$SSX = 1040 - \frac{60^2}{4} = 140$$

Now we calculate the six separate $SSXY$:

$$SSXY_1 = 196 - \frac{12 \times 60}{4} = 16.0$$

$$SSXY_2 = 296 - \frac{18 \times 60}{4} = 26.0$$

$$SSXY_3 = 96 - \frac{6 \times 60}{4} = 6.0$$

$$SSXY_4 = 100 - \frac{6 \times 60}{4} = 10.0$$

$$SSXY_5 = 120 - \frac{8 \times 60}{4} = 0.0$$

$$SSXY_6 = 124 - \frac{7 \times 60}{4} = 19.0$$

So the six slopes are $16/140 = 0.11429$, $26/140 = 0.1857$, $6/140 = 0.04286$, 0 and $19/140 = 0.1357$. Finally we can work out the six SSR:

Clone	SSXY	SSX	b	SSR
A	$196 - \dfrac{60 \times 12}{4} = 16$	$1040 - \dfrac{60^2}{4} = 140$	0.114	1.829
B	$296 - \dfrac{60 \times 18}{4} = 26$	$1040 - \dfrac{60^2}{4} = 140$	0.186	4.829
C	$96 - \dfrac{60 \times 6}{4} = 6$	$1040 - \dfrac{60^2}{4} = 140$	0.043	0.257
D	$100 - \dfrac{60 \times 6}{4} = 10$	$1040 - \dfrac{60^2}{4} = 140$	0.071	0.714
E	$120 - \dfrac{60 \times 8}{4} = 0$	$1040 - \dfrac{60^2}{4} = 140$	0	0
F	$124 - \dfrac{60 \times 7}{4} = 19$	$1040 - \dfrac{60^2}{4} = 140$	0.136	2.579
			Total	10.208

The six component regression sums of squares add up to 10.208. Note that there is substantial variation between the slopes for the different genotypes (b ranges from 0 to 0.186).

The next step is to work out the sum of squares attributable to differences between the slopes. The logic is simple. A single regression slope explained an *SSR* of 7.058 (above). Different slopes, as we have just seen, explained an *SSR* of 10.208. So the difference, attributable to having six slopes instead of one, is $10.208 - 7.058 = 3.149$. Because we have estimated six slopes now, where we had one before, this sum of squares is based on $6 - 1 = 5$ d.f.

Now we can draw all the information together into a single Anova table.

Source	*SS*	d.f.	*MS*	*F*
Genotype	27.87	5	5.574	43.21
Regression	7.059	1	7.059	54.72
Differences in slope	3.149	5	0.63	4.88
Error	1.55	12	$s^2 = 0.129$	
Total	39.63	23		

The Ancova has produced a much better model than either the simple regression or the one-way Anova, and a substantially better model than the two-way Anova. Generally, if the nature of your data allows it, it is a good idea to use analysis of covariance whenever you can.

Analysis of covariance in S-Plus

We could use either **lm** or **aov**; the choice affects only the format of the summary table. We shall use both and compare their output. The model with six different slopes is specified using the asterisk operator:

Growth ~ Genoptype * Photoperiod

An important bit of housekeeping needs to be done before we start. If you have not told it otherwise, S-Plus will assume that you want to use Helmert contrasts (Chapter 18). The output in this chapter, however, has been produced using treatment contrasts (like R would have done), so if you want your output to look like mine, you will need to alter the contrasts option like this:

options(contrasts=c("contr.treatment","contr.poly"))

We call the output model and work like this, starting with **lm**:

```
model<-lm(Growth~Genotype*Photoperiod)
summary(model)
```

Coefficients:

	Estimate	Std. Error	t value	Pr(>|t|)	
(Intercept)	1.28571	0.48865	2.631	0.02193	*
GenotypeB	0.42857	0.69105	0.620	0.54674	
GenotypeC	-0.42857	0.69105	-0.620	0.54674	
GenotypeD	-0.85714	0.69105	-1.240	0.23855	
GenotypeE	0.71429	0.69105	1.034	0.32169	
GenotypeF	-1.57143	0.69105	-2.274	0.04213	*
Photoperiod	0.11429	0.03030	3.771	0.00267	**
GenotypeB.Photoperiod	0.07143	0.04286	1.667	0.12145	
GenotypeC.Photoperiod	-0.07143	0.04286	-1.667	0.12145	
GenotypeD.Photoperiod	-0.04286	0.04286	-1.000	0.33705	
GenotypeE.Photoperiod	-0.11429	0.04286	-2.667	0.02054	*
GenotypeF.Photoperiod	0.02143	0.04286	0.500	0.62612	

The output from Ancova takes a lot of getting used to. The most important thing to remember is that there are 12 rows in the summary table because the model has estimated 12 parameters from the data: 6 intercepts and 6 slopes. With treatment contrasts (as here), the rows of the table are interpreted as follows:

- `Intercept` in row 1 is the *intercept* for Genotype A (first level in the alphabet)

- The next five rows are *differences between intercepts*; for instance, row 2 (labelled `GenotypeB`) is the difference in intercept between Genotypes B and A

- The *slope* is in row 6 (labelled `Photoperiod`); this is the slope of the graph of growth against photoperiod for Genotype A (because this comes first in the alphabet)

- The last 5 rows are *differences between slopes*; for instance, row 7 (labelled `GenotypeB.Photoperiod`) is the difference between the slopes of the graphs for Genotypes B and A

So, for example, the parameterised equation for Genotype D is as follows. The intercept is $1.285\,71 - 0.857\,14 = 0.428\,57$, and the slope is $0.114\,29 - 0.042\,86 = 0.071\,43$. The rightmost column indicates the parameter values that are significantly different from zero (when compared with Genotype A). The table shows that there is one significant interaction term: Genotype E has a significantly different response to Photoperiod compared with the other genotypes; in fact, it didn't respond at all to increasing light exposure. It also shows that Genotype F has a significantly lower intercept than has Genotype A.

This output generated by **lm** is good when you want to look at the individual parameter values. Let's see what happens when we fit the same model using **aov**:

```
model<-aov(Growth~Genotype*Photoperiod)
summary(model)
```

	Df	Sum Sq	Mean Sq	F value	Pr(>F)	
Genotype	5	27.8750	5.5750	43.3611	2.848e-007	***
Photoperiod	1	7.0583	7.0583	54.8981	8.171e-006	***
Genotype:Photoperiod	5	3.1488	0.6298	4.8981	0.0113	*
Residuals	12	1.5429	0.1286			

Here we get a tidy Anova table to compare with the one we calculated earlier by hand, but there is no information on *which* parameters contribute to the significant differences. In practice there is no need to fit two separate models as we did here using **lm** and **aov**. You only need to use one of them, then use **summary.aov** to produce an Anova table or **summary.lm** to produce a list of parameter estimates and standard errors. Try this:

```
summary.aov(model)
summary.lm(model)
```

Ancova with different values of the covariates

The calculations are more complicated if the different factor levels are associated with different values of the covariate. In the last example, all the genotypes were exposed to exactly the same photoperiods, which made the calculations much more straightforward. The next worked example concerns an experiment on the impact of grazing on the seed production of a biennial plant. Forty plants were allocated to two treatments, grazed and ungrazed, and the grazed plants were exposed to rabbits during the first two weeks of stem elongation. They were then protected from subsequent grazing by the erection of a fence and allowed to regrow. Because initial plant size was thought likely to influence fruit production, the diameter of the top of the rootstock was measured before each plant was potted up. At the end of the growing season, the fruit production (dry weight, mg) was recorded on each of the 40 plants, and this forms the response variable in the following analysis.

Ungrazed plants

Fruit	59.77	60.98	14.73	19.28	34.25	35.53	87.73	63.21	24.25
Roots	6.225	6.487	4.919	5.130	5.417	5.359	7.614	6.352	4.975
Fruit	64.34	52.92	32.35	53.61	54.86	64.81	73.24	80.64	18.89
Roots	6.930	6.248	5.451	6.013	5.928	6.264	7.181	7.001	4.426
Fruit	75.49	46.73							
Roots	7.302	5.836							

Grazed plants

Fruit	80.31	82.35	105.1	73.79	50.08	78.28	41.48	98.47	40.15
Roots	8.988	8.975	9.844	8.508	7.354	8.643	7.916	9.351	7.066
Fruit	116.1	38.94	60.77	84.37	70.11	14.95	70.70	71.01	83.03
Roots	10.25	6.958	8.001	9.039	8.910	6.106	7.691	8.515	8.530
Fruit	52.26	46.64							
Roots	8.158	7.382							

The object of the exercise is to estimate the parameters of the minimal adequate model for these data. Here is a plot of the final result so that you can see where we are heading; the triangles are ungrazed plants and the circles are grazed plants.

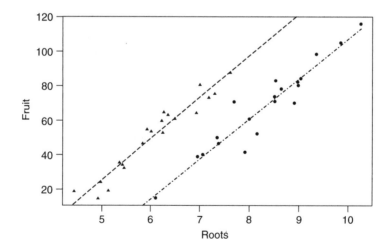

We start by working out the sums, sums of squares and sums of products for the whole data set combined (40 pairs of numbers), and then for each treatment separately (20 pairs of numbers).The data frame is called **ipomopsis**.

```
attach(ipomopsis)
names(ipomopsis)
```

```
[1] "Root"   "Fruit"    "Grazing"
```

First, we'll work out the overall totals based on all 40 data points:

```
sum(Root);sum(Root^2)
```

```
[1] 287.246
[1] 2148.172
```

Start to fill in the table of totals (it helps to be really well organised for these calculations). Check to see where (and why) the sum (287.246) and the sum of squares (2148.172) of the root diameters (the *x* values) have gone in the table:

```
sum(Fruit);sum(Fruit^2)
```

```
[1] 2376.42
[1] 164928.1
```

```
sum(Root*Fruit)
```

```
[1] 18263.16
```

That completes the overall data summary. Now we select only the grazed plant data:

```
sum(Root[Grazing=="Grazed"]);sum(Root[Grazing=="Grazed"]^2)
```

```
[1] 166.188
[1] 1400.834
```

Make sure you see where (and why) these totals go where they go. Now the ungrazed subtotals:

```
sum(Root[Grazing=="Ungrazed"]);sum(Root[Grazing=="Ungrazed"]^2)
```

```
[1] 121.058
[1] 747.3387
```

Now for the data on Fruits for the grazed plants:

```
sum(Fruit[Grazing=="Grazed"]);sum(Fruit[Grazing=="Grazed"]^2)
```

```
[1] 1358.81
[1] 104156.0
```

And the Fruit data from the ungrazed plants:

```
sum(Fruit[Grazing=="Ungrazed"]);sum(Fruit[Grazing=="Ungrazed"]^2)
```

```
[1] 1017.61
[1] 60772.11
```

Finally we want the data for the sums of products for the grazed plants:

```
sum(Root[Grazing=="Grazed"]*Fruit[Grazing=="Grazed"])
```

```
[1] 11753.64
```

and for the ungrazed plants:

```
sum(Root[Grazing=="Ungrazed"]*Fruit[Grazing=="Ungrazed"])
```

```
[1] 6509.522
```

	$\sum$ sums	$\sum^2$ products	$\sum\sum$ totals
x ungrazed	121.058	747.3387	
y ungrazed	1017.61	60 772.11	
xy ungrazed		6 509.522	
$\sum x1 \sum y1$			123 189.8
x grazed	166.188	1 400.834	
y grazed	1358.81	104156.0	
xy grazed		11 753.64	
$\sum x2 \sum y2$			225 817.9
x overall	287.246	2 148.172	
y overall	2376.42	164 928.1	
xy overall		18 263.16	
$\sum x \sum y$			682 617.14

Now we have all the information necessary to carry out the calculations of the corrected sums of squares and products, SSY, SSX and $SSXY$ for the whole data set ($n = 40$) and for the two separate treatments (20 replicates in each).

To get the right answer you will need to be extremely methodical, but there is nothing mysterious or difficult about the process. First, calculate the regression statistics for the whole experiment, ignoring the grazing treatment. Here are the famous five:

```
sum(Root)  sum(Root^2)  sum(Fruit)  sum(Fruit^2)  sum(Root*Fruit)
287.246    2148.172     2376.42     164928.1       18263.16
```

Therefore

$$SST = 164\,928.1 - \frac{2376.42^2}{40} = 23\,743.84$$

$$SSX = 2148.172 - \frac{287.246^2}{40} = 85.4158$$

$$SSXY = 18\,263.16 - \frac{287.246 \times 2376.42}{40} = 1197.731$$

$$SSR = \frac{(1197.731)^2}{85.4158} = 16\,795$$

$$SSE = 23\,743.84 - 16\,795 = 6948.835$$

Next calculate the regression statistics for each of the grazing treatments separately. First, for the grazed plants:

$$SST_g = 104\,156 - \frac{1358.81^2}{20} = 11\,837.79$$

$$SSX_g = 1400.834 - \frac{166.188^2}{20} = 19.9111$$

$$SSXY_g = 11\,753.64 - \frac{1358.81 \times 166.188}{20} = 462.7415$$

$$SSR_g = \frac{(462.7415)^2}{19.9111} = 10\,754.29$$

$$SSE_g = 11\,837.79 - 10\,754.29 = 1083.509$$

so the slope of the graph of Fruit against Root for the grazed plants is given by

$$b_g = \frac{SSXY_g}{SSX_g} = \frac{462.7415}{19.9111} = 23.240$$

Now for the ungrazed plants:

$$SST_u = 60\,772.11 - \frac{1017.61^2}{20} = 8995.606$$

$$SSX_u = 747.3387 - \frac{121.058^2}{20} = 14.58677$$

$$SSXY_u = 6509.522 - \frac{121.058 \times 1017.61}{20} = 350.0302$$

$$SSR_u = \frac{350.0302^2}{14.58677} = 8399.466$$

$$SSE_u = 8995.606 - 8399.466 = 596.1403$$

so the slope of the graph of Fruit against Root for the ungrazed plants is given by

$$b_u = \frac{SSXY_u}{SSX_u} = \frac{350.0302}{14.58677} = 23.996$$

Now add up the regression statistics across the factor levels (grazed and ungrazed):

$$SST_{g+u} = 11\,837.79 + 8995.606 = 20\,833.4$$

$$SSX_{g+u} = 19.9111 + 14.58677 = 34.49788$$

$$SSXY_{g+u} = 462.7415 + 350.0302 = 812.7717$$

$$SSR_{g+u} = 10\,754.29 + 8399.436 = 19\,153.75$$

$$SSE_{g+u} = 1083.509 + 596.1403 = 1684.461$$

The SSR for a model with a single common slope is given by

$$SSR = \frac{(SSXY_{g+u})^2}{SSX_{g+u}} = \frac{812.7717^2}{34.49788} = 19\,148.94$$

and the value of the single common slope is

$$b = \frac{SSXY_{g+u}}{SSX_{g+u}} = \frac{812.7717}{34.49788} = 23.560$$

The difference between the two estimates of SSR ($SSR_{\text{diff}} = 19\,153.75 - 19\,148.94 = 4.81$) is a measure of the significance of the difference between the two slopes estimated separately for each factor level. Finally, SSE is calculated by difference:

$$SSE = SST - SSA - SSR - SSR_{\text{diff}}$$

$$SSE = 23\,743.84 - 2910.44 - 19\,148.94 - 4.81 = 1679.65$$

Now we can complete the Anova table for the full model.

Source	SS	d.f.	MS	F
Grazing	2 910.44	1		
Root	19 148.94	1		
Different slopes	4.81	1	4.81	n.s.
Error	1 679.65	36	46.66	
Total	23 743.84	39		

Degrees of freedom for error are $40 - 4 = 36$ because we have estimated four parameters from the data: two slopes and two intercepts. So the error variance is $46.66 (= SSE/36)$. The difference between the slopes is clearly not significant ($F = 4.81/46.66 = 0.10$) so we can fit a simpler model with a common slope of 23.56. The sum of squares for differences between the slopes (4.81) now becomes part of the error sum of squares.

Source	SS	d.f.	MS	F
Grazing	2 910.44	1	2 910.44	63.9291
Root	19 148.94	1	19 148.94	420.6156
Error	1 684.46	37	45.526	
Total	23 743.84	39		

This is the minimal adequate model. Both of the terms are highly significant and there are no redundant factor levels.

The next step is to calculate the intercepts for the two parallel regression lines. This is done exactly as before, by rearranging the equation of the straight line to obtain $a = y - bx$. For each line we can use the mean values of x and y, with the common slope in each case. Thus

$$a_1 = \overline{Y}_1 - b\overline{X}_1 = 50.88 - 23.56 \times 6.0529 = -91.7261$$

$$a_2 = \overline{Y}_2 - b\overline{X}_2 = 67.94 - 23.56 \times 8.309 = -127.8294$$

This demonstrates that the grazed plants produce, on average, 36.1 fruits *fewer* than the ungrazed plants ($127.83 - 91.73$).

Finally, we need to calculate the standard errors for the common regression slope and for the difference in mean fecundity between the treatments, based on the error variance in the minimal adequate model:

$$s^2 = \frac{1684.46}{37} = 45.526$$

The standard errors are obtained as follows. The standard error of the common slope is found in the usual way:

$$SE_b = \sqrt{\frac{s^2}{SSX}} = \sqrt{\frac{45.526}{19.9111 + 14.45667}} = 1.149$$

The standard error of the intercept of the regression for treatment number 1 (grazed) is also found in the usual way:

$$SE_a = \sqrt{s^2\left[\frac{1}{n} + \frac{(0 - \bar{x})^2}{SSX}\right]} = \sqrt{45.526\left[\frac{1}{20} + \frac{8.3094^2}{34.498}\right]} = 9.664$$

It is clear that the intercept of -127.829 is very significantly less than zero ($t = 127.829/9.664 = 13.2$), suggesting there is a threshold rootstock size before reproduction can begin. Finally, the standard error of the difference between the elevations of the two lines (the grazing effect) is given by

$$SE_{\hat{y}_1 - \hat{y}_2} = \sqrt{s^2\left[\frac{2}{n} + \frac{(\bar{x}_1 - \bar{x}_2)^2}{SSX}\right]}$$

which, substituting the values for the error variance and the mean rootstock sizes of the plants in the two treatments, becomes

$$SE_{\hat{y}_1 - \hat{y}_2} = \sqrt{45.526\left[\frac{2}{20} + \frac{(6.0529 - 8.3094)^2}{34.498}\right]} = 3.357$$

This suggests that any lines differing in elevation by more than about $2 \times 3.357 = 6.66$ mg dry weight would be regarded as significantly different. Thus, the present difference of 36.09 clearly represents a highly significant reduction in fecundity caused by grazing ($t = 10.83$).

Ancova in S-Plus using lm

We now repeat the analysis using **lm**. The response variable is fecundity, and there is one experimental factor (grazing) with two levels (ungrazed and grazed) and one covariate (initial root stock diameter). There are 40 values for each of these variables. You will need to attach the data frame if you did not do so earlier:

```
attach(ipomopsis)
names(ipomopsis)

[1]  "Root"    "Fruit"    "Grazing"
```

The odd thing about these data is that grazing seems to *increase* fruit production, a highly counterintuitive result:

```
tapply(Fruit,Grazing, mean)

Grazed Ungrazed
67.9405  50.8805
```

How could this have come about? We begin by inspecting the data:

```
plot(Root,Fruit)
```

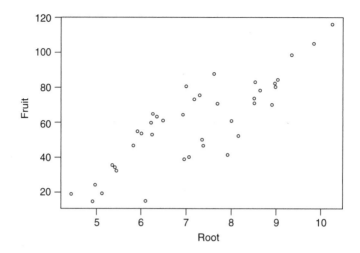

This demonstrates clearly that size matters: the plants that had larger rootstocks at the beginning produced more fruit when they matured. But this plot does not help with the real question, which is, 'How did grazing affect fruit production'? What we need to do is to plot the data separately for the grazed and ungrazed plants. First we plot blank axes:

```
plot(Root,Fruit,type="n")
```

Next we select only those points that refer to ungrazed plants, and add them using **points**:

```
points(Root[Grazing=="Ungrazed"],Fruit[Grazing=="Ungrazed"])
```

Now, using a different plotting symbol (the filled circle, pch=16), we add the points that relate to the grazed plants (use the Up Arrow to edit the last line):

```
points(Root[Grazing=="Grazed"],Fruit[Grazing=="Grazed"],pch=16)
```

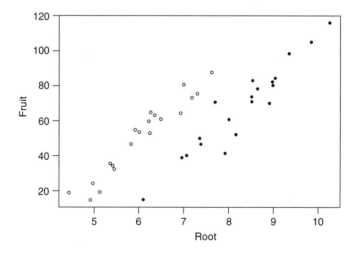

This shows a striking pattern which suggests that the largest plants were allocated to the grazed treatments (the filled symbols). But the plot also indicates that for a *given* rootstock diameter (say 7 mm) the grazed plants produced *fewer* fruits than the ungrazed plants (not more, as a simple comparison of the means suggested). This is an excellent example of where analysis of covariance comes into its own. Here the correct analysis using Ancova completely *reverses* our interpretation of the data.

The analysis proceeds in the following way. We fit the most complicated model first, then simplify it by removing non-significant terms until we are left with a minimal adequate model, in which all the parameters are significantly different from zero. For Ancova, the most complicated model has different slopes and intercepts for each level of the factor. Here we have a two-level factor (grazed and ungrazed) and we are fitting a linear model with two parameters ($y = a + bx$) so the most complicated mode has four parameters (two slopes and two intercepts). To fit different slopes and intercepts we use the asterisk * notation:

```
ancova<-lm(Fruit~Grazing*Root)
```

You should realise that *order matters*. We would get a different output if the model had been written Fruit ~ Root * Grazing (more of this later).

```
summary(ancova)
```

```
Coefficients:
                Estimate Std. Error t value   Pr(>|t|)
(Intercept)     -125.173     12.811  -9.771   1.15e-011 ***
Grazing           30.806     16.842   1.829     0.0757 .
Root              23.240      1.531  15.182   < 2e-016 ***
Grazing.Root       0.756      2.354   0.321     0.7500

Residual standard error: 6.831 on 36 degrees of freedom
Multiple R-Squared: 0.9293,    Adjusted R-squared: 0.9234
F-statistic: 157.6 on 3 and 36 degrees of freedom,
p-value:      0
```

This shows that initial root size has a massive effect on fruit production ($t = 15.182$), but there is no indication of any difference in the slope of this relationship between the two grazing treatments (this is the Grazing by Root interaction with $t = 0.321$, $p \gg 0.05$). The Anova table for the maximal model looks like this:

```
anova(ancova)
```

```
                Df  Sum of Sq    Mean Sq   F Value       Pr(F)
       Grazing   1    2910.44    2910.44   62.3795   0.0000000
          Root   1   19148.94   19148.94  410.4201   0.0000000
  Grazing:Root   1       4.81       4.81    0.1031   0.7499503
     Residuals  36    1679.65      46.66
```

The next step is to delete the non-significant interaction term from the model. We can do this manually or automatically; here we'll do both for the purposes of demonstration. The directive for manual model simplification is **update**. We update the current model (here called ancova) by deleting terms from it. The syntax is important; the punctuation reads "comma tilde dot minus". We define a new name for the simplified model like ancova2:

```
ancova2<-update(ancova, ~ . - Grazing:Root)
```

Now we compare the simplified model with just three parameters (one slope and two intercepts) with the maximal model using **anova** like this:

```
anova(ancova,ancova2)

Analysis of Variance Table

Model 1: Fruit ~ Grazing + Root + Grazing:Root
Model 2: Fruit ~ Grazing + Root

  Res.Df Res.Sum Sq Df  Sum Sq  F  value Pr(>F)
1     36    1679.65
2     37    1684.46 -1   -4.81  0.1031     0.75
```

This says that model simplification was justified because it caused a negligible reduction in the explanatory power of the model ($p = 0.75$; to retain the interaction term in the model we would need $p < 0.05$). The next step in model simplification involves testing whether or not grazing had a significant effect on fruit production once we control for initial root size. The procedure is similar: we make a new model name, say ancova3, and use **update** to remove Grazing from ancova2 like this:

```
ancova3<-update(ancova2, ~ . - Grazing)
```

Now we compare the two models using **anova**:

```
anova(ancova2,ancova3)

Analysis of Variance Table

Model 1: Fruit ~ Grazing + Root
Model 2: Fruit ~ Root

  Res.Df Res.Sum Sq Df    Sum Sq F value      Pr(>F)
1     37     1684.5
2     38     6948.8 -1   -5264.4  115.63  6.107e-013 ***
```

This model simplification is a step too far. Removing the Grazing term causes a massive reduction in the explanatory power of the model, with an F value of 115.63 and a vanishingly small p value. The effect of grazing in reducing fruit production is highly significant and needs to be retained in the model. Thus ancova2 is our minimal adequate model, and we should look at its summary table to compare with our earlier calculations carried out by hand:

```
summary(ancova2)

Coefficients:
              Estimate Std.  Error t value   Pr(>|t|)
(Intercept)  -127.829         9.664 -13.23 1.33e-015 ***
Grazing        36.103         3.357  10.75 6.11e-013 ***
Root           23.560         1.149  20.51  < 2e-016 ***

Residual standard error: 6.747 on 37 degrees of freedom
Multiple R-Squared: 0.9291,    Adjusted R-squared: 0.9252
F-statistic: 242.3 on 2 and 37 degrees of freedom,
p-value:     0
```

You know when you have got the minimal adequate model, because every row of the coefficients table has one or more significance stars (there are three stars in this case, because the effects are all so strong). Unfortunately, the interpretation is not crystal clear from this table because *the variable name, not the level name* appears in the first column, and you might misread this as saying that 'Grazing is associated with +36.103 mg of Fruit production'. This is wrong. It is *the second level of Grazing* (which is 'ungrazed' in the present case) that is associated with this positive difference in intercepts. For a given root size, the grazed plants (factor level 1) produce 36.103 mg of fruit *less* than the ungrazed plants (factor level 2). S-Plus arranges the factor levels in alphabetical order unless you tell it to do otherwise.

anova(ancova2)

	Df	Sum of Sq	Mean Sq	F Value	Pr(F)
Grazing	1	2910.44	2910.44	63.9291	1.397196e-009
Root	1	19148.94	19148.94	420.6156	0.000000e+000
Residuals	37	1684.46	45.53		

These are the values we obtained longhand on p. 292. Now we repeat the model simplification using the automatic model simplification directive called **step**. It couldn't be easier to use. The full model is called ancova:

step(ancova)

This directive causes all the terms to be tested to see whether they are needed in the minimal adequate model. The criterion used is the Akaike information criterion (AIC), more of which later. In the jargon, this is a 'penalised log likelihood'. What this means in simple terms is that it weighs up the inevitable trade-off between degrees of freedom and fit of the model. You can have a perfect fit if you have a parameter for every data point, but this model has zero explanatory power. Thus *deviance goes down as degrees of freedom in the model goes up*. The AIC adds 2 times the number of parameters in the model to the deviance (to penalise it). Deviance, you will recall, is twice the log likelihood of the current model.

Anyway, AIC is a measure of lack of fit; big AIC is bad, small AIC is good. The full model (4 parameters, 2 slopes and 2 intercepts) is fitted first, and AIC calculated as 157.5:

```
Start:   AIC= 157.5
 Fruit ~ Grazing + Root + Grazing:Root
```

Then **step** tries removing the most complicated term (the Grazing by Root interaction):

	Df	Sum of Sq	RSS	AIC
- Grazing:Root	1	4.81	1684.46	155.61
<none>			1679.65	157.50

```
Step:  AIC= 155.61
 Fruit ~ Grazing + Root
```

This has reduced AIC to 155.61 (an improvement, so the simplification is justified):

	Df	Sum of Sq	RSS	AIC
<none>			1684.5	155.6
- Grazing	1	5264.4	6948.8	210.3
- Root	1	19148.9	20833.4	254.2

```
Call:
lm(formula  =  Fruit ~ Grazing + Root)

Coefficients:
(Intercept)          Grazing            Root
   -127.83            36.10            23.56
```

No further simplification is possible (as we saw when we used **update** to remove the Grazing term from the model) because AIC goes up to 210.3 when Grazing is removed and up to 254.2 if Root is removed. Thus, **step** has found the minimal adequate model. It doesn't always, as we shall see later; it is good but not perfect. Again, with this form of output, it is possible to misinterpret the result of Grazing; it would be more informative to have the intercepts labelled by the factor levels grazed and ungrazed, as when we used **tapply** at the beginning of the exercise.

Ancova and experimental design

There is an extremely important general message in this example for experimental design. No matter how carefully we randomise at the outset, our experimental groups are likely to be heterogeneous. Sometimes, as in this case, we may have made initial measurements that we can use as covariates later on, but this will not always be the case. There are bound to be important factors that we did not measure. If we had not measured initial root size in this example, we would have come to entirely the wrong conclusion about the impact of grazing on plant performance.

A far better design for this experiment would have been to measure the rootstock diameters of all the plants at the beginning of the experiment (as was done here), but then to place the plants in matched pairs with similar-sized rootstocks. Then one of the plants is picked at random and allocated to one of the two grazing treatments (e.g. by tossing a coin); the other plant of the pair then receives the unallocated gazing treatment. Under this scheme, the size ranges of the two treatments would overlap, and the analysis of covariance would be unnecessary.

A more complex Ancova

This experiment with Weight as the response variable involved Genotype and Sex as two categorical explanatory variables and Age as a continuous covariate. There are six levels of Genotype and two levels of Sex:

```
attach(Gain)
names(Gain)

[1] "Weight"    "Sex"      "Age"       "Genotype"      "Score"
```

We begin by fitting the maximal model with its 24 parameters, involving different slopes and intercepts for every combination of Sex and Genotype:

```
m1<-lm(Weight~Sex*Age*Genotype)
summary(m1)
```

Coefficients:

	Estimate	Std. Error	t value	Pr(>|t|)	
(Intercept)	7.80053	0.24941	31.276	< 2e-016	***
Sex	-0.51966	0.35272	-1.473	0.14936	
Age	0.34950	0.07520	4.648	4.39e-005	***
GenotypeCloneB	1.19870	0.35272	3.398	0.00167	**

GenotypeCloneC	-0.41751	0.35272	-1.184	0.24429	
GenotypeCloneD	0.95600	0.35272	2.710	0.01023	*
GenotypeCloneE	-0.81604	0.35272	-2.314	0.02651	*
GenotypeCloneF	1.66851	0.35272	4.730	3.41e-005	***
Sex.Age	-0.11283	0.10635	-1.061	0.29579	
Sex.GenotypeCloneB	-0.31716	0.49882	-0.636	0.52891	
Sex.GenotypeCloneC	-1.06234	0.49882	-2.130	0.04010	*
Sex.GenotypeCloneD	-0.73547	0.49882	-1.474	0.14906	
Sex.GenotypeCloneE	-0.28533	0.49882	-0.572	0.57087	
Sex.GenotypeCloneF	-0.19839	0.49882	-0.398	0.69319	
Age.GenotypeCloneB	-0.10146	0.10635	-0.954	0.34643	
Age.GenotypeCloneC	-0.20825	0.10635	-1.958	0.05799	.
Age.GenotypeCloneD	-0.01757	0.10635	-0.165	0.86970	
Age.GenotypeCloneE	-0.03825	0.10635	-0.360	0.72123	
Age.GenotypeCloneF	-0.05512	0.10635	-0.518	0.60743	
Sex.Age.GenotypeCloneB	0.15469	0.15040	1.029	0.31055	
Sex.Age.GenotypeCloneC	0.35322	0.15040	2.349	0.02446	*
Sex.Age.GenotypeCloneD	0.19227	0.15040	1.278	0.20929	
Sex.Age.GenotypeCloneE	0.13203	0.15040	0.878	0.38585	
Sex.Age.GenotypeCloneF	0.08709	0.15040	0.579	0.56616	

```
Residual standard error: 0.2378 on 36 degrees of freedom
Multiple R-Squared: 0.9742, Adjusted R-squared: 0.9577
F-statistic: 59.06 on 23 and 36 degrees of freedom,
p-value:    0
```

Model simplification

There are one or two significant parameters, but it is not at all clear that the three-way or two-way interactions need to be retained in the model. As a first pass, let's use **step** to see how far it gets with model simplification:

```
step(m1)
```

```
Start:  AIC= -155.01
  Weight ~ Sex + Age + Genotype + Sex:Age + Sex:Genotype +
Age:Genotype + Sex:Age:Genotype

                     Df Sum of Sq      RSS       AIC
- Sex:Age:Genotype   5     0.349     2.385   -155.511
<none>                                2.036   -155.007
```

It definitely doesn't need the three-way interaction, despite the effect of Sex.Age.Genotype-CloneC, which gave a significant *t* test on its own. How about the three two-way interactions?

```
Step:  AIC= -155.51
  Weight ~ Sex + Age + Genotype + Sex:Age + Sex:Genotype +
Age:Genotype

                 Df  Sum of Sq    RSS       AIC
- Sex:Genotype   5      0.147    2.532   -161.924
- Age:Genotype   5      0.168    2.553   -161.423
```

```
 - Sex:Age          1       0.049    2.434   -156.292
 <none>                               2.385   -155.511
```

It has left out Sex by Genotype and now assesses the other two:

```
Step:  AIC= -161.92
  Weight ~ Sex + Age + Genotype + Sex:Age + Age:Genotype

                Df  Sum of Sq     RSS        AIC
 - Age:Genotype  5     0.168     2.700   -168.066
 - Sex:Age       1     0.049     2.581   -162.776
 <none>                          2.532   -161.924
```

No need for Age by Genotype. Try removing Sex by Age:

```
Step:  AIC= -168.07
  Weight ~ Sex + Age + Genotype + Sex:Age

              Df  Sum of Sq     RSS        AIC
 - Sex:Age     1     0.049     2.749   -168.989
 <none>                        2.700   -168.066
 - Genotype    5    54.958    57.658      5.612
```

Nothing. What about the main effects?

```
Step:  AIC= -168.99
  Weight ~ Sex + Age + Genotype

             Df  Sum of Sq     RSS        AIC
 <none>                       2.749   -168.989
 - Sex        1    10.374    13.122    -77.201
 - Age        1    10.770    13.519    -75.415
 - Genotype   5    54.958    57.707      3.662
```

They are all highly significant. This is the S-Plus idea of minimal adequate model. Three main effects but no interactions. That is to say, the slope of the graph of weight gain against age does not vary with sex or genotype, but the intercepts *do* vary.

It would be a good idea to look at the Anova table for this model:

```
m2<-aov(Weight~Sex+Age+Genotype)
summary(m2)

Coefficients:
           Df  Sum Sq  Mean Sq  F value    Pr(>F)
Sex         1  10.374   10.374   196.23  < 2.2e-016  ***
Age         1  10.770   10.770   203.73  < 2.2e-016  ***
Genotype    5  54.958   10.992   207.93  < 2.2e-016  ***
Residuals  52   2.749    0.053
```

That certainly looks pretty convincing. What does **lm** produce?

```
summary.lm(m2)

Coefficients:
               Estimate Std. Error t value    Pr(>|t|)
(Intercept)     7.93701    0.10066  78.851    < 2e-016  ***
```

```
Sex                 -0.83161    0.05937 -14.008    < 2e-016 ***
Age                  0.29958    0.02099  14.273    < 2e-016 ***
GenotypeCloneB       0.96778    0.10282   9.412  8.07e-013 ***
GenotypeCloneC      -1.04361    0.10282 -10.149  6.22e-014 ***
GenotypeCloneD       0.82396    0.10282   8.013  1.21e-010 ***
GenotypeCloneE      -0.87540    0.10282  -8.514  1.98e-011 ***
GenotypeCloneF       1.53460    0.10282  14.925    < 2e-016 ***

Residual standard error: 0.2299 on 52 degrees of freedom
Multiple R-Squared: 0.9651,    Adjusted R-squared: 0.9604
F-statistic: 205.7 on 7 and 52 degrees of freedom,
p-value:      0
```

This is where Helmert contrasts would come in handy (Chapter 18). Everything is three-star significantly different from Genotype[1] Sex[1], but it is not obvious that the intercepts for Genotypes B and D need different values (+0.96 and +0.82 above Genotype A with s.e. difference = 0.1028), nor is it obvious that C and E have different intercepts (−1.043 and −0.875). Perhaps we could reduce the number of factor levels of Genotype from the present 6 to 4 without any loss of explanatory power?

Factor level reduction

We create a new categorical variable called newgen with separate levels for clones A and F, and for B and D combined and C and E combined (see p. 263 to revise this technique):

```
newgen<-factor(1+(Genotype=="CloneB")+(Genotype=="CloneD")+
    2*(Genotype=="CloneC")+2*(Genotype=="CloneE")+3*(Genotype=="CloneF"))
```

Then we redo the modelling with newgen (4 levels) instead of Genotype (6 levels):

```
m3<-lm(Weight~Sex+Age+newgen)
```

and check that the simplification was justified:

```
anova(m2,m3)
```

Curve fitting after Ancova

After model simplification, we generally want to draw the fitted model through the data in outline we need to

- use different plotting symbols for each of the factor levels (grazed and ungrazed here)
- use **abline** once for each factor level to draw the lines through the data
- the scatterplot, with different symbols for the grazed (solid) and ungrazed (open) plants we analysed earlier, is on p. 294.

```
plot(Root,Fruit,type="n")
points(Root[Grazing=="Ungrazed"],Fruit[Grazing=="Ungrazed"])
points(Root[Grazing=="Grazed"],Fruit[Grazing=="Grazed",pch=16)
```

To use **abline** to draw the fitted model, we need to provide it with the relevant intercepts and slopes. These are in the **summary.lm** of the model called ancova2:

```
Coefficients:
                  Value   Std. Error   t value   Pr (>|t|)
(Intercept)    -127.8294      9.6641   -13.2272     0.0000
    Grazing      36.1032      3.3574    10.7533     0.0000
       Root      23.5600      1.1488    20.5089     0.0000
```

Let's use a dotted line (lty = 2) for the solid symbols and a solid line (the default) for the open symbols. The minimal model had a common slope for both grazed and ungrazed plants (23.56), so **abline** will take the same slope value for both lines. The intercept was −127.8294 for the grazed plants (line) and 36.1032 higher than this (−127.8294 + 36.1032 = 91.7262) for the ungrazed plants (line):

```
abline(-127.8294,23.56,lty=2)
abline(-127.8294+36.1032,23.56)
```

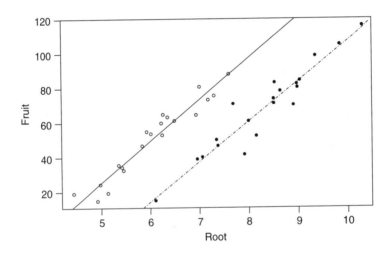

You could write a general ancova plotting function, using subscripts to extract the relevant slopes and intercepts from coef(model) to use in **abline**. Here, for instance, are the slope and two intercepts from our current example:

```
coef(ancova2)
```

```
(Intercept)     Grazing        Root
  -127.8294    36.10325    23.56005
```

which we extract using subscripts like this. The slope is the third element of the coefficients object, so it is extracted like this:

```
as.vector(coef(ancova2)[3])
```

```
[1] 23.56005
```

You would need to take care over the precise value of the subscripts you use, because this would depend on the order of fitting (Grazing + Root is not the same as Root + Grazing), and on the number of parameters in the minimal adequate model (there are three coefficients in this case, because there was a common slope and two intercepts).

Further reading

Huitema, B.E. (1980) *The Analysis of Covariance and Alternatives*. New York, John Wiley.
Neter, J., Kutner, M. *et al.* (1996) *Applied Linear Statistical Models*. New York, McGraw-Hill.
Neter, J., Wasserman, W. *et al.* (1985) *Applied Linear Regression Models*. Homewood, Illinois, Irwin.

17

Model criticism

There is a temptation to become personally attached to a particular model. Statisticians call this 'falling in love with your model'. It is as well to remember the following home truths about models:

- all models are wrong
- some models are better than others
- the correct model can never be known with certainty
- the simpler the model, the better it is

There are several ways that we can improve things if it turns out that our present model is inadequate:

- transform the response variable
- transform one or more of the explanatory variables
- try fitting different explanatory variables if you have any
- use a different error structure
- use non-parametric smoothers instead of parametric functions
- use different weights for different y values

All of these are investigated in the coming chapters. At this point, all we want to do is explain the tools that are used to establish that your model is inadequate. For example, the model may:

- predict some of the y values poorly
- show non-constant variance
- show non-normal errors
- be strongly influenced by a small number of influential data points
- show some sort of systematic pattern in the residuals
- exhibit overdispersion

Scale of measurement

Just as there is no perfect model, so there may be no optimal scale of measurement for a model. Suppose, for example, we had a process that had Poisson errors with multiplicative effects among the explanatory variables. Then one must chose between three different scales, each of which optimises one of three different properties:

- the scale of $\sqrt{y}$ would give constancy of variance

- the scale of $y^{2/3}$ would give approximately Normal errors

- the scale of $\ln y$ would give additivity

Thus any measurement scale is always going to be a compromise, and you should choose the scale that gives the best overall performance of the model.

Residuals

After fitting a model to data, we should investigate how well the model describes the data. In particular, we should look to see if there are any systematic trends in the goodness of fit. For example, does the goodness of fit increase with the observation number, or is it a function of one or more of the explanatory variables? We can work with the raw residuals:

residuals = response variable − fitted values

With Normal errors, the identity link, equal weights and the default scale factor, the raw and standardised residuals are identical (Figure 17.1).

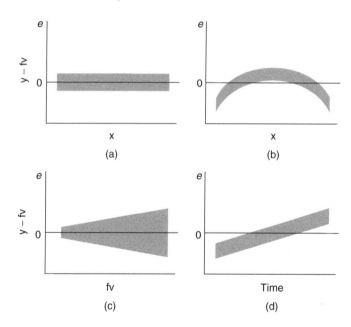

Figure 17.1   Model criticism. (a) The residuals (response variable minus the fitted values, y−fv) should show no patterns when plotted against the explanatory variable x. (b) Curvature in the residuals against the explanatory suggests misspecification of the model (perhaps transformation of y or x is required). (c) The most important assumption is constancy of variance. If, as here, the scatter of the residuals increases with the fitted values, then a different error structure is required (e.g. **glm** with Poisson errors). (d) The residuals are assumed to be constant through time. If, as here, the residuals change through time, then a time series model may be required (Chapters 34 and 35)

For **Poisson** errors, the standardised residuals are

$$\frac{(y - \text{fitted values})}{\sqrt{\text{fitted values}}}$$

For **binomial errors** they are

$$\frac{(y - \text{fitted values})}{\sqrt{\text{fitted values} \times \left[1 - \dfrac{\text{fitted values}}{\text{binomial denominator}}\right]}}$$

And for **Gamma** errors they are

$$\frac{(y - \text{fitted values})}{\text{fitted values}}$$

In general, we can use several kinds of standardised residuals:

$$\text{standardised residuals} = (y - \text{fitted values})\sqrt{\frac{\text{prior weight}}{\text{scale parameter} \times \text{variance funcion}}}$$

Model checking

We should routinely plot the residuals against:

- the fitted values (to look for non-constancy of variance, heteroscedasticity)

- the explanatory variables (to look for evidence of curvature)

- the sequence of data collection (to took for temporal correlation)

- standard normal deviates (to look for non-normality of errors)

Non-constant variance: heteroscedasticity

A good model must also account for the variance–mean relationship adequately and produce additive effects on the appropriate scale (as defined by the link function).

A plot of standardised residuals against fitted values should look like the sky at night, with no trend in the size or degree of scatter of the residuals. A common problem is that the variance increases with the mean, so we obtain an expanding, fan-shaped pattern of residuals.

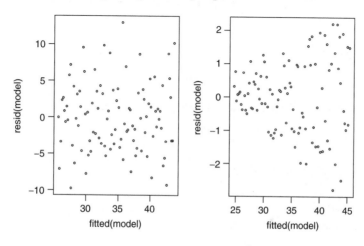

The plot on the left is what we want to see: no trend in the residuals with the fitted values. The plot on the right is a problem. There is a clear pattern of increasing residuals as the fitted values get larger. This is a picture of what *heteroscedasticity* looks like.

Non-normality of errors

Errors may be non-normal for one or several reasons. They may be skew, with long tails to the left or right. Or they may be kurtotic, with a flatter or more pointy top to their distribution. In any case, the theory is based on the assumption of Normal errors, and if the errors are *not* normally distributed, then we shall not know how this affects our interpretation of the data or the inferences we make from it.

It takes considerable experience to interpret the Normal error plots. Here we generate a series of data sets where we introduce different known kinds of non-normal errors. Then we plot them using a simple home-made function called **mcheck** (the name stands for model checking), to see what patterns are generated in Normal plots by the different kinds of non-normality. In real applications we would use the generic plot (model) rather than **mcheck** (see below). First, we write the function **mcheck**. The idea is to produce two plots side by side: a plot of the residuals against the fitted values on the left and a plot of the ordered residuals against the quantiles of the Normal distribution on the right:

```
mcheck <- function (obj, ...) {
  rs<-obj$resid
  fv<-obj$fitted
  par(mfrow=c(1,2))
  plot(fv,rs,xlab="Fitted values",ylab="Residuals")
  abline(h=0, lty=2)
  qqnorm(rs,xlab="Normal scores",ylab="Ordered residuals")
  qqline(rs,lty=2)
  par(mfrow=c(1,1))
  invisible(NULL)
}
```

Note the use of $ to extract the residuals and fitted values from the model object which is passed to the function as obj. The functions **qqnorm** and **qqline** are built-in functions to produce Normal probability plots. It is good programming practice to set the graphics parameters back to their default settings before leaving the function.

We need a vector of x values for the following regression models:

```
x<-0:30
```

Now we manufacture the response variables according to the equation

$$y = 10 + x + \varepsilon$$

where the errors, ε, have zero mean but are taken from different probability distributions in each case.

Normal errors

```
e<-rnorm(31,mean=0,sd=5)
yn<-10+x+e
mn<-lm(yn~x)

mcheck(mn)
```

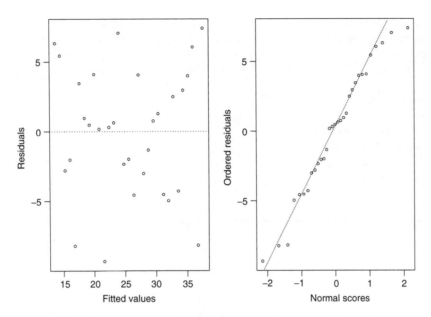

There is no suggestion of non-constant variance (left plot) and the Normal plot (right) is reasonably straight. The judgement as to what constitutes an important departure from normality takes experience, and this is the reason for looking at some distinctly non-normal, but known, error structures next.

Uniform errors

```
eu<-20*(runif(31)-0.5)
yu<-10+x+eu
mu<-lm(yu~x)
mcheck(mu)
```

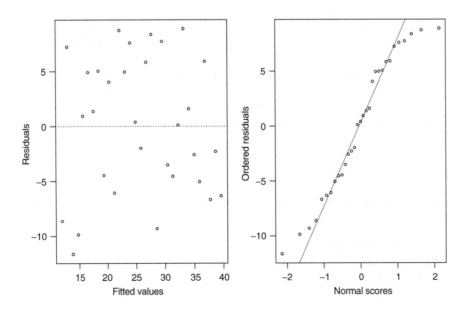

Uniform errors show up as a distinctly S-shaped pattern in the **qqplot** on the right. The fit in the centre is fine, but the largest and smallest residuals are too small (they are constrained in this example to be ±10).

Negative binomial errors

```
enb<-rnbinom(31,2,.3)
ynb<-10+x+enb
mnb<-lm(ynb~x)
mcheck(mnb)
```

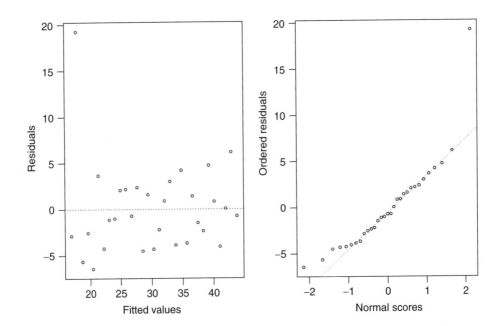

The large negative residuals are all above the line, but the most obvious feature of the plot is the single, very large positive residual (in the top right corner). In general, negative binomial errors will produce a J-shape on the **qqplot**. The biggest positive residuals are much too large to have come from a Normal distribution. These values may turn out to be highly influential (see below).

Gamma errors and increasing variance

The shape parameter is set to 1 and the rate parameter to $1/x$ and the variance increases with the square of the mean:

```
eg<-rgamma(31,1,1/x)
yg<-10+x+eg
mg<-lm(yg~x)
mcheck(mg)
```

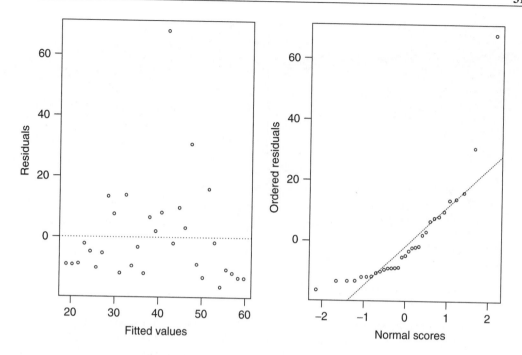

The left-hand plot shows the residuals increasing with the fitted values, and an asymmetry between the size of the positive and negative residuals. The right-hand plot shows that both the large positive and large negative residuals lie above the Normal errors line, demonstrating the presence of outliers in both positive and negative tails of the error distribution.

Influence

One of the commonest reasons for a lack of fit is through the existence of outliers in the data (see p. 238). It is important to understand, however, that a point may *appear* to be an outlier because of misspecification of the model, and not because there is anything wrong with the data.

Take this circle of data that shows no relationship between y and x:

```
x<-c(2,3,3,3,4)
y<-c(2,3,2,1,2)
```

We want to draw two graphs side by side, so we type

```
par(mfrow=c(1,2))
```

and we want them to have the same axis scales, so we specify

```
plot(x,y,xlim=c(0,8,),ylim=c(0,8))
```

Obviously, there is no relationship between y and x in the original data. But let's add an outlier at the point (7,6) using concatenation **c** and see what happens:

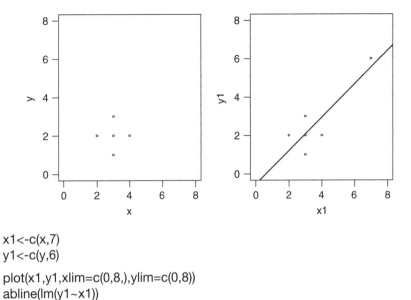

```
x1<-c(x,7)
y1<-c(y,6)

plot(x1,y1,xlim=c(0,8,),ylim=c(0,8))
abline(lm(y1~x1))
```

Now there is a significant regression of y on x. The outlier is said to be highly *influential*. To reduce the influence of outliers, there are a number of modern techniques known as robust regression. To see one of these in action, let's do a straightforward linear regression on these data and print the summary:

```
reg<-lm(y1~x1)
summary(reg)

Call: lm(formula = y1 ~ x1)
Residuals:

       1      2        3       4        5       6
  0.7826  0.913  -0.08696  -1.087  -0.9565  0.4348

Coefficients:

                 Value Std. Error  t value Pr(>|t|)
 (Intercept)   -0.5217   0.9876    -0.5283   0.6253
         x1     0.8696   0.2469     3.5218   0.0244

Residual standard error: 0.9668 on 4 degrees of freedom
Multiple R-Squared: 0.7561
```

The residuals make the important point that analysis of residuals is a very poor way of looking for influence. Precisely because the point (7,6) is so influential, it forces the regression line close to it, hence point number 6 has a small residual (0.4348 is actually the second smallest of all the residuals). The slope of the regression line is 0.8696 with a standard error of 0.2469, and this is significantly different from 0 ($p = 0.0244$) despite the tiny sample size.

Leverage

Points increase in influence to the extent that they lie on their own, a long way from the mean value of x (either left or right). To account for this, measures of leverage for a given data point y are proportional to $(x - \bar{x})^2$. The commonest measure of leverage is

$$h_i = \frac{1}{n} + \frac{(x_i - \bar{x})^2}{\sum (x_j - \bar{x})^2}$$

where the denominator is *SSX*. A good rule of thumb is that a point is highly influential if its h_i satisfies

$$h_i > \frac{2p}{n}$$

where p is the number of parameters in the model. We could easily calculate the leverage value of each point in x_1. It is more efficient, perhaps, to write a general function that could carry out the calculation of h for any vector of x values

```
leverage<-function(x){ 1/length(x)+(x-mean(x))^2/sum((x-mean(x))^2) }
```

Then use the function called leverage on our vector x_1:

```
leverage(x1)
```

```
[1]  0.3478261 0.1956522 0.1956522 0.1956522 0.1739130
[6]  0.8913043
```

This draws attention immediately to the sixth x value: its h value is more than double the next largest. The result is even clearer if we plot the leverage values:

```
plot(leverage(x1))
```

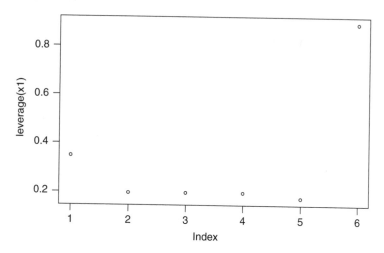

Note that if the **plot** directive has a single argument (as here), then the x values for the plot are taken as the *order* of the numbers in the vector to be plotted (the sequence 1:6 in this case):

The plot would be much easier to interpret if the points were highlighted, perhaps by vertical lines. Also, it would be useful to plot the rule of thumb value of what constitutes an influential point. In this case $p = 2$ and the number of points on the graph is $n = 6$, so a point is influential if $h_i > 0.66$. The improved graph is produced like this:

```
plot(h,type="h")
points(h)
abline(0.66,0,lty=2)
```

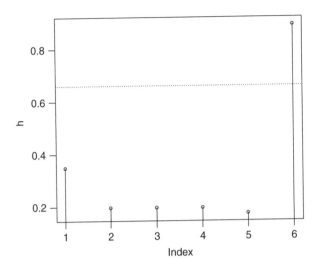

Recall that the straight line drawing function **abline** can be specified in a number of different ways. Here we gave the numeric values of the intercept (0.66) and the slope (0) of the dotted line (lty=2) that we wanted overlaid on the graph.

This is enough to warn us that the point (7,6) could be having a marked effect on the parameter estimates of our model. We can see if this is true by repeating the regression without the point (7,6). There are several ways of doing this. If we know the subscript of the point, [6] in this example, we can drop that point explicitly using the negative subscript convention (see p. 23):

```
reg2<-lm(y1[-6]~x1[-6])

summary(reg2)

Residuals:
          1 2            3  4            5
5.551e-017 1  -1.06e-017 -1 1.715e-017

Coefficients:
             Value Std. Error t value Pr(>|t|)
(Intercept) 2.0000 1.7701      1.1299  0.3407
   x1[-6] 0.0000 0.5774        0.0000  1.0000

Residual standard error: 0.8165 on 3 degrees of freedom
Multiple R-Squared: 0
```

The point (7,6) was indeed highly influential because without it the slope of the graph is zero. Notice that the residuals of the points we created are not exactly zero; they are various numbers times 10^{-17}.

Alternatively, we could use **weights** to 'weight out' the point (7,6) whose influence we want to test. We need to create a vector of weights: 1s for the data we want to include and 0s for the data we want to leave out. In this simple case we could type in the weights directly like this:

```
w<-c(1,1,1,1,1,0)
```

but in general we will want to calculate them on the basis of some logical criterion. A suitable condition for inclusion here would be x1 < 6:

```
w<-(x1<6)

w

[1] T T T T T F
```

Note that when we calculate the weight vector in this way, we get T and F (true and false) rather than 1 and 0, but this works equally well. The new model looks like this:

```
reg3<-lm(y1~x1,weights=w)
summary(reg3)

Call: lm(formula = y1 ~ x1, weights = w)
Residuals:
           1 2           3 4           5
 5.551e-017 1 -1.06e-017 -1 1.715e-017

Coefficients:
             Value Std. Error t value Pr(>|t|)
(Intercept) 2.0000 1.7701      1.1299  0.3407
        x1  0.0000 0.5774      0.0000  1.0000

Residual standard error: 0.8165 on 3 degrees of freedom
Multiple R-Squared:0
F-statistic: 0 on 1 and 3 degrees of freedom, the p-value
is 1

Warning messages:
  1 rows with zero weights not counted in:
```

Finally, we could use **subset** to leave out the points we wanted to exclude from the model fit. Of all the options, this is the most general, and the easiest to use. As with weights, the subset is stated as part of the model specification. It says which points to include, rather than to exclude. Here is the logic to include any points for which x < 6. Here is

```
reg4<-lm(y1~x1,subset=(x1<6))
summary(reg4)

Call: lm(formula = y1 ~ x1,subset = (x1<6))
Residuals:
            1 2          3 4           5
 1.823e-016 1   4.758e-019 -1 -1.03e-016

Coefficients:
             Value Std. Error t value Pr(>|t|)
(Intercept) 2.0000 1.7701      1.1299  0.3407
        x1 0.0000 0.5774       0.0000  1.0000
```

The output is exactly the same as in **reg2** using subscripts and **reg3** using weights.

Akaike information criterion (AIC)

The Akaike information criterion glories under the name of a *penalised log likelihood*. If you have a model for which a log likelihood value can be obtained, then

$$AIC = -2 \times log\ likelihood + 2(p+1)$$

where p is the number of parameters in the model, and 1 is added for the estimated variance (you could call this another parameter if you wanted to). To demystify AIC let's calculate it by hand. We revisit the regression data for which we calculated the log likelihood by hand on p. 121:

```
attach(regression)
names(regression)
```

```
[1] "growth" "tannin"
```

```
growth
```

```
 1  2  3  4  5 6 7 8 9
12 10  8 11  6 7 2 3 3
```

There are nine values of the response variable, Growth, and we calculated the log likelihood as -23.98941 earlier. There was only one parameter estimated from the data for these calculations (the mean value of y), so $p = 1$. This means that AIC should be

$$AIC = -2 \times -23.98941 + 2 \times (1 + 1) = 51.97882$$

Fortunately, we do not need to make these calculations, because there is a built-in function for calculating AIC. It takes a model object as its argument, so we need to fit a one-parameter model to the growth data like this:

```
model<-lm(growth~1)
```

Then we can get AIC directly:

```
AIC(model)
```

```
[1] 51.97882
```

AIC as a measure of the fit of a model

The more parameters there are in the model, the better the fit. You could obtain a perfect fit if you had a separate parameter for every data point, but this model would have absolutely no explanatory power. There is always going to be a trade-off between the goodness of fit and the number of parameters required by parsimony. AIC is useful because it explicitly penalises any superfluous parameters in the model, by adding $2p$ to the deviance.

When comparing two models, the smaller the AIC, the better the fit. This is the basis of automated model simplification using **step**.

You can use the function **AIC** to compare two models, in exactly the same way as you have been using **anova**. Here we revisit the analysis of covariance on p. 295:

```
model.1<-lm(Fruit~Grazing*Root)
model.2<-lm(Fruit~Grazing+Root)
```

```
AIC(model.1, model.2)
```

```
        df      AIC
model.1  5 273.0135
model.2  4 271.1279
```

Because model.2 has the *lower AIC*, we prefer it to model.1. The log likelihood was penalised by $2 \times (4 + 1) = 10$ in model.1 because that model contained 4 parameters (2 slopes and 2 intercepts) and by $2 \times (3 + 1) = 8$ in model.2 because that model had 3 parameters (2 intercepts and 1 common slope). You can see where the two values of *AIC* come from by calculation:

```
-2*logLik(model.1)+2*(4+1)
```

```
[1] 273.0135
```

```
-2*logLik(model.2)+2*(3+1)
```

```
[1] 271.1279
```

Misspecified model

The model may have the wrong terms in it, or the terms may be included in the model in the wrong way. We deal with the selection of terms for inclusion in the minimal adequate model in Chapter 25. Here we simply note that *transformation of the explanatory variables* often produces improvements in model performance. The most frequently used transformations are logs, powers and reciprocals.

Again, in testing for non-linearity in the relationship between y and x, we might add a term in x^2 to the model; a significant parameter in the x^2 term indicates curvilinearity in the relationship between y and x (see p. 317).

A further element of misspecification can occur because of *structural non-linearity*. Suppose, for example, that we were fitting a model of the form

$$y = a + \frac{b}{x}$$

but the underlying process was really of the form

$$y = a + \frac{b}{c + x}$$

then the fit is going to be poor. Of course, if we *knew* that the model structure was of this form, then we could fit it as a non-linear model (p. 413) or as a non-linear mixed effects model (p. 424), but in practice this is seldom the case.

Misspecified error structure

A common problem with real data is that the variance increases with the mean. The assumption so far has been of Normal errors with constant variance at all values of the response variable. For continuous measurement data with non-constant errors, we can specify a generalised linear model (**glm**) with *Gamma errors*. We need only note at this stage that this assumes a *constant coefficient of variation* (see Taylor's power law on p. 56).

With count data, we often assume Poisson errors, but the data may exhibit overdispersion (see below and p. 543), so that the variance is actually greater than the mean (rather than equal to it, as assumed by the Poisson distribution). An important distribution for describing aggregated data is the *negative binomial*. While S-Plus has no direct facility for specifying negative binomial errors, we can use quasi-likelihood to specify the variance function in a **glm** with family = quasi (see p. 544).

Misspecified link function

Although each error structure has a canonical link function associated with it (see p. 505), it is quite possible that a different link function would give a better fit for a particular model specification. For example, in a **glm** with Normal errors we might try a log link or a reciprocal link using **quasi** to improve the fit (examples on p. 509). Similarly, with binomial errors we might try a complementary log-log link instead of the default logit link function (see p. 504).

An alternative to changing the link function is to transform the values of the response variable. Remember that changing the scale of y will alter the error structure (see p. 318). Thus, if you take logs of y and carry out regression with Normal errors, then you will be assuming that the errors in y were lognormally distributed. This may well be a sound assumption, but a bias will have been introduced if the errors really were additive on the original scale of measurement. If, for example, theory suggests that there is an exponential relationship between y and x,

$$y = ae^{bx}$$

then it would be reasonable to suppose that the log of y would be linearly related to x:

$$\ln y = \ln a + bx$$

Now suppose that the errors ε in y are multiplicative with a mean of 0 and constant variance,

$$y = ae^{bx}(1 + \varepsilon)$$

then they will also have a mean of 0 in the transformed model. But if the errors are additive,

$$y = ae^{bx} + \varepsilon$$

then the error variance in the transformed model will depend upon the expected value of y. In a case like this, it is much better to analyse the untransformed response variable and to employ the log link function, because this retains the assumption of additive errors.

When both the error distribution and functional form of the relationship are unknown, there is no single specific rationale for choosing any given transformation in preference to another. The aim is pragmatic, namely to find a transformation that gives:

- constant error variance
- approximately Normal errors
- additivity
- a linear relationship between the response variables and the explanatory variables
- straightforward scientific interpretation

The choice is bound to be a compromise, and as such, is best resolved by quantitative comparison of the deviance produced under different model forms (Chapter 28).

Overdispersion

Overdispersion is the polite statistician's version of Murphy's law: if something can go wrong, it will. Overdispersion can be a problem when working with Poisson or binomial errors, and tends to occur because you have not measured one or more of the factors that turn out to be important. It may also result from the underlying distribution being non-Poisson or non-binomial. This means that the probability you are attempting to model is not constant within each cell, but behaves like a random variable. This, in turn, means that the residual deviance is inflated. In the worst case, all the predictor variables you have measured may turn out to be unimportant so that you have no information at all on any of the genuinely important predictors. In this case the minimal adequate model is just the overall mean, and all your 'explanatory' variables provide no extra information.

The techniques of dealing with overdispersion are discussed in detail in the chapters on Poisson errors (p. 29) and binomial errors (p. 28). Here it is sufficient to point out that there are two general techniques available to us:

- use F tests with an empirical scale parameter instead of chi-square

- use quasi-likelihood to specify a more appropriate variance function

It is important, however, to stress that these techniques introduce another level of uncertainty into the analysis. Overdispersion happens for real, scientifically important reasons, and these reasons may throw doubt upon our ability to interpret the experiment in an unbiased way. It means that something we didn't measure turned out to have an important impact on the results. If we didn't measure this factor, then we have no confidence that our randomisation process took care of it properly and we may have introduced an important bias into the results.

Model checking in S-Plus

The data are on the decay of a biodegradable plastic in soil. The response, y, is the mass of plastic remaining and the explanatory variable, x, is duration of burial:

```
attach(Decay)
names(Decay)
```
```
[1]  "x"  "y"
```
```
plot(x,y,pch=16)
```

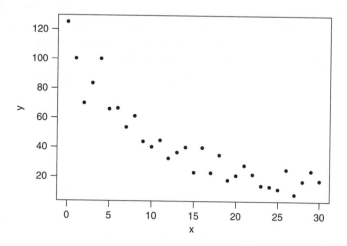

The idea is to fit a linear regression to these data and then use model-checking plots to investigate the adequacy of that model:

```
model<-lm(y~x)
```

The basic model checking could not be simpler; we just say plot (model) like this:

```
plot(model)
```

This one command produces a series of graphs, spread over several pages. First, you get a plot of the residuals against the fitted values.

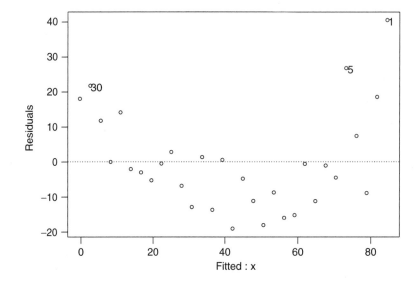

In many ways, this is the most important of all the model-checking plots. We are immediately made aware that all is not well. Remember, this should look like the sky at night, with no patterns of any sort. Here, however, there is a clear U-shaped pattern. For small fitted values of *y*, the residuals are all positive, while for intermediate values of *y*, most of the residuals are negative. This suggests systematic inadequacy in the structure of the model. Perhaps the relationship between *y* and *x* is non-linear rather than linear as we assumed here? The second plot shows the square root of the absolute values of the residuals. This is good for seeing non-constant variance. When there is heterogeneity of variance, then the plot looks like a wedge of cheese—a triangular shape, thicker at one end than the other (usually thicker at the right-hand end see p. 307).

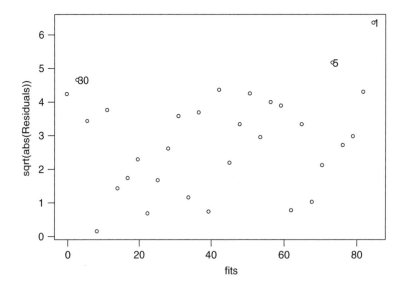

There is not much wrong with this. Our variance is not obviously heterogeneous. Next comes a plot of the data against the fitted values. This should be a straight line with little scatter.

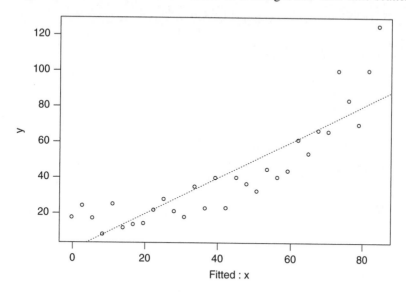

This plot reinforces our impression from plot 1: the relationship between observed and predicted weights is curved, not linear, and the predictions (the fitted values on the *x* axis) are poor everywhere. They are too small for large and small values of *y*, and too big for intermediate values of *y*. The next plot tests the normality of errors assumption; it should be a straight line.

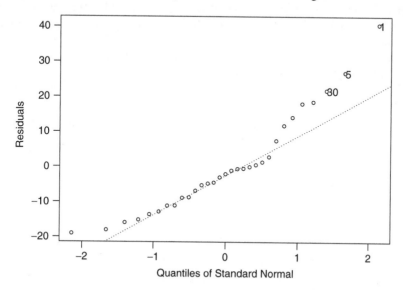

Here again, the model behaves badly for both relatively large values of *y* (right-hand side) and relatively small values of *y* (left-hand side). The residuals are bigger than expected if they were normally distributed. The next graph shows Cook's distance, which highlights the identity of particularly influential data points. Cook's statistic is an attempt to combine leverage and residuals in a single measure. The absolute values of the deletion residuals $|r_i^*|$ are weighted as follows:

$$C_i = |r_i^*| \left(\frac{n-p}{p} \frac{h_i}{1-h_i} \right)^{1/2}$$

see p. 313 and Crawley (1993) for details.

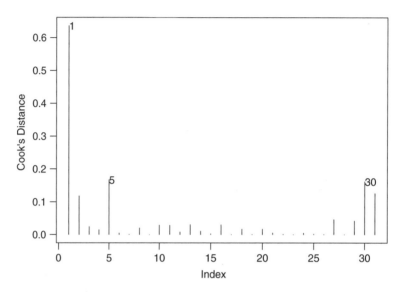

Data points 1, 5 and 30 are singled out as being influential, with point number 1 especially so. When we were happier with other aspects of the model, we would repeat the modelling leaving out each of these points in turn. Alternatively, we could jackknife the data (see p. 195), which involves leaving every data point out, one at a time, in turn. In any event, this is clearly *not* a good model for these data. This analysis is completed in Chapter 22.

Further reading

Aitkin, M., Anderson, D. *et al.* (1989) *Statistical Modelling in GLIM.* Oxford, Clarendon Press.
Crawley, M.J. (1993) *GLIM for Ecologists.* Oxford, Blackwell Science.
Diggle, P.J., Liang, K.-Y. *et al.* (1994) *Analysis of Longitudinal Data.* Oxford, Clarendon Press.

18

Contrasts

Contrasts are the essence of hypothesis testing and model simplification in Anova. They are used to compare means or groups of means with other means or groups of means, in what are known as *single degree of freedom comparisons.* There are two sorts of contrasts we might want to carry out:

- contrasts we had planned to carry out at the experimental design stage (these are referred to as *a priori* contrasts)

- contrasts that look interesting after we have seen the results (these are referred to as *a posteriori* contrasts)

Some people are very snooty about *a posteriori* contrasts, on the grounds that they were *unplanned.* You are not supposed to decide what comparisons to make *after* you have seen the analysis, but scientists do this all the time. You can't change human nature. The key point is that you should only do contrasts *after* the Anova has established that there really are significant differences to be investigated. It is not good practice to carry out tests to compare the largest mean with the smallest mean, if the Anova fails to reject the null hypothesis (tempting though this may be).

Orthogonal designs and non-orthogonal observational data

The data in this book fall into two distinct categories. In the case of planned experiments, all of the treatment combinations are equally represented and, barring accidents, there are no missing values. Such experiments are said to be *orthogonal.* In the case of observational studies, however, we often have no control over the number of individuals for which we have data, or over the combinations of circumstances that are observed. Missing treatment combinations are commonplace, correlations among the explanatory variables are the norm, and the data are said to be non-orthogonal.

This makes an important difference to our statistical modelling because, in orthogonal designs, the variation that is attributed to a given factor is constant, and does not depend upon the order in which that factor is removed from the model. In contrast, with non-orthogonal data, we find that the deviance attributable to a given factor *does* depend upon the order in which the factor is removed from the model. We must be careful, therefore, to judge the significance of factors in non-orthogonal studies, when they are *removed from the maximal model* (i.e. from the model including all the other factors and interactions with which they might be confounded).

For non-orthogonal data, *order matters:* the variation attributed to a factor removed from a full model, will be lower than the variation attributed to that same factor if it is fitted on its own (i.e. added to the null model).

Contrasts: orthogonal and non-orthogonal

There are two important points to understand about contrasts:

- there are a huge number of *possible* contrasts

- there are only $k - 1$ *orthogonal* contrasts

where k is the number of factor levels. Two contrasts are said to be orthogonal to one another if the comparisons are statistically independent. Technically, two contrasts are orthogonal if *the products of their contrast coefficients sum to zero* (we shall see what this means in a moment).

Let's take a simple example. Suppose we have one factor with five levels and the factor levels are called a, b, c, d, and e. Let's start writing down the possible contrasts. Obviously we could compare each mean singly with every other:

$$a \text{ vs. } b, \; a \text{ vs. } c, \; a \text{ vs. } d, \; a \text{ vs. } e, \; b \text{ vs. } c, \; b \text{ vs. } d, \; b \text{ vs. } e, \; c \text{ vs. } d, \; c \text{ vs. } e, \; d \text{ vs. } e$$

but we could also compare pairs of means:

$$\{a, b\} \text{ vs. } \{c, d\}, \; \{a, b\} \text{ vs. } \{c, e\}, \; \{a, b\} \text{ vs. } \{d, e\}, \; \{a, c\} \text{ vs. } \{b, d\}, \; \{a, c\} \text{ vs. } \{b, e\}, etc.$$

or triplets of means:

$$\{a, b, c\} \text{ vs. } d, \; \{a, b, c\} \text{ vs. } e, \; \{a, b, d\} \text{ vs. } c, \; \{a, b, d\} \text{ vs. } e, \; \{a, c, d\} \text{ vs. } b, \; and\ so\ on.$$

or groups of four means

$$\{a, b, c, d\} \text{ vs. } e, \; \{a, b, c, e\} \text{ vs. } d, \; \{b, c, d, e\} \text{ vs. } a, \{a, b, d, e\} \text{ vs. } c, \; \{a, b, c, e\} \text{ vs. } d$$

I think you get the idea. There are absolutely masses of possible contrasts.

In practice we should only compare things once, either directly or implicitly. So the two contrasts

$$a \text{ vs. } b \quad \text{and} \quad a \text{ vs. } c$$

implicitly contrast b vs. c. This means that if we have carried out the two contrasts a vs. b and a vs. c then the third contrast b vs. c is *not* an orthogonal contrast because we have already carried it out, implicitly. Which particular contrasts are orthogonal depends very much on your choice of the first contrast to make. Suppose there were good reasons for comparing a, b, c, e vs. d. For example, d might be the placebo and the other four might be different kinds of drug treatment, so we make this our first contrast. Because $k - 1 = 4$ we only have three possible contrasts that are orthogonal to this. There may be *a priori* reasons to group $\{a, b\}$ and $\{c, e\}$ so we make this our second orthogonal contrast. This means that we have no degrees of freedom in choosing the last two orthogonal contrasts: they have to be a vs. b and c vs. e. Just remember that *with orthogonal contrasts you only compare things once*.

Contrast coefficients

Contrast coefficients are a numerical way of embodying the hypothesis we want to test. The rules for constructing contrast coefficients are straightforward:

- treatments to be lumped together get the same sign (plus or minus)

- groups of means to be contrasted get opposite sign

- factor levels to be excluded get a contrast coefficient of 0

- the contrast coefficients, c, must add up to 0

Suppose that with our five-level factor $\{a, b, c, d, e\}$ we want to begin by comparing the four levels $\{a, b, c, e\}$ with the single level d. All levels enter the contrast, so none of the coefficients is 0. The four terms $\{a, b, c, e\}$ are grouped together so they all get the same sign (minus, for example, although it does not matter which sign is chosen). They are to be compared to d, so it gets the opposite sign (plus, in this case). The choice of what numeric values to give the contrast coefficients is entirely up to you. Most people use whole numbers rather than fractions, but it really doesn't matter. All that matters is that the c sum to 0. The positive and negative coefficients have to add up to the same value. In our example, comparing four means with one mean, a natural choice of coefficients would be -1 for each of $\{a, b, c, e\}$ and $+4$ for d. Alternatively, we could have selected $+0.25$ for each of $\{a, b, c, e\}$ and -1 for d:

Factor level		a	b	c	d	e
Contrast 1 coefficients,	c	-1	-1	-1	4	-1

Suppose the second contrast is to compare a, b with c, e. Because this contrast excludes d, we set its contrast coefficient to 0. a, b get the same sign (say plus) and c, e get the opposite sign. Because the number of levels on each side of the contrast is equal (2 in both cases) we can use the same numeric value for all the coefficients. The value 1 is the most obvious choice (but you could use 13.7 if you wanted to be perverse):

Factor level		a	b	c	d	e
Contrast 2 coefficients,	c	1	1	-1	0	-1

There are only two possibilities for the remaining orthogonal contrasts: a vs. b and c vs. e:

Factor level		a	b	c	d	e
Contrast 3 coefficients,	c	1	-1	0	0	0
Contrast 4 coefficients,	c	0	0	1	0	-1

An example of contrasts in S-Plus

The example comes from a competition experiment in which the biomass of control plants is compared to the biomass of plants grown in conditions where competition was reduced in one of four different ways. There are two treatments in which the roots of neighbouring plants were cut (to 5 cm depth or 10 cm) and two treatments in which the shoots of neighbouring plants were clipped (25% or 50% of the neighbours cut back to ground level).

```
attach(compexpt)
names(compexpt)

[1] "biomass"    "clipping"
```

We begin with data inspection:

```
plot(clipping,biomass)
```

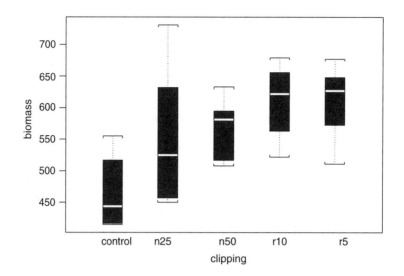

The control plants are much smaller than the rest, but the significance of any differences that there may be between the different competition treatments is unclear. For example, did root pruning lead to higher biomass than shoot pruning as the plot suggests?

First we carry out the analysis of variance to establish whether or not there are any significant differences between the levels of the factor clipping. Recall that the null and alternative hypotheses for Anova are all-embracing, and not particularly informative:

> H_0: *all the means are the same*

> H_1: *at least one of the means is significantly different from the others*

```
model<-aov(biomass~clipping)
summary(model)
```

```
            Df Sum of Sq  Mean Sq  F Value       Pr(F)
 clipping   4    85356.5 21339.12 4.301536 0.008751641
Residuals  25   124020.3  4960.81
```

We can confidently reject the null hypothesis ($p < 0.01$); there are some significant differences in biomass. The experiment consists of five treatments:

```
levels(clipping)
```

```
[1] "control"  "n25"    "n50"    "r10"    "r5"
```

so there are $5 - 1 = 4$ orthogonal contrasts that can be carried out. We choose to carry out the *a priori* contrasts in the following way (although there are many other possible ways of arranging the four orthogonal contrasts):

- compare the controls with the group of reduced-competition treatments

- compare the root-pruned treatments with the shoot-pruned treatments

- compare the two depths of root pruning

- compare the two extents of neighbouring shoot removal

There are sophisticated built-in functions in S-Plus for defining and executing contrasts. The *contrast coefficients are specified as attributes of the factor* clipping like this:

contrasts(clipping)<-cbind(c(4,-1,-1,-1,-1),c(0,1,1,-1,-1),c(0,0,0,1,-1),c(0,-1,1,0,0))

where we bind the relevant contrast vectors together using **cbind**. To inspect the contrasts associated with any factor we just type

contrasts(clipping)

```
          [,1] [,2] [,3] [,4]
control     4    0    0    0
   n25     -1    1    0   -1
   n50     -1    1    0    1
   r10     -1   -1    1    0
    r5     -1   -1   -1    0
```

Notice that all of the column totals sum to zero as required ($\sum c_i = 0$, see above). You can also see that the different contrasts are all *orthogonal* because *the products of their coefficients all sum to zero*. For example, comparing contrasts 1 and 2, we have

$$(4 \times 0) + (-1 \times 1) + (-1 \times 1) + (-1 \times -1) + (-1 \times -1) = -1 + -1 + 1 + 1 = 0$$

Now when we carry out a one-way Anova using the factor clipping, the parameter estimates reflect the differences between the contrasted group means rather than differences between the individual treatment means:

model<-aov(biomass~clipping)

We use the function **summary.lm** to obtain a listing of the coefficients for the four contrasts we have specified:

summary.lm(model)

```
Coefficients:
                 Value Std. Error t value Pr(>|t|)
(Intercept) 561.8000    12.8593 43.6884   0.0000
  clipping1 -24.1583     6.4296 -3.7573   0.0009
  clipping2 -24.6250    14.3771 -1.7128   0.0991
  clipping3   0.0833    20.3323  0.0041   0.9968
  clipping4   8.0000    20.3323  0.3935   0.6973
```

This gives all the information we need. The first row shows the overall mean biomass (561.8). The second row shows clipping contrast 1: the overall mean versus the four competition levels (585.958). The overall mean was 24.1583 g below the mean of the four competition treatments ($561.8 - 585.958 = -24.1583$), and this difference is highly significant ($p = 0.0009$). The third row shows the effect of the second contrast between clipped and root-pruned plants. This contrast is not significant ($p = 0.0991$) despite the fact that the clipped plants produced an average biomass of about 49.25 g less than the root-pruned plants; the coefficient (-24.625) is *half* of the difference between the two groups of means ($561.333 - 610.585$) because all the contrasts are calculated relative to the overall mean (in this case the mean of {b,c,d,e} excluding the controls, 585.96. There is no hint of any difference between the different intensities of root pruning $610.67 - 610.585 = 0.0833(p = 0.9968)$

or shoot clipping $569.333 - 561.333 = 8.0 (p = 0.6973)$, despite the fact that the means differ by $16.0\,g\ (569.333 - 553.333 = 16)$.

You may have noticed that the standard errors are different in four of the five rows of the coefficients table. All the standard errors are based on the same pooled error variance of $s^2 = 4960.813$ (see the Anova table on p. 326). For the grand mean, based on 30 samples, the standard error is $\sqrt{4960.813/30} = 12.859$. The first contrast (row 2) compares a group of four means with the overall mean (a comparison based notionally on $30 \times 4 = 120$ numbers) so the standard error is $\sqrt{4960.813/120} = 6.4296$. The second contrast (row 3) compares the two defoliation means with the two root pruning means (a total of 24 numbers) so the standard error is $\sqrt{4960.813/24} = 14.3771$. The third and fourth contrasts (rows 4 and 5) both involve the comparison of one mean with another (a comparison based on $6 + 6 = 12$ numbers) and therefore they have the same standard error $\sqrt{4960.813/12} = 20.3323$.

Contrast sums of squares by hand

The key point to understand is that *the treatment sum of squares SSA is the sum of all $k - 1$ orthogonal sums of squares*. It is useful to know which of the contrasts contributes most to *SSA*, and to work this out, we compute the contrast sum of squares *SSC* as follows:

$$SSC = \frac{\left(\sum c_i T_i / n_i \right)^2}{\sum c_i^2 / n_i}$$

The significance of a contrast is judged in the usual way by carrying out an F test to compare the contrast variance with the error variance, s^2. Since all contrasts have a single degree of freedom, the contrast variance is equal to *SSC*, so the F test is just

$$F = \frac{SSC}{s^2}$$

where the error variance, s^2, comes from the error mean square column of the Anova table. The contrast is significant (i.e. the two contrasted groups have significantly different means) if the calculated value is larger than the value of F in tables with 1 and $k(n - 1)$ degrees of freedom. We demonstrate these ideas by continuing our example. The five mean biomass values were

```
tapply(biomass,clipping,mean)
    control        n25        n50        r10         r5
  465.1667   553.3333   569.3333   610.6667    610.5
```

We have already established that the contrast between the controls and the other four treatments was highly significant. Here we develop the theme by assessing the significance of the type of competition treatment. The root-pruned plants (r_{10} and r_5) were larger than the shoot pruned plants (n_{25} and n_{50}), suggesting that below ground competition might be more influential than above ground. It remains to be seen whether these differences are significant by using contrasts. To compare defoliation and root pruning (i.e. a comparison of competition for light with below ground competition), the contrast coefficients are

```
        control   n25   n50   r10   r5
  c_i        0    -1    -1     1    1
```

To calculate a new contrast sum of squares, we need the treatment totals, T:

tapply(biomass,clipping,sum)

```
control  n25    n50    r10    r5
2791 3320    3416   3664   3663
```

to which we apply the formula. The controls have zero weight so we ignore them.

$$SSC = \frac{(1/6)^2[(-1 \times 3320) + (-1 \times 3416) + (1 \times 3664) + (1 \times 3663)]^2}{(1/6)[(-1)^2 + (-1)^2 + 1^2 + 1^2]} = \frac{(591/6)^2}{4/6} = 14\,553.38$$

The error variance is 4960.81 (from the Anova table), so the F test for this contrast is

$$F = \frac{14\,553.38}{4960.81} = 2.933\,67$$

Notice that this F value is the square of the t value obtained by contrast number 2 ($1.7128^2 = 2.933\,684$). We need to test the significance of this by comparing our calculated F value with the value in tables with 1 and 25 d.f. We use **qf** for this:

qf(0.95,1,25)

```
[1] 4.241699
```

Our calculated value is less than the value in tables, so this contrast was not significant.

A posteriori model simplification

An alternative to specifying a full set of orthogonal contrasts *a priori*, is to use model simplification. This is typically an *a posteriori* process, dictated by the parameter values and their standard errors. We fit the full model then inspect the parameter values using **summary.lm**:

```
options(contrasts=c("contr.treatment","contr.poly"))
model<-aov(biomass~clipping)
summary.lm(model)

Call: aov(formula = biomass ~ clipping)
Residuals:
   Min     1Q   Median    3Q     Max
 -103.3  -49.67   3.417  43.38   177.7

Coefficients:
                 Value   Std.Error   t value   Pr(>|t|)
(Intercept)    465.1667   28.7542    16.1774    0.0000
clippingn25     88.1667   40.6645     2.1681    0.0399
clippingn50    104.1667   40.6645     2.5616    0.0168
clippingr10    145.5000   40.6645     3.5781    0.0015
clippingr5     145.3333   40.6645     3.5740    0.0015

Residual standard error: 70.43 on 25 degrees of freedom
Multiple R-Squared: 0.4077
```

The *p* values all look significant, but this only means that the four competition treatments are different from the controls. It does not mean that any of the competition treatments is significantly different from the others (this is the downside of using treatment contrasts).

First, let's try lumping the two levels of root pruning together into a single level, because these two means were the most similar (their parameter values are almost identical at 145.5 and 145.333). We need to produce a new factor cc to replace clipping:

```
cc<-clipping
```

but we want it to contain the *same* level for r_{10} and r_5, but retain the other level names unaltered. We use the trick of 'levels gets' for this. The new factor level to replace r_5 and r_{10} is to be called *root*. We need to discover which factor levels in clipping are r_5 and r_{10}:

```
levels(clipping)

[1] "control" "n25"    "n50"    "r10"    "r5"
```

This indicates that we want to replace the fourth and fifth levels of clipping, so we write

```
levels(cc)[4:5]<-"root"
```

To see what this has done, we inspect the contents of cc:

```
cc

 [1]  n25  n25  n25     n25     n25     n25     n50     n50
 [9]  n50  n50  n50     n50     root    root    root    root
[17]  root root control control control control control control
[25]  root root root    root    root    root
```

Note that the levels are changed wherever they occur in the vector cc (e.g. there are values of *root* on either side of the *control* plants in the data frame). Now we fit cc to the model in place of clipping, and ask whether the model simplification is justified, using **anova** to compare the original and simpler models:

```
model2<-aov(biomass~cc)
anova(model,model2)
```

```
Analysis of Variance Table

Response: biomass

    Terms Resid. Df     RSS   Test Df   Sum of Sq       F Value      Pr(F)
1 clipping       25 124020.3
2       cc       26 124020.4 1 vs. 2 -1  -0.08333333  0.00001679832  0.9967623
```

This simplification was justified because the explanatory power of the simpler model is not significantly lower ($p = 0.997$). Next we try replacing the two levels of shoot pruning by a single level. We make a new contrast called dd that has the same values as cc:

```
dd<-cc
```

but we want to replace the factor levels n_{25} and n_{50} by *shoot*:

```
levels(dd)

[1] "control"  "n25"    "n50"    "root"
```

These are the second and third levels in **dd** so we write

```
levels(dd)[2:3]<-"shoot"
levels(dd)
```

```
[1] "control" "shoot"   "root"
```

Now fit **dd** in place of **cc** and compare the two model fits:

```
model3<-aov(biomass~dd)
anova(model2,model3)
```

```
Analysis of Variance Table
```

```
Response: biomass
```

	Terms	Resid. Df	RSS	Test Df	Sum of Sq	F Value	Pr(F)
1	cc	26	124020.4				
2	dd	27	124788.4	1 vs. 2 -1	-768	0.1610057	0.6915111

Again, this model simplification was justified ($p = 0.6915$). Next we compare the two kinds of competition, by lumping together *root* and *shoot* in a new factor called **ee**:

```
ee<-dd
levels(ee)
```

```
[1] "control" "shoot"   "root"
```

```
levels(ee)[2:3]<-"competition"
model4<-aov(biomass~ee)
anova(model3,model4)
```

```
Analysis of Variance Table
```

```
Response: biomass
```

	Terms	Resid. Df	RSS	Test Df	Sum of Sq	F Value	Pr(F)
1	dd	27	124788.4				
2	ee	28	139341.8	1 vs. 2 -1	-14553.38	3.148859	0.08725792

This simplification was also justified (although it is much closer to significance ($p = 0.087$) than the other contrasts we have carried out so far). The final simplification is to lump together *control* and *competition* treatments. This would be a model with only the grand mean (parameter 1) fitted:

```
model5<-aov(biomass~1)
anova(model4,model5)
```

```
Analysis of Variance Table
```

```
Response: biomass
```

	Terms	Resid. Df	RSS	Test Df	Sum of Sq	F Value	Pr(F)
1	ee	28	139341.8				
2	1	29	209376.8		-1 -70035.01	14.07317	0.0008149398

That simplification was certainly *not* justified ($p = 0.0008$). We conclude that model 4 with its two parameters was minimal adequate; all we need to describe the data from this experiment are a mean for the controls, and a mean for all the competition treatments lumped together. Here is the

Anova table for the simplified model:

```
summary(model4)
            Df  Sum of Sq   Mean Sq   F Value          Pr(F)
  ee        1     70035.0   70035.01  14.07317  0.0008149398
  Residuals 28   139341.8    4976.49
```

In this example the kind and the intensity of competition reduction were not important. Note that the error variance in the minimal adequate model ($s^2 = 4976.49$) is not substantially greater than in the maximal model ($s^2 = 4960.81$, see above).

Aliasing

Aliasing occurs when there is no information available on which to base an estimate of a parameter value. Parameters can be aliased for one of two reasons:

- there are no data in the data frame from which to estimate the parameter (e.g. missing values, partial designs or correlation among the explanatory variables)

- the model is structured in such a way that the parameter value cannot be estimated (e.g. over-specified models with more parameters than necessary)

These two kinds of aliasing are called *intrinsic* and *extrinsic*:

- intrinsic aliasing occurs due to the structure of the model

- extrinsic aliasing occurs due to the nature of the data

Intrinsic aliasing with continuous variables

Suppose we are modelling the density of galls on leaves, and we have measurements of the length, breadth and area of each leaf. We might propose to fit the model:

$$galls = a + (b\ log\ length) + (c\ log\ breadth) + (d\ log\ area)$$

with four parameters (a, b, c and d). It turns out that we can only estimate three of the four parameters for the following intrinsic reason:

$$log\ area = e + log\ length + log\ breadth$$

It is worth working through this example, in order to understand the concept of aliasing. Let our three explanatory variables be x_1, x_2 and x_3 so

$$y = a + bx_1 + cx_2 + dx_3$$

and because leaf area is proportional to length times breadth we can write

$$x_3 = e + x_1 + x_2$$

where e is the log of the leaf-shape constant. Now we replace x_3 in the equation, and multiply through by d:

$$y = a + bx_1 + cx_2 + de + dx_1 + dx_2$$

Gathering the terms for x_1 and x_2, we get

$$y = (a + de) + (b + d)x_1 + (c + d)x_2$$

so that the model will only be able to estimate three separate quantities from the data:

$$a + de \quad b + d \quad c + d$$

The fourth parameter is said to be **intrinsically aliased**. As an exercise, you should convince yourself that if leaf breadth is proportional to leaf length (i.e. $x_2 = f + x_1$), then we can estimate only two parameters from the data rather than three.

Intrinsic aliasing with factors

If we had a factor with four levels (say none, light, medium and heavy use) then we could estimate four means from the data, one for each factor level. But the model looks like this:

$$y = \mu + \beta_1 x_1 + \beta_2 x_2 + \beta_3 x_3 + \beta_4 x_4$$

where the x_1 to x_4 are dummy variables having the value 0 or 1 (see p. 215). Clearly there is no point in having five parameters in the model if we can estimate only four independent terms. One of the parameters must be intrinsically aliased.

There are innumerable ways of dealing with this, but three equally logical options are as follows:

- set the grand mean μ to 0, so that β_1 to β_4 are the four individual treatment means

- set the first term β_1 to 0 so that μ is the mean of the first group and the βs are the differences between the first group mean and the other group means

- set the sum of the βs to 0 so that μ is the grand mean and each β is a departure from the grand mean

These options are discussed in detail later on. Former GLIM users will be familiar with the second option. These are known as **treatment contrasts** in S-Plus. The first parameter is a mean and the remaining three parameters are differences between means (all means are compared with the first mean). The default in S-Plus is known as **Helmert contrasts**, which is different from all of these. It has:

- the overall mean as its first parameter

- the difference between the first and the average of the first and second means as its second parameter

- the difference between the average of the first and second means and the average of the first, second and third means as its third parameter

- the difference between the average of the first, second and third means and the overall average as its fourth parameter

Some examples of aliasing

In general, an *aliased* parameter is a parameter whose value cannot be estimated because either the model or the data contain no information on it. Consider the following examples:

- Suppose that in a factorial experiment, all of the animals receiving level 2 of diet (factor A) and level 3 of temperature (factor B) have died accidentally as a result of attack by a fungal pathogen. This particular combination of diet and temperature contributes no data to the response variable, so the interaction term A(2):B(3) cannot be estimated. It is **extrinsically aliased**, and its parameter estimate is set to zero.

- If one continuous variable is perfectly correlated with another variable that has already been fitted to the data (perhaps because it is a constant multiple of the first variable), then the second term is aliased and adds nothing to the model. Suppose that $x_2 = 0.5x_1$ then fitting a model with $x_1 + x_2$ will lead to x_2 being **intrinsically aliased** and given a zero parameter estimate (see the example of galls on leaves, above).

- If all the values of a particular explanatory variable are set to zero for a given level of a particular factor, then that level is **intentionally aliased**. This sort of aliasing is a useful programming trick in Ancova when we wish a covariate to be fitted to some levels of a factor but not to others (see p. 534).

Two-way tables

In two-way tables, the question of aliasing is a little more complicated. If we have replication for every treatment combination then the calculations are reasonably straightforward. Let's say the two factors are nitrogen fertiliser (two N levels; none and regular application rate) and potassium fertilizer (two K levels; none and regular application rate). This design defines a 2×2 table, and the linear model we want to fit is

$$1 + N + K + N{:}K$$

where 1 is the intercept and N:K represents the *interaction* between nitrogen and potassium. There are four cells in the table and hence four means can be estimated. But the linear predictor looks like this:

$$\eta_{ij} = \mu + \alpha_i + \beta_j + \gamma_{ij} + \varepsilon_{ij}$$

where μ is the overall mean, α_i is the effect size for nitrogen, β_j is the effect size for potassium, γ_{ij} is the effect of the interaction between nitrogen and potassium, and ε_{ij} is the error (with mean 0 and variance σ^2). With a 2×2 table, therefore, there are nine terms in the model: the overall mean, plus two α, two β and four γ. Since we can only estimate four quantities, five of the nine parameters must be aliased. We need to impose five *constraints on the estimates* in order to produce uniqueness, but there are lots of ways of dealing with this. These five constraints will leave four estimable quantities: one way of doing this is to make μ equal to the mean yield in cell (1,1), to have a *main effect term* for nitrogen (measured by α), a main effect for potassium (measured by β) and an interaction term (measured by γ). The model now has only four parameters, as required:

$$\eta_{ij} = \mu_{11} + \alpha + \beta + \gamma$$

where μ_{11} is the mean of nitrogen level 1 and potassium level 1, α is added only if $i = 2$, β is added only if $j = 2$ and γ is added only if $i = 2$ *and* $j = 2$. This model has four parameters and therefore includes no aliasing. Suppose that the four means were as follows.

Means	−K	+K
−N	3	4
+N	4.5	6

Then the model terms would be laid out like this.

Model terms	$-K$	$+K$
$-N$	μ_{11}	$\mu_{11} + \beta$
$+N$	$\mu_{11} + \alpha$	$\mu_{11} + \alpha + \beta + \gamma$

And the four parameter estimates would be calculated as follows.

Parameters	$-K$	$+K$
$-N$	$\mu_{11} = 3$	$\beta = 1$
$+N$	$\alpha = 1.5$	$\gamma = 0.5$

Notice that to calculate the mean for the treatment receiving both nitrogen and potassium, you need to add the nitrogen effect (α), the potassium effect (β) and the interaction effect (γ) to the 'intercept' (μ_{11}, the mean of the none:none treatment combination). That is to say, $3 + 1.5 + 1.0 + 0.5 = 6.0$ as required.

Unreplicated two-way tables

The parameter estimates of one-way Anova were easy to understand because they were the k treatment means (or a mean plus some set of $k - 1$ contrasts; see p. 324). With factorial Anova the interpretation is also straightforward, because the parameter estimates are the means of the replicates within each combination of factor levels. A minor blip occurs for cases that are *two-way analyses without replication*. The hard thing to understand is where the estimate of the intercept comes from? We return to the photoperiod example:

```
attach(twoway)
names(twoway)
```

```
[1] "Growth"    "Photoperiod"   "Genotype"
```

Set the **contrasts** to **treatment** (so we can understand them)

```
options(contrasts=c("contr.treatment", "contr.poly"))
```

then fit the model and use **summary.lm** so that we get to see the coefficients:

```
model<-aov(Growth~Photoperiod+Genotype)
summary.lm(model)
```

```
Coefficients:
                        Value   Std. Error   t value   Pr(>|t|)
         (Intercept)   3.1250      0.3400    9.1902     0.0000
      PhotoperiodShort -0.3333      0.3206   -1.0398     0.3149
  PhotoperiodVery.long  0.6667      0.3206    2.0795     0.0551
 PhotoperiodVery.short -0.8333      0.3206   -2.5994     0.0201
             GenotypeB  1.5000      0.3926    3.8203     0.0017
             GenotypeC -1.5000      0.3926   -3.8203     0.0017
             GenotypeD -1.5000      0.3926   -3.8203     0.0017
             GenotypeE -1.0000      0.3926   -2.5469     0.0223
             GenotypeF -1.2500      0.3926   -3.1836     0.0062
```

Now here's the quiz. Where does the value of 3.1250 for the intercept come from? It is going to be useful to have the treatment means to hand:

tapply(Growth,Photoperiod,mean)

```
Long       Short      Very.long  Very.short
  2.5   2.166667      3.166667    1.666667
```

tapply(Growth,Genotype,mean)

```
A     B     C     D    E     F
3   4.5   1.5   1.5    2   1.75
```

The intercept obviously isn't the mean for Genotype A (3.0) or the mean for Long photoperiod (2.5), or indeed the average of the two of these $(3 + 2.5)/2 = 2.75$. So what is it? The thing to appreciate is that *the coefficients are the differences between means*. In particular, they are the differences between *the mean in question and the mean of row 1 or column 1 as appropriate*. Thus, the coefficient for Genotype D (-1.5) is the difference in mean growth between Genotype D and Genotype A, which is $1.5 - 3.0 = -1.5$. Likewise for photoperiod, the parameter value for Very long photoperiod (0.6667) is the difference between mean growth in Very long photoperiod and the mean for Long photoperiod (column 1): $3.166\,66 - 2.5 = 0.666\,67$. Okay, but where does the intercept come from?

The penny drops when you realise that unlike one-way or factorial Anova, where the fitted values are treatment means, in an unreplicated two-way Anova the fitted values are not necessarily equal to *any* of the means. The overall mean is fixed:

mean(Growth)

```
[1] 2.375
```

and the row and column means are fixed (as we saw in the hand calculations earlier). But that is just $1 + (4 - 1) + (6 - 1) = 9$ parameters. We need to predict 24 fitted values using only these 9 parameters. This is how it is done. Look carefully at the following table, where r is a row effect and c is a column effect, and i is the intercept that we are trying to estimate from the data.

	c_1	c_2	c_3	c_4
r_1	i	$i + c_2$	$i + c_3$	$i + c_4$
r_2	$i + r_2$	$i + r_2 + c_2$	$I + r_2 + c_3$	$i + r_2 + c_4$
r_3	$i + r_3$	$i + r_3 + c_2$	$I + r_3 + c_3$	$i + r_3 + c_4$
r_4	$i + r_4$	$i + r_4 + c_2$	$I + r_4 + c_3$	$i + r_4 + c_4$
r_5	$i + r_5$	$i + r_5 + c_2$	$I + r_5 + c_3$	$i + r_5 + c_4$
r_6	$i + r_6$	$i + r_6 + c_2$	$I + r_6 + c_3$	$i + r_6 + c_4$

These effects are the 9 parameter values in the model summary, and they are the differences between the relevant row mean and the mean of row 1, or the difference between the relevant column mean and the mean of column 1. It is important that you see where the following effect sizes come from. The mean of row 1 (A) is 3.0 so

$$r_2 = 1.5, \; r_3 = -1.5, \; r_4 = -1.5, \; r_5 = -1.0 \quad \text{and} \quad r_6 = -1.25$$

The mean of column 1 (Long) is 2.5 so

$$c_2 = -0.333\,33, \quad c_3 = 0.666\,666 \quad \text{and} \quad c_4 = -0.833\,333$$

The idea is to reconstruct the 24 fitted values as the appropriate sums of the 9 parameters. The intercept i appears in all 24 cells of the table. There is no row effect in row 1, because the r parameters are differences from the mean of row 1. There are no column effects in column 1 because the column effects are differences from the mean of column 1. Row effects appear in all four columns (i.e. one for each photoperiod). Column effects appear in all six rows (for each of the genotypes A through F). So the equation for the overall mean is this:

$$\mu = \frac{24(i) + 6(c_2) + 6(c_3) + 6(c_4) + 4(r_2) + 4(r_3) + 4(r_4) + 4(r_5) + 4(r_6)}{24}$$

We know the values of all the parameters except i, so we can rearrange and solve for i:

$$i = \frac{24 \times \mu - [6(c_2) + 6(c_3) + 6(c_4) + 4(r_2) + 4(r_3) + 4(r_4) + 4(r_5) + 4(r_6)]}{24}$$

Now we can calculate i, using the three column differences and the five row differences:

```
(24*2.375-6*(-.33334)-6*.666666-6*(-.83333333)-4*1.5-4*(-1.5)-4*(-1.5)-4*(-1)-4*(-1.25))/24
```

```
[1] 3.125
```

Eventually, we have discovered whence cometh the intercept. Generally, the parameters are much easier to understand than this, because they are the means of n replicates. It was the fact that there was no replication that made these calculations so tedious.

Confounding

The term 'Confounding' means not having every treatment in every block. This sounds like a bad idea at first, but sometimes it is forced upon us. Suppose that we have time to handle only two treatments in a day, but there are six treatments. We don't want to make a 3 day period into a block because we use different batches of animals on every day (nor would this be handy for weekend social life). Efficient use of confounding requires a careful analysis of its advantages and disadvantages. The advantages of confounding are:

- reduction in experimental error arising from the use of smaller and therefore internally more homogeneous blocks

- reduced size of the whole experiment

while the disadvantages are:

- reduction in replication of the confounded treatment comparisons
- increased complexity of calculations

The question that needs to be addressed is whether the reduction in variance per unit that comes from having smaller blocks, more than compensates for the loss of replication due to confounding.

A large factorial experiment (say one with six or seven different factors, each at two levels) may be beyond the resources of the experimenter, or it may be unnecessarily big, in the sense that it would give more precision of the estimates than was really needed. Under these circumstances, fractional replication may be appropriate (Cochran and Cox 1957). Thus, instead of having all 8 of the treatment combinations from a 2^3 factorial in each block, one would have only 4 with half-replication, or 2 with quarter-replication. The difficulty is that experiments with fractional replication are open to misinterpretation because of aliasing. For instance, in a three-factor factorial with two levels of each factor, one half-replicate might have only the three main effects (a), (b) and (c) and the three-way interaction (abc). Then the main effect of A would be calculated as

$$A = (abc) + (a) - (b) - (c)$$

However, the two-way interaction AB would be given by

$$AB = (abc) + (c) - (a) - (b)$$

which you will notice is precisely the same as the formula for the main effect of factor C. Thus C and AB are aliases (Cochran and Cox 1957, p. 244).

The three-way interaction ABC cannot be estimated at all, because

$$ABC = (abc) + (a) + (b) + (c) - (ab) - (ac) - (bc) - (1)$$

and we only have information on the terms with positive signs. In this particular half-replicate we have lost all information on the three-way interaction, and all of the two-factor interactions are inextricably confounded with main effects. Thus, if the experiment suggests that the main effect of A is important, we don't know if this is really due to the effect of A, the interaction between B and C or a combination of both.

Helmert, treatment and sum contrasts

Rather than specify our own contrasts (as we did in the first example), an option in S-Plus allows you to choose one of three standard methods of dealing with contrasts. The default type in S-Plus is Helmert contrasts, while in Genstat and R the default is treatment contrasts. The differences between the three types of contrasts are the cause of untold difficulty for students.

The problem arises because of *overparameterisation* in Anova models (see above). Here we work through a simple example and look in detail at each type of contrast. The response variable is the pile (in mm) exhibited by carpets experiencing four different levels of use: none, light, medium and heavy use. We need to compare the parameter values from each of the three contrast options with each other and with the intrinsically aliased initial Anova model with its five parameters:

$$\hat{y}_i = \mu + \beta_i$$

with its overall mean μ and k differences between means $\beta_i s$ (one for each of the $k = 4$ factor levels). The four treatment means are 13, 12, 9 and 6 under the four levels of use, and the overall mean was 10. Here are four different ways of parameterising the model for this experiment:

	Anova	Sum	Treatment	Helmert
μ	10	10	aliased	10
β_1	3	3	13	-0.5
β_2	2	2	-1	-1.1667
β_3	-1	-1	-4	-1.3333
β_4	-4	aliased	-7	aliased

The full **Anova** in column 1 is overspecified because it has five parameters rather than four. Reading from left to right:

- **Sum** contrast has the overall mean as parameter 1. It has *aliased the mean for the last treatment* (heavy use). The other three parameters are the differences between the grand mean (10) and the means of each of the *first three treatments* (none, light and medium use).

- **Treatment** contrast has *aliased the overall mean*. The mean of the no-use treatment (13) is the first parameter. The remaining three parameters are the following differences between means: no use and light use ($12 - 13 = -1$), no use and medium use ($9 - 13 = -4$), no use and heavy use ($6 - 13 = -7$), respectively.

- **Helmert** contrasts have the overall mean as parameter 1. As with sum contrasts, *the mean for treatment 4* (heavy use) is aliased. The second parameter is the average of the no-use and light use means ($(13 + 12)/2 = 12.5$) minus the no-use mean ($12.5 - 13 = -0.5$). The third parameter is the average of the first three means ($(13 + 12 + 9)/3 = 11.333$) minus the average of the first two means ($11.333 - 12.5 = -1.667$). The fourth parameter is the average of the first four means (10) minus the average of the first three means ($10 - 11.333 = -1.3333$).

The point is that there are lots of different ways of presenting the same information. None of the sensible ways of presenting the information contain any aliased parameters. The best way of presenting the information is a matter of taste. The advantage of the Helmert contrasts is that their p values indicate unequivocally whether a particular contrast is significant, and hence deserves to stay in the model. The disadvantage of Helmert contrasts is that the parameter values are virtually unintelligible.

Note that each of the three different contrast options produces *different estimates for the standard errors* of the parameters. This is because of the different sample sizes used in the calculations of the effect sizes. For instance, in the present example, the standard error for the overall mean is based on 12 numbers, a single mean on 3 numbers, and so on.

Comparison of different contrasts options using S-Plus

```
attach(contrasts)
names(contrasts)
```

```
[1] "pile" "use"
```

The overall mean is obtained first:

```
mean(pile)
```

```
[1] 10
```

then the four treatment means are found like this:

```
tapply(pile,use,mean)
```

```
 A   B  C D
13  12  9 6
```

Because the experiment involves a single factor, and this factor has four levels (none, low, medium and heavy use), the experiment is capable of generating only four parameters—the four treatment means. Any extra parameters that might be estimated from the data (e.g. the overall mean) are said to be **aliased** (see above). The Anova table looks like this:

```
summary(aov(pile~use))
```

```
             Df  Sum of Sq  Mean Sq  F Value
     use      3         90       30       30
Residuals     8          8        1
```

Which shows that the error variance $s^2 = 1$ (which makes interpretation a little easier). The Anova table would be identical for each of the different contrast options.

Now we fit the same model with each of the contrast options in turn, using **lm** rather than **aov** in order to see the parameter estimates rather than the Anova table. The default is Helmert contrasts in S-Plus (but not in R), so we fit these first.

Helmert contrasts

```
options(contrasts=c("contr.helmert","contr.poly"))
```

```
modelH<-lm(pile~use)
summary(modelH)
```

```
Coefficients:
                 Value  Std. Error   t value   Pr(>|t|)
(Intercept)    10.0000      0.2887   34.6410     0.0000
      use1     -0.5000      0.4082   -1.2247     0.2555
      use2     -1.1667      0.2357   -4.9497     0.0011
      use3     -1.3333      0.1667   -8.0000     0.0000
```

The first parameter is the overall mean (10.0). The second parameter is the average of the no-use and light use means $((13 + 12)/2 = 12.5)$ minus the no-use mean $(12.5 - 13 = -0.5)$. The third parameter is the average of the first three means $((13 + 12 + 9)/3 = 11.333)$ minus the average of the first two means $(11.333 - 12.5 = -1.667)$. The fourth parameter is the average of the first four means (10) minus the average of the first three means $(10 - 11.333 = -1.3333)$.

The standard errors are different in every row. In the first row is the *standard error of the overall mean* $\left(\sqrt{(s^2/kn)} = 0.2887\right)$ which is based on $kn = 4 \times 3 = 12$ numbers. In the second row $\left(\sqrt{(s^2/2n)} = 0.4082\right)$, we have a comparison of *a group of two means with a single mean* $(2 \times 1 = 2)$. This is multiplied by the sample size n in the denominator. In the third row $\left(\sqrt{(s^2/6n)} = 0.2357\right)$, we have a comparison of *a group of three means with a group of two means* $(3 \times 2 = 6)$. The standard error in the fourth row $\left(\sqrt{(s^2/12n)} = 0.1667\right)$ is based on a comparison of *a group of four means with a group of three means* $(4 \times 3 = 12)$.

Treatment contrasts

```
options(contrasts=c("contr.treatment","contr.poly"))
```

```
modelT<-lm(pile~use)
summary(modelT)
```

```
Coefficients:
                 Value   Std. Error  t value   Pr(>|t|)
(Intercept)    13.0000     0.5774    22.5167    0.0000
     useB      -1.0000     0.8165    -1.2247    0.2555
     useC      -4.0000     0.8165    -4.8990    0.0012
     useD      -7.0000     0.8165    -8.5732    0.0000
```

The mean of the no-use treatment (13) is the first parameter (Intercept = 13). The remaining three parameters are the differences between means: useB is the difference between the light use and no-use means $(12 - 13 = -1)$, useC is the difference between the medium use and no-use means $(9 - 13 = -4)$, and useD is the difference between the heavy use and no-use means $(6 - 13 = -7)$.

The standard errors are much more straightforward than with Helmert contrasts. The first row shows the standard error of a single mean, $\left(\sqrt{s^2/n} = 0.5774\right)$ and is based on $n = 3$ numbers. The remaining rows have the same value, which is the standard error of the difference between two means $\left(\sqrt{2s^2/n} = 0.8165\right)$; this is based on $2 \times n = 6$ numbers.

Sum contrasts

```
options(contrasts=c("contr.sum","contr.poly"))

modelS<-lm(pile~use)
summary(modelS)

Coefficients:
                Value   Std. Error   t value   Pr(>|t|)
(Intercept)   10.0000     0.2887     34.6410    0.0000
      use1     3.0000     0.5000      6.0000    0.0003
      use2     2.0000     0.5000      4.0000    0.0039
      use3    -1.0000     0.5000     -2.0000    0.0805
```

The first row has the overall mean (10) as parameter 1 (the Intercept). Sum contrasts *alias the mean for the last treatment* (heavy use). The other three parameters are the differences between the grand mean (10) and the means of each of the *first three treatments* (none, light and medium use): $13 - 10 = 3$ for use1, $12 - 10 = 2$ for use2 and $9 - 10 = -1$ for use3.

The standard error in the first row is the *standard error of the overall mean*, $\left(\sqrt{s^2/kn} = 0.2887\right)$, which is based on all $kn = 4 \times 3 = 12$ numbers. The standard errors in the remaining rows $\left(\sqrt{(s^2/kn) + (s^2/2n)} = 0.5\right)$ are based on a comparison between the overall mean (based on kn numbers) and two means (based on $2 \times n$ numbers).

Note that only with treatment contrasts do the parameters appear with the factor level names we gave them (useB, useC and useD). In the other two cases the rows are labelled by their *contrast numbers* (use1, use2 and use3).

Contrasts and the parameters of Ancova models

In analysis of covariance, we estimate a slope and an intercept for each level of one or more factors. Suppose we are modelling growth (the response variable) as a function of sex and age. Sex is a factor with two levels (male and female) and age is a continuous measure. The maximal model therefore has four parameters: two slopes (a slope for males and a slope for females) and two intercepts (one for males and one for females) like this:

$$weight_{male} = a_{male} + b_{male} \times age$$

$$weight_{female} = a_{female} + b_{female} \times age$$

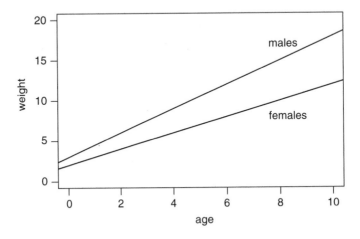

The difficulty arises because there are several different ways of expressing the values of the four parameters in the **summary.lm** table:

- 2 slopes and 2 intercepts (as in the equations above)

- 1 slope and 1 difference between slopes, and 1 intercept and 1 difference between intercepts

- the overall mean slope and the overall mean intercept, and 1 difference between slopes and 1 difference between intercepts

In the second case (two estimates and two differences) a decision needs to be made about which factor level to associate with the estimate, and which level with the difference. For example, should males be expressed as the intercept and females as the difference between intercepts, or vice versa? When the factor levels are unordered (the typical case), then S-Plus takes the factor level that comes first in the alphabet as the estimate and the others are expressed as differences. In our example the parameter estimates would be female, and male parameters would be expressed as differences from the female values, because f comes before m in the alphabet. This should become clear from an example:

```
attach(Ancovacontrasts)
names(Ancovacontrasts)

[1] "weight" "sex"   "age"
```

First we work out the two regressions separately so that we know the values of the two slopes and the two intercepts:

```
lm(weight[sex=="male"]~age[sex=="male"])
```

```
Coefficients:
(Intercept)   age[sex == "male"]
   3.115178              1.560808
```

lm(weight[sex=="female"]~age[sex=="female"])

```
Coefficients:
(Intercept)   age[sex == "female"]
   1.966277                0.9962039
```

So the intercept for males is 3.115 and the intercept for females is 1.966. The difference between the first intercept (female) and the second intercept (male) is therefore

$$3.1152 - 1.96628 = +1.1489$$

Now we can do an overall regression, ignoring gender:

lm(weight~age)

```
Coefficients:
(Intercept)          age
   2.540728 1.278506
```

This tells us that the average intercept is 2.541 and the average slope is 1.279.

Next we can carry out an analysis of covariance and compare the output produced by each of the three different contrast options allowed by S-Plus: **Helmert** (the default), **treatment** (the default in R and in GLIM) and **sum**:

options(contrasts=c("contr.helmert", "contr.poly"))

The Ancova estimates separate slopes and intercepts for each sex because we use the * operator:

lm(weight~age*sex)

```
Coefficients:
(Intercept)          age          sex      age:sex
   2.540728 1.278506 0.5744508 0.2823018
```

Let's see if we can work out what the four parameter values represent. The first parameter 2.5407 (labelled Intercept) is the intercept of the overall regression, ignoring sex (see above). The parameter labelled age (1.2785) is a *slope* because age is our continuous explanatory variable. Again, you will see that it is the slope for the regression of weight against age, ignoring sex. The parameter labelled sex (0.5744) must have something to do with intercepts because sex is our categorical variable. If we want to reconstruct the second intercept (for males) we need to add 0.5744 to the overall intercept: $2.5407 + 0.5744 = 3.1151$. To get the intercept for females we need to subtract 0.5744: $2.5407 - 0.5744 = 1.9663$. The parameter labelled age:sex (0.2823) is the difference between the overall mean slope (1.279) and the male slope: $1.2785 + 0.2823 = 1.5608$. To get the slope of weight against age for females, we need to subtract the interaction term from the age term: $1.2785 - 0.2823 = 0.9962$.

The advantage of Helmert contrasts is in hypothesis testing, because it is easy to see which terms we need to retain in a simplified model by inspecting their significance levels in the **summary.lm** table. The disadvantage is that it is hard to reconstruct the slopes and the intercepts from the estimated parameter values (see also p. 340). Let's repeat the analysis using **treatment contrasts** as used by R and by GLIM:

```
options(contrasts=c("contr.treatment", "contr.poly"))
lm(weight~age*sex)
```
```
Coefficients:
(Intercept)          age        sex    age:sex
   1.966277 0.9962039 1.148902 0.5646037
```

The Intercept (1.9662) is now the intercept for females (because f comes before m in the alphabet). The age parameter (0.9962) is the slope of the graph of weight against age for females. The sex parameter (1.1489) is the difference between the (female) intercept and the male intercept $(1.966\,277 + 1.148\,902 = 3.1151)$. The age:sex interaction term is the difference between slopes of the female and male graphs $(0.9962 + 0.5646 = 1.5608)$. So with treatment contrasts, the parameters (in order 1 to 4) are an intercept, a slope, a difference between two intercepts, and a difference between two slopes. Many people are more comfortable with this method of presentation than they are with Helmert contrasts.

Finally, we look at the third option, which is **sum contrasts**:

```
options(contrasts=c("contr.sum", "contr.poly"))
```
```
lm(weight ~ age*sex)
```
```
Coefficients:
(Intercept)          age        sex    age:sex
   2.540728 1.278506 -0.5744508 -0.2823018
```

The first two terms are the same as those produced by Helmert contrasts: the overall intercept and slope of the graph relating weight to age ignoring sex. The sex parameter (-0.5744) is *sign reversed* compared with the Helmert option; it shows how to calculate the female (the first) intercept from the overall intercept $2.5407 - 0.5746 = 1.9661$. The interaction term also has reversed sign; to get the slope for females, add the interaction term to the slope for age: $1.2785 - 0.2823 = 0.9962$.

Further reading

Sokal, R.R. and Rohlf, F.J. (1995) *Biometry: The Principles and Practice of Statistics in Biological Research.* San Francisco, W.H. Freeman.

19

Split-plot Anova

The key features of split-plot experiments are:

- different treatments are applied to plots of different sizes

- a separate Anova table is produced for every different plot size

- a different error variance is used to testing hypotheses about treatments that are applied to different plot sizes

It is somewhat unfortunate that they are called split-plot experiments (from their origin in agricultural field trials), because they are of very wide applicability in other kinds of work. Lots of experiments that are really split plots, are published as if they were straightforward Anovas. Failure to spot that an experiment is a split plot is one of the commonest causes of pseudoreplication (see p. 66).

Why would you design an experiment as a split plot, rather than say a randomised block? The answer is usually very pragmatic: because it is impractical to apply some of the treatments to very small plots. This is commonly the case in agricultural experiments; it is easy to apply fertiliser or seed to small plots, but it is very difficult to irrigate small plots effectively, or to plough them or harvest them using large machinery. In medical work, it may be too risky to employ several different protocols in the same hospital, so different protocols are randomly allocated to whole hospitals, while other treatments can easily be administered to different wards within a hospitals, and still other treatments can be allocated at random to different patients within a ward. In a physiology experiment, there may only be one controlled environment cabinet for each temperature, and all the different drug treatments at a given temperature must therefore be applied in the same cabinet.

The statistical problem that arises if you fail to spot that an experiment is a split plot is pseudoreplication. The smaller plots are pseudoreplicates when it comes to testing hypotheses about the treatments that are applied to larger plots. This leads to the wrong error variance and the wrong number of degrees of freedom being used to test hypotheses. This can lead to both Type I and Type II errors:

- you will reject the null hypothesis wrongly when the error variance is smaller than the error variance you ought to be using

- you will accept the null hypothesis wrongly when the error variance is larger than the error variance you ought to be using

Pseudoreplication matters because *it contravenes the fundamental assumption of independence of errors*. Small plots that are spatially or temporally close to one another are likely to show high positive covariance.

The really important thing to remember about split-plot experiments are that:

- we need to draw up as many Anova tables as there are plot sizes

- the error term in each table is *the interaction between blocks and all treatments that are applied to that plot size or larger*

What all these things mean will only become clear from an example. You will need to work through this slowly and painstakingly if you want to understand it.

Split-plot Anova by hand

This experiment, replicated in four blocks, involves the yield of cereals in a factorial experiment with three treatments, each applied to plots of different sizes. The largest plots (half of each block) were irrigated or not, because of the practical difficulties of watering a large number of small plots. Next the irrigated plots were split into three smaller split plots and seeds were sown at different densities. Again, because the seeds were machine sown, larger plots were preferred. Finally, each sowing density plot was split into three small split-split plots and fertilisers were applied by hand (N alone, P alone and N + P together). The yield data look like this.

	Control			Irrigated		
	N	NP	P	N	NP	P
Block A	81	93	92	78	122	98
	90	107	95	80	100	87
	92	92	89	121	119	110
Block B	74	74	81	136	132	133
	83	95	80	102	105	109
	98	106	98	99	123	94
Block C	82	94	78	119	136	122
	85	88	88	60	114	104
	112	91	104	90	113	118
Block D	85	83	89	116	133	136
	86	89	78	73	114	114
	79	87	86	109	126	131

We begin by calculating *SST* in the usual way:

$$CF = \frac{\left[\sum y\right]^2}{abcd} = \frac{7180^2}{72} = 716\,005.6$$

$$SST = \sum y^2 - \frac{\left[\sum y\right]^2}{abcd} = 739\,762 - CF = 23\,756.44$$

The block sum of squares, *SSB*, is straightforward. All we need are the four block totals:

$$SSB = \frac{\sum B^2}{acd} - CF = \frac{1746^2 + 1822^2 + 1798^2 + 1814^2}{18} - CF = 194.444$$

The irrigation main effect is calculated in a similar fashion:

$$SSI = \frac{\sum I^2}{bcd} - CF = \frac{3204^2 + 3976^2}{4 \times 3 \times 3} - CF = 8277.556$$

This is where things get different. In a split-plot experiment, *the error term is the interaction between blocks and all factors at that plot size or larger.* Because irrigation treatments are applied to the largest plots, the error term is the block by irrigation interaction. We need to calculate an interaction subtotal table (see Chapter 15 for an introduction to interaction sums of squares):

```
    control irrigated
A      831       915
B      789      1033
C      822       976
D      762      1052
```

The large plot error sum of squares *SSBI* is therefore

$$SSBI = \frac{\sum Q^2}{cd} - SSB - SSI - CF$$

$$SSBI = \frac{831^2 + 915^2 + \cdots + 1052^2}{3 \times 3} - SSB - SSI - CF = 1411.778$$

At this point we can draw up the Anova table for the largest plots. There were only 8 large plots in the whole experiment, so there are just 7 degrees of freedom in total. Block has 3 d.f., irrigation has 1 d.f., so the error variance has only $7 - 3 - 1 = 3$ d.f.

Source	SS	d.f.	MS	F	p
Block	194.44	3			
Irrigation	8277.556	1	8277.556	17.59	0.025
Error	1411.778	3	470.593		

Block is a random effect so we are not interested in testing hypotheses about differences between block means (Chapters 20 and 35). We have not written in a row for the totals, because in due course, we want the subtotals and the degrees of freedom to add up correctly across the three different Anova tables.

Now we move on to consider the sowing density effects. At this split-plot scale we are interested in the main effects of sowing density, and the interaction between sowing density and irrigation. The error term will be the block:irrigation:density interaction:

$$SSD = \frac{2467^2 + 2226^2 + 2487^2}{4 \times 2 \times 3} - CF = 1758.361$$

For the irrigation:density interaction we need the table of subtotals:

```
          high  low medium
  control 1006 1064   1134
irrigated 1461 1162   1353
```

$$SSID = \frac{1006^2 + \cdots + 1353^2}{4 \times 3} - SSI - SSD - CF = 2747.028$$

The error term for the split plots is the block:irrigation:density interaction, and we need the table of subtotals:

	high		low		medium	
	control	irrigated	control	irrigated	control	irrigated
A	266	298	292	267	273	350
B	229	401	258	316	302	316
C	254	377	261	278	307	321
D	257	385	253	301	252	366

$$SSBID = \frac{226^2 + 298^2 + \cdots + 252^2 + 366^2}{3} - SSBI - SSID - SSB - SSI - SSD - CF$$

$$= 2787.944$$

At this point we draw up the second Anova table for the split plots.

Source	SS	d.f.	MS	F	p
Density	1758.361	2	879.181	3.784	0.053
Irrigation:Density	2747.028	2	1373.514	5.912	0.016
Error	2787.944	12	232.329		

There are $4 \times 2 \times 3 = 24$ of these plots so there are 23 d.f. in total. We have already used up 7 in the first Anova table: there are 2 for density, 2 for irrigation:density and hence $23 - 7 - 2 - 2 = 12$ d.f. for error. The F test uses our new, smaller estimate of error variance (232.3 instead of 470.6) to test hypotheses about the effects of sowing density. There is a significant interaction between irrigation and density, so we take no notice of the apparently non-significant main effect of density.

Finally, we move on to the smallest, split-split plots. We first calculate the main effect of fertiliser in the familiar way:

$$SSF = \frac{\sum F^2}{abc} - CF = \frac{2230^2 + 2536^2 + 2414^2}{2 \times 4 \times 3} - CF = 1977.444$$

Now the irrigation:fertiliser interaction; the interaction subtotals are

	N	NP	P
control	1047	1099	1058
irrigated	1183	1437	1356

$$SSIF = \frac{1047^2 + 1099^2 + \cdots + 1356^2}{4 \times 3} - SSI - SSF - CF = 953.444$$

and the density:fertiliser interaction is

	N	NP	P
high	771	867	829
low	659	812	755
medium	800	857	830

$$SSDF = \frac{771^2 + 867^2 + \cdots + 830^2}{4 \times 2} - SSD - SSF - CF = 304.889$$

The final three-way interaction is irrigation:density:fertiliser and we calculate it like this:

	N			NP			P		
	high	low	medium	high	low	medium	high	low	medium
control	322	344	381	344	379	376	340	341	377
irrigated	449	315	419	523	433	481	489	414	453

$$SSIDF = \frac{322^2 + 344^2 + \cdots + 414^2 + 453^2}{4} - SSID - SSIF$$
$$- SSDF - SSI - SSD - SSF - CF = 234.722$$

The rest is easy. The error sum of squares is just the remainder when all the calculated sums of squares are subtracted from the total:

$$SSE = SST - SSB - SSI - SSD - SSF - SSIB - SSID - \cdots - SSIDF = 3108.833$$

Technically, this is the block:irrigation:density:fertiliser interaction. There are 72 plots at this scale and we have used up $7 + 16 = 23$ degrees of freedom in the first two Anova tables. In the last Anova table, fertiliser has 2 d.f., the fertiliser:irrigation interaction has 2 d.f., the fertiliser:density interaction has 4 d.f. and the irrigation:density:fertiliser interaction has a further 4 d.f.. This leaves $71 - 23 - 2 - 2 - 4 - 4 = 36$ d.f. for error. At this point we can draw up the third and final Anova table.

Source	SS	d.f.	MS	F	p
Fertiliser	1977.444	2	988.722	11.449	0.000 14
Irrigation:Fertiliser	953.444	2	476.722	5.52	0.0081
Density:Fertiliser	304.889	4	76.222	0.883	n.s.
Irrigation:Density:Fertiliser	234.722	4	58.681	0.68	n.s.
Error	3108.833	36	86.356		

We are able to test hypotheses about the effects of fertiliser using a much smaller error variance than for either of the two larger plot sizes (86.4 compared with 232.3 and 470.6). The three-way interaction was not significant, nor was the two-way interaction between density and fertiliser. The two-way interaction between irrigation and fertiliser, however, was highly significant and, not surprisingly, there was a highly significant main effect of fertiliser.

Obviously, you would not want to have to do calculations like this by hand every day. Fortunately, the computer will do it for you:

```
attach(splityield)
names(splityield)

[1] "yield"  "block"  "irrigation" "density"  "fertilizer"

model<-aov(yield~irrigation*density*fertilizer+Error(block/irrigation/density/fertilizer))
```

The model is long, but not particularly complicated. Note the two parts: the model formula (the factorial design irrigation*density*fertilizer), and the error structure (with plot sizes listed left to right from largest to smallest, separated by slash / operators). You don't actually need to put the name of the smallest plots ('fertiliser') in the error statement (you might like to experiment to see what difference leaving it out makes). The main replicates are blocks, and these provide the estimate of the error variance for the largest treatment plots (irrigation). We use asterisks in the model formula because these are fixed effects (i.e. their factor levels *are* informative; see p. 215).

>   summary(model)

This produces a series of Anova tables, one for each plot size, starting with the largest plots (block), then looking at irrigation within blocks, then density within irrigation within block, then finally fertiliser within density within irrigation within block. Notice that the error degrees of freedom are correct in each case (e.g. there are only 3 d.f. for error in assessing the irrigation main effect, but it is nevertheless significant, $p = 0.025$).

```
Error: block
        Df   Sum Sq  Mean Sq
block    3  194.444   64.815

Error: block:irrigation
            Df Sum Sq Mean Sq F value  Pr(>F)
irrigation   1 8277.6  8277.6  17.590 0.02473 *
Residuals    3 1411.8   470.6

Error: block:irrigation:density
                   Df  Sum Sq Mean Sq F value  Pr(>F)
density             2 1758.36  879.18  3.7842 0.05318 .
irrigation:density  2 2747.03 1373.51  5.9119 0.01633 *
Residuals          12 2787.94  232.33

Error: block:irrigation:density:fertilizer
                            Df  Sum Sq Mean Sq F value    Pr(>F)
fertilizer                   2 1977.44  988.72 11.4493 0.0001418 ***
irrigation:fertilizer        2  953.44  476.72  5.5204 0.0081078 **
density:fertilizer           4  304.89   76.22  0.8826 0.4840526
irrigation:density:fertilizer 4 234.72   58.68  0.6795 0.6106672
Residuals                   36 3108.83   86.36
```

There are lots of significant interactions. The best way to understand these is to use the **interaction.plot** function (Chapter 15). The variables are listed in a non-obvious order: first the factor to go on the x axis, then the factor to go as different lines on the plot, then the response variable. There are three plots to look at so we make a 2×2 plotting area:

```
par(mfrow=c(2,2))
interaction.plot(fertilizer,density,yield)
interaction.plot(fertilizer,irrigation,yield)
interaction.plot(density,irrigation,yield)
par(mfrow=c(1,1))
```

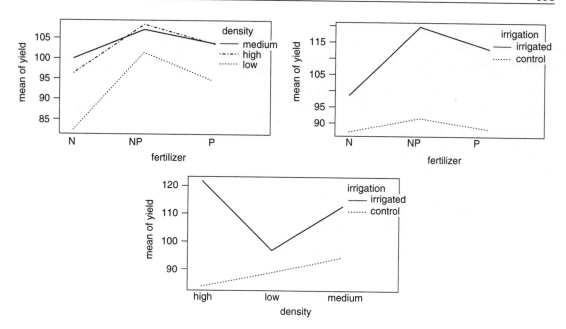

Obviously, the really strong interaction is that between irrigation and density, with a reversal of the high to low density difference on the irrigated and control plots. Interestingly, this is not the most significant interaction. That honour goes to the fertiliser:irrigation interaction (top right graph).

With nested designs, the **model.tables** can be quite complex:

```
        model.tables(model,se=T)

Tables of effects

irrigation
control irrigated
-10.722    10.722

density

  high      low medium
3.0694 -6.9722 3.9028

fertilizer
       N      NP        P
-6.8056 5.9444 0.86111

  irrigation:density
Dim 1: irrigation
Dim 2: density

             high      low  medium
  control -8.2361   6.6389  1.5972
irrigated  8.2361  -6.6389 -1.5972
```

```
    irrigation:fertilizer
Dim 1: irrigation
Dim 2: fertilizer
                    N        NP        P
   control   5.0556  -3.3611  -1.6944
 irrigated  -5.0556   3.3611   1.6944
    density:fertilizer
Dim 1: density
Dim 2: fertilizer
                  N        NP        P
   high   0.3889  -0.3611  -0.0278
    low  -3.5694   2.8056   0.7639
 medium   3.1806  -2.4444  -0.7361

    irrigation:density:fertilizer
Dim 1: irrigation
Dim 2: density
Dim 3: fertilizer

, , N

                high     low  medium
   control  -1.9722   2.6528  -0.6806
 irrigated   1.9722  -2.6528   0.6806

, , NP

                high     low  medium
   control  -0.0556   0.6944  -0.6389
 irrigated   0.0556  -0.6944   0.6389

, , P

                high     low  medium
   control   2.0278  -3.3472   1.3194
 irrigated  -2.0278   3.3472  -1.3194

Standard errors of effects
irrigatn density fertilizer irrig:den irrig:fert den:fert irrig:den:fert
  3.6155 3.1113     1.8969    4.4001     2.6826    3.2855       4.6464
 36.0000 24.0000   24.0000   12.0000    12.0000    8.0000       4.0000
```

The replication on which each standard error is based appears beneath the relevant mean.

A complex split-split-split-split plot field experiment

This example explains the analysis of a field experiment on the impact of grazing and plant competition on the yield of forage biomass (measured as harvested dry matter in tonnes per hectare). To understand what is going on, you will need to study the experimental layout quite carefully.

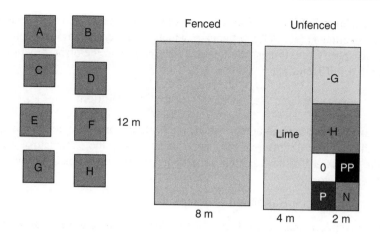

There are eight large plots (A to H), forming the blocks of the experiment, each measuring 12 m × 16 m. Two treatments are applied at this scale: plus and minus insecticide spray, and plus and minus mollusc pellets. Each of the four treatment combinations is replicated twice, and applied at random to one of the eight blocks. Each block is split in half in a randomly selected direction (up and down, or left to right) and rabbit fencing is allocated at random to half of the block. Within each split plot the plot is split again, and one of two liming treatments is allocated at random. Within each split-split plot the area is divided into three, and one of three plant competition treatments is allocated at random (control, minus grass using a grass-specific herbicide, or minus herb using a herb-specific herbicide). Finally, within each split-split-split plot the area is divided into four and one of four nutrient treatments is applied at random (plus and minus nitrogen and plus and minus phosphorus). The whole design therefore contains

$$8 \times 2 \times 2 \times 3 \times 4 = 384$$

of the smallest (2 m × 2 m) plots. The plot sizes for the different treatments are as follows:

- insecticide or molluscicide   16 m × 12 m
- rabbit grazing   12 m × 8 m
- lime   12 m × 4 m
- plant competition   4 m × 4 m
- nutrients   2 m × 2 m

The data frame looks like this:

```
attach(splitplot)
names(splitplot)
```

```
[1] "Block"   "Insect"   "Mollusc"  "Rabbit"  "Lime"   "Competition"
[7] "Nutrient"   "Biomass"
```

The **response variable** is Biomass, Block is a **random effect** and the other variables are all **fixed effects** applied as treatments (at random of course, which is a bit confusing). As before, analysing the data requires us to specify two things: the treatment structure and the error structure. The treatment structure is simply a full factorial:

Insect*Mollusc*Rabbit*Lime*Competition*Nutrient

specified using the * operator to indicate that all main effects and all interactions are to be fitted. The error term shows how the different plot sizes are related to the explanatory variables. We list, from left to right, the names of the variables relating to progressively smaller plots, with each name separated by the / operator:

Block/Rabbit/Lime/Competition

There are four variable names in the Error function because there are five different sizes of plots (see above); we can specify if we want to that the smallest plots are the Nutrient treatments, but we don't have to. The analysis is run by combining the treatment and error structure in a single **aov** function. It may not run on your machine because of memory limitations but here is the S-Plus output anyway:

```
model<-aov(Biomass~Insect*Mollusc*Rabbit*Lime*Competition*Nutrient
          +Error(Block/Rabbit/Lime/Competition))
   summary(model)
```

Error: Block

	Df	Sum of Sq	Mean Sq	F Value	Pr(F)
Insect	1	414.6085	414.6085	34.27117	0.0042482
Mollusc	1	8.7458	8.7458	0.72292	0.4430877
Insect:Mollusc	1	11.0567	11.0567	0.91394	0.3932091
Residuals	4	48.3915	12.0979		

Error: Rabbit %in% Block

	Df	Sum of Sq	Mean Sq	F Value	Pr(F)
Rabbit	1	388.7935	388.7935	4563.592	0.0000003
Insect:Rabbit	1	0.4003	0.4003	4.698	0.0960688
Mollusc:Rabbit	1	0.0136	0.0136	0.160	0.7096319
Insect:Mollusc:Rabbit	1	0.2477	0.2477	2.908	0.1633515
Residuals	4	0.3408	0.0852		

Error: Lime %in% (Block/Rabbit)

	Df	Sum of Sq	Mean Sq	F Value	Pr(F)
Lime	1	86.63703	86.63703	1918.264	0.0000000
Insect:Lime	1	0.03413	0.03413	0.756	0.4100144
Mollusc:Lime	1	0.12197	0.12197	2.701	0.1389385
Rabbit:Lime	1	0.14581	0.14581	3.228	0.1100955
Insect:Mollusc:Lime	1	0.05160	0.05160	1.143	0.3163116
Insect:Rabbit:Lime	1	0.00359	0.00359	0.079	0.7852903
Mollusc:Rabbit:Lime	1	0.09052	0.09052	2.004	0.1945819
Insect:Mollusc:Rabbit:Lime	1	0.46679	0.46679	10.335	0.0123340
Residuals	8	0.36131	0.04516		

Error: Competition %in% (Block/Rabbit/Lime)

	Df	Sum of Sq	Mean Sq	F Value	Pr(F)
Competition	2	214.4145	107.2073	1188.317	0.0000000
Insect:Competition	2	0.1502	0.0751	0.832	0.4442496
Mollusc:Competition	2	0.1563	0.0782	0.866	0.4300752
Rabbit:Competition	2	0.1981	0.0991	1.098	0.3457568
Lime:Competition	2	0.0226	0.0113	0.125	0.8825194
Insect:Mollusc:Competition	2	0.4132	0.2066	2.290	0.1176343
Insect:Rabbit:Competition	2	0.1221	0.0611	0.677	0.5153674
Mollusc:Rabbit:Competition	2	0.0221	0.0111	0.123	0.8850922
Insect:Lime:Competition	2	0.0527	0.0263	0.292	0.7487901
Mollusc:Lime:Competition	2	0.0296	0.0148	0.164	0.8493921
Rabbit:Lime:Competition	2	0.0134	0.0067	0.074	0.9286778
Insect:Mollusc:Rabbit:Competition	2	0.0307	0.0154	0.170	0.8442710
Insect:Mollusc:Lime:Competition	2	0.0621	0.0311	0.344	0.7112350
Insect:Rabbit:Lime:Competition	2	0.3755	0.1878	2.081	0.1413456
Mollusc:Rabbit:Lime:Competition	2	0.5007	0.2504	2.775	0.0773730
Insect:Mollusc:Rabbit:Lime:Competition	2	0.0115	0.0057	0.064	0.9385470
Residuals	32	2.8870	0.0902		

Error: Within

	Df	Sum of Sq	Mean Sq	F Value	Pr(F)
Nutrient	3	1426.017	475.3389	6992.589	0.0000000
Insect:Nutrient	3	0.213	0.0711	1.046	0.3743566
Mollusc:Nutrient	3	0.120	0.0400	0.589	0.6231103
Rabbit:Nutrient	3	0.087	0.0291	0.428	0.7332115
Lime:Nutrient	3	0.035	0.0116	0.171	0.9156675
Competition:Nutrient	6	0.724	0.1207	1.775	0.1081838
Insect:Mollusc:Nutrient	3	0.112	0.0373	0.549	0.6497462
Insect:Rabbit:Nutrient	3	0.783	0.2611	3.840	0.0110859
Mollusc:Rabbit:Nutrient	3	0.929	0.3096	4.554	0.0044319
Insect:Lime:Nutrient	3	0.059	0.0196	0.289	0.8335956
Mollusc:Lime:Nutrient	3	0.476	0.1585	2.332	0.0766893
Rabbit:Lime:Nutrient	3	0.376	0.1252	1.842	0.1422169
Insect:Competition:Nutrient	6	0.556	0.0927	1.363	0.2333048
Mollusc:Competition:Nutrient	6	0.347	0.0578	0.851	0.5327535
Rabbit:Competition:Nutrient	6	0.431	0.0718	1.057	0.3914693
Lime:Competition:Nutrient	6	0.451	0.0752	1.106	0.3616015
Insect:Mollusc:Rabbit:Nutrient	3	0.411	0.1370	2.016	0.1143309
Insect:Mollusc:Lime:Nutrient	3	0.697	0.2324	3.418	0.0190776
Insect:Rabbit:Lime:Nutrient	3	0.435	0.1450	2.132	0.0987222
Mollusc:Rabbit:Lime:Nutrient	3	0.055	0.0184	0.270	0.8468838
Insect:Mollusc:Competition:Nutrient	6	0.561	0.0936	1.376	0.2279809
Insect:Rabbit:Competition:Nutrient	6	0.728	0.1213	1.784	0.1062479
Mollusc:Rabbit:Competition:Nutrient	6	0.427	0.0712	1.047	0.3974232
Insect:Lime:Competition:Nutrient	6	0.258	0.0430	0.632	0.7043702
Mollusc:Lime:Competition:Nutrient	6	0.417	0.0695	1.023	0.4128486
Rabbit:Lime:Competition:Nutrient	6	0.406	0.0676	0.994	0.4315034
Insect:Mollusc:Rabbit:Lime:Nutrient	3	0.350	0.1166	1.715	0.1665522
Insect:Mollusc:Rabbit:Competition:Nutrient	6	0.259	0.0431	0.635	0.7023702
Insect:Mollusc:Lime:Competition:Nutrient	6	0.403	0.0672	0.988	0.4355882
Insect:Rabbit:Lime:Competition:Nutrient	6	0.282	0.0470	0.692	0.6565638
Mollusc:Rabbit:Lime:Competition:Nutrient	6	0.355	0.0592	0.870	0.5183944
Insect:Mollusc:Rabbit:Lime:Competition:Nutrient	6	0.989	0.1648	2.424	0.0291380
Residuals	144	9.789	0.0680		

Notice that you get five separate Anova tables, one for each different plot size. It is the number of plot sizes, not the number of treatments, that determines the shape of the split-plot Anova table. The number of Anova tables would not have changed if we had specified the Nutrient treatment (4 levels) as a 2×2 factorial with Nitrogen and Phosphorus each as two-level treatments. Because Insecticide and Molluscicide were both applied at the same plot size (whole blocks) they appear in the same Anova table.

Interpretation of output tables like this requires a high level of serenity. The first thing to do is to check that the degrees of freedom have been handled properly. There were 8 blocks, with 2 replicates of the large-plot factorial experiment of plus and minus insects and plus and minus molluscs. This means that there are 7 d.f. in total, and so with 1 d.f. for Insect, 1 d.f. for Mollusc and 1 d.f. for the Insect:Mollusc interaction, there should be $7 - 1 - 1 - 1 = 4$ d.f. for error. This checks out in the top Anova table where Error is labelled as Block. For the largest plots, therefore, the error variance is 12.1. We can now assess the significance of these treatments that were applied to the largest plots. As ever, we begin with the interaction. This is clearly not significant, so we can move on to interpreting the main effects. There is no effect of mollusc exclusion on biomass, but insect exclusion led to a significant increase in mean biomass ($p = 0.0042$).

The second largest plots were those with or without fences to protect them from rabbit grazing. To check the degrees of freedom, we need to work out the total number of rabbit-grazed and fenced plots. There were 8 blocks, each split in half, so there are 16 plots. We have already used 7 d.f. for the insect by mollusc experiment (above) so there are $16 - 7 - 1 = 8$ d.f. remaining. Rabbit grazing has 1 d.f. and there are three interaction terms, each with 1 d.f. (Rabbit:Insect, Rabbit:Mollusc and Rabbit:Insect:Mollusc). This means that these terms should be assessed by an error variance that has $8 - 1 - 3 = 4$ d.f. This also checks out. The error variance is 0.085, and shows that there are no significant interactions between rabbit grazing and invertebrate herbivores, but there is a highly significant main effect of rabbit grazing.

The third largest plots were either limed or not limed. In this case there are 8 d.f. for error, and we discover the first significant interaction is Insect:Mollusc:Rabbit:Lime ($p = 0.012$). Like all high-order interactions, this is extremely complicated to interpret. It means that the three-way interaction between Insect:Mollusc:Rabbit works differently on limed and unlimed plots. It would be unwise to overinterpret this result without further experimentation focused on the way this interaction might work. There is a highly significant main effect of lime.

The fourth largest plots received one of three plant competition treatments: control, minus grass or minus herb. There are 96 competition plots, and we have used 31 d.f. so far on the larger plots, so there should be $96 - 31 - 1 = 64$ d.f. at this level. With 32 d.f. for main effects and interactions, that leaves 32 d.f. for error. This checks out, and the error variance is 0.09. There are no significant interaction terms, but competition had a significant main effect on biomass.

The fifth largest plots (the smallest at $2\,\text{m} \times 2\,\text{m}$) received one of four nutrient treatments: plus or minus nitrogen and plus or minus phosphorus. All the remaining degrees of freedom can be used at this scale, leaving 144 d.f. for error, and an error variance of 0.068. There are several significant interactions: the six-way Insect:Mollusc:Rabbit:Lime:Competition:Nutrient ($p = 0.029$) a four-way, Insect:Mollusc:Lime:Nutrient ($p = 0.019$) and two three-way interactions, Insect:Rabbit:Nutrient ($p = 0.011$) and Mollusc:Rabbit:Nutrient ($p = 0.004$). At this point I can confide in you. I made up these results, so I know that all of these small-plot interactions are due to chance alone.

This raises an important general point. In big, complicated experiments like this, it is sensible to use a very low value of α in assessing the significance of high-order interactions (e.g. $\alpha = 0.01$ or $\alpha = 0.005$). This compensates for the fact that you are doing a vast number of hypothesis tests, and has the added bonus of making the results much more straightforward to write up. It is a trade-off, because you do not want to be so harsh that you miss scientifically important and potentially very interesting interactions.

We can inspect the interactions in two ways. Tables of interaction means can be produced using **tapply**:

```
tapply(Biomass,list(Mollusc,Rabbit,Nutrient),mean)


, , N
           Fenced    Grazed
Pellets 6.963431  4.984890
  Slugs 7.322273  5.298524


, , NP
           Fenced    Grazed
Pellets 9.056132  6.923177
  Slugs 9.257630  7.324827


, , O
           Fenced    Grazed
Pellets 3.873364  2.069350
  Slugs 4.282690  2.149054


, , P
           Fenced    Grazed
Pellets 5.066604  2.979994
  Slugs 5.351919  3.344672
```

Better still, use **interaction.plot** to inspect the interaction terms, two at a time. Take the Mollusc:Rabbit:Nutrient interaction ($p = 0.004$) whose means we have just calculated. We expect that the interaction plot will show non-parallelness of some form or other. We divide the plotting space into four, then make three separate interaction plots:

```
par(mfrow=c(2,2))
interaction.plot(Rabbit,Mollusc,Biomass)
interaction.plot(Rabbit,Nutrient,Biomass)
interaction.plot(Nutrient,Mollusc,Biomass)
```

It is quite clear from the plots that the interaction, though statistically significant, is not biologically substantial. The plots are virtually parallel in all three panels. The graphs demonstrate clearly the interaction between nitrogen and phosphorus (bottom left) but this is not materially altered by either mollusc or rabbit grazing.

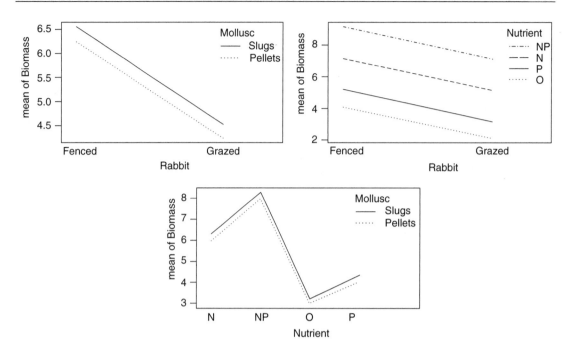

Doing the wrong 'factorial' analysis of multi-split-plot data

Here's how *not* to do it. With data like this there are always a lot of different ways of getting the model and/or the error structure wrong. The commonest mistake is to treat the data as a full factorial, like this:

```
model<-aov(Biomass~Insect*Mollusc*Rabbit*Lime*Competition*Nutrient)
```

It looks perfectly reasonable, and indeed, this is the experimental design intended. But it is rife with pseudoreplication. Ask yourself how many independent plots had insecticide applied to them or not. The answer is 8. The analysis carried out here assumes there were 384! And what about plant competition treatments? The answer is 96, but this analysis assumes 384. And so on. Let's see the consequences of the pseudoreplication. It thinks the error variance is 0.3217 for testing all the interactions and main effects. The only significant interaction is insecticide by molluscicide and this appears to have an F ratio of 34.4 on d.f. = 1192. All six main effects appear to be significant at $p < 0.000\,001$.

```
summary(model)
```

	Df	Sum of Sq	Mean Sq	F Value	Pr(F)
Insect	1	414.609	414.6085	1288.743	0.0000000
Mollusc	1	8.746	8.7458	27.185	0.0000005
Rabbit	1	388.793	388.7935	1208.502	0.0000000
Lime	1	86.637	86.6370	269.297	0.0000000
Competition	2	214.415	107.2073	333.236	0.0000000
Nutrient	3	1426.017	475.3389	1477.514	0.0000000
Insect:Mollusc	1	11.057	11.0567	34.368	0.0000000
Insect:Rabbit	1	0.400	0.4003	1.244	0.2660546
Mollusc:Rabbit	1	0.014	0.0136	0.042	0.8371543

Insect:Lime	1	0.034	0.0341	0.106	0.7450036
Mollusc:Lime	1	0.122	0.1220	0.379	0.5388059
Rabbit:Lime	1	0.146	0.1458	0.453	0.5016203
Insect:Competition	2	0.150	0.0751	0.233	0.7920629
Mollusc:Competition	2	0.156	0.0782	0.243	0.7845329
Rabbit:Competition	2	0.198	0.0991	0.308	0.7353323
Lime:Competition	2	0.023	0.0113	0.035	0.9654345
Insect:Nutrient	3	0.213	0.0711	0.221	0.8817673
Mollusc:Nutrient	3	0.120	0.0400	0.124	0.9455545
Rabbit:Nutrient	3	0.087	0.0291	0.090	0.9652343
Lime:Nutrient	3	0.035	0.0116	0.036	0.9907599
Competition:Nutrient	6	0.724	0.1207	0.375	0.8942470
Insect:Mollusc:Rabbit	1	0.248	0.2477	0.770	0.3813061
Insect:Mollusc:Lime	1	0.052	0.0516	0.160	0.6892415
Insect:Rabbit:Lime	1	0.004	0.0036	0.011	0.9160383
Mollusc:Rabbit:Lime	1	0.091	0.0905	0.281	0.5964109
Insect:Mollusc:Competition	2	0.413	0.2066	0.642	0.5272844
Insect:Rabbit:Competition	2	0.122	0.0611	0.190	0.8272856
Mollusc:Rabbit:Competition	2	0.022	0.0111	0.034	0.9662288
Insect:Lime:Competition	2	0.053	0.0263	0.082	0.9214289
Mollusc:Lime:Competition	2	0.030	0.0148	0.046	0.9550433
Rabbit:Lime:Competition	2	0.013	0.0067	0.021	0.9794192
Insect:Mollusc:Nutrient	3	0.112	0.0373	0.116	0.9506702
Insect:Rabbit:Nutrient	3	0.783	0.2611	0.811	0.4889277
Mollusc:Rabbit:Nutrient	3	0.929	0.3096	0.962	0.4116812
Insect:Lime:Nutrient	3	0.059	0.0196	0.061	0.9802371
Mollusc:Lime:Nutrient	3	0.476	0.1585	0.493	0.6877598
Rabbit:Lime:Nutrient	3	0.376	0.1252	0.389	0.7609242
Insect:Competition:Nutrient	6	0.556	0.0927	0.288	0.9421006
Mollusc:Competition:Nutrient	6	0.347	0.0578	0.180	0.9820982
Rabbit:Competition:Nutrient	6	0.431	0.0718	0.223	0.9688921
Lime:Competition:Nutrient	6	0.451	0.0752	0.234	0.9651323
Insect:Mollusc:Rabbit:Lime	1	0.467	0.4668	1.451	0.2298613
Insect:Mollusc:Rabbit:Competition	2	0.031	0.0154	0.048	0.9534089
Insect:Mollusc:Lime:Competition	2	0.062	0.0311	0.097	0.9079804
Insect:Rabbit:Lime:Competition	2	0.376	0.1878	0.584	0.5588583
Mollusc:Rabbit:Lime:Competition	2	0.501	0.2504	0.778	0.4606637
Insect:Mollusc:Rabbit:Nutrient	3	0.411	0.1370	0.426	0.7346104
Insect:Mollusc:Lime:Nutrient	3	0.697	0.2324	0.722	0.5397982
Insect:Rabbit:Lime:Nutrient	3	0.435	0.1450	0.451	0.7171854
Mollusc:Rabbit:Lime:Nutrient	3	0.055	0.0184	0.057	0.9820420
Insect:Mollusc:Competition:Nutrient	6	0.561	0.0936	0.291	0.9407903
Insect:Rabbit:Competition:Nutrient	6	0.728	0.1213	0.377	0.8930313
Mollusc:Rabbit:Competition:Nutrient	6	0.427	0.0712	0.221	0.9695915
Insect:Lime:Competition:Nutrient	6	0.258	0.0430	0.134	0.9918574
Mollusc:Lime:Competition:Nutrient	6	0.417	0.0695	0.216	0.9713330
Rabbit:Lime:Competition:Nutrient	6	0.406	0.0676	0.210	0.9733124
Insect:Mollusc:Rabbit:Lime:Competition	2	0.011	0.0057	0.018	0.9823388
Insect:Mollusc:Rabbit:Lime:Nutrient	3	0.350	0.1166	0.362	0.7802436
Insect:Mollusc:Rabbit:Competition:Nutrient	6	0.259	0.0431	0.134	0.9917702
Insect:Mollusc:Lime:Competition:Nutrient	6	0.403	0.0672	0.209	0.9737285
Insect:Rabbit:Lime:Competition:Nutrient	6	0.282	0.0470	0.146	0.9896249
Mollusc:Rabbit:Lime:Competition:Nutrient	6	0.355	0.0592	0.184	0.9810233
Insect:Mollusc:Rabbit:Lime:Competition:Nutrient	6	0.989	0.1648	0.512	0.7987098
Residuals	192	61.769	0.3217		

The problems are of two kinds. For the largest plots, the error variance is underestimated because of the pseudoreplication, so things appear to be significant which actually are not. For example, the correct analysis shows that the insect by mollusc interaction is not significant ($p = 0.39$) but the wrong analysis suggests that it is highly significant ($p = 0.00000$). The other problem is of the opposite kind. Because this analysis does not factor out the large between-plot variation at the larger scale, the error variance for testing small plot effects is much too big. In the correct, split-plot analysis, the small-plot error variance is very small (0.068) compared with the pseudoreplicated, small-plot error variance of 0.32 (above).

Further reading

Neter, J., Kutner, M. *et al.* (1996) *Applied Linear Statistical Models.* New York, McGraw-Hill.

Sokal, R.R. and Rohlf, F.J. (1995) *Biometry: The Principles and Practice of Statistics in Biological Research.* San Francisco, W.H. Freeman.

20

Nested designs and variance components analysis

Up to this point, we have treated all categorical explanatory variables as if they were the same. This is certainly what R.A. Fisher had in mind when he invented the analysis of variance in the 1920s. It was Eisenhart (1947) who realised that there were actually two fundamentally different sorts of categorical explanatory variables; he called these **fixed effects** and **random effects**:

- fixed effects influence only the *mean* of y
- random effects influence only the *variance* of y

A random effect should be thought of as coming from a population of effects: the existence of this population is an extra assumption. We speak of *prediction of random effects*, rather than estimation: we *estimate* fixed effects from data, but we intend to make predictions about the population from which our random effects were sampled. Fixed effects are unknown constants to be estimated from the data. Random effects govern the variance–covariance structure of the response variable.

The key point is that *observations affected by random effects are not independent*. Observations that contain the same random effect are correlated, and this contravenes one of the fundamental assumptions of standard statistical models—*independence of errors*.

The distinction between fixed and random effects is best seen by an example. In most mammal species the categorical variable sex has two levels: male and female. For any individual that you find, the knowledge that it is, say, female conveys a great deal of extra information about the individual, and this information draws on experience gleaned from many other individuals that were female. A female will have a whole set of attributes (associated with her being female) no matter what population that individual was drawn from. Take a different categorical variable like genotype. If we have two genotypes in a population, we might label them A and B. If we take two more genotypes from a *different* population, we might label them A and B as well. In a case like this, the label A does not convey any information at all about the genotype, other than that it is different from genotype B. In the case of sex, the factor level (male or female) is informative: sex is a fixed effect. In the case of genotype, the factor level (A or B) is uninformative: genotype is a random effect. Random effects have factor levels that are drawn from a large (potentially very large) population in which the individuals differ in many ways, but *we do not know exactly how or why they differ*. In the case of sex we know that males and females are likely to differ in characteristic and predictable ways. To get a feel for the difference between fixed effects and random effects, here are some more examples.

Fixed effects	Random effects
Drug administered or not	Genotype
Insecticide sprayed or not	Brood
Nutrient added or not	Block within a field
One country versus another	Split plot within a plot
Male or female	Factory
Wet versus dry	Individual
Light versus shade	Family
Old versus young	Parent

Hierarchical sampling

Many kinds of statistical work involve studies carried out at a range of spatial scales. For example, in medical parasitology we might investigate worm burdens in people as:

- individuals

- families

- villages

- districts

- countries

Naturally, the questions we ask at different scales would concern different processes. For instance, differences between individuals within families might involve genetics, behaviour patterns, sex and age, whereas differences between countries might involve such factors as sanitation, climate, geology, average income, and so on. The key point is that *variability accumulates as spatial scale increases*; thus variation between families within villages, and variation between villages within districts, all contribute to variation in worm burdens measured at the scale of the country as a whole.

Model I and Model II Anova

In Model I, the differences between the means are ascribed entirely to the fixed treatment effects, so any data point can be decomposed like this:

$$y_{ij} = \mu + \alpha_i + \varepsilon_{ij}$$

It is the overall mean μ plus a treatment effect due to the ith level of the fixed effect α_i plus a deviation ε_{ij} that is unique to that individual, and is drawn from a Normal distribution with mean 0 and variance σ_e^2. The variance is assumed to be constant across all the levels, which are assumed to differ only in their means. Because variance is constant, there is only one error term, based on σ_e^2.

Model II is subtly different. Here the model is written

$$y_{ij} = \mu + U_i + \varepsilon_{ij}$$

which looks as if it is *exactly* the same as Model I, just replacing α_i by U_i. The distinction is this. In Model I the experimenter either *made* the treatments like they were (e.g. by the random allocation of treatments to individuals), or selected the categories *because* they were like they were, i.e. fixed and clearly distinctive. The main interest is in the values of the differences between the means of the various factor levels. In Model II the factor levels are different from one another, but the experimenter did not *make* them different. They were selected (perhaps *because* they were different, perhaps not), but they came from a much larger pool of factor levels that exhibits variation beyond the control of the experimenter and beyond their immediate interest (e.g. six genotypes were selected to work with out of a pool of perhaps many thousands of genotypes). We call it random variation, and call such factors random effects. The main interest is not in differences between the means of the different factor levels, but rather in the way that the factor influences the *variance* of the response.

The important point is that because the random effects U_i come from a large population, there is not much point in concentrating on estimating means of our small subset, k, of factor levels, and no point at all in comparing individual pairs of means for different factor levels. Much better to recognise them for what they are, random samples from a much larger population, and to concentrate on their variance σ_U^2. This is the *added* variation caused by differences between the levels of the random effects. Model II Anova is all about estimating the size of this variance, and working out its percentage contribution to the overall variation.

The fundamental difference between Model I and Model II, therefore, is that Model I has constant variance, hence a single error term based on σ_e^2. In Model II we are not interested in the means of the random effects, and use these instead to estimate extra variances, σ_U^2, one for each spatial scale. The random variables U_i and e_{ij} are distributed like this:

$$U_i \quad \text{independent} \quad N(0, \sigma_U^2)$$

$$e_{ij} \quad \text{independent} \quad N(0, \sigma_e^2)$$

$$\{U_i\} \quad \text{independent of} \quad \{e_{ij}\}$$

Statistically, the main concern is in estimating and testing the variances σ_U^2 and σ_e^2.

The issues involved in Model II fall into two broad categories:

- questions about experimental design and the management of experimental error, e.g. where does most of the variation occur, and where would increased replication be most profitable?

- questions about hierarchical structure, and the relative magnitude of variation at different levels within the hierarchy, e.g. studies on nested spatial scales.

Calculations involved in nested analysis

Most Anova models are based on the assumption that there is a single error term. But in nested experiments, where the data are gathered at two or more different spatial scales, there is *a different error variance for each different plot size*.

We take an example from Sokal and Rohlf (1995). The experiment involved a simple one-factor Anova with three treatments given to six rats. The analysis was complicated by the fact that three preparations were taken from the liver of each rat, and two readings of glycogen content were taken from each preparation. This generated 6 pseudoreplicates per rat to give a total of 36 readings in all. Clearly, it would be a mistake to analyse these data as if they were a straightforward one-way Anova, because that would give us 33 degrees of freedom for error. In fact, since there are only two rats in each treatment, we have only one degree of freedom per treatment, giving a correct total of only 3 d.f. for error.

The variance is likely to be different at each level of this nested analysis because:

- the readings differ because of variation in the glycogen detection method within each liver sample (measurement error)

- the pieces of liver may differ because of heterogeneity in the distribution of glycogen within the liver of a single rat

- the rats will differ from one another in their glycogen levels because of sex, age, size, genotype, etc.

- rats allocated different experimental treatments may differ as a result of the fixed effects of the clinical treatment.

If all we want to test is whether the experimental treatments have affected the glycogen levels, then we are not interested in liver bits within rat's livers, or in preparations within liver bits. We could add all the pseudoreplicates together, and analyse the six averages. This would have the virtue of showing what a tiny experiment this really was (see below). But to analyse the full data set, we must proceed as follows.

The only trick is to ensure the factor levels are set up properly. There were 3 treatments, so we make a treatment factor T with 3 levels. While there were 6 rats in total, there were only 2 in each treatment, so we declare rats as a factor R with 2 levels (not 6). There were 18 bits of liver in all, but only 3 per rat, so we declare liver bits as a factor L with 3 levels (not 18):

```
attach(rats)
names(rats)

[1]  "Glycogen"  "Treatment"  "Rat"  "Liver"

tapply(Glycogen,Treatment,mean)

        1         2         3
140.5000  151.0000  135.1667
```

There are substantial differences between the treatment means, and our job is to say whether these differences are statistically significant or not. Because the three factors Treatment, Rat and Liver have numeric factor levels, we must declare them to be factors before beginning the modelling:

```
Treatment<-factor(Treatment)
Rat<-factor(Rat)
Liver<-factor(Liver)
```

Because the sums of squares in nested designs are so confusing on first acquaintance, it is important to work through a simple example like this one by hand. The first two steps are easy, because they relate to the **fixed effect**, which is Treatment. We calculate *SST* and *SSA* in the usual way (Chapter 15). We need the sum of squares and the grand total of the response variable Glycogen:

```
sum(Glycogen);sum(Glycogen^2)

[1]  5120
[1]  731508
```

$$SST = \sum y^2 - \frac{\left[\sum y\right]^2}{36}$$

$$SST = 731\,508 - \frac{5120^2}{36} = 3330.222$$

For the treatment sum of squares, SSA, we need the three treatment totals:

tapply(Glycogen,Treatment,sum)

```
   1    2    3
1686 1812 1622
```

Each of these was the sum of 12 numbers (2 preparations $\times$ 3 liver bits $\times$ 2 rats) so we divide the square of each subtotal by 12 before subtracting the correction factor:

$$SSA = \frac{\sum T^2}{12} - \frac{\left[\sum y\right]^2}{36}$$

sum(tapply(Glycogen,Treatment,sum)^2)/12-5120^2/36

```
[1] 1557.556
```

So the Error sums of squares must be $SST - SSA = 3330.222 - 1557.556 = 1772.666$. If this is correct, then the Anova table must be like this.

Source	SS	d.f.	MS	F	Critical F
Treatment	1557.556	2	778.778	14.497	3.28
Error	1772.66	33	53.717		
Total	3330.222	35			

qf(.95,2,33)

```
[1] 3.284918
```

The calculated value is much larger than the value in tables, so treatment has a highly significant effect on liver glycogen content. Wrong! We have made the classic mistake of pseudoreplication. We have counted all 36 data points as if they were replicates. The definition of replicates is that *replicates are independent of one another*. Two measurements from the same piece of rat's liver are clearly *not* independent. Nor are measures from three regions of the same rat liver. It is the rats that are the replicates in this experiment and there are only 6 of them in total. So the correct total degrees of freedom is 5, not 35, and *there are $5 - 2 = 3$ degrees of freedom for error, not 33.*

There are lots of ways of getting this analysis wrong, but only one way of getting it right. There are three spatial scales to the measurements (rats, liver bits from each rat and preparations from each liver bit); hence there must be three different error variances in the analysis. Our first task is to compute the sum of squares for differences between the rats. It is easy to find the sums:

tapply(Glycogen,list(Treatment,Rat),sum)

```
    1   2
1 795 891
2 898 914
3 806 816
```

Note the use of Treatment and Rat to get the totals; it is *wrong* to do the following:

```
tapply(Glycogen,Rat,sum)
```

```
   1      2
2499   2621
```

because the rats are numbered 1 and 2 within each treatment. This is very important. There are 6 rats in the experiment, so we need 6 rat totals. Each rat total is the sum of 6 numbers (3 liver bits and 2 preparations per liver bit). So we square the 6 rat totals, add them up and divide by 6:

```
sum(tapply(Glycogen,list(Treatment,Rat),sum)^2)/6
```

```
[1]  730533
```

But what now? Our experience so far tells us simply to subtract the correction factor. But this is wrong. We use *the sum of squares from the spatial scale above* as the correction term at any given level. This may become clearer from the worked example.

What about the sum of squares for liver bits? There are 3 per rat, so there must be 18 liver-bit totals in all. We can inspect them like this:

```
tapply(Glycogen,list(Treatment,Rat,Liver),sum)
```

261	298	256	283	278	310
302	306	296	294	300	314
259	278	276	277	271	261

We need to square these, add them up and divide by 2. Why 2? Because each of these totals is the sum of the two measurements of each preparation from each liver bit.

```
sum(tapply(Glycogen,list(Treatment,Rat,Liver),sum)^2)/2
```

```
[1]  731127
```

So we have the following uncorrected sums of squares.

Source	Uncorrected *SS*
Treatment	729 735.3
Rats	730 533
Liver bits	731 127
Total	731 508

The key to understanding nested designs is to understand the next step. Up to now we have used the correction factor

$$CF = \frac{\left[\sum y\right]^2}{\sum n}$$

to determine the corrected sum of squares in all cases. We used it correctly to determine the treatment sum of squares, earlier in this example. With nested factors, however, we don't do this. We use the *uncorrected sum of squares from the next spatial scale larger than the scale in question*:

$$SS_{\text{Rats}} = \frac{\sum R^2}{6} - \frac{\sum T^2}{12}$$

Where R is a rat total and T is a treatment total. The sum of squares of rat totals is divided by 6 because each total was the sum of 6 numbers. The sum of squares of treatment totals is divided by 12 because each treatment total was the sum of 12 numbers (2 rats each generating 6 numbers). So, if L is the sum of the two liver-bit preparations, and y is an individual preparation (the 'sum of one number' if you like), we get

$$SS_{\text{Liver Bits}} = \frac{\sum L^2}{2} - \frac{\sum R^2}{6}$$

$$SS_{\text{Preparations}} = \frac{\sum y^2}{1} - \frac{\sum L^2}{2}$$

We can now compute the numerical values of the nested sums of squares:

$$SS_{\text{Rats}} = 730\,533 - 729\,735.3 = 797.7$$

$$SS_{\text{Liver bits}} = 731\,127 - 730\,533 = 594.0$$

$$SS_{\text{Preparations}} = 731\,508 - 731\,127 = 381.0$$

Finally, we can fill in the Anova table correctly, taking account of the nesting and the pseudoreplication.

Source	SS	d.f.	MS	F	Critical F
Treatment	1557.556	2	778.778	2.929	9.552094
Rats in Treatments	797.7	3	265.9	5.372	3.490295
Liver bits in Rats	594.0	12	49.5	2.339	2.342067
Readings in Liver bits	381.0	18	21.1666		
Total	3330.222	35			

There are several important things to see in this Anova table:

- you use the mean square from the spatial scale immediately below in testing significance. (You do *not* use 21.166 as we would have done up to now.) So the F ratio for treatment is $778.778/265.9 = 2.929$ on 2 and 3 d.f., which is a long way short of significance. The critical value is 9.55; compare this with what we did wrong earlier.

- a different recipe is used in nested designs to compute the degrees of freedom. Treatment and Total degrees of freedom are the same as usual but the others are different.

- there are 6 rats in total but there are not 5 d.f. for rats; there are 2 rats per treatment, so there is 1 d.f. for rats within each treatment. There are 3 treatments so there are $3 \times 1 = 3$ d.f. for rats within treatments.

- there are 18 liver bits in total but there are not 17 d.f. for liver bits; there are 3 bits in each liver, so there are 2 d.f. for liver bits within each liver. Since there are 6 livers in total (one for each rat) there are $6 \times 2 = 12$ d.f. for liver bits within rats.

- there are 36 preparations but there are not 35 d.f. for preparations; there are 2 preparations per liver bit, so there is 1 d.f. for preparation within each liver bit. There are 18 liver bits in all (3 from each of 6 rats) and so there are $1 \times 18 = 18$ d.f. for preparations within liver bits.

- using the three different error variances to carry out the appropriate F tests, we learn that there are no significant differences between treatments but there are significant differences between rats, and between parts of the liver within rats (i.e. small-scale spatial heterogeneity in the distribution of glycogen within livers).

Nested analysis in S-Plus

The new concept is that we include an **Error** term in the **aov** model formula to show:

- how many error terms are required (as many as there are plot sizes)

- what is the hierarchy of plot sizes (biggest on the left, smallest on the right)

- in the model formula (after the tilde ~) we use the slash operator / instead of *

- the Error term follows a plus sign + and is enclosed in round brackets

- the plots, ranked by their relative sizes, are separated by slash operators like this:

```
model<-aov(Glycogen~Treatment/Rat/Liver+Error(Treatment/Rat/Liver))
summary(model)
Error: Treatment
            Df  Sum Sq Mean Sq
Treatment   2 1557.56  778.78

Error: Treatment:Rat
               Df Sum Sq Mean Sq
Treatment:Rat   3 797.67  265.89

Error: Treatment:Rat:Liver
                   Df Sum Sq Mean Sq
Treatment:Rat:Liver 12  594.0    49.5

Error: Within
            Df Sum Sq Mean Sq F value Pr(>F)
Residuals 18 381.00   21.17
```

These are the correct mean squares, as we can see from our longhand calculations earlier. The F test for the effect of treatment is $778.78 / 265.89 = 2.93$ (n.s.), for differences between rats within treatments it is $265.89 / 49.5 = 5.37$ ($p < 0.05$) and for liver bits within rats $F = 49.5/21.17 = 2.24$ ($p = 0.05$).

In general, we use the slash operator in **model formulas** where the variables are random effects (i.e. where the factor levels are uninformative), and the asterisk operator in model formulas where the variables are fixed effects (i.e. the factor levels are informative). Knowing that a rat is number 2 tells us nothing about that rat. Knowing a rat is male tells us a lot about that rat. In **error formulas** we **always use the slash operator** (never the asterisk operator) to indicate the order of 'plot sizes': the largest plots are on the left of the list and the smallest on the right (see below).

The wrong analysis in S-Plus

Here is what *not* to do:

```
model2<-aov(Glycogen~Treatment*Rat*Liver)
```

The model has been specified as if it were a full factorial with no nesting and no pseudoreplication. Note that the structure of the data frame allows this mistake to be made (a very common problem where there is pseudoreplication). A summary of the model fit looks like this:

```
summary(model2)
```

	Df	Sum Sq	Mean Sq	F value	Pr(>F)	
Treatment	2	1557.56	778.78	36.7927	4.375e-07	***
Rat	1	413.44	413.44	19.5328	0.0003308	***
Liver	2	113.56	56.78	2.6824	0.0955848	.
Treatment:Rat	2	384.22	192.11	9.0761	0.0018803	**
Treatment:Liver	4	328.11	82.03	3.8753	0.0192714	*
Rat:Liver	2	50.89	25.44	1.2021	0.3235761	
Treatment:Rat:Liver	4	101.44	25.36	1.1982	0.3455924	
Residuals	18	381.00	21.17			

This says that there was an enormously significant difference between the treatment means, rats were significantly different from one another and there was a significant interaction between treatments and rats. Wrong! The analysis is flawed because it is based on the assumption that there is only one error variance and that its value is 21.17. This value is actually the measurement error; that is to say, the variation between one reading and another from the *same* piece of liver. For testing whether the treatment has had any effect, it is the rats that are the replicates, and there were only six of them in the whole experiment.

Avoiding the mistake of pseudoreplication

The idea is to get rid of the pseudoreplication by averaging over the liver bits and preparations for each rat. We need to create a new vector, *ym*, of length 6 containing the mean glycogen levels of each rat. You can see how this works as follows:

```
tapply(Glycogen,list(Treatment,Rat),mean)
```

```
          1         2
1  132.5000  148.5000
2  149.6667  152.3333
3  134.3333  136.0000
```

We make this into a vector for use in the model like this:

```
ym<-as.vector(tapply(Glycogen,list(Treatment,Rat),mean))
```

We also need a new vector, *tm*, of length 6 to contain a factor for the three treatment levels:

```
tm<-factor(as.vector(tapply(as.numeric(Treatment),list(Treatment,Rat),mean)))
```

Now we can do the Anova:

```
summary(aov(ym~tm))
            Df  Sum Sq Mean Sq F value Pr(>F)
tm           2 259.593 129.796   2.929 0.1971
Residuals    3 132.944  44.315
```

This gives us the same (correct) value of the F test as when we did the full nested analysis. The sums of squares are different, of course, because we are using 6 mean values rather than 36 raw values of glycogen to carry out the analysis. We conclude correctly that treatment had no significant effect on liver glycogen content.

Variance components analysis

Variance components analysis is an extremely useful technique for summarising the results of a hierarchical sampling exercise, when all (or most) of the explanatory variables are *random effects*. It shows the proportion of total variation in the response variable that enters at each scale of the hierarchy (e.g. at each spatial scale). As you go up the hierarchy, the errors accumulate. At the smallest scale, the only variance component comes from differences between the replicates (e.g. measurement error). At the next scale up, we have differences between the replicates plus differences between the individuals. At the next scale up, variation is due to differences between the replicates plus differences between the individuals, plus differences between locations. And so on.

As we go up the hierarchy something else is happening. The number of numbers on which each variance estimate is based is increasing. In our rat example earlier, there were 2 readings per liver bit, 6 readings per rat, 12 readings per treatment.

In a three-level variance components analysis, where there are k levels of the fixed effect, a numbers per level at the largest scale, b numbers per level at the intermediate scale, and n replicates per level at the smallest scale, then the mean square at each level is estimated as follows.

Source	d.f.	Estimated variances
Treatments	$k - 1$	
A	$k(a - 1)$	$\sigma^2 + n\sigma_B^2 + bn\sigma_A^2$
B in A	$ka(b - 1)$	$\sigma^2 + n\sigma_B^2$
Residual	$kab(n - 1)$	σ^2
Total	$kabn - 1$	

Make sure you understand how the degrees of freedom differ from those in a standard Anova. Also, note the way that the replicates enter the variances in the rightmost column. Variance components analysis aims to tease apart the values of σ^2, σ_B^2, and σ_A^2:

- σ^2 is variation from measurement to measurement at the smallest spatial scale

- $\sigma^2 + n\sigma_B^2$ is variation from measurement to measurement *plus* variation from unit to unit at the first spatial scale up (we multiply by n because there are only $1/n$ as many numbers at this scale than at the scale below)

- $\sigma^2 + n\sigma_B^2 + bn\sigma_A^2$ is variation from measurement to measurement, *plus* variation from unit to unit at the first scale up, *plus* variation from one unit to another at the next scale up (we multiply by bn because there are only $1/bn$ as many of these numbers)

- and so on

The variance components are calculated like this. The first variance component is easy. It is just the residual mean square $\sigma^2 = SSE/kab(n-1)$.

The second variance component reflects heterogeneity from one unit to another at the first scale up from the bottom. The Anova table gives us MSB and we obtain the variance component σ_B^2 simply by rearrangement:

$$\sigma^2 + n\sigma_B^2 = MSB$$

$$n\sigma_B^2 = MSB - \sigma^2$$

$$\sigma_B^2 = \frac{MSB - \sigma^2}{n}$$

The third variance component reflects differences between units at the second scale up from the bottom plus the other two variance components. We know MSA from the Anova table $= \sigma^2 + n\sigma_B^2 + bn\sigma_A^2$. We also know that $\sigma^2 + n\sigma_B^2 = MSB$ (above) so we can get by subtraction:

$$\sigma_A^2 = \frac{MSA - MSB}{bn}$$

Variance components analysis by hand

The idea is straightforward, but the computation for complicated data sets can be cumbersome. If there is unequal replication, the answers cannot be calculated by hand, and have to be found by iteration:

1. fill out the nested Anova table in which the sums of squares have been corrected using the sum of squares from the next scale up (i.e. don't use the standard correction factor)

2. subtract the mean square from the scale below

3. divide by the number of numbers on which the mean at that level was based

4. this is the variance component at that level

We return to the rats data. The nested Anova table looks like this.

Source	SS	d.f.	MS	F	Critical F
Treatment	1557.556	2	778.778	2.929	9.552094
Rats in Treatments	797.7	3	265.9	5.372	3.490295
Liver bits in Rats	594.0	12	49.5	2.339	2.342067
Readings in Liver bits	381.0	18	21.1666		
Total	3330.222	35			

The first variance component is extracted directly; it is the residual mean square $MSE = \sigma^2 = 21.1666$.

The second variance component is obtained from the difference between the mean squares at the smallest and next smallest scales, $\sigma^2 + n\sigma_B^2$:

$$\sigma_B^2 = \frac{MSB - \sigma^2}{n} = \frac{49.5 - 21.16667}{2} = 14.16667$$

The third variance component is obtained from the difference between the second and third scales:

$$\sigma_A^2 = \frac{MSA - MSB}{bn} = \frac{265.8889 - 49.5}{6} = 36.0648$$

Varcomp in S-Plus

Before **varcomp** will work, the relevant explanatory variables need to be defined as random effects. This is a two-stage process:

- first declare the variables to be factors

- then define the factors to be random effects

For the rats data, we would proceed like this:

```
attach(rats)
names(rats)

[1] "Glycogen" "Treatment" "Rat" "Liver"

Treatment<-factor(Treatment)
Rat<-factor(Rat)
Liver<-factor(Liver)
```

Now that the three categorical explanatory variables have been defined as factors, we can distinguish between the **fixed effects** (Treatment) and the **random effects** (Rat and Liver) like this:

```
is.random(Rat)<-T
is.random(Liver)<-T
```

Variance components analysis, **varcomp**, looks very like the **aov** model (see above):

```
varcomp(Glycogen~Treatment/Rat/Liver)

Variances:
Rat %in% Treatment Liver %in% (Treatment/Rat) Residuals
        36.06481                    14.16667  21.16667
```

as we calculated by hand (above).

You should appreciate that the value given by **varcomp** depends on the method of estimation used. Variance components analysis has more than its share of jargon.

Acronym	Meaning
MINQUE	minimum norm quadratic unbiased estimation
MIVQUE	minimum variance quadratic unbiased estimation
REML	restricted maximum likelihood estimation
ML	maximum likelihood estimation
BLUP	best linear unbiased predictor
Winsorization	the 'winsor' method produces robust estimates of the variance components after the data have been 'cleaned' (Burns 1992)
Variograms	correlations of random processes

What the biases are, and what they mean, are described and discussed by Rao (1997). The default method in S-Plus is minque0. In order that you can compare them, here are the results obtained using all the different methods: minque0, maximum likelihood, restricted maximum likelihood, and winsor for the rats data:

varcomp(Glycogen~Treatment/Rat/Liver,method="minque0")

```
Rat %in% Treatment Liver %in% (Treatment/Rat) Residuals
      36.6481                     14.16667   21.16667
```

varcomp(Glycogen~Treatment/Rat/Liver,method="ml")

```
Rat %in% Treatment Liver %in% (Treatment/Rat) Residuals
      13.90741                     14.16666   21.16667
```

varcomp(Glycogen~Treatment/Rat/Liver,method="reml")

```
Rat %in% Treatment Liver %in% (Treatment/Rat) Residuals
      36.06481                     14.16667   21.16667
```

varcomp(Glycogen~Treatment/Rat/Liver,method="winsor")

```
Rat %in% Treatment Liver %in% (Treatment/Rat) Residuals
      36.06482                     14.16667   21.16667
```

All methods give the same variance components at the two smallest scales. Maximum likelihood gives a much smaller estimate than the other three methods at the largest scale (this is because maximum likelihood does not make any allowance for the degrees of freedom that are used up by fixed effects). The differences in the estimates produced by the different methods only become large when there is unequal replication.

One thing that people find difficult and confusing is the appearance of negative variance components in the analysis. These come about when the mean square for the higher level is less than the mean square for the level below, i.e. when the higher level happens to vary less than random samples typically do when drawn from a single Normal population. Accurate estimates of the higher-level variance components require very large samples. To work out how many negative variance components to expect in a given circumstance, we can apply the following reasoning. Negative estimates come about when the F ratio is less than 1. The probability of observing an F ratio <1 is

$$\text{probability that } F < \frac{1}{1 + 2(\sigma_A^2/\sigma^2)}$$

So if $\sigma_A^2 = 25$ and $\sigma^2 = 100$ this is $1/(1 + 50/100) = 1/1.5 = 0.6667$. This F value has 9 d.f. in the numerator and 10 d.f. in the denominator. To work out its probability we use **pf**, the cumulative probability function of the F distribution (see p. 470):

pf(0.66667,9,10)

```
[1] 0.2768787
```

So for this example we expect to get negative variance components in roughly one-quarter of cases.

One of the most useful applications of variance components is in designing efficient experiments. In the rats example (above) the largest variance component was 36.065. It seems clear, therefore, that the number of rats should be increased and the number of liver bits decreased if this is possible without increasing the total cost of the experiment.

An example from genetics

In this experiment, one male is mated with four different females. Two offspring are born from each mating, and the intensity of the colour of their eyes is measured. The question is how much of the variation in eye colour is attributable to differences between females and how much to differences between males?

```
attach(flies)
names(flies)
```

```
[1] "eye" "female" "male"
```

In the data frame, males and females are both identified by numeric factor levels, so we begin by declaring them to be categorical rather than continuous variables:

```
female<-factor(female)
male<-factor(male)
```

Next we declare them to be random effects like this:

```
is.random(female)<-T
is.random(male)<-T
```

The variance components analysis is straightforward:

```
varcomp(eye~male/female)
```

```
Variances:
      male female %in% male Residuals
   17.70644          94.94236   1.301667
```

There is very little variation from offspring to offspring within a litter (1.30), and variation from female to female within males is much greater (94.94) than from male to male (17.71).

Because this design was balanced, with equal replication, we can see how the variance components were calculated by fitting a correctly specified nested **aov**, in which females are nested within males for the error term:

```
model<-aov(eye~male/female+Error(male/female))
summary(model)
```

```
Error: male
        Df Sum of Sq  Mean Sq
male  2   665.6758 332.8379
```

```
Error: female %in% male
                    Df Sum of Sq   Mean Sq
female %in% male  9  1720.677 191.1864
```

```
Error: Within
            Df Sum of Sq Mean Sq F Value Pr(F)
Residuals 12     15.62 1.301667
```

The residual (1.30) is the same in both models, as usual. The variance component for females within males is

```
(191.1864 - 1.301667) / 2
```

```
[1]  94.94237
```

We divide by 2 because there were 2 offspring per female. The variance component for males is

(332.8379 - 191.1864) / 8

```
[1] 17.70644
```

We divide by 8 because there were 8 offspring per male (4 females per male and 2 offspring per female). With unbalanced designs, the variance components are calculated iteratively, and will be different from those given by the analysis of variance.

Geneticists use variance components analysis to estimate the *heritability*, h^2, of traits as follows:

$$h^2 = \frac{4\sigma_a^2}{\sigma^2 + \sigma_a^2}$$

Further reading

Rao, P.S.R.S. (1997) *Variance Components Estimation: Mixed Models, Methodologies and Applications*. London, Chapman & Hall.

Searle, S.R., Casella, G. *et al.* (1992) *Variance Components*. New York, John Wiley.

Sokal, R.R. and Rohlf, F.J. (1995) *Biometry: The Principles and Practice of Statistics in Biological Research*. San Francisco, W.H. Freeman and Company.

21

Graphs, functions and transformations

It is useful to be able to sketch the shape of a graph when your are presented with an equation. Likewise, it is useful to be able to look at a scatterplot of some data, and suggest a range of mathematical functions that might be reasonable models for describing the data. This pictorial key is intended to direct you to some appropriate functions.

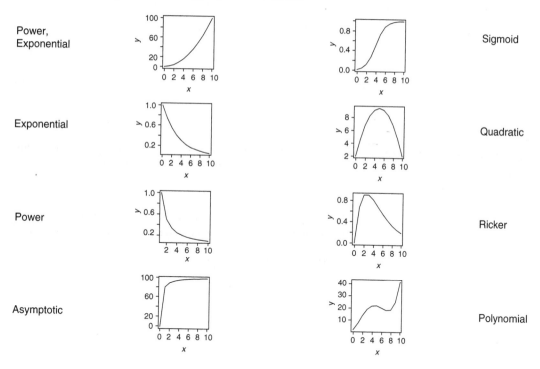

For the kinds of functions you will meet in statistical computing there are only three mathematical rules that you need to learn:

- power: x^b the explanatory variable is raised to a power b

- exponent: e^x the explanatory variable appears as a power (in this case of $e = 2.71828$)

- logarithm: $\log x$ all our logs are to the base e, so when we write $\log x$ it is the same as $\ln x$.

It is also useful to remember a handful of mathematical facts that are useful for working out **behaviour at the limits**. We would like to know what happens to y when x gets very large (e.g. $x \to \infty$) and what happens to y when x goes to zero (i.e. the intercept, if there is one). These are the most important rules:

- anything to the power zero is 1: $x^0 = 1$

- 1 raised to any power is still 1: $1^x = 1$

- infinity plus 1 is infinity: $\infty + 1 = \infty$

- one over infinity (the reciprocal of infinity, ∞^{-1}) is zero: $1/\infty = 0$

- a number bigger than 1 raised to the power infinity is infinity: $1.2^\infty = \infty$

- a fraction (e.g. 0.99) raised to the power infinity is zero: $0.99^\infty = 0$

- negative powers are reciprocals: $x^{-b} = 1/x^b$

- fractional powers are roots: $x^{1/3} = \sqrt[3]{x}$

- the base of natural logarithms is e $= 2.718\,28$ so $e^\infty = \infty$

- $e^{-\infty} = 1/e^\infty = 1/\infty = 0$

You should be able to work out the behaviour of y at $x = 0$ and $x = \infty$ for all of the following examples.

Power functions

In a power function the explanatory variable x is raised to a power. For example, x^2 is a power function where the power is 2. In more general terms, we present a power function as

$$y = ax^b$$

where the two parameters, a and b, describe the initial slope and the shape of the curve, respectively. It is easy to linearise a power function by taking logs of both sides. All you need to remember is that

the log of x^b is b times $\log x$

So, starting from the left, by saying 'the log of y is $\log y$', etc., we get

$$\log y = \log a + b \log x$$

and if we replace $\log y$ by Y, $\log a$ by A and $\log x$ by x, this becomes

$$Y = A + bX$$

which you will recognise as the equation of a straight line. Log transformation of both axes has linearised the power function. The slope of the transformed graph is b (the power) and the intercept of the transformed graph is $A = \log a$.

The power function is so useful because it can take such a wide variety of shapes. Let's deal with two rather trivial cases first. What if $b = 0$, so $y = ax^0$. You will recall that any number to power 0 is 1, so this is the same as saying $y = a \times 1$, which is $y = a$. This means that y is a constant and the graph is a horizontal straight line. Suppose that the constant $a = 0.5$. We start by drawing the axes:

```
plot(c(0,1),c(0,1),type="n",xlab="x",ylab="y")
```

then add a line with intercept 0.5 and slope 0 using **abline** like this:

    abline(0.5,0)

With **abline** the first parameter is a (the intercept) and the second is b (the slope), so we get this graph.

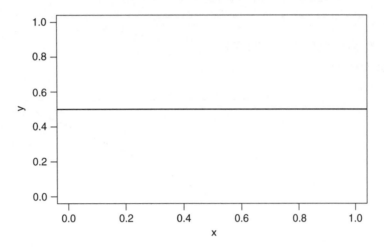

The other simple case arises when the power $b = 1$. Now $y = ax^1$ which is just $y = ax$, and this is a straight line passing through the origin with slope a:

    plot(c(0,1),c(0,1),type="n",xlab="x",ylab='y")
    abline(0,0.5)

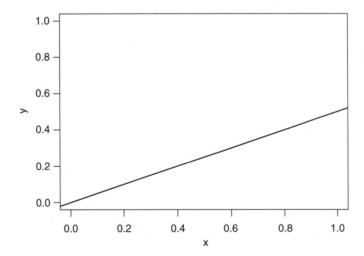

A more interesting family of power functions are the convex and concave curves that have $0 < b < 1$ and $b > 1$ respectively. For **fractional powers** (like square root where $b = 0.5$) we get convex curves like this, whose **slope declines** as x gets bigger. To draw curves we need to generate a series of x values that will produce a smooth-looking curve when plotted out. A good rule of thumb is that about 100 segments will look smooth when plotted:

```
x<-seq(0,1,0.01)
```

Suppose that we want to plot the function for $b = 0.7$. We need to calculate the y values as a function of x; that is to say, x raised to the power (^) of 0.7:

```
y<-x^0.7
```

And finally we plot y against x using type = "l":

```
plot(x,y,type="l")
```

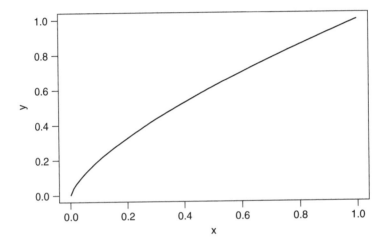

When $b > 1$ the slope of the graph *increases* as x increases, and the graph is concave to the y axis. Here is the power function for $b = 3$:

```
y<-x^3
plot(x,y,type="l")
```

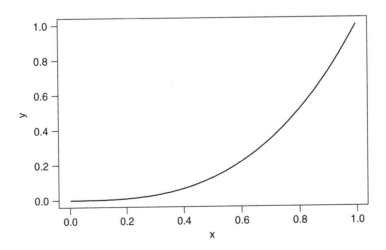

The fifth and final shape that can be taken by a power function is when the power b is negative. *Negative powers are reciprocals*, so instead of writing

$$y = ax^{-b}$$

we might just as easily write

$$y = \frac{a}{x^b}$$

They are exactly the same. The point is that y gets smaller as x gets bigger. The graph has a **negative slope**. Note especially what happens when x is zero:

```
x<-seq(0.1,1,0.01)
y<-x^-3
plot(x,y,type="l")
```

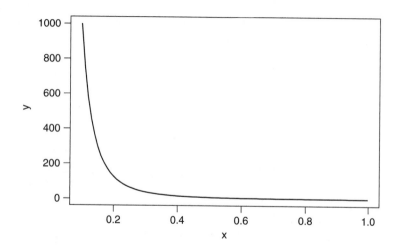

The previous three power functions all went through the origin, but this one clearly does not. Division of a by zero gives **plus infinity** ($+\infty$), which means the graph has no intercept. We say that y goes asymptotically to infinity as x approaches zero. But it is still easy to draw the graph, so long as we don't try to include the point $(0, y)$.

Parameter estimation

How do we work out the parameter values of a power function if we know the values of two points on the line? Suppose that we know the graph is a power function and that it passes through the point (1,10) and the point (10,2). We know that a power function is linearized by log-log transformation, so the points (log 1, log 10) and (log 10, log 2) must lie on a straight line. The slope of that line is the change in log y divided by the change in log x, and this is the value of parameter b (see above):

$$\text{slope} = b = \frac{\log 2 - \log 10}{\log 10 - \log 1} = \frac{0.693\,1472 - 2.302\,585}{2.302\,585 - 0} = \frac{-1.609\,438}{2.302\,585} = -0.698\,97$$

We now use either one of the known points on the line to obtain a:

$$a = \frac{y}{x^b}$$

Let's use the point (1,10) so $a = 10/1^b = 10/1 = 10$ and the parameterised equation is

$$y = 10x^{-0.69897}$$

Polynomial functions

Polynomial functions are functions in which x appears several times, each time raised to a different power. They are useful for describing curves with humps, inflections or local maxima like these.

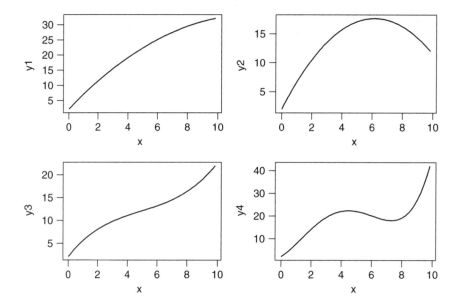

The top left panel shows a decelerating positive function, modelled by the quadratic

$$y = 2 + 5x - 0.2x^2$$

Making the coefficient of the x^2 term larger produces a curve with a hump:

$$y = 2 + 5x - 0.4x^2$$

Cubic polynomials can show points of inflection, as in the lower left panel:

$$y = 2 + 4x - 0.6x^2 + 0.4x^3$$

Finally, polynomials containing powers of 4 are capable of producing curves with local maxima, as in the lower right panel:

$$y = 2 + 4x + 2x^2 - 0.6x^3 + 0.04x^4$$

Inverse polynomials

An important class of models for which Gamma errors are appropriate are the *inverse polynomials*. The simplest of these is the inverse linear polynomial

$$\frac{1}{y} = a + b\frac{1}{x}$$

Which is an asymptotic curve familiar to scientists under a variety of names, including Briggs–Haldane, Michaelis–Menten and Holling's disc equation, to name just three from physics, biochemistry and ecology respectively (see p. 387). Taking reciprocals gives

$$y = \frac{x}{b + ax}$$

The curve rises with an initial slope of $1/b$, and asymptotes at $1/a$. The general formulation of the inverse polynomial looks like this:

$$\frac{x}{y} = a + bx + cx^2 + dx^3 + \cdots + zx^n$$

and can have a wide range of shapes, including asymmetrical humped curves. Here are two contrasting inverse polynomials:

```
par(mfrow=c(1,2))
x<-seq(0,14,.01)
Y<-x/(1+x)
plot(x,y,type="l")
y<-1/(x-2+4/x)
plot(x,y,type="l")
```

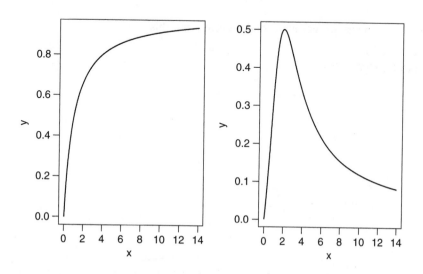

The reciprocal link function is appropriate for these models and, indeed, the reciprocal is the canonical link function for the Gamma error distribution.

Exponential functions

Exponential functions are an important family of two-parameter distributions; they are extremely useful for describing ubiquitous processes like **growth** and **decay**:

$$y = ae^{bx}$$

where e^x (base e to the power x) is read as 'e to the x'. This means raise the number $2.718\,28$ to the power of the number that is the explanatory variable. When the parameter b is positive we get exponential growth, and when b is negative we get exponential decay.

For positive values of b, exponential growth looks like this:

```
x<-seq(0,5,0.1)
y<-exp(x)
plot(x,y,type="l")
```

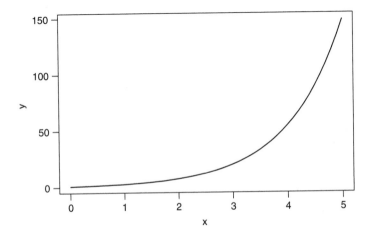

For negative exponents we get exponential decay, like this:

```
y<-exp(-x)
plot(x,y,type="l")
```

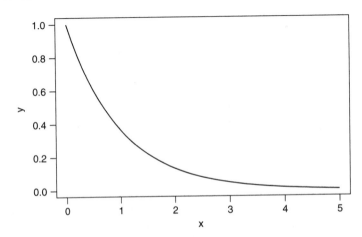

Note that unlike the negative power function, the negative exponential *does* have an intercept; the graph cuts the y axis at $y = a$ when $x = 0$ (e.g. the initial amount of a decaying substance), because $e^0 = 1$.

To linearise an exponential relationship we take logs of both sides. Note that 'e to the power' also means antilog. So $\log(e^x)$ is like saying 'the log of the antilog of x'. Obviously, this is just x itself. So remember that $\log(e^x) = x$. Taking logs of both sides gives

$$\log y = \log a + bx$$

This is linear if we plot $\log y$ against x. Note the difference compared with power functions, where we logged **both axes** in order to linearise the model.

Suppose we know that an exponential curve passes through the points $(0,15)$ and $(12,2)$; what is its equation? We need to log transform the y values but not the x values (see above). The slope of the transformed function is b, and this is

$$\text{slope} = b = \frac{\log 2 - \log 15}{12 - 0} = \frac{-2.014\,903}{12} = -0.167\,9086$$

The graph has an intercept (at $x = 0$) of $y = 15$, so we can immediately write the fully parameterised equation:

$$y = 15e^{-0.1679086x}$$

Logarithmic functions

Some important processes are logarithmic, like this:

$$y = \log(a + bx)$$

which is linearised by **taking antilogs of both sides**:

$$e^y = a + bx$$

For the linearised graph, we would plot antilog of y on the y axis against untransformed x on the x axis. When b is positive the logarithmic function looks like this:

```
y<-log(0.1+x)
plot(x,y,type="l")
```

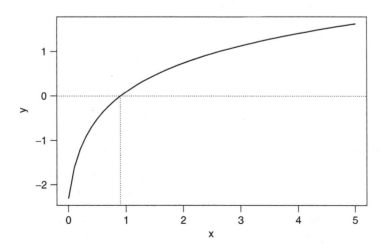

Note the existence of a **threshold value** of x (in this example $x = 0.9$), below which the value of the response variable, y, is negative. We can add dotted lines to the plot to show the threshold at $x = 0.9$:

```
abline(0,lty=2)
lines(c(0.9,0.9),c(-2.5,0),lty=2)
```

For negative values of b, the relationship looks like this:

```
y<-log(5-x)
plot(x,y,type="l")
```

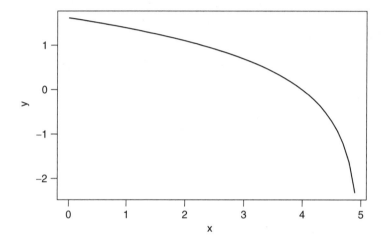

You need to be careful with this function because there is a risk that you might try to take the log of a negative number (such things don't exist). In this example, x must not exceed 5.

Suppose that a logarithmic function passes through the points (2,2) and (5,1); what are the parameter values for a and b? We need to antilog the y values to calculate the slope:

$$\text{slope} = b = \frac{e^1 - e^2}{5 - 2} = \frac{-4.670\,774}{3} = -1.556\,925$$

We can get a from either of the points, say (5,8):

$$a = e^y - bx = e^2 + 1.556\,925 \times 2 = 10.502\,91$$

so the parameterised equation is

$$y = \exp(10.502\,91 - 1.556\,925x)$$

Asymptotic relationships

Many dynamic processes are asymptotic; the response variable increases with x at first, but for high values of x, further increases in x do not lead to any further increase in y; a familiar case is the law of diminishing returns. Examples are reaction rates as a function of enzyme concentration, photosynthetic rate as a function of light intensity, predation rate as a function of prey availability, profit as a function of investment, etc. The two most commonly used asymptotic relationships are the **asymptotic exponential** and the **Michaelis–Menten equation**. The Michaelis–Menten equation is the easiest to parameterise from data, because it is straightforward to linearise.

Michaelis–Menten equation

This famous asymptotic relationship is written as follows:

$$y = \frac{ax}{1 + bx}$$

To see how this model behaves, try putting $x = 0$, then $x = \infty$. For $x = 0$ we have

$$y = \frac{0}{1 + 0} = \frac{0}{1} = 0$$

so the graph passes through the origin. For $x = \infty$ we have

$$y = \frac{\infty}{1 + \infty} = \frac{\infty}{\infty}$$

Now you might imagine that this should be 1 (the infinities cancel, you might think). But no. For cases like this, where we end up with infinity over infinity, we need to use a special rule known as L'Hospital's rule. **The limit of the ratio is given by the ratio of the derivatives.** The numerator is ax so its derivative is a. The denominator is $1 + bx$, so its derivative is $0 + b = b$. So the ratio of the derivatives is a/b. This is the asymptotic value of y; no matter how large x gets, y gets no bigger than a/b. Thus for $a = 400$ and $b = 4$ the asymptote is $400/4 = 100$ and the function looks like this:

```
x<-seq(0,1,0.01)
y<-400*x/(1+4*x)
plot(x,y,type="l")
```

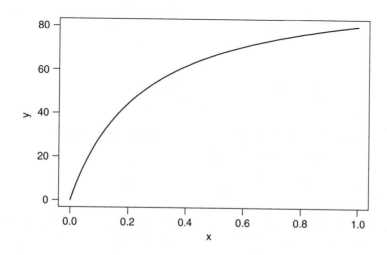

In order to estimate the values of the parameters a and b from data, we need to linearise the relationship. This time, taking logs doesn't work:

$$\log y = \log a + \log x - \log(1 + bx)$$

which doesn't help because 'we can't take the log of plus'. This is *not* a linear function.

But all is not lost, because we can try other tricks. What happens if we try the **reciprocal transformation**, for instance? To do this we ask, What is $1/y$ a function of?

$$\frac{1}{y} = \frac{1 + bx}{ax}$$

which, at first glance, isn't a big help. But we can separate the terms on the right because they have a common denominator. Then we can cancel some of the x terms like this:

$$\frac{1}{y} = \frac{1}{ax} + \frac{bx}{ax} = \frac{1}{ax} + \frac{b}{a}$$

so if we put $y = 1/y$, $x = 1/x$, $A = 1/a$ and $C = b/a$ then we see that

$$Y = AX + C$$

which is linear, C is the intercept and A is the slope. So to estimate the values of a and b from data, we would transform both x and y to reciprocals, plot a graph of $1/y$ against $1/x$, carry out a linear regression, then back-transform to get

$$a = \frac{1}{A}$$

$$b = aC$$

Suppose that we knew that the graph passed through the two points (0.2, 44.44) and (0.6, 70.59). How do we work out the values of the parameters a and b? First we calculate the four reciprocals. The slope of the linearised function, A, is the change in $1/y$ divided by the change in $1/x$:

```
(1/44.44 - 1/70.59)/(1/0.2 - 1/0.6)
```

```
[1] 0.002500781
```

so $a = 1/A = 1/0.0025 = 400$. Now we rearrange the equation and use the point ($x = 0.2$, $y = 44.44$) to get the value of b:

$$b = \frac{1}{x}\left(\frac{ax}{y} - 1\right) = \frac{1}{0.2}\left(\frac{400 \times 0.2}{44.44} - 1\right) = 4$$

Alternatively we could use the intercept of the transformed graph, which is C. This is $y - AX$:

```
(1/44.44) - 0.0025 *(1/0.2)
```

```
[1] 0.01000225
```

then $b = aC = 400 \times 0.01 = 4$.

With a scatterplot of many values for x and y we could estimate the parameters from a linear regression of $1/y$ against $1/x$ (a double reciprocal transformation).

Asymptotic exponential

Another important function in statistical modelling where there are diminishing returns involves the asymptotic exponential function

$$y = a(1 - e^{-bx})$$

To see how this behaves, look at the values of y when $x = 0$ and $x = \infty$. At $x = 0$ we have $-b$ times zero, which is zero; and e to the power zero is 1. One minus one is zero, and zero times a is zero. So the function passes through the origin.

When x is very large we have minus b times infinity, which is minus infinity; and 2.718 26 to the power minus infinity is zero. Remember the minus power means reciprocal, so 1/e to the power $+\infty$ is $1/\infty$, which is zero. If you don't know it already, it is well worth memorising this:

$$e^{-\infty} = \frac{1}{e^{\infty}} = \frac{1}{\infty} = 0$$

So $1 - 0$ is 1 and $1 \times a$ is a. No matter how large x becomes, y never gets any bigger than a. For the parameter values $a = 100$ and $b = 0.1$ (both are always positive) we get

```
x<-0:100
y<-100*(1-exp(-0.1*x))
plot(x,y,type="l")
```

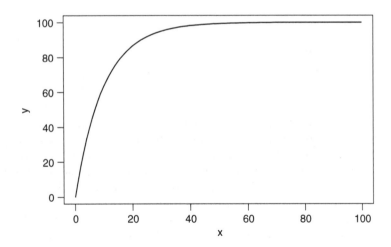

Once x gets bigger than about 70, y has reached its asymptotic value of 100.

This equation cannot readily be linearised unless we know the value of the asymptote, a, in advance, and this would seldom if ever be the case, so Michaelis–Menten is better if we intend to use linear regression to estimate the parameters. Non-linear estimation of the asymptotic exponential is covered in Chapter 23.

Sigmoid functions

A range of well-known processes start out exponential but then become asymptotic; growth of a population of bacteria in a culture vessel would be a good example. For such processes we need a different kind of growth function, a function that is S-shaped. The best known S-shaped function is the two-parameter logistic, whose equation is

$$y = \frac{\exp(a + bx)}{1 + \exp(a + bx)}$$

Suppose that $a = 1$ and $b = 0.1$. The x values go all the way from $-\infty$ to $+\infty$. The point of inflection of our curve is at $x = 50$, so we write the function using $x - 50$ like this:

```
x<-0:100
y<-exp(1+0.1*(x-50))/(1+exp(1+0.1*(x-50))
plot(x,y,type="l")
```

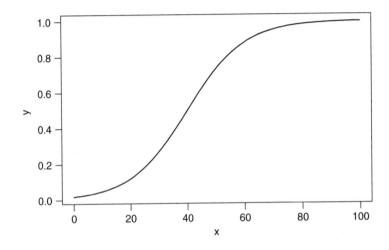

The equation looks a bit daunting, but it is actually quite easy to linearise. First, note that y varies between 0 and 1. This function is often used to describe proportion data (e.g. the proportion of a population that dies over winter, the proportion of a share portfolio that increases in value). If y is the proportion that die, then $1 - y$ is obviously the proportion that survive. The ratio $y/(1 - y)$ is what bookmakers refer to as odds. So if a horse is thought to have a 33.3% chance of winning a race, then $y = 0.33$ and $1 - y = 0.67$ and the odds are 0.33/0.67; one chance of winning to two chances of losing. The odds are therefore 2 to 1. To make a consistent profit, the bookie would have to believe that the horse had a real chance of winning of somewhat less than 33.3% (e.g. real odds of, say, 3 to 1).

Let us express the logistic in terms of odds, i.e. as the ratio $y/(1 - y)$.

First, notice that $1 - y$ is actually rather simple; by making a common denominator of $1 + e^{a+bx}$ we get

$$1 - y = \frac{1 + e^{a+bx}}{1 + e^{a+bx}} - \frac{e^{a+bx}}{1 + e^{a+bx}} = \frac{1}{1 + e^{a+bx}}$$

which is actually a simpler equation than for y. Since the odds are the ratio $y/(1 - y)$, we have

$$\frac{y}{1 - y} = \frac{e^{a+bx}}{1 + e^{a+bx}} \frac{1 + e^{a+bx}}{1} = e^{a+bx}$$

It is now straightforward to linearise this by taking logs of both sides:

$$\log\left(\frac{y}{1-y}\right) = a + bx$$

This says that if we plot log odds against x, the relationship will be linear. For our graph, $a = 1$ and $b = 0.1$ (x scaled as $x - 50$). Note that when $y = 0.5$ then $1 - y = 0.5$ as well, so the log odds will be $\log 1 = 0$; the straight line graph will cut the x axis at $50 - a/b = 40$:

```
plot(x,log(y/(1-y)),type="l")
abline(0,lty=2)
```

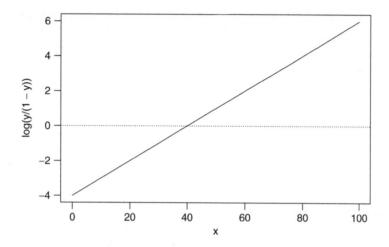

Sigmoid

Another commonly used three-parameter model for an S-shaped growth curve is a reparameterised version of the logistic which looks like this:

$$y = \frac{a}{1 + be^{-cx}}$$

where a is the asymptotic value of y, b is related to the intercept, and c is the exponential growth rate at small values of x. To check the meaning of the parameters, we can ask what values of y are observed at the limits of x. For $x = 0$ we have $-c$ times zero, which is zero, e^0 is 1, and $b \times 1$ is b. So at $x = 0$ the intercept is

$$y = \frac{a}{1+b}$$

At the other extreme we put $x = \infty$. Now $-c$ times infinity is minus infinity, $e^{-\infty}$ is 0, and $b \times 0$ is 0. One plus zero is one, and a divided by one is a. So as x tends to infinity, y tends to the asymptotic value a.

An important application of this equation is in ecology where y is population size at time x. The slope of the graph (the derivative) is the **instantaneous, per capita rate of reproduction**:

$$\frac{dy}{dx} = cy\left(\frac{a-y}{a}\right)$$

This represents the simplest possible model of density dependence. When population size y is low, then $(a - y)/a$ is close to 1, and the rate of population change is cy; this is exponential growth:

$$\frac{dy}{dx} = cy$$

The per capita rate of reproduction declines linearly as density increases (e.g. as a result of competition for food), and $dy/dx \propto (a - y)$. Once population density has risen to a, the rate of change in population becomes zero (since $a - y = 0$ at this point), and population size y remains constant:

$$\frac{dy}{dx} = \frac{cy(a - a)}{a} = 0$$

Population growth is greatest at intermediate population size. To find the maximum of a function:

- find the derivative of the function

- set the derivative to zero

- solve for the variable of interest

In this case we want to find the derivative of $cy(a - y)/a$ with respect to y.

Multiplying out the bracket gives $cy - cy^2/a$. The derivative of this with respect to y is therefore $c - 2cy/a = c(1 - 2y/a)$. Set this to zero and divide both sides by c:

$$1 - 2y/a = 0 \quad \text{so} \quad 2y/a = 1 \quad \text{and} \quad y = \frac{a}{2}$$

The maximum growth rate therefore occurs when population size y is at half its carrying capacity $= a/2$ (where the parameter a is the asymptotic population size). To find the time, x, at which maximum population growth occurs (the point of inflection) we need to substitute the value $y = a/2$ in the original equation, and rearrange to get $x = \log b/c$.

We plot the sigmoid growth curve like this for $a = 100$, $b = 9$ and $c = 0.07$:

```
x<-0:100
y<-100/(1+9*exp(-0.07*x))
plot(x,y,type="l")
```

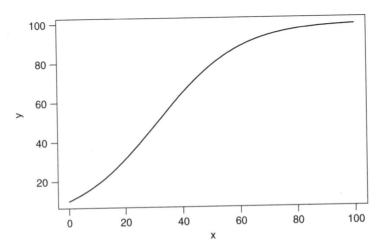

The parameter values are used to work out the maximum growth rate, which occurs when population $y = a/2 = 100/2 = 50$. Visual inspection suggests this occurs about time $x = 40$. To get the exact answer, we evaluate $x = (\log b)/c = (\log 9)/0.07 = 31.39$. The maximum population growth is $cy(a - y)/a$ evaluated at $y = a/2$. Replacing y by $a/2$ and rearranging gives the maximum population growth as $ca/4$, which is $0.07 \times 100/4 = 1.75$ individuals per individual per unit time.

Gompertz function

Our previous S-shaped curves were symmetrical but sometimes we need an asymmetrical S-shaped function. The best known of these is the Gompertz curve, named after the famous human demographer and one of the first actuaries (see p. 614). It is a three-parameter model:

$$y = ae^{be^{cx}}$$

This double exponential can take a variety of different shapes, depending upon the **signs** of b and c. It can give faster than exponential increase (when b and c are both positive), or exponential decline to an asymptote of a (when b is positive and c negative). For a **negative sigmoid**, b is negative (in this case -1) and c is positive (here 0.02):

$$y = 100e^{-e^{0.02x}}$$

We plot the curve like this, using a new range of x values:

```
x<- -200:100
y<-100*exp(-exp(0.02*x))
plot(x,y,type="l")
```

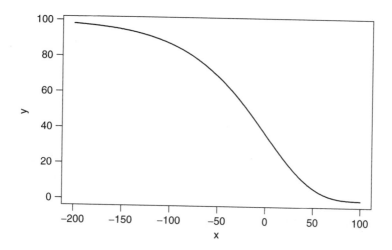

For a **positive sigmoid** Gompertz curve, both parameters are negative. Here is an example:

$$y = 50e^{-5e^{-0.08x}}$$

```
x<-0:100
y<-50*exp(-5*exp(-0.08*x))
plot(x,y,type="l")
```

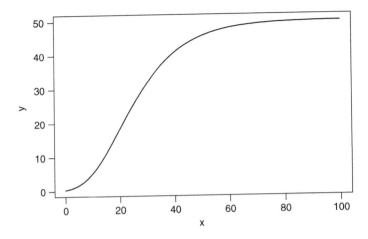

Table 21.1 Derivatives of important mathematical functions

Function	Derivative
$y = a$	0
$y = a + bx$	b
$y = ax^b$	abx^{b-1}
$y = ae^{bx}$	abe^{bx}
$y = a(1 - e^{-bx})$	abe^{-bx}
$y = \dfrac{ax}{1 + bx}$	$\dfrac{a}{1 + bx} - \dfrac{abx}{(1 + bx)^2}$
$y = \dfrac{a}{1 + be^{-cx}}$	$cy\left[\dfrac{a - y}{a}\right]$
$y = \dfrac{\exp(a + bx)}{1 + \exp(a + bx)}$	$\dfrac{b\exp(a + bx)}{1 + \exp(a + bx)} - \dfrac{b\exp\ (2a + 2bx)}{(1 + \exp(a + bx))^2}$
$y = \log x$	$\dfrac{1}{x}$
$y = \log(a + bx)$	$\dfrac{b}{a + bx}$
$y = a\exp(b\exp(cx))$	$abc\exp(cx + b\exp(cx))$

You should check the behaviour at the limits for the other sign combinations of b and c.

Transformations of the response and explanatory variables

We have seen the use of transformation to linearise the relationship between the response and the explanatory variables:

- $\log y$ against x for exponential relationships
- $\log y$ against $\log x$ for power functions
- e^y against x for logarithmic relationships

- $1/y$ against $1/x$ for asymptotic relationships
- $\log[p/(1-p)]$ against x for proportion data

Other transformations are useful for variance stabilisation

- $\sqrt{y}$ to stabilise the variance for count data
- arcsin y to stabilise the variance of percentage data

Sometimes, however, it may not be obvious what transformation to use. In these circumstances, the Box–Cox transformation offers a useful empirical solution.

Box–Cox transformations

The idea is to find the power transformation, λ (lambda), that maximises the likelihood when a specified set of explanatory variables is fitted to

$$\frac{y^{\lambda} - 1}{\lambda}$$

as the response. The value of λ can be positive or negative, but it can't be zero (you would get a zero-divide error when the formula was applied to the response variable, y). For $\lambda = 0$ the Box–Cox transformation is *defined* as $\log y$.

Suppose that $\lambda = -1$. The formula now looks like this:

$$\frac{y^{-1} - 1}{-1} = \frac{1/y - 1}{-1} = 1 - \frac{1}{y}$$

This quantity is then regressed against the explanatory variables and the log likelihood computed. We start by loading the MASS library of Venables and Ripley:

```
library(MASS)
```

We work with the timber data.

```
attach(timber)
names(timber)
```

```
[1]   "volume"   "girth"   "height"
```

Suppose that we want to know the best transformation of timber volume when we use log(girth) and log(height) of the timber as explanatory variables. The **boxcox** function is very easy to use; just specify the model formula, and the default options take care of everything else:

```
boxcox(volume~log(girth)+log(height))
```

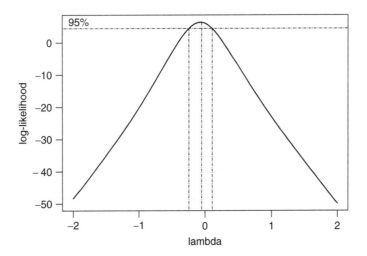

It is clear that the optimal value of λ is close to zero (i.e. the log transformation of y). We can zoom in to get a more accurate estimate by specifying our own, non-default, range of λ values. It looks as if it would be sensible to plot from -0.5 to $+0.5$:

```
boxcox(volume~log(girth)+log(height),lambda=seq(-0.5,0.5,0.01))
```

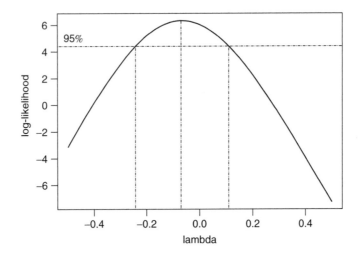

The log likelihood is maximised at λ about -0.08, but the log likelihood for $\lambda = 0$ is very close to the maximum. $\lambda = 0$ gives a much more straightforward interpretation, so we would go with that, and model log(volume) as a function of log(girth) and log(height).

What if we had not log transformed the explanatory variables? What would have been the optimal transformation of volume in that case? To find out, we rerun the **boxcox** function, simply changing the model formula like this:

```
boxcox(volume~girth+height)
```

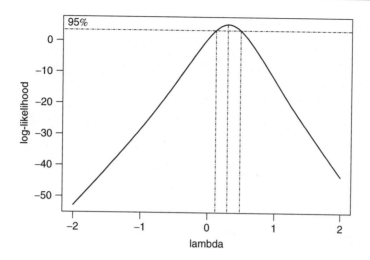

We can zoom in from 0.1 to 0.6 like this:

boxcox(volume~girth+height,lambda=seq(0.1,0.6,0.01))

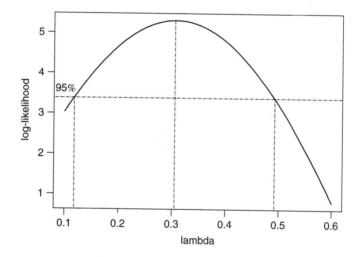

This suggests that the cube root transformation would be best ($\lambda = 1/3$). Again, this accords with dimensional arguments, since the response and explanatory variables would all have dimensions of length in this case (i.e. the cube root of volume would be measured in metres, as would both girth and height).

Self-starting functions

When we come to fit non-linear functions to data (Chapters 22 and 23) we are normally required to work out an initial guess for the value of each of the parameters. As we have seen, this can be both time-consuming and error-prone. To save time, S-Plus provides self-starting versions of many of the most popular non-linear functions. These work out their own starting values for the parameters from the data, as follows.

SSasymp: asymptotic exponential model

$$y = a - be^{-cx}$$

In the self-starting version:

- $a = $ Asym, the horizontal asymptote on the right-hand side
- $b = $ Asym −R0 where R0 is the intercept (the response when x is zero)
- $c = $ exp(lrc) where lrc is the natural logarithm of the rate constant

The equation is written like this:

Asym+(R0-Asym)*exp(-exp(lrc)*input).

Examples of this function are shown on p. 415. Options allow one to specify the function with an offset on the x values **SSasympOff**, and a the two-parameter form that passes through the origin **SSasympOrig**:

$$y = a(1 - e^{-cx})$$

SSbiexp: biexponential model

A useful four-parameter non-linear function, **SSbiexp** is the sum of two exponential functions of x:

$$y = ae^{bx} + ce^{dx}$$

Various shapes depend upon the signs of the parameters b, c and d.

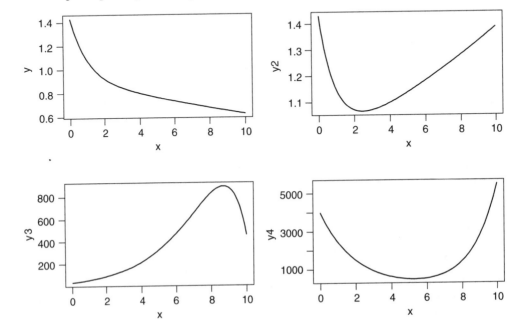

- the upper left panel shows c positive, b and d negative
- the upper right panel shows c and d positive, b negative
- the lower left panel shows c and d negative, b positive
- the lower right panel shows c and b negative, d positive

The self-starting function is parameterised as follows:

A1*exp(-exp(lrc1)*input)+A2*exp(-exp(lrc2)*input).

where A1 is the multiplier of the first exponential, lrc1 is the natural logarithm of the rate constant of the first exponential, A2 is the multiplier of the second exponential and lrc2 is the natural logarithm of the rate constant of the second exponential.

When b, c and d are all negative, this function is known as the first-order compartment model in which a drug administered at time 0 passes through the system with its dynamics affected by elimination, absorption and clearance. It has a self-starting version **SSfol** with parameters

Dose*exp(lKe+lKa-lCl)*(exp(-exp(lKe)*input)
-exp(-exp(lKa)*input))/(exp(lKa) - exp(lKe))

where Dose is the initial dose, lKe is the natural logarithm of the elimination rate constant, lKa is the natural logarithm of the absorption rate constant, lCl is the natural logarithm of the clearance rate, and input is the vector of values for the x axis. For the model $y = 40e^{-0.5x} - 5e^{-0.7x}$, the function looks like this.

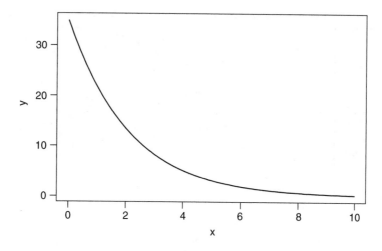

Three-parameter logistic model

$$y = \frac{a}{1 + e^{b(c-x)}}$$

This has a self-starting version **SSlogis** with

Asym/(1+exp((xmid-input)/scal))

where Asym is the asymptote, xmid is the x value at the inflection point of the curve, and scal is the rate parameter on the input axis.

Four-parameter logistic model

The four-parameter version has asymptotes at the left-hand (a) and right-hand (b) ends of the x axis and scales (c) the response to x about the midpoint (d) where the curve has its inflection.

$$y = a + \frac{b - a}{1 + e^{c(d-x)}}$$

It has a self-starting version **SSfpl**:

A+(B-A)/(1+exp((xmid-input)/scal))

where A is the horizontal asymptote on the left-hand side (for very small values of x), B is the horizontal asymptote on the right-hand side (for very large values of x), xmid is the value of x at the inflection point of the curve, and scal is a scale parameter on the x axis. The graph shows the function

$$y = 20 + \frac{100}{1 + e^{0.8 \times (3-x)}}$$

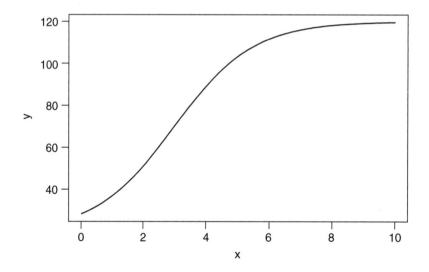

Negative sigmoid curves have the parameter $c < 0$ as in the following graph for the function

$$y = 20 + \frac{100}{1 + e^{-0.8 \times (3-x)}}$$

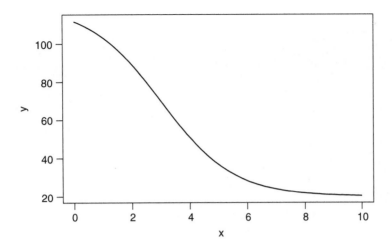

Michaelis−Menten model

The Michaelis−Menten model has a self-starting version **SSmicmen**:

> Vm*input/(K+input)

where two parameters are Vm, the asymptotic value of the response, and K, which is the x value at which half the maximum response is attained. In the field of enzyme kinetics this is called the Michaelis parameter (see p. 387).

Further reading

Abramowitz, M. and Stegun, I. (1964) *Handbook of Mathematical Functions*. Washington, DC, National Bureau of Standards.

Atkinson, A. C. (1985) *Plots, Transformations, and Regression*. Oxford, Clarendon Press.

Box, G. E. P. and Cox, D. R. (1964) An analysis of transformations. *Journal of the Royal Statistical Society, Series B* **26**: 211−246.

Caroll, R. J. and Ruppert, D. (1988) *Transformation and Weighting in Regression*. New York, Chapman & Hall.

Chambers, J. M., Cleveland, W. S. *et al.* (1983) *Graphical Methods for Data Analysis*. Belmont, California, Wadsworth.

22

Curve fitting and piecewise regression

Suppose that initial data inspection indicated that the relationship between the response variable and a continuous explanatory variable was clearly not linear. For example, in Chapter 17 we investigated the dry mass remaining after different periods of time in a decomposition experiment:

```
attach(Decay)
names(Decay)
```

```
[1] "x" "y"
```

```
plot(x,y)
```

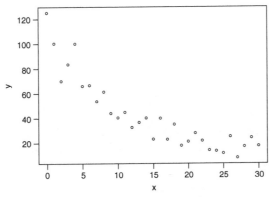

A straight-line relationship was not appropriate for these data (see p. 319). For one thing, the fitted model would predict negative dry matter once time (x) was larger than 30 or so. Let's put a straight line through the data using **abline** just to get a better impression of the curvature:

```
abline(lm(y~x),lty=2)
```

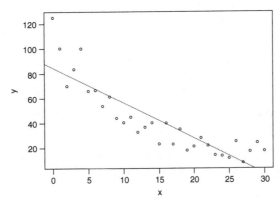

What we see are 'groups of residuals'. Below about $x = 5$ most of the residuals are positive, as is the case for the residuals for x bigger than about 25. In between, most of the residuals are negative. This is what we mean by 'evidence of curvature'. Or more forthrightly, 'systematic inadequacy of the model'. The diagnostic plots from a linear regression gave further support to the inadequacy of the linear models for these data (Chapter 17).

We replot the data with y on a log scale:

plot(x,y,log="y")

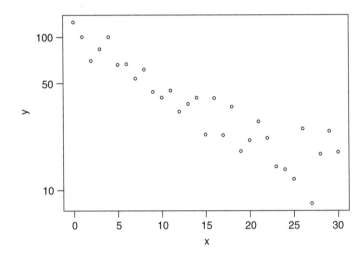

That is much straighter, but now a new problem has arisen: the variance in y apparently increases as y gets smaller (the scatter of the data around the line is fan-shaped). We shall deal with that later. For the moment, instead of

y ~ x

we shall fit

log(y) ~ x

Let's call the new model object transformed like this:

```
transformed <- lm(log(y)~x)
summary(transformed)

Coefficients:
                Estimate  Std.Error  t value    Pr(>|t|)
(Intercept)     4.547386   0.100295     45.34    < 2e-016  ***
x              -0.068528   0.005743    -11.93    1.04e-012  ***

Residual standard error: 0.286 on 29 degrees of freedom
Multiple R-Squared: 0.8308
```

As before, all the parameters are highly significant, and the value of r^2 is higher. But does the model behave any better? Let's look at the diagnostic plots:

plot(transformed)

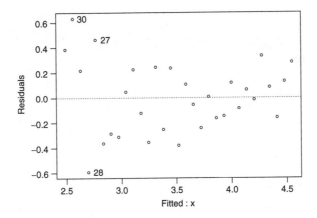

We have cured the U-shaped residuals, but at the price of introducing non-constancy of variance (in the jargon, this is the dreaded heteroscedasticity). What about the normality of the errors ? This is the fourth plot.

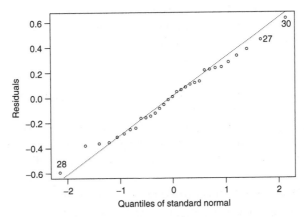

Not perfect, but much better than before. In this case, transformation of the response variable has solved most of our difficulties, as we might have expected with an essentially exponential process such as decay (Chapter 21).

Curve fitting after regression

Generally, you would want to plot your model as a smooth curve of fitted values through the scatterplot of data. You use the **predict** function for this. There are three steps:

- use **seq** to generate a series of values for the x axis (they should cover the same range as the data, and have at least 100 increments so that the curve of the fitted values looks reasonably smooth when it is plotted)

- use **predict** to calculate the fitted values

- use **lines** to overlay the smooth curve of the fitted values on the scatterplot of data

Let's remind ourselves what the raw data look like, to see the range of x values:

```
plot(x,y)
```

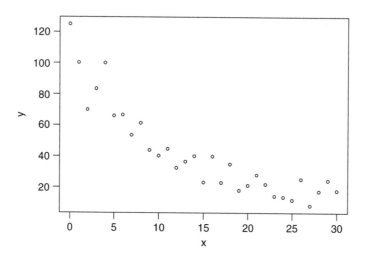

So we want our generated x values to go from 0 to 30. When the x values are supplied to the **predict** function *they must have exactly the same name as the explanatory variable used in the model* called transformed. Now there's a slight problem here because x is the name of the vector containing our data points, and we don't want to mess with this. The solution is a follows. Generate the values for drawing the fitted curve under a *different* name, say *smoothx*, like this:

```
smoothx<-seq(0,30,0.1)
```

Then tell **predict** that the data are called x rather than *smoothx* by using a **list** like this list(x=smoothx) inside the **predict** function.

The last issue to be confronted concerns the scale of measurement. The model has log y as the response variable, but we want to plot a curve of y against x (not a straight line of log y against x). To do this, we need to back-transform the values produced by **predict** using **exp** (because these values are in logs):

```
smoothy<-exp(predict(transformed,list(x=smoothx)))
```

Now we can draw the fitted model through the data, using **lines**:

```
lines(smoothx,smoothy)
```

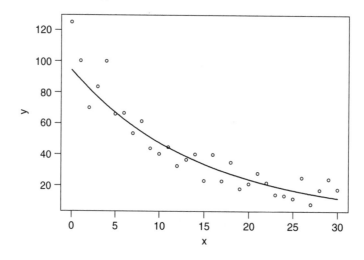

Piecewise regression

This kind of regression fits different functions over different ranges of the explanatory variable. For example, it might fit different linear regressions to the left- and right-hand halves of a scatterplot. Two important questions arise in piecewise regression:

- how many segments to divide up the line into
- where to position the break points on the x axis

Suppose we want to do the simplest piecewise regression, using just two linear segments. The question then becomes where to break up the x values. A simple, pragmatic view is to divide the x values at the point where the piecewise regression best fits the response variable. Let's take an example using a **glm** where the response is a count (the number of species recorded) and the explanatory variable is the log of the area searched for the species:

```
attach(sasilwood)
ames(sasilwood)
```

```
[1] "Species" "Area"
```

A quick scatterplot suggests that the relationship between log(Species) and log(Area) is not linear (as required by the log-linear **glm** with Poisson errors):

```
plot(log(Area),log(Species))
```

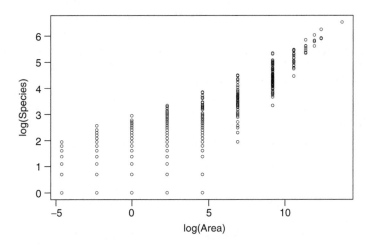

The slope appears to be shallower at small scales than at large. The overall **glm** regression highlights this at the model-checking stage:

```
model<-glm(Species~log(Area),poisson)
summary(model)
```

```
Coefficients:
                Value   Std. Error    t value
(Intercept) 1.8079559 0.011819952   152.9580
  log(Area) 0.2977006 0.001433644   207.6531
```

```
(Dispersion Parameter for Poisson family taken to be 1)

      Null Deviance: 62756.63 on 1486 degrees of freedom

  Residual Deviance: 7583.043 on 1485 degrees of freedom
```

Note the high degree of overdispersion, 7583 on 1485 d.f. (see p. 541). The model checking shows that the residuals are very badly behaved:

plot(model)

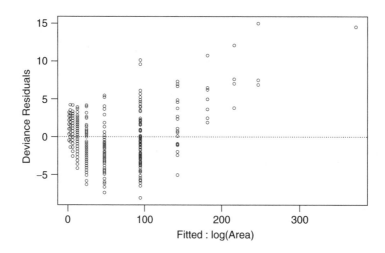

All of the deviance residuals for the larger scales are positive. So we decide to fit two separate straight lines, one for small scales and another for large. But where to make the break? Suppose we put it in the middle. We need to count the number of discrete values of Area, perhaps using **table** to do this:

table(Area)

```
0.01 0.1    1   10 100 1000 10000 40000 90000 160000 250000 1000000
 346 345  259  239  88   67   110    18     7      4      3       1
```

There are 12 levels of Area, so we could split the data into 6 and 6. Areas less than $10\,000\,\text{m}^2$ would make up the left-hand graph, while areas greater than $1000\,\text{m}^2$ make up the right-hand graph. Replication is highly asymmetric between the two graphs, but we ignore this for the moment. Piecewise regression is extremely simple in S-Plus: we just include the logical statement as part of the model formula. In the present example, the two logical statements are (Area < 10000) to define the left-hand regression and (Area > 1000) to define the right-hand regression. The piecewise model looks like this:

model<-glm(Species~log(Area)*(Area<10000)+log(Area)*(Area>1000),poisson)

Note the use of the multiply operator * so that when the logical expression is false, the model entry evaluates to zero, hence it is not fitted to the response (see p. 334). The output needs a little getting used to:

summary(model)

```
Call: glm(formula = Species ~ log(Area) * (Area < 10000) +
log(Area) * (Area > 1000),  family = poisson)

Coefficients: (2 not defined because of singularities)
                              Value    Std. Error     t value
           (Intercept)    0.9138669   0.029540887    30.93566
              log(Area)   0.3506907   0.003103151   113.01119
         Area < 10000     1.0321647   0.029540887    34.94021
          Area > 1000           NA            NA          NA
  log(Area):(Area < 10000) -0.1416193  0.003103151   -45.63726
   log(Area):(Area > 1000)        NA            NA          NA

(Dispersion Parameter for Poisson family taken to be 1)

   Null Deviance: 62756.63 on 1486 degrees of freedom

Residual Deviance: 5332.982 on 1483 degrees of freedom
```

First things first. The residual deviance is down to 5332.98, a great improvement over the linear regression we began with (where the deviance was 7583.043) but still overdispersed. The top message indicates that there are only 4 parameters (not 6) to be estimated (we have intentionally created a singularity in the piecewise regression between Area = 1000 and Area = 10000; the aliased parameters show up as NA).

The intercept of 0.913 8669 is for the *right-hand graph*, whose slope (labelled log(Area)) is 0.350 6907. The other two parameters (labelled Area < 10000 and log(Area):(Area < 10000) respectively) are the difference between the two intercepts and the difference between the two slopes. So the left-hand graph has intercept = 0.913 8669 + 1.032 1647 = 1.946 032 and slope = 0.350 6907 − 0.141 6193 = 0.209 0714. You might like to confirm this by fitting the two regressions separately, using the **subset** function (see below). Note that because we are using a log link, these intercepts are in units of species, not log(species).

But is this 6:6 split the best possible piecewise regression? One way to tackle this question is to try putting the break point in all possible places, and pick the break that produces the best fit; in this case the minimal residual deviance. We do this kind of thing using *loops* and *subscripts*.

First make an ordered vector *aa* containing all the different values of Area using **unique** like this:

```
aa<-sort(unique(Area))
```

The most extreme piecewise regression would use the leftmost two levels of *aa* for one of the regressions, and the remaining data for the right-hand part of the regression. Then the leftmost three levels for the left-hand graph, and so on. In general, we put the break point between aa[i] and aa[i+1] where *i* starts at 2 and goes up to 8 like this:

```
d<-numeric(7)
for (i in 2:8) {
model<-
    glm(Species~(Area<aa[i+1])*log(Area)+(Area>aa[i])*log(Area),poisson)
    d[i-1]<-deviance(model)    }
```

To see the results, we can plot the deviances *d* in a simple indexed series like this:

```
plot(d,type="b")
```

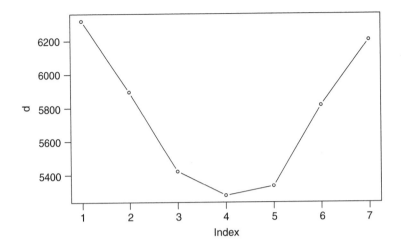

It shows the clear U-shape in the deviances, as the fit gets better and better up to index 4, then progressively worse as the break point is put too far to the right. Index 4 corresponds to $i = 5$ (see above), which means that the best break point is between Area $= 100$ and Area $= 1000$. Our original attempt, dividing the data 6:6 was quite close to the best, but a split of 5:7 gives a slightly lower deviance (5276.947 compared with 5332.98 that we got earlier).

We can compare the two pieces of the regression visually like this:

```
model1<-lm(log(Species)~log(Area),subset=(Area<1000))
model2<-lm(log(Species)~log(Area),subset=(Area>100))
plot(log(Area),log(Species))
abline(model1)
abline(model2)
```

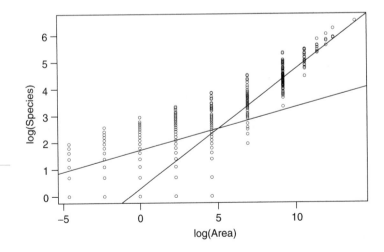

The slope of the relationship over the lowest five areas is significantly shallower than the slope over the largest seven areas.

Further reading

Atkinson, A.C. (1985) *Plots, Transformations, and Regression*. Oxford, Clarendon Press.

Ross, G.J.S. (1990) *Nonlinear Estimation*. New York, Springer-Verlag.

Wetherill, G.B., Duncombe, P. *et al.* (1986) *Regression Analysis with Applications*. London, Chapman & Hall.

23

Non-linear regression

In many cases there simply *is* no transformation that will linearise the relationship between y and x. Then we have several options, including:

- non-linear regression (**nls**)

- non-linear mixed effects models (**nlme**)

- polynomial regression (**lm** using x plus powers of x)

- non-parametric smoothing models (**loess**)

- generalised additive models (**gam**)

- tree models (**tree**)

- non-parametric rank correlation (**cor**)

Here we deal with non-linear parametric regression. The other topics are covered in separate chapters (loess, gam, tree, non-linear mixed effects) or as parts of other chapters (polynomials, rank correlation).

Up to now S-Plus knew exactly what model we wanted to fit *because the model was linear*. So when we said y ~ x + z, it knew that we wanted to estimate the three parameters in the linear model $y = a + bx + cz$. With non-linear regression, S-Plus has no way of knowing which of the many possible models we want to fit to the data—we need to tell it.

Also, we need to give S-Plus a helping hand to get the procedure started, by providing some rough estimates for the parameter values. The novel components of non-linear regression are therefore:

- explicit statement of the equation as part of the model formula

- provision of a list of starting values for the parameter estimates

The data show jaw size as a function of age in deer. The aim is to fit an asymptotic exponential to the data by obtaining non-linear least squares estimates of the parameters a, b and c in

$$y = a - b\,e^{-cx}$$

```
attach(asymptotic)
names(asymptotic)
```

```
[1] "x" "y"
```

We produce a scatterplot to begin with:

```
plot(x,y)
```

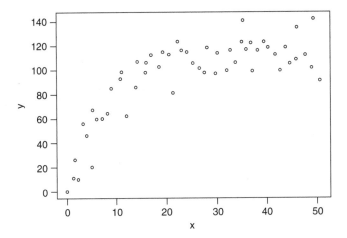

When we come to fit the model, we shall specify the equation explicitly like this:

$$y \sim a\text{-}b\text{*}exp(\text{-}c\text{*}x)$$

First, we need to work out initial estimates for the three parameter values. A good way to start is to ask about the behaviour of the equation at the limits, i.e. work out the values of y when x is 0 and when x is ∞. Putting $x = 0$ we have

$$y = a - b\,e^{-c \times 0} = a - b\,e^0 = a - b \times 1 = a - b$$

so if we can estimate the value of the intercept from the scatterplot, this will give us $a - b$. It is roughly 10, say. Now we find the value of y when x is infinity:

$$y = a - b\,e^{-c \times \infty} = a - b\,e^{-\infty} = a - b \times 0 = a - 0 = a$$

So a is the asymptotic value of y, and the scatterplot suggests this is about 100. This, in turn, means that $b = 100 - 10 = 90$. The last parameter, c, is a bit more tricky. The best plan is to rearrange the equation to get c as a function of the two variables and the other parameters:

$$b\,e^{-cx} = a - y \quad \text{so} \quad e^{-cx} = \frac{a - y}{b}$$

Taking logs gives

$$\ln(e^{-cx}) = \ln\left[\frac{a - y}{b}\right] = -cx \quad \text{so} \quad c = -\frac{1}{x}\ln\left(\frac{a - y}{b}\right)$$

We have already estimated a and b so we need only to pick a point on the line (preferably on the ascending part of the curve) to get a value for x and y. It looks as if $x = 10$, $y = 80$ would be reasonable. So $c \approx -\ln((100 - 80)/90)/10 \approx 0.15$.

These three estimates ($a = 100$, $b = 90$, $c = 0.15$) need to be specified in the **start** function. The simplest way to do this is with a **list**.

The model is fitted to the data using **nls** by combining the model formula and the start values like this:

model<-nls(y~a-b*exp(-c*x),start=list(a=100,b=90,c=0.15))

The model is summarised in the usual way:

```
summary(model)
```

```
Formula: y ~ a - b * exp (- c * x)
```

```
Parameters:
        Value Std. Error  t value
a 115.251000  2.9137200 39.55470
b 118.690000  7.8928400 15.03770
c   0.123517  0.0171061  7.22061
```

```
Residual standard error: 13.2082 on 51 degrees of freedom
```

which gives the parameter estimates (headed Value), their standard errors and an indication of the significance of their differences from zero (t value); all ours are highly significantly different from zero, but more on this below. So the parameterised non-linear model is

$$y = 115.251 - 118.69\,e^{-0.123517x}$$

We draw the fitted values through the scatterplot as before. First we generate a fine-grained sequence of x values to cover the range over which we want the graph to be plotted:

```
xv<-seq(0,50,0.1)
```

Then the curve is drawn using **lines**, with **predict** used to generate the y values to match each x value (see p. 239):

```
lines(xv,predict(model,list(x=xv)))
```

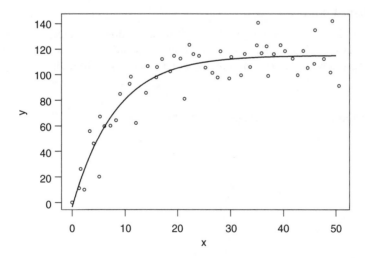

The fit is very good, but the question of model simplification now arises. All of our parameters were significantly different from zero, but that does not necessarily mean we need to keep all three parameters in a minimal adequate model, because the parameters might not be significantly different from each other. It is possible, for instance, that the graph passes through the origin, so the two-parameter model

$$y = a(1 - e^{-cx})$$

might fit the data just as well, in which case parsimony requires that we accept the simpler model. We fit the simpler model like this:

model2<-nls(y~a*(1-exp(-c*x)),start=list(a=100,c=0.15))

summary(model2)

```
Parameters:
         Value   Std. Error   t value
a  115.58100    2.8437400    40.6441
c    0.11881    0.0123307     9.6353

Residual standard error: 13.104 on 52 degrees of freedom
```

The residual standard error (13.104) is actually smaller than with the three-parameter model (13.2082), so the model simplification was clearly justified. We check this by comparing the two models using **anova** in the normal way:

anova(model,model2)

```
Analysis of Variance Table

Response: y

                  Terms Resid. Df      RSS  Test  Df  Sum of Sq      F Value       Pr(F)
1    a - b * exp (- c * x)     51 8897.300     1
2    a * (1 - exp (- c * x))   52 8929.143     2  -1  -31.84329  0.1825281  0.6710079
```

The difference between the two models is very small ($p = 0.671$) so we conclude that the model simplification was justified, and go with the two-parameter model:

$$y = 115.581(1 - e^{-0.11881x})$$

Models for ∩-shaped data

Several of the humped models we might want to fit to data cannot be linearised, hence linear models cannot be used to estimate their parameter values and standard errors. Here are two examples. First is the Ricker curve (named after the famous Canadian fisheries biologist):

$$y = ax\,e^{-bx}$$

which is a two-parameter model. It looks like this for $a = 1$ and $b = 0.08$.

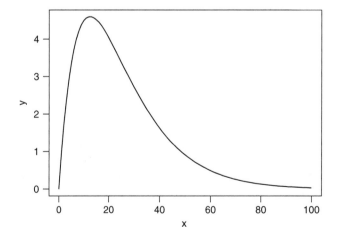

This function is much used in fisheries research to describe the relationship between stock density (x) and recruitment (y). It cannot be linearised because the equation contains both x and e^x on the right-hand side (it is transcendental).

Another useful non-linear relationship is the three-parameter model

$$y = \frac{ax}{(1 + cx)^b}$$

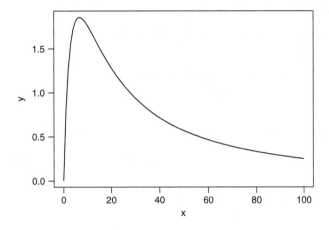

This function is widely used to describe the dynamics of competition in single-species populations, where y is population at time $t + 1$ and x is population at time t. This graph has $a = 1$, $b = 2.5$ and $c = 0.1$.

In practice we would have a scatterplot of y against x and we would want to estimate the parameter values and their standard errors:

```
attach(fishery)
names(fishery)
```

```
[1] "xv"  "yv"
```

```
plot(xv,yx)
```

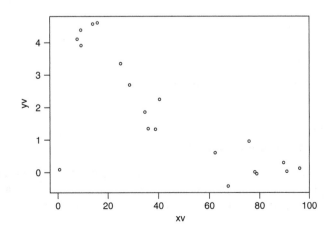

Now we write the equation in the model formula and provide a list of guesstimates for the starting values of the parameters. Suppose we decide to fit a Ricker curve. We need to get rough starting values for the two parameters, a and b. The graph rose to about $y = 4$ over the first 4 units of x, so a guess of $a \approx 1$ is reasonable ($e^{-bx} \approx 1$ when x is small). The second parameter measures the decay of the curve from its peak towards the right-hand side. This declined from 4 to 1 over the range of x from 20 to 60, so an estimate of $(\log 4)/40 = 0.035$ should work.

```
model<-nls(yv~a*xv*exp(-b*xv),start=list(a=1,b=0.035))
summary(model)
```

```
Formula: yv ~ a * xv * exp (- b * xv)

Parameters:
        Value   Std. Error   t value
a   0.9710130   0.06323430   15.3558
b   0.0803303   0.00339969   23.6287

Residual standard error: 0.368357 on 18 degrees of freedom
```

We can draw the smooth curve through the data, using **lines** with **predict** from the non-linear model:

```
lines(1:100,predict(model,list(xv=1:100)))
```

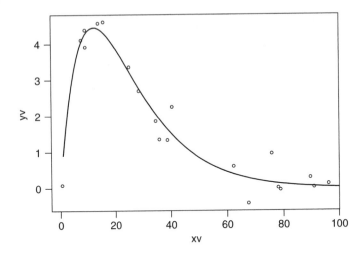

The best-fit equation for this model for these data is therefore

$$y = 0.97x\, e^{-0.08x}$$

Non-linear regression with grouped data

It is often necessary to fit a family of curves, rather than one curve on its own. The function for this is **nlsList**. This example comes from experimental work on reaction rate (y) as a function of enzyme concentration (x) in different strains of mice (a categorical variable with five levels). The data look like this:

```
attach(reaction)
names(reaction)
```

```
[1] "strain" "enzyme" "rate"
```

```
plot(enzyme,rate)
```

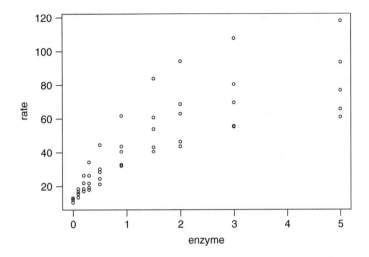

There is a lot of scatter, but the process is clearly asymptotic. We shall fit the model

$$y = c + \frac{ax}{1 + bx}$$

An initial guess of $c = 10$ seems sensible. The asymptote $c + a/b$ is roughly 90, so we need to guess at a (the initial slope). Over the first interval of x (0 to 1), y increased by about 20, so let's put $a = 20$. This means that $b = 0.25$ because a/b is 80. Now we can fit the model and summarise the parameter estimates:

```
model<- nls(rate~c+a*enzyme/(1+b*enzyme),start=c(a=20,b=0.25,c=10))
summary(model)
```

```
Formula: rate ~ c + (a * enzyme)/(1 + b * enzyme)
```

```
Parameters:
      Value Std. Error  t value
a   51.4881  15.131700  3.40267
b    0.5020   0.206599  2.42983
c   10.9516   3.988260  2.74595
```

```
Residual standard error: 13.1649 on 47 degrees of freedom
```

So the parameterised model (ignoring the strains of mice) is

$$y = 10.95 + \frac{51.49x}{1 + 0.502x}$$

We can inspect the fit of the model to the data using **predict**:

```
xv<-seq(0,5,0.01)
yv<-predict(model,list(enzyme=xv))
lines(xy,yv)
```

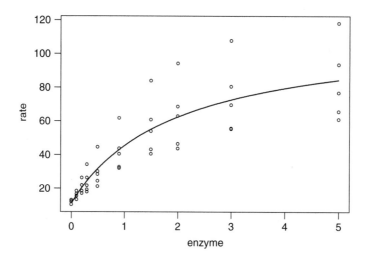

The trend looks okay but there is a great deal of scatter. Perhaps the fit would be improved if we took account of the differences between the different strains of mice using **nlsList**. It is used like **nls** but there are two new features:

- we use the conditioning operator | to define the factor on which the model is to be fitted

- we supply the name of the data frame containing the factor levels

```
model2<- nlsList(rate~
c+a*enzyme/(1+b*enzyme)|strain,data=reaction,start=c(a=20,b=0.25,c=10))
```

```
summary(model2)
```

```
Call:
 Model: rate ~ c + (a * enzyme)/(1 + b * enzyme) |
strain
 Data: reaction

Coefficients:
   a
       Value Std. Error      t value
A 51.79702    4.093751   12.652704
B 26.05968    3.063577    8.506294
C 51.87215    5.087172   10.196657
D 94.46586    5.814211   16.247408
E 37.51030    4.840817    7.748754
   b
         Value    Std. Error     t value
A 0.4238518    0.04971589   8.525479
```

```
B 0.2802579    0.05761729   4.864129
C 0.5585563    0.07413151   7.534668
D 0.6560854    0.05207563  12.598704
E 0.5253571    0.09354990   5.615795
     c
      Value Std. Error    t value
A 11.46506    1.194152   9.601008
B 11.73294    1.120460  10.471544
C 10.53145    1.254956   8.391886
D 10.40912    1.294460   8.041290
E 10.30131    1.240668   8.303032

Residual standard error: 1.81625 on 35 degrees of freedom
```

The t tests compare the parameter estimates with zero, but we are more interested in whether the parameter estimates differ from genotype to genotype. Some idea of significance can be gained by asking which parameter estimates differ by more than 2 standard errors.

Starting with the estimates of c, the reaction rate with no extra enzyme, it is clear that there is no significant difference between the five genotypes (range = 1.4, s.e. = 1.3). There does appear to be significant variation among the b values (range = 0.37, s.e. = 0.08), and substantial variation among the a values (range = 66, s.e. = 5.0 approximately).

Multiple curve drawing

We have estimated a lot of parameters from a fairly meagre data frame: 3 parameters for each of 5 genetic strains, giving 15 in all. We can inspect their values using **coef**:

```
coef(model)
```

```
          a          b         c
A 51.79702  0.4238518  11.46506
B 26.05969  0.2802579  11.73294
C 51.87215  0.5585563  10.53145
D 94.46586  0.6560854  10.40912
E 37.51030  0.5253571  10.30131
```

where the rows are the genotypes and the columns are the parameters of the model. Each parameter is accessible by the use of subscripts. For instance, if we want parameter b for genotype D (which is 0.656 0854) this is in row 4 column 2, and we extract it like this:

```
coef(model)[4,2]
```

```
[1] 0.6560854
```

This allows us to automate the production of fitted lines (one for each strain) to overlay on the scatterplot of rate against enzyme. First, we need to generate a fine-grained set of enzyme values over the range 0 to 5:

```
e<-seq(0,5,0.01)
```

We plot the scattergraph again

```
plot(enzyme,rate)
```

then, in a loop, draw the fitted lines for each genotype like this:

```
for (i in 1:5) {
r<-coef(model2)[i,3]+e*coef(model2)[i,1]/(1+e*coef(model2)[i,2])
lines(e,r,lty=i)}
```

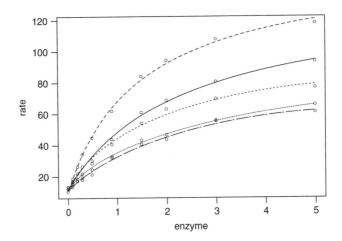

The alternative is to use **predict** to generate the fitted values for plotting. To do this longhand is a rather roundabout procedure, because we need to create a data frame to contain not just the x values for calculating the smooth curve, but also a vector of identifiers for the genetic strain associated with each of the five curves (it is much easier to use **augPred** with a trellis plot, as described on p. 426). First we create a long vector to contain the five sets of x values:

```
es<-rep(seq(0,5,0.01),5)
```

Now we generate 501 repeats of each of the strain identifiers in another long vector:

```
ss<-rep(levels(strain),rep(501,5))
```

Next combine these into a data frame called df in which each vector is to be known by exactly the same name as it was known in the model fit (enzyme and strain respectively):

```
df<-data.frame(strain=ss,enzyme=es)
```

Now we use **predict** with the model name and the data frame name as its two arguments. This produces an object called *fit2* which is of length 501:

```
fit2<-predict(model2,df)
```

To plot these fitted values through the scatterplot using different line types for each strain, we need to make a vector of the unique values of strain. This will control the selection of fitted values from the model that are to be plotted on each line:

```
st<-unique(strain)
```

Now we loop round, once for each strain, and draw the **lines** from the object *fit2* produced by **predict**:

```
plot(enzyme,rate)

for (i in 1:5) lines(es[ss==st[i]],fit2[ss==st[i]],lty=i+1)
```

The key is to use the levels of strain, *st*, as subscripts on the long vectors (*es* and *ss*) that contain the *x* values and *strain* identifiers that were used in computing the long vector of fitted values (*fit2*). Note that on this second graph we did not use the solid line lty=1 for strain 1, but fitted a variety of dotted lines (line types 2 to 6) instead.

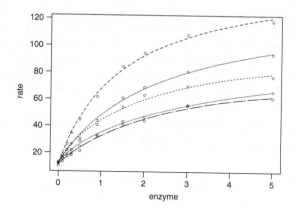

There is a much easier way to plot fitted curves through scatterplots using panel plots with non-linear mixed effects models. But here is a non-parametric smoother fitted to the reaction rates for each strain of mice separately:

```
xyplot(rate ~ enzyme | strain,
        panel = function(x, y) {
            panel.grid(h = 2)
            panel.xyplot(x, y,pch=16,col=1)
            panel.loess(x, y, span = 1)
    })
```

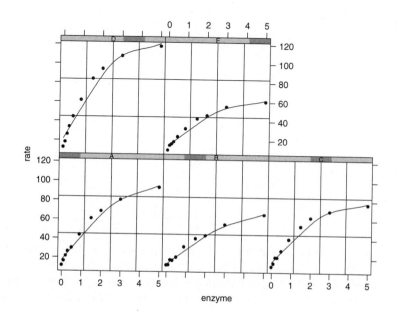

Note that the panels are ranked from lower left to upper right, in alphabetical order of the factor levels (A to E in this case).

Non-linear mixed effects models

Commonly we have repeated measures on growing individuals, but we wish to use these repeated measures to estimate the parameters of a more sophisticated growth model (e.g. a model that asymptotes at some maximum size). The parameters of the growth model are often functions of experimental treatments (fixed effects) or of blocks (random effects). The models are typically based on sound theory of the mechanisms involved, and the parameters generally have a clear physical meaning.

There are two important kinds of non-linear mixed effects models that you might want to fit:

- designed experiments, where a continuous explanatory variable is a *fixed* effect that has been allocated at random to individuals, and each individual is measured only once at the end of the experiment

- growth experiments, where a continuous explanatory variable is a *random* effect like time, and where there are repeated measurements on the same individuals

Rather than estimate all the parameters separately for every individual (as we could do if we chose to use **nlsList**, see p. ●●●), we might prefer to economise on parameters, and to estimate the variance in the parameter values using mixed effects techniques. The background to mixed models is presented in Chapter 35.

Fixed effects as the explanatory variable

Plant height was measured as a function of soil phosphorus concentration for five different genotypes:

```
attach(phosphorus)
names(phosphorus)

[1] "Genotype" "P"        "Height"
```

We intend to fit a Michaelis–Menten function separately for each genotype:

$$y = c + \frac{ax}{1 + bx}$$

This is a non-linear mixed effects model with Genotype as the grouping variable. The bulk of the function **nlme** is similar to the straightforward non-linear model **nls** (see p. 414). The difference is that we need to specify three items:

- fixed effects

- random effects

- grouping

The parameters of the model are assumed to be fixed effects and random effects, and if the list of random effects is omitted, the random effects are assumed to be identical to the fixed effects. The form of the expression is a formula (parameters $a + b + c$ in this case) then ~ (tilde) then another expression. A 1 on the right-hand side of the formula (as here) indicates a single fixed effect for the corresponding parameters. As with **nls** models, the name of the data frame is provided along with estimates of the starting values for each of the parameters (see p. 414).

```
mixed <- nlme(Height~c+a*P/(1+b*P),
fixed=a+b+c~1,
random=a+b+c~1 | Genotype,
data=phosphorus,
start=c(a=20,b=0.25,c=10))

summary(mixed)

Nonlinear mixed-effects model fit by maximum likelihood
  Model: Height ~ C + (a * P)/(1 + b * P)
 Data: phosphorus
        AIC        BIC     logLik
  253.4741   272.5943   -116.737

Random effects:
 Formula: list (a ~ 1, b ~ 1, c ~ 1)
 Level: Genotype
 Structure: General positive-definite
            StdDev     Corr
      a 22.9141004 a          b
      b  0.1132708  0.876
      c  0.4244736 -0.534 -0.876
Residual  1.7105836

Fixed effects: a + b + c ~ 1
      Value Std.Error DF  t-value  p-value
a  51.59883  10.74086 43   4.80398  <.0001
b   0.47666   0.05880 43   8.10647  <.0001
c  10.98535   0.55667 43  19.73414  <.0001
 Correlation:
        a       b
b   0.843
c  -0.313  -0.544

Standardized Within-Group Residuals:
       Min         Q1        Med        Q3        Max
 -1.792171 -0.6578614 0.05630396 0.7429433 2.026849

Number of Observations: 50
Number of Groups: 5
```

To see the estimates of the parameter values, we inspect the coefficients:

```
mixed$coefficients

$fixed:
        a         b         c
51.59883 0.4766555 10.98535

$random:
$random$Genotype:
```

```
            a              b              c
A    1.598693   -0.03410015    0.2535983
B  -23.591969   -0.15285712    0.5664573
C   -1.951207    0.04280621   -0.3173121
D   41.444787    0.16735922   -0.3306736
E  -17.500304   -0.02320816   -0.1720698
```

Note that the fixed effects are the overall mean values for a, b and c while the random effects are *deviations* from the fixed means for each of the five genotypes. The random effects have means of zero. Thus, Genotype A has a bigger than average value of parameter $c(+0.2536)$ while Genotype D has a smaller than average value (-0.3307) compared with the mean value of 10.985 (the fixed effect for parameter c).

Plotting the fitted model

The function **augPred** is very useful here in saving us the need to generate the values for the x axes for each Genotype necessary to produce the smooth curves:

    plot(augPred(mixed))

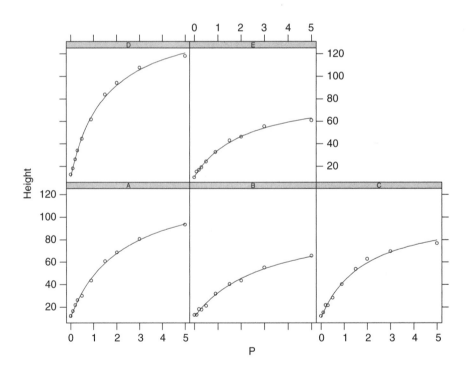

Growth curves are produced for the five genotypes in alphabetical sequence from lower left to upper right.

Random effects as the explanatory variable

In the second kind of application we have several measurements on the same individual (time series), and we want to fit a specified non-linear function to the repeated measures, to describe the process of growth. The model might have fixed effects (e.g different experimental treatments), and we might want to model the correlation structure of the repeated measures (the temporal pseudoreplicates). The example concerns the radial growth of fungal colonies on medium in petri dishes:

```
attach(nonlinear)
names(nonlinear)
```

```
[1] "time"     "dish"     "isolate"  "diam"
```

To use the mixed effects models most effectively, we create a **groupedData** object which shows the time series model diam ~ time and the structure of the blocks (dishes within isolates). There are fixed effects (isolate) and random effects (time and dish):

```
growth<-groupedData(diam~time|dish)
plot(growth)
```

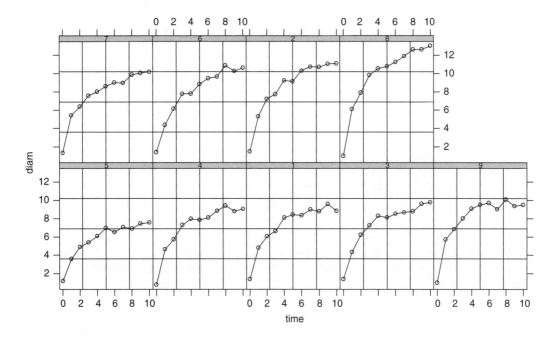

The fitted lines here are obtained by joining the dots using dots (type="b"). We shall fit smooth, parameterised non-linear growth functions in due course. Recall that, in general, the **nlme** model has the following components:

- model formula
- fixed
- random
- correlation
- data
- start

We need to specify the structure of the non-linear model (say the three-parameter Michaelis–Menten) in which the parameters are to be our fixed and random effects, and (optionally) specify the correlation structure of the repeated measures (here we assume they are autoregressive with lag 1 by using corAR1( ) as the argument for correlation). Finally, we need to provide an initial guess at the values of each of the three parameter values at the start:

```
model<-nlme(diam~a+b*time/(1+c*time),
fixed=a+b+c~1,
random=a+b+c~1,
data=growth,
correlation=corAR1( ),
start=c(a=0.5,b=5,c=0.5))

summary(model)
Nonlinear mixed-effects model fit by maximum
  likelihood
  Model: diam ~ a + (b * time) / (1 + c * time)

 Data: growth
       AIC       BIC      logLik
 129.7678 158.3141 -53.88388

Random effects:
 Formula: list (a ~ 1, b ~ 1, c ~ 1)
 Level: dish
 Structure: General positive-definite
           StdDev       Corr
    a   0.1008073   a       b
    b   1.2051205  -0.556
    c   0.1094426  -0.958  0.771
Residual  0.3150598
```

There is rather little variation in parameters a or c (s.d. values of 0.10 and 0.11 respectively), but substantial variation in b (s.d. = 1.205). There is very little autocorrelation structure to the time series:

```
Correlation Structure: AR(1)
 Parameter estimate(s):
        Phi
-0.03211362
```

Phi is the estimated correlation of the errors from reading to reading within the same individual colony (see p. 685); its very low value indicates there is little effect of temporal autocorrelation in this example (see below):

```
Fixed effects: a + b + c ~ 1
        Value  Std.Error  DF   t-value    p-value
 a   1.288065  0.1086142  88  11.85909    <.0001
 b   5.215433  0.4739829  88  11.00342    <.0001
 c   0.498234  0.0450352  88  11.06320    <.0001

Number of Observations: 99
Number of Groups: 9
```

So the maximum likelihood equation is this:

$$diam = 1.288\,065 + \frac{5.215\,433 \times time}{1 + 0.498\,234 \times time}$$

for the data set as a whole. We can use **anova** to compare this model with model2 that contains no temporal autocorrelation (you can create model2 as an exercise):

anova(model,model2)

	Model	df	AIC	BIC	logLik	Test	L.Ratio	p-value
model1	1	11	129.7677	158.3141	-53.88387			
model2	2	10	127.7965	153.7477	-53.89827	1 vs 2	0.0288028	0.8652

There is clearly no justification for including a correlation structure in this case; the simpler model2 with its 10 parameters is preferred. If we had not used a mixed effects model, then the non-linear fits would have used up a whopping 27 degrees of freedom (3 for each of the 9 dishes). To see the different parameters estimated for each of the dishes, we could inspect the coefficients:

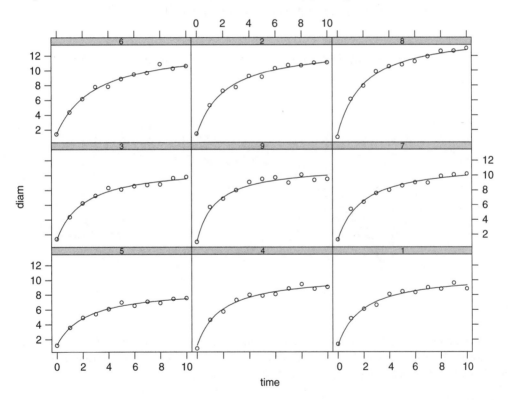

coef(model)

```
5 1.288606 3.349391 0.4394615
4 1.235797 5.074612 0.5373240
1 1.252696 5.009867 0.5212936
3 1.285710 4.843153 0.4885687
9 1.112243 7.166793 0.7056759
```

```
7  1.272391  5.362449  0.5159102
6  1.434626  4.057521  0.3399603
2  1.347895  5.441997  0.4555013
8  1.362622  6.633112  0.4804087
```

The dishes (column 1) are ranked from smallest to largest on the basis of their final diameters. It is extremely straightforward to plot the model through the data for each of the 9 dishes separately using **augPred** to generate all the necessary *x* values and factor levels automatically:

```
plot(augPred(model))
```

If we want all the plots in a single frame, we can do it like this. Begin by generating values for the explanatory variables to be used in **predict**: xv gets the values for time (0 to 10) and dl gets the labels for dish (1 to 9 each repeated 101 times):

```
xv<-seq(0,10,0.1)
xv<-c(xv,xv,xv,xv,xv,xv,xv,xv,xv)
dl<-rep(1:9,rep(101,9))
```

And then plot the nine growth curves through the scatterplot:

```
plot(time,diam)
```

```
for (i in 1:9) lines(xv[dl==i],predict(model,data.frame(dish=dl,time=xv))[dl==i])
```

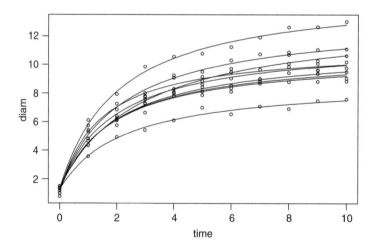

The overall fit is excellent. Our next task would be model simplification. Do we need separate graphs for each of the nine dishes? Would a three-level factor on isolates describe the data equally well? Are all three isolates significantly different from one another? And so on.

Further reading

Davidian, M. and Giltinan, D.M. (1995) *Nonlinear Models for Repeated Measurement Data*. London, Chapman & Hall.

Diggle, P.J., Liang, K.-Y. *et al.* (1994) *Analysis of Longitudinal Data*. Oxford, Clarendon Press.

McCulloch, C.E. and Searle, S.R. (2001) *Generalized, Linear and Mixed Models*. New York, John Wiley.

Pinheiro, J.C. and Bates, D.M. (2000) *Mixed-effects Models in S and S-Plus*. New York, Springer-Verlag.

Ross, G.J.S. (1990) *Nonlinear Estimation*. New York, Springer-Verlag.

24

Multiple regression

When there is more than one continuous explanatory variable, a great many choices need to be made. Among other things, we need to decide on these four items to include in the model:

- the number of explanatory variables
- the identity of the explanatory variables
- the number and identity of interaction terms
- the number, kind and identity of non-linear terms

It is likely that many of the explanatory variables are correlated with each other, and so *the order in which variables are deleted from the model* will influence the explanatory power attributed to them.

The thing to remember about multiple regression is that, in principle, there is no end to it. The number of combinations of interaction terms and non-linear terms is endless. There are some simple rules (like parsimony) and some automated functions (like **step**) to help. But, in principle, you could spend a very great deal of time in modelling a single data frame. There are no hard and fast rules about the best way to proceed, but we shall typically use simplification of a complex model by stepwise deletion: non-significant terms are left out, significant terms are added back (Chapter 25).

At the data inspection stage, there are many more kinds of plots we could do:

- plot the response against each of the explanatory variables separately
- plot all the variables against one another (e.g. **pairs**)
- plot the response against pairs of explanatory variables in 3D plots
- plot the response against one explanatory variable for different combinations of other explanatory variables (e.g. conditioning plots, **coplot**)

At the modelling stage, we need to choose between:

- multiple regression
- mixed effects models
- non parametric surface-fitting (local regression or loess)
- additive models (with perhaps a mix of parametric and non-parametric terms)
- tree models
- multivariate techniques

At the model-checking stage, we need to be particularly concerned with the extent of correlations between the explanatory variables.

We shall begin with an extremely simple example with just two explanatory variables. The response variable is the volume of utilisable timber in harvested trunks of different lengths and diameters:

```
attach(timber)
names(timber)

[1]  "volume"  "girth"  "height"
```

We begin by comparing different kinds of plot. It is a good idea to start by plotting the response against each of the explanatory variables in turn, and by looking at the extent to which the explanatory variables are correlated with each other. The multipanel **pairs** function is excellent for this:

```
pairs(timber)
```

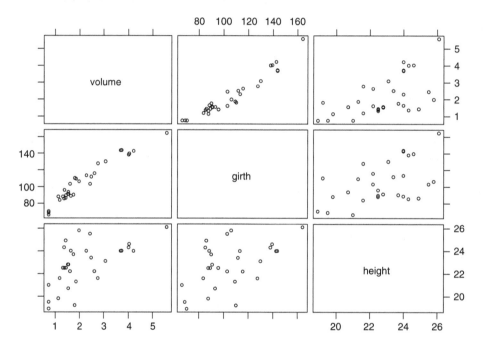

It is clear that girth is a much better predictor of the volume of usable timber than is height (compare the middle and right-hand panels in the top row). This is what we might have expected, because it is very easy to imagine two tree trunks of exactly the same length that differ enormously in the volume of timber they continue. The long thin one might contain no useful timber at all, but you could build a whole house out of the stout one. The bottom right quartet of panels show there is a positive correlation between the two explanatory variables, but trunks of the same girth appear to differ more in height than trunks of the same height differ in girth.

Although experts always advise against it, you might well be tempted to look at a **3D scatterplot** of the data. The reason experts advise against it is a good one; they argue that it is hard to interpret the data correctly from a perspective plot, and the whole idea of plotting is to aid interpretation. The problem with 3D scatterplots is that it is impossible to know where a point is located, because you don't know whether it is in the foreground or the background. Plots showing 3D *surfaces* are another matter altogether; they can be very useful in interpreting the models that emerge from multiple regression analysis. For what it's worth, here is the 3D scatterplot of the timber data using the function **cloud**:

cloud(volume~girth*height)

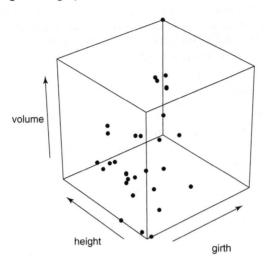

Perhaps the most useful kinds of plot for this purpose are **conditioning plots**. These allow you to see whether the response depends on one explanatory variable in different ways at different levels of other explanatory variables. It takes two dimensional slices through a potentially multidimensional volume of parameter space, and these are often much more informative than the unconditioned scatterplots produced by **pairs**. Here is volume against girth, conditioning on height. Note the use of a model formula in the plotting directive:

coplot(volume~girth|height)

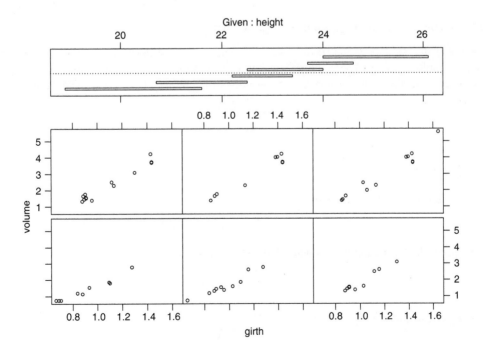

Much of the scatter that was such a feature of the **pairs** plot has gone. Within a height class, the relationship between volume and girth is relatively close. What seems to be happening is that the *slope* of the relationship depends upon the height, getting steeper for taller trees. This is a very informative plot. It tells us that both girth and height are likely to be required in the minimal model and that there may be an interaction between the two explanatory variables.

The multiple regression model

The assumptions are the same as with simple linear regression. The explanatory variables are assumed to be measured without error, the errors are normally distributed, the errors are confined to the response variable, and the variance is constant. The model for a multiple regression with two explanatory variables (x_1 and x_2) looks like this:

$$y_i = \beta_0 + \beta_1 x_{1,i} + \beta_2 x_{2,i} + \varepsilon_i$$

The ith data point, y_i, is determined by the levels of the two continuous explanatory variables $x_{1,i}$ and $x_{2,i}$ by the model's three parameters (the intercept β_0 and the two slopes β_1 and β_2), and by the residual ε_i of point i from the fitted surface. More generally, the model is presented like this:

$$y_i = \sum \beta_i x_i + \varepsilon_i$$

where the summation term is called the **linear predictor**, and can involve many explanatory variables, non-linear terms and interactions. The question that immediately arises is, How are the values of the parameters determined? In order to see this, we return to the simple linear regression that we investigated earlier, and recast that example in terms of matrix algebra. Once the matrix notation is reasonably clear, we can then extend it to deal with an arbitrarily large number of explanatory variables.

Working through a linear regression in matrix form

The best way to learn how matrix algebra can generalise our ideas about regression is to work through a simple example where the data are already familiar. The example involved weight gain of caterpillars fed diets with different concentrations of tannin (Chapter 14). The famous five and sample size were

$$\sum x \quad \sum x^2 \quad \sum y \quad \sum y^2 \quad \sum xy \quad n$$
$$36 \quad\quad 204 \quad\quad 62 \quad\quad 536 \quad\quad 175 \quad\quad 9$$

Here are the three steps involved:

1. Display the model in matrix form:

$$\mathbf{Y} = \mathbf{X}\boldsymbol{\beta} + \boldsymbol{\varepsilon}$$

2. Determine the least squares estimate of $\boldsymbol{\beta}$, which we call **b**:

$$\mathbf{b} = (\mathbf{X}'\mathbf{X})^{-1}\mathbf{X}'\mathbf{Y}$$

3. Carry out the analysis of variance:

$$\mathbf{b}'\mathbf{X}'\mathbf{Y}'$$

We look at each of these in turn. The response vector **Y**, the unit vector **1** and the error vector ε are simple $n \times 1$ column vectors, **X** is an $n \times 2$ matrix and **b** is a 2×1 vector of coefficients:

$$\mathbf{Y} = \begin{bmatrix} 12 \\ 10 \\ 8 \\ 11 \\ 6 \\ 7 \\ 2 \\ 3 \\ 3 \end{bmatrix} \quad \mathbf{X} = \begin{bmatrix} 1 & 0 \\ 1 & 1 \\ 1 & 2 \\ 1 & 3 \\ 1 & 4 \\ 1 & 5 \\ 1 & 6 \\ 1 & 7 \\ 1 & 8 \end{bmatrix} \quad \varepsilon = \begin{bmatrix} \varepsilon_1 \\ \varepsilon_2 \\ \varepsilon_3 \\ \varepsilon_4 \\ \varepsilon_5 \\ \varepsilon_6 \\ \varepsilon_7 \\ \varepsilon_8 \\ \varepsilon_9 \end{bmatrix} \quad \mathbf{1} = \begin{bmatrix} 1 \\ 1 \\ 1 \\ 1 \\ 1 \\ 1 \\ 1 \\ 1 \\ 1 \end{bmatrix} \quad \mathbf{b} = \begin{bmatrix} b_0 \\ b_1 \end{bmatrix}$$

The vectors **y** and **1** are entered into the computer like this:

```
y<-c(12,10,8,11,6,7,2,3,3)
one<-rep(1,9)
```

The sample size is given by $\mathbf{1'1}$ (transpose of **1** times **1**):

```
t(one) %*% one

          [,1]
[1,]        9
```

The only hard part concerns the matrix of the explanatory variables. This has one more column than the number of explanatory variables. The extra column is a column of 1s on the left-hand side of the matrix. Because the x values are evenly spaced, we can generate them rather than type them in (use lower case x):

```
x<-0:8
x

[1]  0  1  2  3  4  5  6  7  8
```

We manufacture the matrix x by using **cbind** to tack a column of 1s in front of the vector of x values. Note that the single number 1 is *coerced* into length n to match **X** (use upper case X):

```
X<-cbind(1,x)

X

          x
[1,]  1   0
[2,]  1   1
[3,]  1   2
[4,]  1   3
[5,]  1   4
[6,]  1   5
[7,]  1   6
[8,]  1   7
[9,]  1   8
```

Thus, all of the matrices are of length $n = 9$, except for $\boldsymbol{\beta}$ which is length $k + 1$, where $k =$ number of explanatory variables (1 in this example).

The transpose of a matrix (denoted by a prime) is the matrix obtained by writing the rows as columns in the order in which they occur, so the columns all become rows. A 9×2 matrix **X** transposes to a 2×9 matrix **X**$'$ (called Xp for 'x prime') like this. The transpose function is **t**:

```
Xp <- t(X)

Xp
      [,1]  [,2]  [,3]  [,4]  [,5]  [,6]  [,7]  [,8]  [,9]
        1     1     1     1     1     1     1     1     1
  x     0     1     2     3     4     5     6     7     8
```

This is useful for regression, because

$$\sum y^2 = y_1^2 + y_2^2 + \cdots + y_n^2 = \mathbf{Y'Y}$$

```
t(y) %*% y

      [,1]
[1,]   536
```

$$\sum y = n\overline{y} = y_1 + y_2 + \cdots + y_n = \mathbf{1'Y}$$

```
t(one) %*% y

      [,1]
[1,]    62
```

$$\left(\sum y \right)^2 = \mathbf{Y'11'Y}$$

```
t(y) %*% one %*% t(one) %*% y

       [,1]
[1,]   3844
```

For the matrix of explanatory variables, we see that **X**$'$**X** gives a 2×2 matrix containing n, $\sum x$ and $\sum x^2$. The numerical values are easy to find using matrix multiplication %*%:

```
Xp %*% X

             x
       9    36
  x   36   204
```

Note that **X**$'$**X** (a 2×2 matrix) is completely different from **XX**$'$ (a 9×9 matrix). The matrix **X**$'$**Y** gives a 2×1 matrix containing $\sum y$, and the sum of products $\sum xy$:

```
Xp %*% y

      [,1]
        62
  x    175
```

Using the beautiful symmetry of the Normal equations

$$b_0 n + b_1 \sum x = \sum y$$

$$b_0 \sum x + b_1 \sum x^2 = \sum xy$$

we can now write the regression directly in matrix form as

$$\mathbf{X'Xb = X'Y}$$

because we already have the necessary matrices to form the left- hand right-hand sides. To find the least squares parameter values **b**, we need to divide both sides by $\mathbf{X'X}$. This involves calculating the **inverse** of the $\mathbf{X'X}$ matrix. The inverse exists only when the matrix is square and when its determinant is non-singular. The inverse contains $-\bar{x}$ and $\sum x^2$ as its terms, with SSX as the denominator:

$$(\mathbf{X'X})^{-1} = \begin{bmatrix} \dfrac{\sum x^2}{n \sum (x-\bar{x})^2} & \dfrac{-\bar{x}}{\sum (x-\bar{x})^2} \\ \dfrac{-\bar{x}}{\sum (x-\bar{x})^2} & \dfrac{1}{\sum (x-\bar{x})^2} \end{bmatrix}$$

When every element of a matrix has a common factor, it can be taken outside the matrix. Here the term $1/(n \times SSX)$ can be taken outside to give

$$(\mathbf{X'X})^{-1} = \dfrac{1}{n \sum (x-\bar{x})^2} \begin{bmatrix} \sum x^2 & -\sum x \\ -\sum x & n \end{bmatrix}$$

Computing the numerical value of this is easy using the matrix function **ginverse**:

```
XM<-Xp %*% X
ginverse(XM)
```

```
                  [,1]              [,2]
[1,]    0.37777778    -0.06666667
[2,]   -0.06666667     0.01666667
```

Now we can solve the Normal equations:

$$(\mathbf{X'X})^{-1}(\mathbf{X'X})\mathbf{b} = (\mathbf{X'X})^{-1}\mathbf{X'Y}$$

using the fact that $(\mathbf{X'X})^{-1}(\mathbf{X'X}) = \mathbf{I}$ to obtain the important general result:

$$\mathbf{b = (X'X)^{-1}(X'Y)}$$

In our example, we have $(\mathbf{X'X})^{-1}$ as

```
                  [,1]              [,2]
[1,]    0.37777778    -0.06666667
[2,]   -0.06666667     0.01666667
```

and $\mathbf{X'Y}$ as

```
         [,1]
          62
  x      175
```

so **b** is found by the three-matrix product

```
b<-ginverse(XM) %*% Xp %*% y
b
           [,1]
[1,]    11.755556
[2,]    -1.216667
```

which you will recognise from our earlier hand calculations as the intercept and slope respectively.

The **Anova computations** are as follows. The correction factor is, $CF = \mathbf{Y'11'Y}/n$:

```
CF<-t(y) %*% one %*% t(one) %*% y / 9
CF
           [,1]
[1,]    427.1111
```

SST is $\mathbf{Y'Y}-CF$:

```
t(y) %*% y - CF
           [,1]
[1,]    108.8889
```

SSR is $\mathbf{b'X'Y}-CF$:

```
t(b) %*% t(X) %*% y - CF
           [,1]
[1,]    88.81667
```

and SSE is $\mathbf{Y'Y}-\mathbf{b'X'Y}$:

```
t(y) %*% y - t(b) %*% t(X) %*% y
           [,1]
[1,]    20.07222
```

You should check these figures against the hand calculations on p. 233. Obviously, this is not a sensible way to carry out a single linear regression, but it demonstrates how to generalise the calculations for cases that have two or more continuous explanatory variables.

Working through a multiple regression by hand

To show how multiple regression works, we shall go through the simple example from the timber data longhand. We have two explanatory variables (girth and height) and the response variable is the volume of usable timber from logs with this girth and height. Here is the data set:

```
timber

      volume   girth    height
  1   0.7458   66.23     21.0
  2   0.7458   68.62     19.5
  3   0.7386   70.22     18.9
  4   1.1875   83.79     21.6
  5   1.3613   85.38     24.3
```

```
 6   1.4265    86.18      24.9
 7   1.1296    87.78      19.8
 8   1.3179    87.78      22.5
 9   1.6365    88.57      24.0
10   1.4410    89.37      22.5
11   1.7524    90.17      23.7
12   1.5206    90.97      22.8
13   1.5496    90.97      22.8
14   1.5424    93.36      20.7
15   1.3831    95.76      22.5
16   1.6075   102.94      22.2
17   2.4475   102.94      25.5
18   1.9841   106.13      25.8
19   1.8610   109.32      21.3
20   1.8030   110.12      19.2
21   2.4982   111.71      23.4
22   2.2954   113.31      24.0
23   2.6285   115.70      22.2
24   2.7734   127.67      21.6
25   3.0847   130.07      23.1
26   4.0116   138.05      24.3
27   4.0333   139.64      24.6
28   4.2216   142.84      24.0
29   3.7292   143.63      24.0
30   3.6930   143.63      24.0
31   5.5757   164.38      26.1
```

Note that in the data frame, girth is measured in centimetres and height in metres. To keep things simple, we convert girth to metres as well (volume is in cubic metres already):

girth<-girth/100

The response variable (volume) is renamed y (just to save on typing):

y<-volume

The vector of explanatory variables $\mathbf{X}$ is made by **cbind** like this:

X<-cbind(1,girth,height)

and its transpose $\mathbf{X'}$ is entered as Xp:

Xp<-t(X)

We obtain the sums of x and the sums of squares of x from $\mathbf{X'X}$:

Xp %*% X

```
                          girth        height
          31.0000      32.77230      706.8000
 girth    32.7723      36.52706      754.6654
height   706.8000     754.66542    16224.6600
```

This reads as follows. Top left is $n = 31$, the number of trees. The second row of the first column shows the sum of the girths $= 32.7723$, and the third row is the sum of the heights $= 706.8$. The second row of column 2 shows the sum of the squares of the girths $= 36.527\,06$, and the third row the sum of the products girth $\times$ height $= 754.665\,42$ (the symmetry has this figure in the second row of the third column as well). The last figure in the bottom right is the sum of the squares of the heights $= 16\,224.66$.

We get the sums of products from $\mathbf{X'Y}$

```
Xp %*% y

                  [,1]
             67.72630
    girth     80.24607
    height  1584.99573
```

The top number is the sum of the timber volumes $\sum y = 67.7263$. The second number is the sum of the products volume $\times$ girth $= 80.246\,07$ and the last number is the sum of the products volume $\times$ height $= 1584.99\,573$.

Now we need the inverse of $\mathbf{X'X}$:

```
ginverse(Xp%*% X)

              [,1]            [,2]            [,3]
[1,]     4.9519523      0.35943351     -0.23244197
[2,]     0.3594335      0.72786995     -0.04951388
[3,]    -0.2324420     -0.04951388      0.01249064
```

which shows all of the corrected sums of squares of the explanatory variables. For instance, the top left-hand number 4.951 9523 is

$$\frac{\sum g^2 \sum h^2 - \left(\sum gh\right)^2}{c - d}$$

where g stands for girth and h for height, c is $n \sum g^2 \sum h^2 + 2 \sum g \sum h \sum gh$ and d is $n \left(\sum gh^2\right) + \left(\sum g^2\right) \sum h^2 + \left(\sum h\right)^2 \sum g^2$; see Draper and Smith (1981) for details of how to calculate determinants for matrices larger than 2×2.

Finally, we compute the parameter values, $\mathbf{b}$:

```
ginverse(Xp %*% X) %*% Xp %*% y

                [,1]
[1,]     -4.19899732
[2,]      4.27251096
[3,]      0.08188343
```

and compare the vector $\mathbf{b}$ with the parameter estimates obtained by multiple regression:

```
lm(volume~girth+height)

Coefficients:
  (Intercept)       girth        height
    -4.198997    4.272511    0.08188343
```

We are hugely relieved to find they are the same. The first element of **b** is the intercept (-4.199), the second is the slope of the graph of volume against girth (4.27) and the third is the slope of the graph of volume against height (0.082).

To finish, we compute the sums of squares for the Anova table, starting with the correction factor:

```
CF<-t(volume) %*% t(one) %*% one %*% volume / 31
CF
```

```
           [,1]
[1,]   147.963
```

The total sum of squares is

```
sst<-t(volume) %*% volume - CF
sst
```

```
          [,1]
[1,]   42.50408
```

The regression sum of squares is

```
ssr<-t(b) %*% Xp %*% volume - CF
ssr
```

```
          [,1]
[1,]   40.29156
```

The error sum of squares is

```
sse<-t(volume) %*% volume - t(b) %*% Xp %*% volume
sse
```

```
          [,1]
[1,]   2.212518
```

These check out with sums of squares in the Anova produced by the **aov** fit:

```
model<-aov(volume~girth+height)
summary(model)
```

	Df	Sum of Sq	Mean Sq	F Value	Pr(F)
girth	1	39.75477	39.75477	503.1070	0.00000000
height	1	0.53679	0.53679	6.7933	0.01449791
Residuals	28	2.21252	0.07902		

Note that *SSR* (40.291 56) is the sum of the individual regression sums of squares for girth and height (39.754 77 + 0.536 79). A more thorough analysis of these data, employing a more realistic scale of measurement, is described later.

Regression with many explanatory variables

Real multiple regression studies can be immensely time-consuming. The problem is that there are a vast number of combinations of explanatory variables, non-linear terms and interaction terms that we might want to try out. Almost invariably, the number of combinations of explanatory variables will exceed the number of data points by a substantial margin. This, in turn, means that we typically have to select terms to include in the model and leave other terms out without knowing whether or not the omitted ones have important interactions with the included ones. As a good rule of thumb, you do not want to have more than about $n/3$ parameters in your initial model in a multiple regression. This means that if you have 30 data points, you do not want to include more than 10 parameters at any one time. So if you have 6 explanatory variables, each with a quadratic term (6 more parameters) and all 15 pairwise interactions, you are in difficulties (27 parameters and only 30 data points). Our example is only a little easier than this; we have 41 data points in an air pollution study:

```
attach(Pollute)
names(Pollute)
```

```
[1]  "Pollution"  "Temp"  "Industry"  "Population"  "Wind"  "Rain"
[7]  "Wetdays"
```

As ever, we begin with data inspection, in this case using **pairs** to plot all the variables against one another:

```
pairs(Pollute)
```

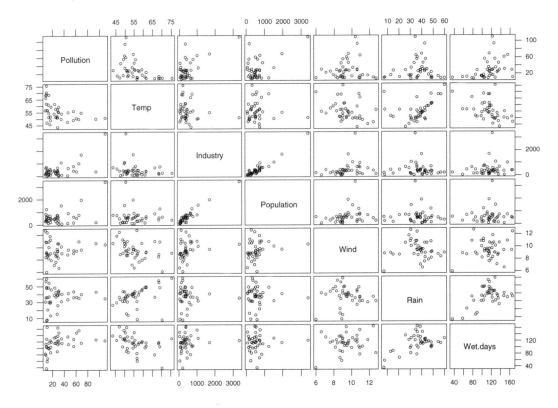

We start by doing the simplest possible multiple regression analysis. We fit only the main effects (no interactions) and no non-linear terms. Call this model1

```
model1<-lm(Pollution~Temp+Industry+Population+Wind+Rain+Wetdays)
```

```
summary(model1)
```

```
Coefficients:
                Estimate Std. Error t value   Pr(>|t|)
(Intercept)    111.59357   47.11131   2.369   0.023673  *
Temp            -1.26510    0.61796  -2.047   0.048427  *
Industry         0.06542    0.01567   4.175   0.000196  ***
Population      -0.03975    0.01506  -2.640   0.012430  *
Wind            -3.17961    1.81397  -1.753   0.088643  .
Rain             0.50507    0.36118   1.398   0.171054
Wetdays         -0.04911    0.16123  -0.305   0.762554
```

This says that Industry is very significant, Temperature and Population are significant, Wind is of marginal significance and both Rain and Wetdays are not significant. We could use **update** to remove terms one at a time by hand, starting with Wetdays (the least significant term in the current model) like this:

model2<-update(model1,~. -Wetdays)

Then compare the original and simplified models using **anova**:

anova(model1,model2)

```
Analysis of Variance Table

    Res.Df Res.Sum Sq Df Sum Sq F value Pr(>F)
1      34      7221.1
2      35      7240.8 -1  -19.7  0.0928 0.7626
```

Life, however, is short. Let's try to automate this procedure, using **step**:

step(model1)

```
Start:  AIC = 226.02
  Pollution ~ Temp + Industry + Population + Wind + Rain + Wetdays
             Df Sum of Sq      RSS     AIC
- Wetdays     1      19.7   7240.8   224.1
<none>                      7221.1   226.0
- Rain        1     415.3   7636.4   226.3
- Wind        1     652.5   7873.7   227.6
- Temp        1     890.1   8111.3   228.8
- Population  1    1480.1   8701.3   231.7
- Industry    1    3702.1  10923.3   241.0
Step:   AIC=  224.13
  Pollution ~ Temp + Industry + Population + Wind + Rain
             Df  Sum of Sq      RSS     AIC
<none>                       7240.8   224.1
- Wind        1      632.9   7873.7   225.6
- Rain        1      777.1   8017.9   226.3
- Population  1     1485.4   8726.2   229.8
- Temp        1     1538.6   8779.4   230.0
- Industry    1     3700.4  10941.3   239.1
Call:
lm(formula = Pollution ~ Temp + Industry + Population + Wind + Rain)
```

```
Coefficients:
(Intercept)      Temp    Industry   Population      Wind      Rain
  100.67844   -1.12664    0.06541    -0.03982   -3.08730   0.41741
```

Remember, **step** works by trying to find the minimum value of Akaike's information criterion (AIC); see Chapter 17. After deleting Wetdays (as we did by hand, above) it thinks that Wind is the next least significant term, but leaving it out causes an *increase* in AIC (from 224.1 to 225.6), so automatic model simplification stops there. Let's see if we agree that Wind should stay in the model. We remove it from model2:

```
model3<-update(model2,~. -Wind)
anova(model2,model3)
```

```
Analysis of Variance Table

Model 1: Pollution ~ Temp + Industry + Population + Wind + Rain
Model 2: Pollution ~ Temp + Industry + Population + Rain
    Res.Df Res.Sum Sq Df Sum Sq F value  Pr(>F)
1       35      7240.8
2       36      7873.7 -1 -632.9  3.0591 0.08905
```

This is quite typical of **step**. It is rather generous at leaving terms in the model. With a p value of 0.089 we would certainly leave Wind out.

```
summary(model3)
```

This says that Rain is the least significant remaining factor; we try removing it:

```
model4<-update(model3,~. -Rain)
anova(model3,model4)
```

```
Analysis of Variance Table

Model 1: Pollution ~ Temp + Industry + Population + Rain
Model 2: Pollution ~ Temp + Industry + Population
    Res.Df Res.Sum Sq Df Sum Sq F value Pr(>F)
1       36      7873.7
2       37      8462.5 -1 -588.8  2.6919 0.1096
```

We would leave Rain out as well.

```
summary(model4)
```

Now Temp looks to be non-significant:

```
model5<-update(model4,~. -Temp)
anova(model4,model5)
```

```
Analysis of Variance Table

Model 1: Pollution ~ Temp + Industry + Population
Model 2: Pollution ~ Industry + Population
    Res.Df Res.Sum Sq Df Sum Sq F value Pr(>F)
1       37      8462.5
2       38      9058.8 -1 -596.4  2.6074 0.1149
```

We don't need Temp either.

summary(model5)

```
Coefficients:
              Estimate Std. Error  t value   Pr(>|t|)
(Intercept)   26.42480    3.83294    6.894   3.42e-008  ***
Industry       0.08296    0.01469    5.648   1.73e-006  ***
Population    -0.05707    0.01428   -3.996   0.000286   ***
```

It certainly looks as if we do need the remaining two terms. But let's remove a significant term just to demonstrate what happens:

model6<-update(model5,~. -Population)
anova(model5,model6)

```
Analysis of Variance Table

Model 1: Pollution ~ Industry + Population
Model 2: Pollution ~ Industry
  Res.Df Res.Sum Sq Df  Sum Sq  F value    Pr(>F)
1     38     9058.8
2     39    12864.9 -1 -3806.1  15.966   0.000286 ***
```

This simplification was not justified, because it caused a highly significant increase in deviance ($p = 0.000\,286$). Would we have got a different result if we had compared all our terms with the maximal model rather than with the current part-simplified model? This is easy to check by deleting the terms from model1 rather than models model2, model3 or model4. For Temp, we have

model7<-update(model1,~. -Temp)

anova(model1,model7)

```
Analysis of Variance Table

Model 1: Pollution ~ Temp + Industry + Population + Wind
+ Rain + Wetdays
Model 2: Pollution ~ Industry + Population + Wind + Rain
+ Wetdays
   Res.Df Res.Sum Sq Df Sum Sq F value   Pr(>F)
1     34     7221.1
2     35     8111.3 -1 -890.1  4.1911   0.04843  *
```

Yes. It does make a difference. If we had tested Temp against the maximal model we would have retained it. We deleted it because it was not significant when removed from model4. So *fitting sequence matters*.

Now we need to address the thorny problem of interactions and non-linear terms. There are 6 explanatory variables and so there are $5 + 4 + 3 + 2 + 1 = 15$ two-way interactions that we might want to fit. Remember that there are only 40 d.f. in total. Likewise we might want to test for curvature in each of the explanatory variables by fitting quadratic terms ($Temp^2$, $Rain^2$, etc.). That's another 6 parameters.

Our rule of thumb is not to try to fit more than $n/3$ parameters at any one time. In our case that is $41/3 = 13.6$. We have 6 main effects and 6 quadratics, so that only leaves room for 1 interaction out of the 15. Which one? Which one indeed! Or we could try testing for curvature first, and then test for interactions with only the significant curvatures in the model. That rules out the possibility of finding interactions between curvatures, of course. There is no answer. What I intend to do is look at the six curvature terms first. If there are none, then I shall go on and look at the interactions. First, calculate the six quadratic terms:

```
t2<-Temp^2
i2<-Industry^2
p2<-Population^2
w2<-Wind^2
r2<-Rain^2
d2<-Wetdays^2
```

then fit them along with the six linear effects

```
model10<-lm(Pollution~Temp+Industry+Population+Wind
          +Rain+Wetdays+t2+i2+p2+w2+r2+d2)
summary(model10)
```

```
Coefficients:
              Estimate  Std. Error   t value  Pr(>|t|)
(Intercept) -6.641e+001  2.234e+002    -0.297   0.76844
Temp         5.814e-001  6.295e+000     0.092   0.92708
Industry     8.123e-002  2.868e-002     2.832   0.00847   **
Population  -7.844e-002  3.573e-002    -2.195   0.03662   *
Wind         3.172e+001  2.067e+001     1.535   0.13606
Rain         1.155e+000  1.636e+000     0.706   0.48575
Wetdays     -1.048e+000  1.049e+000    -0.999   0.32615
t2          -1.297e-002  5.188e-002    -0.250   0.80445
i2          -1.969e-005  1.899e-005    -1.037   0.30862
p2           2.551e-005  2.158e-005     1.182   0.24714
w2          -1.784e+000  1.078e+000    -1.655   0.10912
r2          -9.714e-003  2.538e-002    -0.383   0.70476
d2           4.555e-003  3.996e-003     1.140   0.26398
```

That is pretty conclusive; there is absolutely no evidence for any curvature in any of the explanatory variables. Now for the interactions. As a purely pragmatic approach, I have randomised the 15 two-way interactions and I'm going to fit them five at a time in three different models. If you wanted to, you could write a function to repeatedly randomise the combinations in which the interactions were fitted; I leave this to you as an exercise if you are interested. The thing to remember about multiple regression is that, in principle, there is no end to it. The number of combinations of interaction terms and non-linear terms is essentially endless.

```
interactions<-c("ti","tp","tw","tr","td","ip","iw","ir","id","pw","pr","pd","wr","wd","rd")
```

```
sample(interactions)
```

```
[1] "iw" "pr" "wr" "pw" "tw" "pd" "id" "ip" "rd" "ir"
    "wd" "tp" "td" "tr" "ti"
```

Now add the first five of these interactions to the model without the quadratic terms (model1):

```
model12<-update(model1,~. +
Industry:Wind+Population:Rain+Wind:Rain+Population:Wind+Temp:Wind)
summary(model12)
```

```
Coefficients:
                  Estimate   Std. Error  t value   Pr(>|t|)
(Intercept)      6.575e+002  1.553e+002    4.235   0.000211 ***
Temp            -8.882e+000  2.541e+000   -3.495   0.001546 **
Industry        -2.761e-001  1.488e-001   -1.855   0.073755 .
Population       2.748e-001  1.342e-001    2.048   0.049722 *
Wind            -5.758e+001  1.612e+001   -3.572   0.001260 **
Rain            -3.879e+000  1.461e+000   -2.655   0.012758 *
Wetdays         -1.020e-001  1.401e-001   -0.728   0.472641
Temp.Wind        7.143e-001  2.650e-001    2.696   0.011572 *
Industry.Wind    3.535e-002  1.504e-002    2.351   0.025729 *
Population.Wind -3.335e-002  1.342e-002   -2.486   0.018941 *
Population.Rain  1.104e-004  7.712e-004    0.143   0.887152
Wind.Rain        5.375e-001  1.741e-001    3.088   0.004409 **
```

So there are four significant two-way interactions (interestingly, all four involve Wind). Now for the next five randomly selected interactions:

> model13<-update(model1,~. +Population:Wetdays+Industry:Wetdays+
> Industry:Population+Rain:Wetdays+Industry:Rain)

> summary(model13)

None of those interactions were significant. And the last five:

> model14<-update(model1,~. +Wind:Wetdays+Temp:Population+
> Temp:Wetdays+Temp:Rain+Temp:Industry)
> summary(model14)

```
Coefficients:
                   Estimate   Std. Error  t value   Pr(>|t|)
(Intercept)      167.007869   173.916362    0.960    0.3449
Temp              -2.321804     2.175943   -1.067    0.2948
Industry           0.263254     0.159050    1.655    0.1087
Population        -0.224795     0.125983   -1.784    0.0848 .
Wind              -4.436409     7.413031   -0.598    0.5542
Rain               5.575865     2.140227    2.605    0.0143 *
Wetdays           -2.545766     1.680962   -1.514    0.1407
Temp.Industry     -0.003682     0.002998   -1.228    0.2293
Temp.Population    0.003406     0.002325    1.465    0.1537
Temp.Rain         -0.095239     0.040302   -2.363    0.0250 *
Temp.Wetdays       0.044441     0.025050    1.774    0.0865 .
Wind.Wetdays       0.029151     0.063735    0.457    0.6508
```

So the temperature by rain interaction looks important. Next we make a model containing all the significant interactions we have discovered and see if we can simplify it:

> model15<-update(model1,~. + Industry:Wind+Wind:Rain+Population:Wind
> +Temp:Wind+Temp:Rain)

> summary(model15)

```
Coefficients:
                   Estimate  Std. Error  t value  Pr(>|t|)
(Intercept)       750.20370   195.08940    3.845  0.000608  ***
Temp              -10.02009     2.77107   -3.616  0.001122  **
Industry           -0.29055     0.14930   -1.946  0.061398  .
Population          0.29361     0.13145    2.234  0.033384  *
Wind              -62.12415    15.65758   -3.968  0.000437  ***
Rain               -5.60375     2.94317   -1.904  0.066877  .
Wetdays            -0.13701     0.14886   -0.920  0.364951
Temp.Wind           0.73799     0.24183    3.052  0.004832  **
Temp.Rain           0.01814     0.02684    0.676  0.504426
Industry.Wind       0.03679     0.01508    2.439  0.021078  *
Population.Wind    -0.03483     0.01349   -2.581  0.015186  *
Wind.Rain           0.62656     0.21155    2.962  0.006049  **
```

The temperature by rain interaction does not survive being included with the other three interactions, despite the low p value it showed on its own ($p = 0.025$). I don't think that I need to go on. You get the idea. If you really care about the data, this can be a very time-consuming business.

Common problems arising in multiple regression

- differences in the measurement scales of the explanatory variables, leading to large variation in the sums of squares and hence to an ill-conditioned matrix

- multicollinearity in which there is a near-linear relation between two of the explanatory variables, leading to unstable parameter estimates (ill-conditioning)

- rounding errors during the fitting procedure

- non-independence of groups of measurements

- temporal or spatial correlation among the explanatory variables

- pseudoreplication

- a host of others difficulties

See Wetherill *et al.* (1986) for a detailed discussion of these problems. Also, don't forget the other powerful tools like **tree models** and **neural networks** for dealing with multiple explanatory variables.

Further reading

Belsley, D.A., Kuh,E. *et al.* (1980) *Regression Diagnostics: Identifying Influential Data and Sources of Collinearity.* New York, John Wiley.

Draper, N.R. and Smith,H. (1981) *Applied Regression Analysis.* New York, John Wiley.

Neter, J., Wasserman, W. *et al.* (1985) *Applied Linear Regression Models.* Homewood, Illinois, Irwin.

Weisberg, S. (1985) *Applied Linear Regression.* New York, John Wiley.

Wetherill, G.B., Duncombe, P. *et al.* (1986). *Regression Analysis with Applications.* London, Chapman & Hall.

25

Model simplification

The *principle of parsimony* (Occam's razor) requires that the model should be as simple as possible. This means that the model should not contain any redundant parameters or factor levels. We achieve this by fitting a maximal model then simplifying it by following one or more of these steps:

- remove non-significant interaction terms

- remove non-significant quadratic or other non-linear terms

- remove non-significant explanatory variables

- group together factor levels that do not differ from one another

- in Ancova, set non-significant slopes of continuous explanatory variables to zero

subject, of course, to the caveats that the simplifications make good scientific sense, and do not lead to significant reductions in explanatory power.

Model	Interpretation
Saturated model	One parameter for every data point
	Fit: perfect
	Degrees of freedom: none
	Explanatory power of the model: none
Maximal model	Contains all (p) factors, interactions and covariates that might be of any interest. Many of the model's terms are likely to be insignificant
	Degrees of freedom: $n - p - 1$
	Explanatory power of the model: it depends
Minimal adequate model	A simplified model with $0 \leq p' \leq p$ parameters
	Fit: less than the maximal model, but not significantly so
	Degrees of freedom: $n - p' - 1$
	Explanatory power of the model: $r^2 = SSR/SST$
Null model	Just one parameter, the overall mean $\overline{y}$
	Fit: none; $SSE = SST$ Degrees of freedom: $n - 1$
	Explanatory power of the model: none

The steps involved in model simplification

There are no hard and fast rules, but the procedure laid out below works well in practice. With large numbers of explanatory variables, and many interactions and non-linear terms, the process of model simplification can take a very long time. But this is time well spent because it reduces the risk of overlooking an important aspect of the data. It is important to realise that there is no guaranteed way of finding all the important structures in a complex data frame.

Step	Procedure	Explanation
1	Fit the maximal model	Fit all the factors, interactions and covariates of interest. Note the residual deviance. If you are using Poisson or binomial errors, check for overdispersion and rescale if necessary
2	Begin model simplification	Inspect the parameter estimates using **summary**. Remove the least significant terms first, using **update −**, starting with the highest-order interactions
3	If the deletion causes an insignificant increase in deviance	Leave that term out of the model Inspect the parameter values again Remove the least significant term remaining
4	If the deletion causes a significant increase in deviance	Put the term back in the model using **update +**. These are the statistically significant terms as assessed by deletion from the maximal model
5	Keep removing terms from the model	Repeat steps 3 or 4 until the model contains nothing but significant terms. This is the minimal adequate model If none of the parameters is significant, then the minimal adequate model is the null model

Model simplification in multiple regression

This example involves the body weight of desert rodents as a function of four continuous explanatory variables: the amount of rainfall, the abundance of predators, the cover of perennial vegetation and the amount of seed production.

```
attach(seeds)
names(seeds)
[1] "rodent"  "rain"  "predators"  "cover"  "seed"
```

Preliminary data inspection uses **pairs** to plot each variable against every other:

```
pairs(seeds)
```

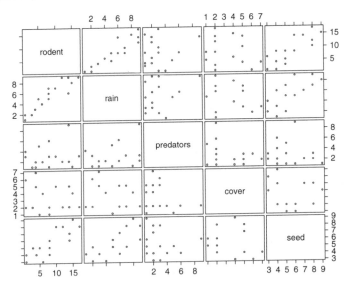

There is a close correlation between rodent weight and rainfall, and a reasonable correlation between weight and seed production but no obvious effect of predators or vegetation cover. Among the explanatory variables, there is a positive correlation between rainfall and seed production (not surprisingly, perhaps, in a desert ecosystem).

There are only 16 data points, so our rule of thumb (p. 442) does not allow us to fit any non-linear or interaction terms if we want to include all four explanatory variables:

model1<-lm(rodent~rain+predators+cover+seed)

To suppress the correlations between parameters and make the output much shorter, use cor=F like this:

summary(model1,cor=F)

```
Coefficients:
                 Value  Std. Error  t value  Pr(>|t|)
  (Intercept)  -4.2977      2.5205  -1.7051    0.1162
         rain   1.3651      0.3752   3.6381    0.0039
    predators  -0.0032      0.3105  -0.0104    0.9919
        cover   0.2333      0.2957   0.7887    0.4469
         seed   0.7796      0.5030   1.5498    0.1495
```

The least significant term is predators (it has the highest p value), so we try deleting that variable first:

model2<-update(model1,~. -predators)

then compare the two models using AIC (see p. 315):

AIC(model1,model2)

```
        df       AIC
model1   6  73.21943
model2   5  71.21959
```

The simpler model2 had the lower AIC so we accept the simplification. Now try deleting vegetation cover (the next least significant term). We could delete it from the maximal model1 or from the simplified model2. Here we do both:

model3<-update(model1,~. -cover)
model4<-update(model2,~. -cover)
AIC(model1,model3,model2,model4)

```
        df       AIC
model1   6  73.21943
model3   5  72.09964
model2   5  71.21959
model4   4  70.16667
```

In both cases the simplification is justified, so we shall leave out both predators and cover. Next we test the deletion of seed production; again we delete from both the maximal model (model1) and the current model (model4):

```
model5<-update(model1,~. -seed)
model6<-update(model4,~. -seed)
AIC(model1,model5,model4,model6)
```

```
         df      AIC
model1   6  73.21943
model5   5  74.37939
model4   4  70.16667
model6   3  72.78143
```

In both cases, AIC goes up, suggesting that this simplification is unjustified, and that seed ought to be added back into the model. We inspect the parameter values of model4:

```
summary(model4)
```

```
Coefficients:
                  Value Std. Error t value Pr(>|t|)
(Intercept)  -3.3756      1.4780  -2.2840   0.0398
       rain   1.2960      0.2818   4.5996   0.0005
       seed   0.8152      0.3910   2.0847   0.0574
```

It is typical of AIC to be rather generous at leaving terms in the model. The t test for the inclusion of seed is non-significant because p is greater than 0.05 (only a little bigger, but bigger nonetheless). My inclination in these circumstances is to leave out seed; after all, we have done lots of tests and we are trying to work at $\alpha = 0.05$. In simplifying complicated regression models, I tend to work at $\alpha = 0.01$. So now we abandon **AIC** and use **anova** instead. Our minimal adequate model has a single explanatory variable and explains 84% of the variation in rodent growth:

```
summary(model6)
```

```
Call: lm(formula = rodent ~ rain)
Coefficients:
                  Value Std. Error t value Pr(>|t|)
(Intercept)  -1.2657      1.1989  -1.0557   0.3090
       rain   1.7471      0.2009   8.6982   0.0000
Residual standard error: 2.085 on 14 degrees of freedom
Multiple R-Squared: 0.8439
```

Note that if you fit seed production to the null model on its own, it appears to be highly significant:

```
summary(lm(rodent~seed))
```

```
Coefficients:
                  Value Std. Error t value Pr(>|t|)
(Intercept)  -4.0939      2.2956  -1.7834   0.0962
       seed   2.1966      0.3912   5.6154   0.0001
Residual standard error: 2.926 on 14 degrees of freedom
Multiple R-Squared: 0.6925
```

This highlights the reason why deletion is so much better than addition in model testing. Seed production is correlated with rainfall, but rainfall is a much better predictor of rodent growth than seed production. When the explanatory variables are correlated *order matters*; an insignificant explanatory

variable will often appear to be significant when it is added to mode
significant variables with which it is correlated.

Deletion of the intercept in simple linear regression

Sometimes you might want to remove the intercept from a regre
graph through the origin (the intercept is parameter 1). You w
doing this, of course, because it is not obviously a sensible th
had a single explanatory variable x, and you wanted to compare

$$y = a + bx$$

$$y = bx$$

Note that deletion of the intercept using ~. -1 does *not* rotate the regression line about the point
$(\bar{x}, \bar{y})$ until it passes through the origin (this is a common misconception). We can demonstrate this
as follows. Take the following data

```
x<-c(0,1,2,3,4,5)
y<-c(2,1,1,5,6,8)
```

and fit a two-parameter linear regression as model1:

```
model1<-lm(y~x)
model1

Coefficients:
 (Intercept)     x
   0.3333333  1.4
```

The intercept is 0.333 33 and the slope is 1.4. We know (from p. 226) that this regression line is
defined as passing through the point $(\bar{x}, \bar{y})$, but we should check:

```
mean(x); mean(y)

[1] 2.5
[1] 3.833333
```

We check that the model goes through the point (2.5,3.8333) using **predict** like this:

```
predict(model1,list(x=2.5))

      1
3.833333
```

So that's all right then. Now we remove the intercept using **update** to fit a single-parameter model
that is forced to pass through the origin:

```
model2<-update(model1,~. -1)
model2

Coefficients:
      x
1.490909
```

the model formula has been altered by update. The slope of model2 is steeper (1.49 with 1.40), but does the line still pass through the point $(\bar{x}, \bar{y})$? We can test this using

predict(model2,list(x=2.5))

```
          1
3.727273
```

No, it doesn't. The single slope parameter is estimated from

$$y_i = \beta x_i + \varepsilon_i$$

in which the least squares estimate of β is b where

$$b = \frac{\sum xy}{\sum x^2}$$

instead of $SSXY/SSX$ when two parameters are estimated from the data (Chapter 14). This graph does not pass through the point $(\bar{x}, \bar{y})$.

Because update(model1,~.-1) causes the regression line to be rotated away from its maximum likelihood position, this will inevitably cause an increase in the residual deviance. If the increase in deviance is significant, as judged by a likelihood ratio test, then the simplification is unwarranted, and the intercept should be added back to the model. Forcing the regression to pass through the origin may also cause problems with non-constancy of variance. Removing the intercept is generally not recommended unless we have confidence in the linearity of the relationship over the whole range of the explanatory variable (i.e. all the way from the origin up to the maximum value of x).

Model simplification in Anova

In general we shall tend to begin the process of model simplification by removing high-order inter-action terms from a maximal model. If removal of the high-order interaction terms shows them to be non-significant, we move on to test the low-order interactions and then main effects. The justification for deletion is made with the current model *at the level in question*. Thus in considering whether or not to retain a given second-order interaction, it is deleted with all other second-order interactions (plus any significant higher-order terms that do not contain the variables of interest) still included in the model. If deletion leads to a significant increase in deviance, the term must retained, and the interaction added back into the model. In considering a given first-order interaction, all other first-order interactions plus any significant higher-order interactions are included at each deletion. And so on.

Main effects which figure in significant interactions should not be deleted. If you do try to delete them there will be no change in deviance and no change in degrees of freedom, because the factor will have been aliased (see p. 332); the deviance and degrees of freedom will simply be transferred to one of the surviving interaction terms. It is a moot point whether block effects should be removed during model simplification. In the unlikely event that block effects were *not* significant (bearing in mind that everything varies, and so insignificant block effects are likely to be the exception rather than the rule), then you should compare your conclusions with and without the block terms in the model. If the conclusions differ, then you need to think very carefully about why, precisely, this has happened. The balance of opinion is that block effects are best left unaltered in the minimal adequate model.

Collapsing factor levels

It often happens that a categorical explanatory variable is significant, but not all of the factor levels are significantly different from one another. We may have a set of *a priori* contrasts that guide model selection. Often, however, we simply want to get rid of factor levels that are redundant. A frequent outcome during Anova that, say, the 'low' and 'medium' levels of a treatment are not significantly different from one another, but both differ from the 'high' level of treatment. Collapsing factor levels involves calculating a new factor that has the same value for 'low' and 'medium' levels, and another level for 'high'. This new two-level factor is then added to the model at the same time as the original three-level factor is removed. The increase in deviance is noted. If the change is not significant, then the simplification is justified and the new two-level factor is retained. If a significant increase in deviance occurs, then the original three-level factor must be restored to the model:

```
attach(Levels)
names(Levels)

[1] "yield" "level"

model<-aov(yield~level)
summary(model)

           Df Sum of Sq  Mean Sq F Value        Pr(F)
    level   2  28.13333 14.06667 13.1875 0.0009349689
Residuals  12  12.80000  1.06667
```

A highly significant effect of level is clear. We might leave it at this, but in the interests of parsimony, it would be sensible to ask whether we really need to retain all three levels of the factor. We should tabulate the treatment means:

```
tapply(yield,level,mean)

  A   B    C
7.2 7.4 10.2
```

The means of levels A and B are very close to one another. Perhaps we can get almost the same explanatory power with a two-level factor (level2) that has level1 for A and for B and level2 for C:

```
level2<-factor(1+(level=="C"))
level2

[1] 1 1 1 1 1 1 1 1 1 1 2 2 2 2 2
```

Note the use of logical arithmetic (Chapter 2) to get the second level of level2; level=="C" evaluates to 1 when true and to zero when false. Now we can remove level from model and replace it with level2 using **update** like this:

```
model2<-update(model, ~. -level + level2)
```

To see if the model simplification was justified, we compare the simpler model2 with the original model using **anova**:

```
anova(model,model2)

Analysis of Variance Table
Response: yield
```

```
     Terms Resid. Df  RSS    Test Df Sum of Sq F Value    Pr(F)
1 level           12 12.8
2 level2           13 12.9 1 vs. 2 -1      -0.1 0.09375 0.76471
```

The simplification is vindicated by the very high *p* value. Had *p* been less than 0.05 we would return the three-level factor to the model, but there is no justification for keeping the three-level factor in this case.

Simplifying factor levels for Anova using levels

Instead of using logical arithmetic to create simplified models with reduced numbers of factor levels, you can use the **levels** function to replace factor levels with new text.

Suppose that the values of *y* are these:

```
y<-rep(0:7,c(210,99,32,12,4,1,5,2))
```

```
table(y)
```

```
   0   1   2   3  4  5  6  7
 210  99  32  12  4  1  5  2
```

As you see, there are eight levels of *y* (the numbers 0 through 7) and we turn *y* into a factor like this:

```
yf<-factor(y)
```

Unless we say otherwise, there are as many values of **levels** are there are distinct values of the factor:

```
levels(yf)
```

```
[1] "0" "1" "2" "3" "4" "5" "6" "7"
```

We want to create a factor with just five levels rather than eight, in which the largest category is 'four or more' which we want to label as '4+' (there would be 12 values for this level, $4 + 1 + 5 + 2$). All we do is this. We want to leave the first four levels as they are (0, 1, 2, 3), but change all of the other levels to 4+. So we drop all the values of yf that have one of the first four levels, using negative subscripts [-(1:4)] (note it is their *subscripts* that we use, *not* their values), and assign '4+' to all the remaining levels:

```
levels(yf)[-(1:4)]<-"4+"
```

and inspect the result:

```
levels(yf)
```

```
[1] "0" "1" "2" "3" "4+"
```

The vector of values in yf now contains the factor level '4+' wherever it previously had either a 4, 5, 6 or 7. We can see this using **table**:

```
table(yf)
```

```
   0   1   2   3 4+
 210  99  32  12 12
```

Another way to simplify factor levels is to turn the vector into a two-level factor with values either true or false. Suppose we wanted to simplify y even further, so that it was either '0' or 'greater than 0'. This is what we do:

```
yf<-factor(y>0)

levels(yf)

[1] "FALSE" "TRUE"
```

During model simplification, we often want to lump two or more factor levels together, but retain other levels as distinct. In a trial with four drugs, for example, the initial Anova suggests that A and D are alike in their effects, but B and C are different.

This was the original set of factor levels (with slightly unequal replication):

```
drug<-factor(rep(c("A","B","C","D"),c(10,9,11,12)))
```

In this case we want to assess whether a new, three-level factor would describe the data equally well. We want to keep the second and third factor levels as they are (B and C), but change the first and fourth (A and D) so that they are the same (remember that factor levels are in alphabetical order unless you override this). Let's call the new amalgamated factor level AD and the new factor drug2. We compute drug2 like this:

```
drug2<-drug
levels(drug2)[-(2:3)]<-"AD"
```

so the new three-level factor has the following replication:

```
table(drug2)

AD  B   C
22  9  11
```

Now suppose that further model simplification suggested that B and C might not be significantly different from one another. Then we would reduce the three-level factor drug2 to a two-level factor called drug3 like this:

```
drug3<-drug2
levels(drug3)[-1]<-"BC"

table(drug3)

AD  BC
22  20
```

bearing in mind that the already amalgamated factor AD is the first level because of alphabetic priority. More examples using 'levels gets' are in Chapter 18.

Offsets in model simplification

Sometimes you will want to compare a model where you specify one or more of the model's parameter values with a model where S-Plus has estimated the values of the parameters from the data using maximum likelihood. Typical cases are where theory predicts one of the parameter values.

In this example we have gathered data on the basal diameters of 10 herbaceous plants (x, cm) and measured the dry mass of seeds produced by the same plant later in the season (y, g):

```
attach(seedoutput)
names(seedoutput)
```

```
[1] "diam" "seed"
```

We begin by plotting seed output against diameter, and log(seed output) against log(diameter) on adjacent graphs:

```
par(mfrow=c(1,2))
plot(diam,seed)
plot(log(diam),log(seed))
```

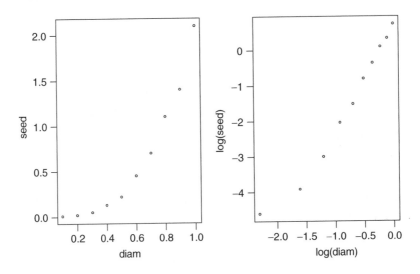

The relationship looks very straight indeed on a log-log scale, except for the plant with the lowest diameter, which lies well above where it might have been expected to fall based on an extrapolation of the linear part. We shall carry out the regression with and without this point, to determine how influential it is in determining parameter estimates:

```
model<-glm(log(seed)~log(diam))
summary(model)
```

```
Coefficients:
                Value Std. Error    t value
(Intercept) 0.458045  0.1726347   2.653261
   log(diam) 2.477983  0.1637777  15.130156
```

Now refit the model omitting the suspicious-looking point using **subset**. We don't know its subscript, but we do know from the plot (above) that it is the only value with log(diam) < -2.0

```
model2<-glm(log(seed)~log(diam),subset=(log(diam)>-2))
summary(model2)
```

```
Coefficients:
                Value Std. Error   t value
(Intercept) 0.6792402 0.05430173  12.50863
   log(diam) 2.9466384 0.06759064  43.59536
```

This confirms our guess about the influence of the large positive residual of the smallest plant. The slope of the log-log plot is 2.478 with the smallest plant included, but 2.947 with it excluded. This difference is highly significant, giving a t test of $(2.947 - 2.478)/0.1638 = 2.863$.

Inspection of the parameter estimates now suggests a simplification that requires the use of offsets. The slope looks to be sufficiently close to 3.0 that constraining it to be 3 would not cause a significant change in deviance. This would make reasonable biological sense, because x is measured in centimetres and seed production is proportional to total shoot mass and this, in turn, is proportional to shoot volume (i.e. to cubic centimetres). This is how we proceed. We calculate a new vector containing three times the log of diameter:

```
values<-3*log(diam)
model3<-glm(log(seed)~1+offset(values),subset=(log(diam)>-2))
```

Now compare the fit of the models with and without the offset:

```
anova(model2,model3)
```

```
Analysis of Deviance Table
Response: y
            Terms Resid. Df Resid. Dev Test Df      Deviance   F Value      Pr(F)
1           log(diam)      7 0.07358445
2 1 + offset(values)      8 0.08013644         -1 -0.006551991 0.623283 0.4557347
```

Changing the slope of the log-log plot from 2.947 to 3 caused no significant reduction in explanatory power. Technically, **glm** has fitted this model:

$$\eta = \text{offset} + \sum \beta_i x_i$$

subtracting the offset from the linear predictor and estimating the remaining parameters by maximum likelihood. We need only estimate the intercept now, because we have specified the rest of the model in the offset. The new estimate for the intercept is

```
model3
```

```
Coefficients:
 (Intercept)
   0.7125548
```

We need to know the antilog of this:

```
exp(model3$coeff)
```

```
(Intercept)
   2.039194
```

which is very close to 2.0. We therefore try a further simplification, this time specifying the entire model in a new offset called full:

```
full<-log(2)+3*log(diam)
```

Now we fit the fully specified equation. We don't want the **glm** to estimate an intercept, so we specify -1 in the model formula after the **offset**:

```
model4<-glm(log(seed)~offset(full)-1,subset=(log(diam)>-2))
summary(model4)

Coefficients:
NULL

Degrees of Freedom: 9 Total; 9 Residual
Residual Deviance: 0.08352634
Warning messages:
   This model has zero rank --- no summary is provided in:
summary.glm(model4)
```

Again, there is an inevitable increase in deviance, but the change is very small (from 0.080 14 to 0.083 526 34) and this is not significant. The error variance is marginally smaller than it was before we used the offset, because the increase in the degrees of freedom has more than compensated for the increase in deviance. From this we conclude that the data are just as well described by the model

$$y = 2x^3$$

as by the initial regression model using the maximum likelihood estimates of the parameters

$$y = 1.927\,23x^{2.947}$$

We can now give some more thought to the data point that we omitted. We saw that the seed production from the smallest plant (which had a y value of 0.01) was highly influential. On closer inspection of the data, a possible reason for its influence emerges. Notice that all the data on seed weights are given to only 2 decimal places. Since the predicted value of y at $x = 0.1$ would be substantially below 0.01 g the outlier may have arisen from a measurement precision error or a data entry error. In this case it turns out from inspection of the lab notebook that the y value had been rounded up to the smallest value possible (e.g. from 0.002 to 0.01) at the data entry stage, since to round it down to zero would suggest that the plant produced no seed at all, and the computer operator assumed that number had to fit into a field with space for only 2 decimal places. On a log scale this change to the data makes an enormous difference. Let's compare the overall fit of the two models with and without the data point in question, and plot observed and fitted seed weights on the original axes.

In the first case we just compute $y = 2x^3$ in the **lines** directive. In the second case we use **predict** with the original model to generate the lines, using the range of diameters generated in xp to produce the predictions:

```
xp<-seq(0.02,1,.02)
par(mfrow=c(1,2))
plot(diam,seed)
xp<-seq(0,1,.02)
yp<-2*xp^3
lines(xp,yp)
plot(diam,seed)
lines(xp,exp(predict(model,list(diam=xp))))
```

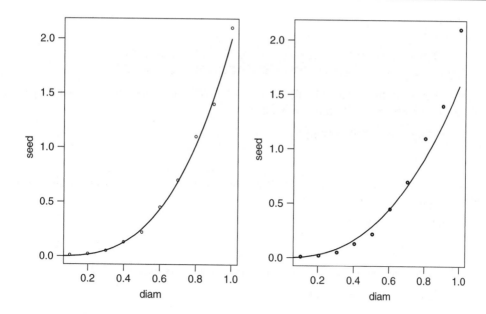

The fit of the fully specified model is excellent except for the largest plant, where the fitted value slightly underestimates the fecundity. Omission of the smallest plant makes little practical difference to the fit on a linear scale. This point did, however, have an enormous influence on the parameter estimates, reducing the slope of the log-log plot from 2.947 to 2.478. The effect can be seen by comparing the very poor fit of the right-hand graph (with the point included) with the excellent fit of the left-hand graph (point excluded). The moral is clear. Make sure that the decimal precision is correct when working with log transformed data, and always test the influence of outlying points.

Allometry

Sometimes there may be a clear theoretical prediction about one or more of the parameter values. In a study of density-dependent mortality in plant populations, for example, we might wish to test whether the self-thinning rule applied to a particular set of data relating mean plant weight (y) to mean plant density (x). Yoda's rule states that the relationship is allometric with $y = ax^{-3/2}$. We might wish to test whether our estimated exponent is significantly different from this. One way would be to do a t test, comparing our estimate to -1.5 using the standard error from the summary table. An alternative is so specify the exponent as $-3/2$ in an offset and compare the fit of this model with our initial model in which the maximum likelihood estimate of the exponent was used:

```
attach(Yoda)
names(Yoda)
```

```
[1] "density" "meansize"
```

The theory is a power law relationship, so the appropriate linear model is log y against log x. We fit the model using **glm** rather than **lm** because **glm** allows us to use offsets and **lm** does not:

```
model<-glm(log(meansize)~log(density))
```

```
summary(model)
```

```
Coefficients:
                   Value Std. Error    t value
  (Intercept)  12.213579 0.26622433   45.87702
  log(density) -1.529393 0.06442428  -23.73938
```

The slope is close to the predicted value of $-3/2$ but is it close enough? A t test suggests that it is definitely not significantly different from -1.5:

(1.529393-1.5)/0.06442482

```
[1]  0.4562372
```

To find the probability of $t = 0.456$ with 19 d.f. we use **pt**, the cumulative probability of Student's t distribution, like this:

pt(1-0.4562372,19)

```
[1]  0.7035357
```

Here is the same test, but using offsets to specify the slope as exactly $-3/2$:

values<- - 1.5*log(density)

The trick is that we specify the **offset** as part of the model formula like this (much as we specified the error term in a nested design). Now, the only parameter estimated by maximum likelihood is the intercept (coefficient 1):

model2<-glm(log(meansize)~1+offset(values))

Now we compare the two models using **anova**:

anova(model,model2,test="F")

```
Analysis of Deviance Table
Response: log(meansize)
          Terms Resid. Df Resid. Dev Test Df    Deviance   F Value       Pr(F)
1     log(density)     19   8.781306
2 1 + offset(values)   20   8.877508      -1 -0.09620223 0.2081515   0.6533922
```

Our conclusion is the same in both cases, even though the p values differ slightly (0.704 vs. 0.653). These data conform very precisely with the allometry predicted by Yoda's rule.

Caveats

Model simplification is an important process but it should not be taken to extremes. For example, the interpretation of deviances and standard errors produced with fixed parameters that have been estimated from the data should be undertaken with caution. Again, the search for 'nice numbers' should not be pursued uncritically. Sometimes there are good scientific reasons for using a particular number (e.g. a power of 0.66 in an allometric relationship between respiration and body mass). It is much more straightforward, for example, to say that yield increases by 2 kg per hectare for every extra unit of fertiliser, than to say that it increases by 1.947 kg. Similarly, it may be preferable to say that the odds of infection increase 10-fold under a given treatment, than to say that the logits increase by 2.321; without model simplification this is equivalent to saying that there is a 10.186-fold increase in the odds. It would be absurd, however, to fix on an estimate of 6 rather than 6.1 just because 6 is a whole number.

Summary

Remember that *order matters*. If your explanatory variables are correlated with each other, then the significance you attach to a given explanatory variable will depend upon whether you delete it from a maximal model or add it to the null model. Always test by model simplification and you won't fall into this trap.

The fact that we have laboured long and hard to include a particular experimental treatment does not justify the retention of that factor in the model if the analysis shows it to have no explanatory power. Anova tables are often published containing a mixture of significant and non-significant effects. This is not a problem in orthogonal designs, because sums of squares can be unequivocally attributed to each factor and interaction term. But as soon as there are missing values or unequal weights, then it is impossible to tell how the parameter estimates and standard errors of the significant terms would have been altered if the non-significant terms had been deleted. The best practice is this:

- say whether your data are orthogonal or not

- present a minimal adequate model

- give a list of the non-significant terms that were omitted, and the deviance changes that resulted from their deletion

The reader can then judge for themselves the relative magnitude of the non-significant factors, and the importance of correlations between the explanatory variables.

The temptation to retain terms in the model that are 'close to significance' should be resisted. The best way to proceed is this. If a result would have been *important* if it had been statistically significant, then it is worth repeating the experiment with higher replication and/or more efficient blocking, in order to demonstrate the importance of the factor in a convincing and statistically acceptable way.

26

Probability distributions

We use probability distributions in statistical computing in four main ways:

- for hypothesis testing
- calculating probabilities from continuous distributions
- calculating probabilities from discrete distributions
- in simulations, to generate different distributions of random numbers

S-Plus has a wide range of built-in probability distributions, for each of which four functions are available:

- **d** is the probability density function
- **p** is the cumulative probability
- **q** is the quantiles of the distribution
- **r** is random numbers generated from the distribution.

Each letter can be prefixed to the S-Plus function names in Table 26.1 (e.g. **dbeta**).

Normal distribution

The Normal distribution is central to the theory of parametric statistics, and gets Chapter 7 to itself. Here we simply illustrate the four built-in probability functions in S-Plus in the context of the Normal distribution.

The density function **dnorm** has a value of z (a quantile) as its argument. Optional arguments specify the mean and standard deviation (default is the standard Normal with mean $= 0$ and s.d. $= 1$). Values of z outside the range -3.5 to $+3.5$ are very unlikely:

```
z<-seq(-3.5,3.5,.01)
fz<-dnorm(z)
plot(z,fz,type="l",ylab="Probability density")
```

Table 26.1   The probability distributions supported by S-Plus. The meanings of the parameters are explained in the text

S-Plus function	Distribution	Parameters
beta	Beta	shape1, shape2
binom	Binomial	sample size, probability
cauchy	Cauchy	location, scale
exp	Exponential	rate (optional)
chisq	Chi-square	degrees of freedom
f	Fisher's F	df1, df2
gamma	Gamma	shape
geom	Geometric	probability
hyper	Hypergeometric	m, n, k
lnorm	Lognormal	mean, s.d.
logis	Logistic	location, scale
nbinom	Negative binomial	size, probability
norm	Normal	mean, s.d.
pois	Poisson	mean
stab	Stable	index, skewness
t	Student's t	degrees of freedom
unif	Uniform	minimum, maximum (optional)
weibull	Weibull	shape
wilcox	Wilcoxon rank sum	m, n

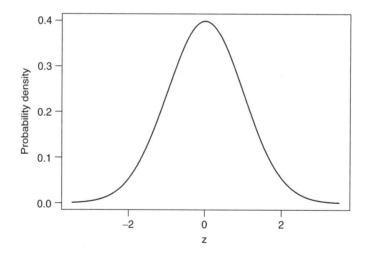

The probability function **pnorm** also has a value of z (a quantile) as its argument. Optional arguments specify the mean and standard deviation (default is the standard Normal with mean $= 0$ and s.d. $= 1$). It shows the cumulative probability of a value of z less than or equal to the value specified, and is an S-shaped curve:

```
pz<-pnorm(z)
plot(z,pz,type="l",ylab="Probability")
```

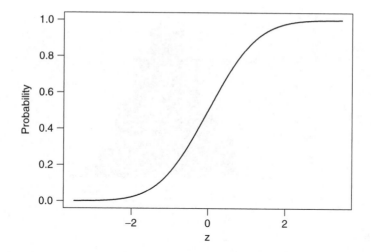

Quantiles of the Normal distribution **qnorm** have a cumulative probability as their argument. They perform the opposite function of **pnorm**, returning a value of z when provided with a probability:

```
p<-seq(0,1,0.01)
z<-qnorm(p)
plot(p,z,type="l",ylab="Quantile (z)")
```

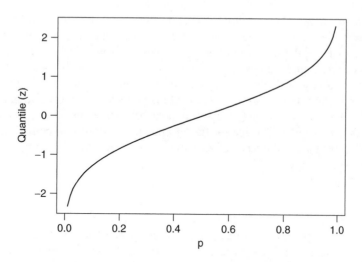

The Normal distribution random number generator **rnorm** produces random real numbers from a distribution with specified mean and standard deviation. The first argument is the number of numbers that you want to be generated; here are 1000 random numbers with mean = 60 and standard deviation = 10:

```
y<-rnorm(1000,60,10)
hist(y)
```

The four functions work in similar ways with all the other probability distributions.

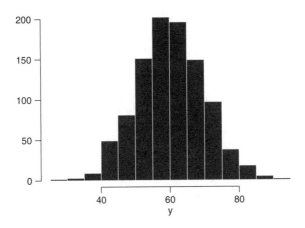

Hypothesis testing

The main distributions used in hypothesis testing are

- chi-square, for testing hypotheses involving count data

- F, in analysis of variance for comparing two variances

- t, in small-sample work for comparing two parameter estimates

Chi-square, Fisher's F and Student's t distributions tell us *the size of the test statistic that could be expected by chance alone* when nothing was happening (i.e. when the null hypothesis was true). Given the rule that a big value of the test statistic tells us that something *is* happening, and hence that the null hypothesis is false, these distributions define what constitutes a big value of the test statistic. For instance, if we are doing a chi-square test, and our test statistic is 14.3 on 9 d.f., we need to know whether this is a large value (meaning the null hypothesis is false) or a small value (meaning the null hypothesis is true). In the old days we would have looked up the value in chi-squares tables. We would have looked in the row labelled 9 (the degrees of freedom row) and the column headed $\alpha = 0.05$. This is the conventional value for the acceptable probability of committing a Type I error; that is to say, we allow a 1 in 20 chance of rejecting the null hypothesis when it is actually true (see p. 55). Nowadays, we just type

```
1-pchisq(14.3,9)
```
```
[1] 0.1120467
```

This indicates that 14.3 is actually a relatively small number when we have 9 d.f. We would conclude that the null hypothesis is true (nothing is happening), because a value of chi-square as large as 14.3 has greater than 11% probability of arising by chance alone. We would want the probability to be less than 5% before we rejected the null hypothesis. So how large would the test statistic need to be before we would reject the null hypothesis? We use **qchisq** to answer this. Its two arguments are $(1 - \alpha)$ and the number of degrees of freedom:

```
qchisq(.95,9)
```
```
[1] 16.91898
```

So the test statistic would need to be larger than 16.92 in order for us to reject the null hypothesis when there were 9 d.f.

We could use **pf** and **qf** in an exactly analogous manner for Fisher's F. Thus, the probability of getting a variance ratio of 2.85 by chance alone when the null hypothesis is true, given that we have

8 d.f. in the numerator and 12 d.f. in the denominator, is just under 5% (i.e. the value is just large enough to allow us to reject the null hypothesis):

```
1-pf(2.85,8,12)
```
```
[1] 0.04992133
```

Note that with **pf**, degrees of freedom in the numerator (8) come first in the list of arguments, followed by degrees of freedom in the denominator (12).

Similarly with Student's t statistics and **pt** and **qt**. For instance, the value of t in tables for a two-tailed test at $\alpha/2 = 0.025$ with d.f. = 10 is

```
qt(.975,10)
```
```
[1] 2.228139
```

Chi-square distribution

The chi-square distribution is perhaps the best known of all the statistical distributions, introduced to generations of schoolchildren in their geography lessons, and comprehensively misunderstood thereafter. It is a special case of the Gamma distribution (p. 487) characterised by a single parameter—the number of degrees of freedom. The mean is equal to the degrees of freedom v (pronounced 'new'), and the variance is equal to $2v$. The density function looks like this:

$$f(x) = \frac{1}{2^{v/2}\Gamma(v/2)} x^{v/2-1} e^{-x/2}$$

where Γ is the gamma function (see p. 15). The chi-square distribution is important because many quadratic forms follow it under the assumption that the data follow the Normal distribution. In particular, the sample variance is a scaled chi-square variable. Likelihood ratio statistics are also approximately distributed as a chi-square (see the F distribution, above).

When the cumulative probability is used, an optional third argument can be provided to describe non-centrality. If the non-central chi-square is the sum of v independent normal random variables, then the non-centrality parameter is equal to the sum of the squared means of the Normal variables. Here are the cumulative probability plots for a non-centrality parameter (ncp) based on three normal means (of 1, 1.5 and 2) and another with four means and ncp = 10:

```
par(mfrow=c(1,2))
x<-seq(0,30,.25)
plot(x,pchisq(x,3,7.25),type="l",ylab="p(x)",xlab="x")
plot(x,pchisq(x,5,10),type="l",ylab="p(x)",xlab="x")
```

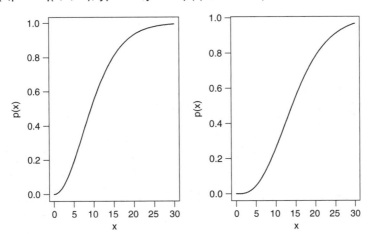

The cumulative probability on the left has 3 d.f. and ncp $= 1^2 + 1.5^2 + 2^2 = 7.25$, while the distribution on the right has 4 d.f. and ncp $= 10$; note the longer left-hand tail at low probabilities.

Chi-square is also used to establish confidence intervals for sample variances. The quantity

$$\frac{(n-1)s^2}{\sigma^2}$$

is the degrees of freedom $(n-1)$ multiplied by the ratio of the sample variance s^2 to the unknown population variance σ^2. This is chi-square distributed, so we can establish a 95% confidence interval for σ^2 as follows:

$$\frac{(n-1)s^2}{\chi^2_{1-\alpha/2}} \leq \sigma^2 \leq \frac{(n-1)s^2}{\chi^2_{\alpha/2}}$$

Suppose the sample variance $s^2 = 10.2$ on 8 d.f. Then the interval on σ^2 is given by

```
8*10.2/qchisq(.975,8)
```
```
[1]  4.65367
```
```
8*10.2/qchisq(.025,8)
```
```
[1]  37.43582
```

which means that we can be 95% confident that the population variance lies in the range $4.65 \leq \sigma^2 \leq 37.44$.

Fisher's *F* distribution

Fisher's F is the famous variance ratio test that occupies the penultimate column of every Anova table. The ratio of treatment variance to error variance follows the F distribution, and you will often want to use the quantile **qf** to look up critical values of F. You specify, in order, the probability of your one-tailed test (this will usually be 0.95), then the two degrees of freedom: numerator first, then denominator. So the 95% value of F with 2 and 18 d.f. is

```
qf(.95,2,18)
```
```
[1]  3.554557
```

This is what the density function of F looks like for 2 and 18 d.f. (left) and 6 and 18 d.f. (right):

```
x<-seq(0.05,4,0.05)
plot(x,df(x,2,18),type="l",ylab="f(x)",xlab="x")
plot(x,df(x,6,18),type="l",ylab="f(x)",xlab="x")
```

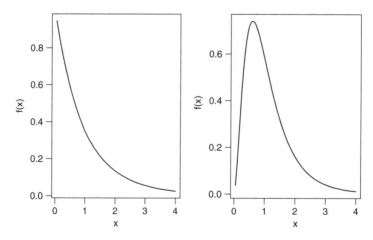

The F distribution is a two-parameter distribution defined by the density function

$$f(x) = \frac{r\Gamma\left(\frac{1}{2}(r+s)\right)}{s\Gamma\left(\frac{1}{2}r\right)\Gamma\left(\frac{1}{2}s\right)} \frac{(rx/s)^{r/2-1}}{[1+(rx/s)]^{(r+s)/2}}$$

where r is the degrees of freedom in the numerator and s is the degrees of freedom in the denominator. The distribution is named after R.A. Fisher, the father of Anova and principal developer of quantitative genetics. It is central to hypothesis testing, because of its use in *assessing the significance of the differences between two variances*. The test statistic is calculated by dividing the larger variance by the smaller variance. The two variances are significantly different when this ratio is larger than the critical value of Fisher's F. The degrees of freedom in the numerator and in the denominator allow the calculation of the critical value of the test statistic. When there is a single degree of freedom in the numerator, the distribution is equal to the square of Student's t, i.e. $F = t^2$. Thus, while the rule of thumb for the critical value of t is 2, so the rule of thumb for $F = t^2 = 4$. To see how well the rule of thumb works, we can plot critical F against d.f. in the numerator:

```
df<-seq(1,30,.1)
plot(df,qf(.95,df,30),type="l",ylab="Critical F")
lines(df,qf(.95,df,10),lty=2)
```

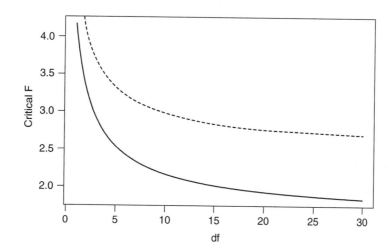

You see that the rule of thumb (critical $F = 4$) quickly becomes much too large once the d.f. in the numerator (on the x axis) is larger than 2. The solid line shows the critical values of F when the denominator has 30 d.f. and the dashed line shows the case in which the denominator has 10 d.f.

The shape of the density function of the F distribution depends on the degrees of freedom in the numerator:

```
x<-seq(0.01,3,0.01)
plot(x,df(x,1,10),type="l",ylim=c(0,1),ylab="f(x)")
lines(x,df(x,2,10),lty=6)
lines(x,df(x,5,10),lty=2)
lines(x,df(x,30,10),lty=3)
```

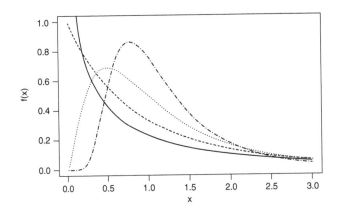

The probability density $f(x)$ declines monotonically when the numerator has 1 d.f. or 2 d.f., but rises to a maximum for d.f. of 3 or more (5 and 30 are shown here). All the graphs have 10 d.f. in the denominator.

Student's t distribution

The famous Student's t distribution was first published by W.S. Gosset in 1908 under the pseudonym of 'Student' because his then employer, the Guinness Brewing Company in Dublin, would not permit employees to publish under their own names. It is a one-parameter model with density function

$$f(x) = \frac{\Gamma\left(\frac{1}{2}(r+1)\right)}{(\pi r)^{1/2}\Gamma\left(\frac{1}{2}r\right)}\left(1 + \frac{x^2}{r}\right)^{-(r+1)/2}$$

where $-\infty < x < +\infty$. This looks very complicated, but if all the constants are stripped away, you can see just how simple the underlying structure really is:

$$f(x) = \left(1 + x^2\right)^{-1/2}$$

We can plot this for values of x from -3 to $+3$ as follows:

```
x<-seq(-3,3,0.01)
fx<-(1+x^2)^(-0.5)
plot(x,fx,type="l")
```

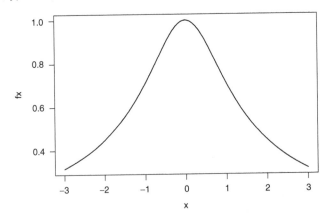

The main thing to notice is how fat the tails of the distribution are, compared with the Normal distribution. The plethora of constants is necessary to scale the density function so that its integral is 1. If we define U as

$$U = \frac{(n-1)}{\sigma^2} s^2$$

then this is chi-square distributed on $n-1$ d.f. (see above). Now define V as

$$V = \frac{n^{1/2}}{\sigma} (\bar{y} - \mu)$$

and note that this is normally distributed with mean 0 and standard deviation 1 (the standard Normal distribution), so

$$T = \frac{V}{[U/(n-1)]^{1/2}}$$

is the ratio of a Normal distribution and a chi-square distribution. You might like to compare this with the F distribution (above), which is the ratio of two chi-square distributed random variables.

At what point does the rule of thumb for Student's $t = 2$ break down so seriously that it is actually misleading? To find this out, we need to plot the value of Student's t against sample size (actually against degrees of freedom) for small samples. We use **qt** (quantile of t) and fix the probability at the two-tailed value of 0.975:

plot(1:30,qt(0.975,1:30),type="l",ylab="Student's t value",xlab="d.f.")

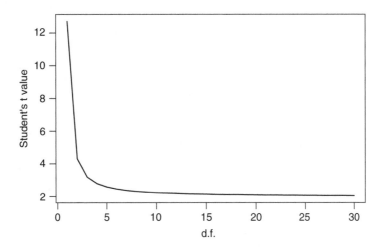

As you see, the rule of thumb only becomes really hopeless for degrees of freedom less than about 5 or so. For most practical purposes, $t \approx 2$ really is a good working rule of thumb.

DISCRETE PROBABILITY DISTRIBUTIONS

Binomial distribution

The binomial distribution is a discrete one-parameter distribution in which p describes the probability of success in a binary trial. The probability of x successes out of n attempts is given by multiplying together:

- the probability of obtaining one specific realisation

- the number of ways of getting that realisation

Suppose we are interested in fish parasites. If we catch 4 fish we want to know the probability that none of them will be parasitised, the probability that 1 of them, 2 of them, 3 of them or all 4 of them will be parasitised. Let us say that, on average, 1 fish in 10 in our particular lake is parasitised. This means that the probability of 'success' (i.e. obtaining a parasitised fish in a random sample) is $p = 0.1$. This immediately defines the probability of 'failure' (i.e. the probability of obtaining a clean, unparasitised fish) as $q = 1 - p = 0.9$. It is clear that the probability that all 4 fish are unparasitised is $0.9^4 = 0.6561$ because there is only 1 way that this outcome can occur. Similarly, the (low) probability that all 4 fish are parasitised is $0.1^4 = 0.0001$. The less extreme cases are a little more tricky, because we need to allow for the fact that there are several ways of getting each outcome. For 1 parasitised (P) and 3 unparasitised (U), for instance, we could get

PUUU

UPUU

UUPU

UUUP

so that the probability of one particular outcome (say PUUU) $= 0.1 \times 0.9^3 = 0.0729$ needs to be multiplied by 4 to get the probability of getting any 1 out of 4. For 2 parasitised and 2 unparasitised there are even more possibilities:

PPUU

PUPU

PUUP

UUPP

UPUP

UPPU

so the probability of any one realisation (say PPUU) $= 0.1^2 \times 0.9^2 = 0.0081$ needs to be multiplied by 6 to get the probability of obtaining any 2 from 4. The are 4 ways of getting 3 from 4, so the probability of this is $4 \times 0.1^3 \times 0.9 = 0.0036$.

This technique of counting up the number of combinations would be exceptionally cumbersome and error-prone for larger samples. We need a way of generalising the number of ways of getting x items out of n items. The answer is the combinatorial formula:

$$\binom{n}{x} = \text{ways of getting } x \text{ out of } n = \frac{n!}{x!(n-x)!}$$

where ! means factorial. For instance, $5! = 5 \times 4 \times 3 \times 2 = 120$. This formula has immense practical utility. It shows you at once, for example, how unlikely you are to win the National Lottery in which you are invited to select 6 numbers between 1 and 49. We can use the built-in **factorial** function for this:

```
factorial(49)/(factorial(6)*factorial(49-6))
```

```
[1] 13983816
```

which is roughly a 1 in 14 million chance of winning the jackpot. You are more likely to die between buying your ticket and hearing the outcome of the draw!

The general form of the binomial distribution is given by

$$p(x) = \binom{n}{x} p^x (1-p)^{n-x}$$

where the 'n above x' in parentheses is the conventional abbreviation for the combinatorial formula. The mean of the binomial distribution is np and the variance is $np(1-p)$.

Since $(1-p)$ is less than 1 it is obvious that *the variance is less than the mean* for the binomial distribution (except, of course, in the trivial case when $p = 0$ and the variance is 0). It is easy to visualise the distribution for particular values of n and p: for our fish we have

```
p<-0.1
n<-4
x<-0:n
px<-factorial(n)/(factorial(x)*factorial(n-x))*p^x*(1-p)^(n-x)
barplot(px,names=as.character(x))
```

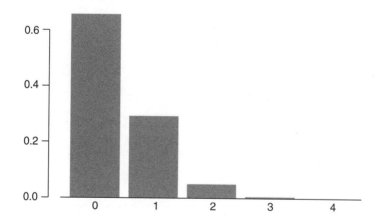

The four distribution functions available for the binomial in S-Plus (density, cumulative probability, quantiles and random generation) are used like this:

```
dbinom(x, size, prob)
```

The **density function** shows the probability for the specified count x (e.g. the number of parasitised fish) out of a sample of $n = size$ with probability of success $= prob$. For our example, $size = 4$ and $prob = 0.1$, so a graph of probability density against number of parasitised fish can be obtained like this:

```
plot(0:4,dbinom(0:4,4,0.1),type="l",xlab="x",ylab="p(x)")
```

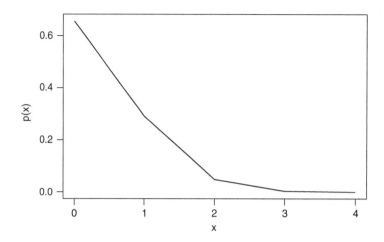

The most likely number of parasitised fish is zero. Note that we can generate the sequence of x values we want to plot (0:4 in this case) *inside* the density function.

The **cumulative probability** shows the integral of the density curve, plotting cumulative probability against the number of successes, for a sample of $n = size$ and probability $= prob$:

```
plot(0:4,pbinom(0:4,4,0.1),type="l",xlab="x",ylab="p(x)")
```

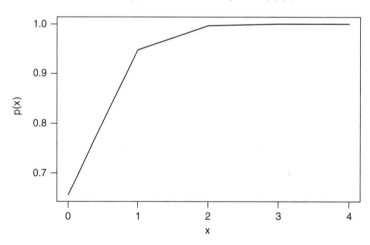

This says that the probability of getting 2 or fewer parasitised fish out of a sample of 4 is very close to 1.

The **quantiles** function asks, With specified probability p (often 0.025 and 0.975 for two-tailed 95% tests) what is the expected number of fish to be caught in a sample of $n = size$ and a probability $= prob$? So for our example, the two-tailed lower and upper 95% expected catches of parasitised fish are

```
qbinom(.025,4,0.1)
```

```
[1]  0
```

```
qbinom(.975,4,0.1)
```

```
[1]  2
```

which means that with 95% certainty we shall catch between 0 and 2 parasitised fish out of 4 if we repeat the sampling exercise. We are very unlikely to get 3 or more parasitised fish out of a sample of 4 if the proportion parasitised really is 0.1.

Random numbers are generated from the binomial distribution like this. The first argument is the number of random numbers we want. The second argument is the sample size ($n = 4$) and the third is the probability of success ($p = 0.1$).

```
rbinom(10,4,0.1)
```
```
[1] 0 0 0 0 0 1 0 1 0 1
```

Here we repeated the sampling of 4 fish 10 times. We got 1 parasitised fish out of 4 on 3 occasions, and 0 parasitised fish on the remaining 7 occasions. We never caught 2 or more parasitised fish in any of these samples of 4.

Bernoulli distribution

The Bernoulli distribution is a special case if the binomial in which the sample size $n = 1$ and x takes one of only two values, $x = 0$ or $x = 1$:

$$p(x) = p^x(1 - p)^{1-x}$$

so $p(0) = 1 - p$ and $p(1) = p$.

Geometric distribution

Suppose that a series of independent Bernoulli trials with probability p are carried out at times 1, 2, 3, etc. Now let W be the waiting time until the first success occurs. So

$$P(W > x) = (1 - p)^x$$

which means that

$$P(W = x) = P(W > x - 1) - P(W > x)$$

The density function is therefore

$$f(x) = p(1 - p)^{x-1}$$

```
fx<-dgeom(0:20,0.2)
plot(0:20,fx,type="l")
```

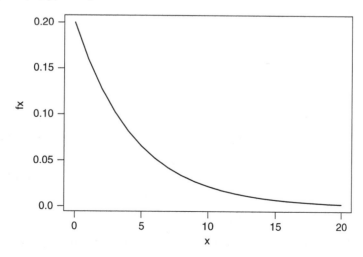

For the geometric distribution:

- the mean is $\dfrac{(1-p)}{p}$

- the variance is $\dfrac{(1-p)}{p^2}$

The geometric has a very long tail. Here are 100 random numbers from a geometric distribution with $p = 0.1$; the mode is 0 but outlying values as large as 43 and 51 have been generated:

```
table(rgeom(100,0.1)
```

```
 0   1   2   3   4   5   7   8   9   10  11  12  13  14  15  16  17  18  19  21
13   8   9   6   4  12   5   4   3    4   3   6   1   1   3   1   1   3   1   1
22  23  25  26  30  43  51
 1   2   1   3   2   1   1
```

Hypergeometric distribution

'Balls in urns' are the classic sort of problem solved by this distribution. The density function of the hypergeometric is

$$f(x) = \frac{\binom{b}{x}\binom{N-b}{n-x}}{\binom{N}{n}}$$

Suppose there are N coloured balls in the statistician's famous urn, b of them are blue and $r = N - b$ of them are red. A sample of n balls is removed from the urn; this is sampling *without replacement*. Now $f(x)$ gives the probability that x of these n balls are blue.

The built-in function for the hypergeometric is used like this:

```
dhyper(q, b,r,n)
rhyper(m, b,r,n)
```

where

q is the vector of values of a random variable representing the number of blue balls out of a sample of size n drawn from an urn containing b blue balls and r red ones

b is the number of blue balls in the urn; this could be a vector with non-negative integer elements

r is the number of red balls in the urn $= N - b$; this could also be a vector with non-negative integer elements

n is the number of balls drawn from an urn with b blue and r red balls; this can be a vector like b and r

p is the vector of probabilities with values between 0 and 1

m is the number of hypergeometrically distributed random numbers to be generated

Let the urn contain $N = 20$ balls, of which 6 are blue and 14 are red. We take a sample of $n = 5$ balls so x could be 0, 1, 2, 3, 4 or 5 of them blue, although since the proportion blue is only 6/20, the higher frequencies are most unlikely.

So our example can be evaluated like this:

```
ph<-numeric(6)
for(i in 0:5) ph[i]<-dhyper(i,6,14,5)
barplot(ph,names=as.character(0:5))
```

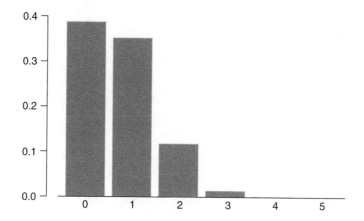

We are very unlikely to get more than 2 red balls out of 5. The most likely outcome is that we get 0 or 1 red ball out of 5. We can simulate a set of Monte Carlo trials of size 5. Here are the numbers of red balls obtained in 20 realisations of our example:

```
rhyper(20,6,14,5)
```

```
[1] 1 2 1 0 1 0 2 1 2 2 1 2 3 2 2 1 2 1 2 2
```

The binomial distribution is a limiting case of the hypergeometric distribution which arises as N, b and r approach infinity in such a way that b/N approaches p, and r/N approaches $1 - p$ (see p. 475). This is because as the numbers get large, the fact that we are *sampling without replacement* becomes irrelevant. The binomial distribution assumes sampling *with* replacement from a finite population, or sampling without replacement from an infinite population.

Multinomial distribution

Suppose there are t possible outcomes from an experimental trial, and the outcome i has probability p_i. Now allow n independent trials where $n = n_1 + n_2 + \cdots + n_t$ and ask what is the probability of obtaining the vector of N_i occurrences of the ith outcome:

$$P(N_i = n_i) = \frac{n!}{n_1!n_2!n_3!\ldots n_t!} p_1^{n_1} p_2^{n_2} p_3^{n_3} \cdots p_t^{n_t}$$

where i goes from 1 to t. Take an example with three outcomes, where the first outcome is twice as likely as the other two ($p_1 = 0.5$, $p_2 = 0.25$ and $p_3 = 0.25$). We do four trials with $n_1 = 6$, $n_2 = 5$, $n_3 = 7$ and $n_4 = 6$ so $n = 24$. We need to evaluate the formula for i equal to 1, 2 and 3 (because there are three possible outcomes). It is sensible to start by writing a function called **multi** to carry out the calculations for any numbers of successes a, b and c for the three outcomes given our three probabilities 0.5, 0.25 and 0.25:

```
multi<-function(a,b,c) {
        factorial(a+b+c)/(factorial(a)*factorial(b)*factorial(c))*.5^a*.25^b*.25^c}
```

Now put the function in a loop to work out the probability of getting the required patterns of success, **psuc**, for the three outcomes. We illustrate just one case, in which the third outcome is fixed at four successes. This means that the first and second cases vary stepwise between 19 and 1 and between 9 and 11 respectively:

```
psuc<-numeric(11)
for (i in 0:10) psuc[i]<- multi(19-i,1+i,4)
barplot(psuc,names=as.character(0:10))
```

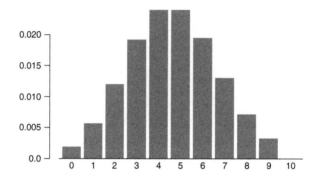

The most likely outcome here is that we would get $19 - 4 = 15$ or $19 - 5 = 14$ successes of type 1 in a trial of size 24 with probabilities 0.5, 0.25 and 0.25 when the number of successes of the third case was 4 out of 24. You can easily modify the function to deal with other probabilities and other numbers of outcomes.

Poisson distribution

The Poisson is one of the most useful and important of the discrete probability distributions for describing count data. It is a one-parameter distribution with the interesting property that *the variance is equal to the mean.*

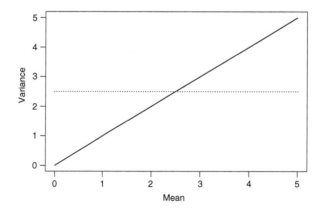

The assumption in normal parametric statistics is that the variance is constant (the horizontal dotted line). A great many processes show variance increasing with the mean, often faster than linearly (see negative binomial distribution). The density function of the Poisson shows the probability of obtaining a count of x when the mean count per unit is λ:

$$p(x) = \frac{e^{-\lambda}\lambda^x}{x!}$$

The zero term of the Poisson (the probability of obtaining a count of zero) is obtained by setting $x = 0$:

$$p(0) = e^{-\lambda}$$

simply the antilog of minus the mean. Given $p(0)$ it is clear that $p(1)$ is just

$$p(1) = p(0)\lambda$$

and any subsequent probability is readily obtained by multiplying the previous probability by the mean and dividing by the count:

$$p(x) = p(x-1)\frac{\lambda}{x}$$

Functions for the density, cumulative distribution, quantiles and random number generation of the Poisson distribution are obtained by

```
dpois(x, lambda)
ppois(q, lambda)
qpois(p, lambda)
rpois(n, lambda)
```

The Poisson distribution holds a central position in three quite separate areas of statistics:

- in the description of random spatial point patterns (see p. 709)
- as the frequency distribution of counts of rare but independent events (see p. 538)
- as the error distribution in **glm** for count data (see p. 537)

If we want 600 simulated counts from a Poisson distribution with a mean of, say, 0.90 blood cells per slide, we just type

```
count<-rpois(600,0.9)
```

We can use **table** to see the frequencies of each count generated:

```
table(count)
```

```
   0     1     2     3    4    5
 259   199   108    29    4    1
```

or **hist** to see a histogram of the counts:

```
hist(count,breaks = - 0.5:6.5)
```

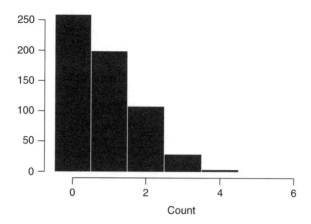

Note the use of the vector of break points on integer increments from -0.5 to 6.5 to create the bins for the histogram bars.

Negative binomial distribution

A discrete, two-parameter distribution, the negative binomial distribution is useful for describing the distribution of count data, where the variance is often much greater than the mean. The two parameters are the mean μ and the clumping parameter k. The smaller the value of k, the greater the degree of clumping. The density function is

$$p(x) = \left(1 + \frac{\mu}{k}\right)^{-k} \frac{(k + x - 1)!}{x!(k - 1)!} \left(\frac{\mu}{\mu + k}\right)^{x}$$

The zero term is found by setting $x = 0$ and simplifying:

$$p(0) = \left(1 + \frac{\mu}{k}\right)^{-k}$$

and successive terms in the distribution can be computed iteratively from

$$p(x) = p(x - 1) \left(\frac{k + x - 1}{x}\right) \left(\frac{\mu}{\mu + k}\right)$$

An initial estimate of the value of k can be obtained from the sample mean and variance:

$$k \approx \frac{\mu^2}{s^2 - \mu}$$

Since k cannot be negative, it is clear that the negative binomial distribution should not be fitted to data where the variance is less than the mean.

The maximum likelihood estimate of k is found numerically, by iterating progressively more fine-tuned values of k until the left- and right-hand sides of the following equation are equal:

$$n \ln \left(1 + \frac{\mu}{k}\right) = \sum_{x=0}^{\max} \left(\frac{A_{(x)}}{k + x}\right)$$

where the vector $A_{(x)}$ contains the total frequency of values *greater* than x.

There is another, quite different way of looking at the negative binomial distribution. Here the response variable is the waiting time W_r for the rth success:

$$f(x) = \binom{x-1}{r-1} p^r (1-p)^{x-r}$$

Note that x starts at r and increases from there (obviously, the rth success cannot occur before the rth attempt). The function **dnbinom** represents the number of failures x (e.g. tails in coin tossing) before *size* successes (or heads in coin tossing) are achieved, when the probability of a success (or of a head) is *prob*:

dnbinom(x, size, prob)

Suppose we are interested in the distribution of waiting times until the fifth success occurs in a negative binomial process with $p = 0.1$. We start the sequence of x values at 5:

plot(5:100,dnbinom(5:100,5,0.1),type="l",xlab="x",ylab="f(x)")

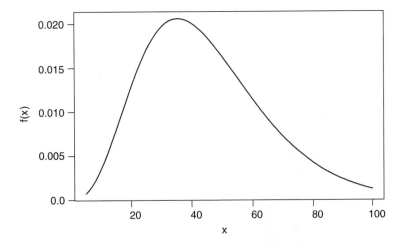

This shows that the most likely waiting time for the fifth success, when probability of a success is 1/10, is about 31 trials after the fifth trial (i.e. on trial 36). Note that the negative binomial distribution is quite strongly skewed to the right.

It is easy to generate negative binomial data using the random number generator:

rnbinom(n, size, prob)

The number of random numbers required is n. When the second parameter, *size*, is set to 1 the distribution becomes the geometric (see above). The final parameter *prob* is the probability of success per trial, p. Here we generate 100 counts with a mean of 0.6:

count<-rnbinom(100,1,0.6)

We can use **table** to see the frequency of the different counts:

table(count)

```
 0   1    2   3   5   6
65  18   13   2   1   1
```

It is sensible to check that the mean really is 0.6 (or very close to it):

```
mean(count)
```

```
[1]  0.61
```

The variance will be substantially greater than the mean:

```
var(count)
```

```
[1]  1.129192
```

and this gives an estimate for k of $0.61^2/(1.129 - 0.61) = 0.717$.

An example of aggregated count data

The data show the number of spores counted on 238 buried glass slides. There are two questions:

- are these data well described by a negative binomial distribution?
- if so, what is the maximum likelihood estimate of the aggregation parameter k?

```
x<-0:12
freq<-c(131,55,21,14,6,6,2,0,0,0,0,2,1)
barplot(freq,names=as.character(x),ylab="frequency",xlab="spores")
```

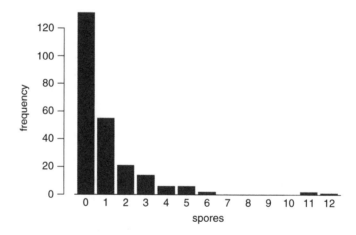

We start by looking at the variance/mean ratio of the counts. We cannot use mean and variance directly, because our data are frequencies of counts, rather than counts themselves. This is easy to rectify; we use **rep** to create a vector of counts in which each count (x) is repeated the relevant number of times (*freq*). Now we can use **mean** and **var** directly:

```
y<-rep(x,freq)
```

```
mean(y)
```

```
[1]  1.004202
```

```
var(y)
```

```
[1]  3.075932
```

   This shows that the data are highly aggregated (the variance/mean ratio is roughly 3 compared with the 1 it would be if the data were Poisson distributed). Our rough estimate of k is

```
mean(y)^2/(var(y)-mean(y))
```
```
[1]  0.4867531
```

Here is a function that takes a vector of frequencies of counts x (between 0 and $length(x) - 1$) and computes the maximum likelihood estimate of the aggregation parameter, k:

```
kfit <- function(x)
{
    lhs<-numeric()
    rhs<-numeric()
    y <- 0:(length(x) - 1)
    j <- 0:(length(x) - 2)
    m <- sum(x * y)/(sum(x))
    s2 <- (sum(x * y^2) - sum(x * y)^2/sum(x))/(sum(x)- 1)
    k1 <- m^2/(s2 - m)
    a<-numeric(length(x)-1)
    for(i in 1:(length(x) - 1)) a[i] <- sum(x [- c(1:i)])
    i<-0
    for (k in seq(k1/1.2,2*k1,0.001) ) {
    i<-i+1
    lhs[i] <- sum(x) * log(1 + m/k)
    rhs[i] <- sum(a/(k + j))
      }
    k<-seq(k1/1.2,2*k1,0.001)
    plot(k, abs(lhs-rhs),xlab="k",ylab="Difference",type="l")

    d<-min(abs(lhs-rhs))
    sdd<-which(abs(lhs-rhs)==d)
    k[sdd]
}
```

We can try it out with our spore count data:

```
kfit(freq)
```
```
[1]  0.5826276
```

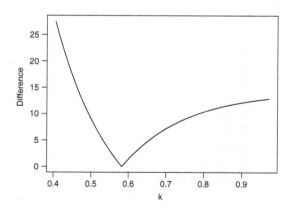

The minimum difference is close to zero and occurs at about $k = 0.55$. The printout shows the maximum likelihood estimate of k to be 0.582 (to the 3 decimal places we simulated; the last 4 decimals (6276) are irrelevant and would not be printed in a more polished function).

How would a negative binomial distribution with a mean of 1.0042 and a k value of 0.582 describe our count data? The expected frequencies are obtained by multiplying the probability density (above) by the total sample size (238 slides in this case).

$$\text{nb<-238*(1+1.0042/0.582)^(-0.582)*factorial(.582+(0:12)-1)/}$$
$$\text{(factorial(0:12)*factorial(0.582-1))*(1.0042/(1.0042+0.582))^(0:12)}$$

We shall compare the observed and expected frequencies using **barplot**. We need to alternate the observed and expected frequencies. There are three steps to the procedure

- concatenate the observed and expected frequencies in an alternating sequence
- create a list of labels to name the bars (alternating blanks and counts)
- produce a legend to describe the different bar colours

The concatenated list of frequencies (called *both*) is made like this, putting the 13 observed counts (*freq*) in the odd-numbered bars and the 13 expected counts (*nb*) in the even-numbered bars (note the use of modulo %% to do this):

```
both<-numeric(26)
both[1:26 %% 2 != 0]<-freq
both[1:26 %% 2 == 0]<-nb
```

The names are the counts (0:12), and these are to be applied only to the even-numbered bars (the odd-numbered bars are to be named with blanks ""):

```
labs<-character(26)
labs[1:26%%2==0]<-as.character(0:12)
```

Now we can draw and label the combined barplot:

```
barplot(both,col=rep(c(1,0),13),ylab="frequency",names=labs)
```

The **key** function creates a legend to show which bars are the observed (solid in this case) and which are the expected, negative binomial frequencies (open):

```
key(text=c("Expected","Observed"),rectangle=list(size=3,col=0:1),border=T)
```

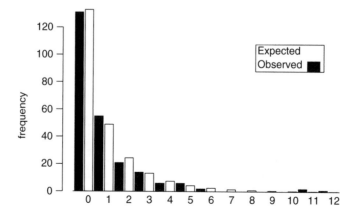

The fit is very close, so we can be reasonably confident in describing the observed counts as negative binomially distributed. The tail of the observed distribution is rather fatter than the expected negative binomial tail, so we might want to measure the lack of fit between observed and expected distributions. A simple way to do this is to use Pearson's chi-square, taking care to use only those cases where the expected frequency nb is greater than 5:

```
sum((((freq-nb)^2/nb)[nb>5])
```

```
[1] 1.634975
```

This is based on five legitimate comparisons:

```
sum(nb>5)
```

```
[1] 5
```

and hence on $5 - p - 1 = 2$ d.f. (because we have estimated $p = 2$ parameters from the data in estimating the expected distribution; the mean and k of the negative binomial). Our calculated value of chi-square $= 1.63$ is much smaller than the value in tables:

```
qchisq(0.95,2)
```

```
[1] 5.991465
```

so we accept that our data are not significantly different from a negative binomial with mean $= 1.0042$ and $k = 0.582$.

Wilcoxon rank sum statistic

The function **wilcox** calculates the distribution of the Wilcoxon rank sum statistic (also known as Mann–Whitney), and returns values for the exact probability at discrete values of q:

```
dwilcox(q, m, n)
```

where q is a vector of quantiles, m is the number of observations in sample x (a positive integer not greater than 50), and n is the number of observations in sample y (also a positive integer not greater than 50). The Wilcoxon rank sum statistic is the sum of the ranks of x in the combined sample c (x, y). The Wilcoxon rank sum statistic takes on values W between these limits:

$$\frac{m(m + 1)}{2} \leq W \leq \frac{m(m + 2n + 1)}{2}$$

This statistic can be used for a non-parametric test of location shift between the parent populations x and y.

CONTINUOUS PROBABILITY DISTRIBUTIONS

Gamma distribution

The Gamma distribution is useful for describing a wide range of processes where the data have positive skew (i.e. non-normal data with a long tail on the right). It is a two-parameter distribution, where the parameters are traditionally known as shape and rate. Its density function is

$$f(x) = \frac{1}{\beta^\alpha \Gamma(\alpha)} x^{\alpha - 1} e^{-x/\beta}$$

where α is the shape parameter, $\alpha\beta$ is the mean and $\alpha\beta^2$ is the variance. Special cases of the Gamma distribution are the **exponential** (mean $= \beta$, variance $= \beta^2$, and $\alpha = 1$) and **chi-square** (shape $=$ degrees of freedom/2, mean $=$ d.f. and variance $= 2$ d.f.).

To see the effect of the shape parameter on the probability density, we can plot the Gamma distribution for different values of shape and rate over the range 0.01 to 4:

```
x<-seq(0.01,4,.01)
par(mfrow=c(2,2))
y<-dgamma(x,.5,.5)
plot(x,y,type="l")
y<-dgamma(x,.8,.8)
plot(x,y,type="l")
y<-dgamma(x,2,2)
plot(x,y,type="l")
y<-dgamma(x,10,10)
plot(x,y,type="l")
```

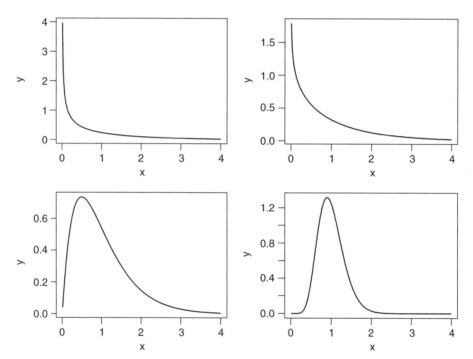

The graphs show different values of α:0.5, 0.8, 2 and 10. Note how $\alpha < 1$ produces monotonic declining functions and $\alpha > 1$ produces humped curves that pass through the origin, with the degree of skew declining as α increases.

The **mean** of the distribution is *shape/rate*, the **variance** is *shape/rate²*, and the **skewness** is $2/\sqrt{shape}$. Thus, if you know the mean and the variance, then

$$rate = \frac{mean}{variance}$$

$$shape = rate \times mean$$

We can answer questions like this: What value is 95% quantile expected from a Gamma distribution of *mean* = 2 and *variance* = 3? This implies that *rate* is 0.6666 and *shape* is 1.333 33, so

```
qgamma(0.95,0.666666,1.333333)
```

```
[1] 1.732095
```

An important use of the Gamma distribution is in describing continuous measurement data that are *not* normally distributed. Here is an example where body mass data for 200 fishes is plotted as a histogram and a Gamma distribution with the same mean and variance is overlayed as a smooth curve:

```
attach(fishes)
names(fishes)
```

```
[1] "mass"
```

First we calculate the two parameter values for the Gamma distribution:

```
rate<-mean(mass)/var(mass)
shape<-rate*mean(mass)
```

```
rate
```

```
[1] 0.8775119
```

```
shape
```

```
[1] 3.680526
```

We need to know the largest value of mass, in order to make the bins for the histogram:

```
max(mass)
```

```
[1] 15.53216
```

Now we can plot the histogram, using break points at 0.5 to get integer-centred bars up to a maximum of 16.5 to accommodate our biggest fish:

```
hist(mass,breaks=-0.5:16.5)
```

The density function of the Gamma distribution is overlaid using **lines** like this:

```
lines(seq(0.01,15,0.01),length(mass)*dgamma(seq(0.01,15,0.01),shape,rate))
```

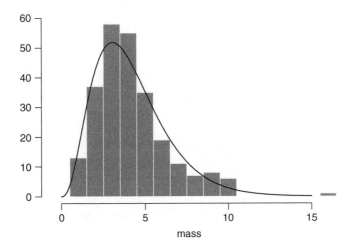

The fit is much better than when we tried to fit a Normal distribution to these same data earlier (see p. 127).

Exponential distribution

The exponential distribution is a one-parameter distribution that is a special case of the Gamma distribution. Much used in survival analysis, its density function is given on p. 612 and its use in survival analysis is explained on p. 616. The random number generator of the exponential is useful for Monte Carlo simulations of time to death when the hazard (the instantaneous risk of death) is constant with age. You specify the hazard, which is the reciprocal of the mean age at death:

```
rexp(15,0.1)
```

```
 [1]   9.811752  5.738169 16.261665 13.170321   1.114943
 [6]   1.986883  5.019848  9.399658 11.382526   2.121905
[11]  10.941043  5.868017  1.019131 13.040792 38.023316
```

These are 15 random lifetimes with an expected value of $1/0.1 = 10$ years; they give a sample mean of 9.66 years.

Beta distribution

The Beta distribution has two positive constants, a and b, and x is bounded, $0 \le x \le 1$:

$$f(x) = \frac{\Gamma(a+b)}{\Gamma(a)\Gamma(b)} x^{a-1}(1-x)^{b-1}$$

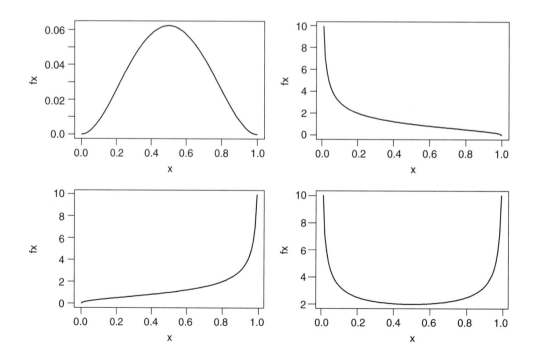

The important point is whether the parameters are greater or less than 1. When both are >1 we get an n-shaped curve which becomes more skew as $b > a$ (top left). If $0 < a < 1$ and $b > 1$ then the density is negative (top right), while for $a > 1$ and $0 < b < 1$ the density is positive (bottom left). The function is u-shaped when both a and b are positive fractions. If $a = b = 1$ then we obtain the uniform distribution on [0,1].

Here are 20 random numbers from the Beta distribution with shape parameters 2 and 3:

```
rbeta(20,2,3)
```

```
 [1]  0.5820844  0.5150638  0.5420181  0.1110348  0.5012057  0.3641780
 [7]  0.1133799  0.3340035  0.2802908  0.3852897  0.6496373  0.3377459
[13]  0.1743189  0.4568897  0.7343201  0.3040988  0.5670311  0.2241543
[19]  0.6358050  0.5932503
```

Cauchy distribution

The Cauchy distribution is a long-tailed two-parameter distribution, characterised by a location parameter and a scale parameter. It is real-valued, symmetric about location, and is a curiosity in that it has long enough tails that *the expectation does not exist*. The default distribution is the same as Student's t distribution with one degree of freedom. The harmonic mean of a variable with positive density at 0 is typically distributed as Cauchy, and the Cauchy distribution also appears in the theory of Brownian motion (e.g. random walks). The density function for $-\infty < x < \infty$ is

$$f(x) = \frac{1}{\pi \ (1 + x^2)}$$

```
par(mfrow=c(1,2))
plot(-200:200,dcauchy(-200:200,0,10),type="l",ylab="p(x)",xlab="x")
plot(-200:200,dcauchy(-200:200,0,50),type="l",ylab="p(x)",xlab="x")
```

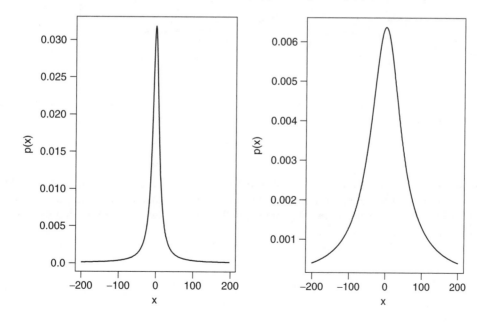

Note the very long, fat tail of the Cauchy distribution The left-hand density function has scale $= 10$ and the right-hand plot has scale $= 50$; both have location $= 0$.

Lognormal distribution

The lognormal distribution takes values on the positive real line. If the logarithm of a lognormal deviate is taken, the result is a normal deviate, hence the name. Applications for the lognormal include the distribution of particle sizes in aggregates, flood flows, concentrations of air contaminants, and failure times. The hazard function of the lognormal is increasing for small values and then decreasing. A mixture of heterogeneous items that individually have monotone hazards can create such a hazard function.

Density, cumulative probability, quantiles and random generation for the lognormal distribution employ the function **dlnorm** like this:

```
dlnorm(x, meanlog=0, sdlog=1)
```

The mean and standard deviation are optional, with default mean $= 0$ and s.d. $= 1$:

```
par(mfrow=c(1,1))
plot(seq(0,10,0.05),dlnorm(seq(0,10,0.05)),type="l",xlab="x",ylab="Log Normal f(x)")
```

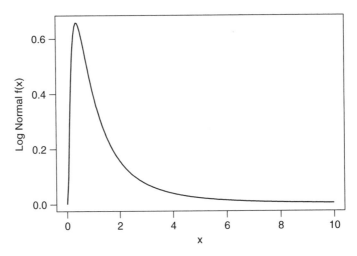

The extremely long tail and exaggerated positive skew are characteristic of the lognormal distribution. Logarithmic transformation followed by analysis with Normal errors is often appropriate for data such as these.

Logistic distribution

The logistic is the canonical link function in **glm** with binomial errors and is described in detail in Chapter 28 on the analysis of proportion data. The cumulative probability is a symmetrical S-shaped distribution that is bounded above by 1 and below by 0. There are two ways of writing the cumulative probability equation:

$$p(x) = \frac{e^{a+bx}}{1 + e^{a+bx}}$$

$$p(x) = \frac{1}{1 + \beta e^{-\alpha x}}$$

The great advantage of the first form is that it linearises under the log odds transformation (see p. 390) so that

$$\ln\left(\frac{p}{q}\right) = a + bx$$

where p is the probability of success and $q = (1 - p)$ is the probability of failure. The logistic is a unimodal, symmetric distribution on the real line with tails that are longer than the Normal distribution. It is often used to model growth curves, but has also been used in bioassay studies and other applications. A motivation for using the logistic with growth curves is because the logistic distribution function $f(x)$ satisfies this condition: the derivative of $f(x)$ with respect to x is proportional to $[f(x) - A][B - f(x)]$ with $A < B$. The interpretation is that the rate of growth is proportional to the amount already grown, times the amount of growth that is still expected.

```
par(mfrow=c(1,2))
plot(seq(-5,5,0.02),dlogis(seq(-5,5,.02)),type="l",ylab="Logistic f(x)")
plot(seq(-5,5,0.02),dnorm(seq(-5,5,.02)),type="l",ylab="Normal f(x)")
```

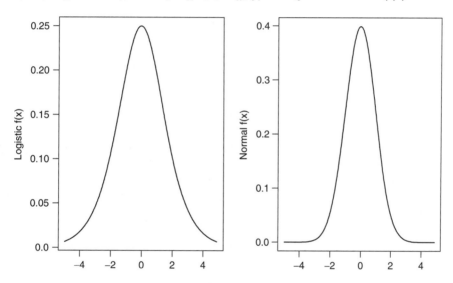

Here the logistic density function **dlogis** (left) is compared with an equivalent Normal density function **dnorm** (right) using the defaults mean $= 0$ and s.d. $= 1$ in both cases. Note the much fatter tails of the logistic (still substantial probability at ± 4 standard deviations. Note also the difference in the scales of the two y axes (0.25 for the logistic, 0.4 for the Normal).

Stable distributions

Stable distributions are of considerable mathematical interest. Each stable distribution is the limit distribution of a suitably scaled sum of independent and identically distributed random variables. Statistically, they are used mostly when an example of a very long-tailed distribution is required. S-Plus provides a random number generator to generate n random numbers from a stable distribution, but no density, probability or quantile functions are available:

```
rstab(n, index, skewness=0)
```

The index is a number from the interval (0, 2], where index = 2 corresponds to the Normal, and 1 to the Cauchy. Smaller values mean longer tails. Negative values of skewness correspond to skewness to the left (the median is smaller than the mean, if it exists), and positive values correspond to skewness to the right (the median is larger than the mean). The absolute value of skewness should not exceed 1.

hist(rstab(1000,1.5,0.6))

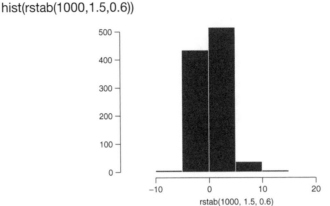

Uniform distribution

The uniform distribution is what the random number generator in your calculator hopes to emulate. The idea is to generate numbers between 0 and 1 where every possible real number on this interval has exactly the same probability of being produced. If you have thought about this, it will have occurred to you that there is something wrong here. Computers produce numbers by following recipes. If you are following a recipe then the outcome is predictable. If the outcome is predictable then how can it be random? This raises the question of what, exactly, a computer-generated random number is. The answer turns out to be scientifically very interesting. The notion of random-looking determinism is the idea that underpins **chaos** theory. Some reasonably good random number generators were based on the simple quadratic map (May 1973) that can generate a chaotic time series from the simple difference equation

$$N_{t+1} = \lambda N_t (1 - N_t)$$

when λ is larger than about 3.3. Here are some time series for different values of λ.

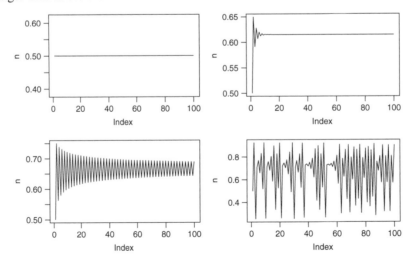

The top left panel has $\lambda = 2$, top right has $\lambda = 2.6$, bottom left has $\lambda = 3$ and bottom right has $\lambda = 3.7$. As λ is increased the dynamics change from stable equilibrium, to damped oscillations, through persistent cycles to chaos. The graphics were produced like this:

```
par(mfrow=c(2,2))
n[1]<-.5
for (i in 2:100) n[i]<-n[i-1]*2*(1-n[i-1])
plot(n,type="l")
⋮
n[1]<-.5
for (i in 2:100) n[i]<-n[i-1]*3.7*(1-n[i-1])
plot(n,type="l")
```

The period-doubling route to chaos is illustrated further on p. 50.

```
par(mfrow=c(1,1))
```

Modern random number generators are very good. Here is the outcome of the S-Plus function **runif** throwing a six-sided die 10 000 times; the histogram ought to be flat:

```
x<-ceiling(runif(10000)*6)
table(x)

   1     2     3     4     5     6
1632  1677  1659  1723  1601  1708

hist(x,breaks=0.5:6.5)
```

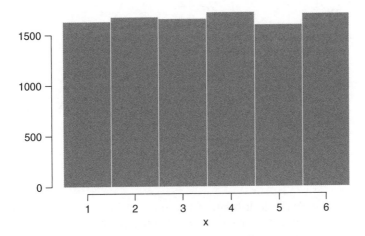

This is remarkably close to theoretical expectation, reflecting the very high efficiency of modern pseudorandom number generators.

Weibull distribution

The origin of the Weibull distribution is in *weakest link analysis*. If there are r links in a chain, and the strengths of each link Z_i are independently distributed $(0, \infty)$, then the distribution of weakest links $V = \min(Z_j)$ approaches the Weibull distribution as the number of links increases.

The Weibull is a two-parameter model that has the exponential distribution as a special case. Its value in demographic studies and survival analysis is that it allows for the death rate to increase or to decrease with age, so that all three kinds of survivorship curve can be analysed (see p. 611). The density, hazard and survival functions with $\lambda = \mu^{-\alpha}$ are

$$f(t) = \alpha\lambda t^{\alpha-1}e^{-\lambda t^\alpha}$$

$$h(t) = \alpha\lambda t^{\alpha-1}$$

$$S(t) = e^{-\lambda t^\alpha}$$

The mean of the Weibull distribution is $\Gamma(1 + \alpha^{-1})\mu$ and the parameter α describes the shape of the hazard function (the background to determining the likelihood equations is given by Aitkin *et al.* (1989, pp. 281–283). For $\alpha = 1$ (the exponential distribution) the hazard is constant, while for $\alpha > 1$ the hazard increases with age, and for $\alpha < 1$ the hazard decreases with age.

Because the Weibull, lognormal and log-logistic all have positive skewness, it will be difficult to discriminate between them with small samples. This is an important problem, because each distribution has differently shaped hazard functions, and therefore it will be hard to discriminate between different assumptions about the age specificity of death rates. In survival studies, parsimony requires that we fit the exponential rather than the Weibull unless the shape parameter α is significantly different from 1.

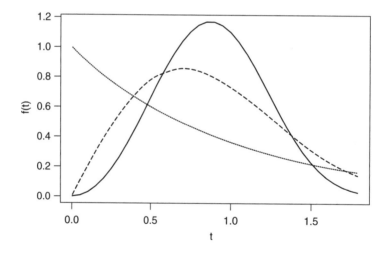

Here is a family of three Weibull distributions with $\alpha = 1$, $\alpha = 2$ and $\alpha = 3$ (dotted, dashed and solid lines, respectively). Note that for large values of α the distribution becomes symmetrical, while for $\alpha \leq 1$ the distribution has its mode at $t = 0$.

```
a<-3
l<-1
t<-seq(0,1.8,.05)
ft<-a*l*t^(a-1)*exp(-l*t^a)
plot(t,ft,type="l")
a<-1
ft<-a*l*t^(a-1)*exp(-l*t^a)
```

```
lines(t,ft,type="l",lty=2)
a<-2
ft<-a*l*t^(a-1)*exp(-l*t^a)
lines(t,ft,type="l",lty=3)
```

Further reading

Grimmett, G.R. and Stirzaker, D.R. (1992) *Probability and Random Processes*. Oxford, Clarendon Press.
Johnson, N.L. and Kotz, S. (1970) *Continuous Univariate Distributions*, Volume 2. New York, John Wiley.

27

Generalised linear models

So far we have assumed that the variance is constant and that the errors are normally distributed. For many kinds of data, one or both of these assumptions is wrong, and we need to be able to deal with this if our analysis is to be unbiased and our interpretations scientifically correct. In count data, for example, where the response variable is an integer and there are often lots of zeros in the data frame, the variance may increase linearly with the mean. With proportion data, where we have a count of the number of failures of an event as well as the number of successes, the variance will be a ∩-shaped function of the mean. Where the response variable follows a Gamma distribution (as in data on time to death) the variance increases faster than linearly with the mean. So our assumption has been like the top left panel (where variance is constant) but the data are often like one of the other three panels.

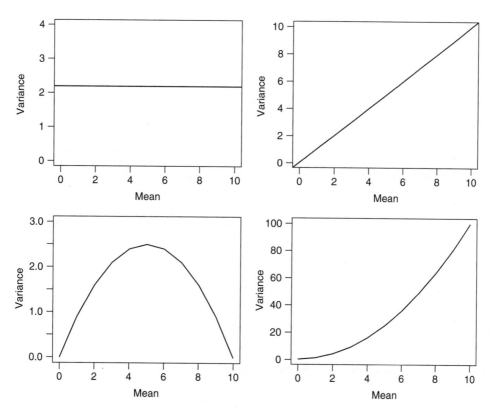

The way to deal with all these problems in a single theoretical framework was discovered by John Nelder, who christened the technique generalised linear models (GLMs), often shortened to 'glims'.

The directive to fit a generalised linear model in S-Plus is **glm**. It is used in exactly the same way as the model-fitting directives that are now familiar to you, **aov** and **lm**. There are yet more ways of fitting models that you can discover later if you want to: mixed effects models (**lme**), generalised additive models (**gam**), non-parametric surface fitting models (**loess**), tree models (**tree**), and so on.

A common misconception about generalised linear models is that linear models involve a straight-line relationship between the response variable and the explanatory variables. This is *not* the case, as you can see from these two linear models:

```
par(mfrow=c(1,2))
x<-seq(0,10,0.1)
plot(x,1+x-x^2/15,type="l")
plot(x,3+0.1*exp(x),type="l")
```

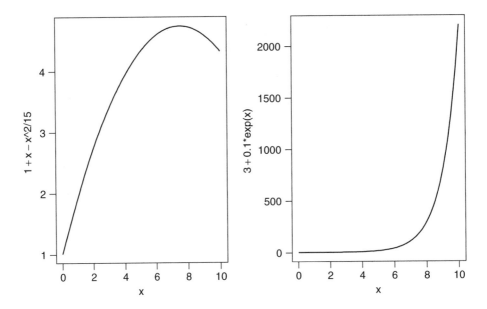

The definition of a linear model is an equation that contains mathematical variables, parameters and random variables that is *linear in the parameters and in the random variables*. What this means is that if a, b and c are parameters then obviously

$$y = a + bx$$

is a linear model. But so is

$$y = a + bx - cx^2$$

because x^2 can be replaced by z, which gives a linear relationship

$$y = a + bx + cz$$

And so is

$$y = a + b\,e^x$$

because we can create a new variable $z = e^x$, so that

$$y = a + bz$$

Some models are non-linear but can be readily linearised by transformation. For example

$$y = \exp(a + bx)$$

is non-linear, but on taking logs of both sides, it becomes

$$\ln y = a + bx$$

This kind of relationship is handled in a GLM by specifying the log link (see below). Again, the much used asymptotic relationship (known in different disciplines as Michaelis–Menten, Briggs–Haldane or Holling's disc equation) is given by

$$y = \frac{x}{b + ax}$$

and is non-linear, but it is readily linearised by taking reciprocals:

$$\frac{1}{y} = a + b\frac{1}{x}$$

Generalised linear models handle this family of equations by a transformation of the explanatory variable ($z = 1/x$) and using the reciprocal link (often, for data like this, associated with Gamma rather than Normal errors).

Other models are *intrinsically non-linear* because there is no transformation that can linearise them in all the parameters. Some important examples include the hyperbolic function

$$y = a + \frac{b}{c + x}$$

and the asymptotic exponential

$$y = a(1 - b\,e^{-cx})$$

where both models are non-linear unless the parameter c is known in advance. In cases like this, a GLM is unable to estimate the full set of parameters, and we need to resort to non-linear modelling (Chapter 23).

Generalised linear models

A generalised linear model has three important properties:

- the error structure
- the linear predictor
- the link function

These are all likely to be unfamiliar concepts. The ideas behind them are straightforward, however, and it is worth learning what each of the concepts involves.

The error structure

Up to this point, we have dealt with the statistical analysis of data with Normal errors. In practice, however, many kinds of data have non-normal errors:

- errors that are strongly skewed
- errors that are kurtotic
- errors that are strictly bounded (as in proportions)
- errors that cannot lead to negative fitted values (as in counts)

In the past, the only tools available to deal with these problems were transformation of the response variable or the adoption of non-parametric methods. A GLM allows the specification of a variety of different error distributions:

- Poisson errors, useful with count data
- binomial errors, useful with data on proportions
- Gamma errors, useful with data showing a constant coefficient of variation
- exponential errors, useful with data on time to death (survival analysis)

The error structure is defined by means of the **family** directive, used as part of the model formula like this:

$$\text{glm}(y \sim z, \text{family} = \text{poisson})$$

which means that the response variable y has Poisson errors. Or

$$\text{glm}(y \sim z, \text{family} = \text{binomial})$$

which means that the response is binary, and the model has binomial errors. As with previous models, the explanatory variable z can be continuous (leading to a regression analysis) or categorical (leading to an Anova-like procedure called analysis of deviance; see below).

The linear predictor

The structure of the model relates each observed y value to a predicted value. The predicted value is obtained *by transformation of the value emerging from the linear predictor*. The linear predictor, η (eta), is a linear sum of the effects of one or more explanatory variables, x_j:

$$\eta_i = \sum_{j=1}^{p} x_{ij} \beta_j$$

where the x are the values of the p different explanatory variables, and the β are the (usually) unknown parameters to be estimated from the data. The right-hand side of the equation is called the *linear structure*.

There are as many terms in the linear predictor as there are parameters, p, to be estimated from the data. With a simple regression, the linear predictor is the sum of two terms, the intercept and the slope. With a one-way Anova with four treatments, the linear predictor is the sum of four terms

— a mean for each level of the factor. If there are covariates in the model, they add one term each to the linear predictor (the slope of the relationship). Each interaction term in a factorial Anova adds one or more parameters to the linear predictor, depending upon the degrees of freedom of each factor (e.g. three extra parameters for the interaction between a two-level factor and a four-level factor $(2 - 1) \times (4 - 1) = 3$).

Fitted values

To determine the fit of a given model, a GLM evaluates the linear predictor for each value of the response variable, then compares the predicted value with a *transformed* value of y. The transformation to be employed is specified in the link function (see below). The fitted value is computed by applying the reciprocal of the link function, in order to get back to the original scale of measurement of the response variable. With a log link, the fitted value is the antilog of the linear predictor; with a reciprocal link, the fitted value is the reciprocal of the linear predictor.

The link function

One of the difficult things to grasp about GLMs is the relationship between the values of the response variable (as measured in the data and predicted by the model in fitted values) and the linear predictor. The thing to remember is that the *link function relates the mean value of y to its linear predictor*. In symbols, this means that

$$\eta = g(\mu)$$

which is simple but needs thinking about. The linear predictor, η (eta), emerges from the linear model as a sum of the terms for each of the p parameters. *This is not a value of y* (except in the special case of the *identity link* that we have been using implicitly up to now). The value of η is obtained by transforming the value of y by the link function, and the predicted value of y is obtained by applying the inverse link function to η.

The most frequently used link functions are shown below. An important criterion in the choice of link function is to ensure that the fitted values stay within reasonable bounds. We would want to ensure, for example, that counts were all greater than or equal to 0 (negative count data would be nonsense). Similarly, if the response variable were the proportion of animals that died, then the fitted values would have to lie between 0 and 1 (fitted values greater than 1 or less than 0 would be meaningless). In the first case, a log link is appropriate because the fitted values are antilogs of the linear predictor, and all antilogs are greater than or equal to 0. In the second case, the logit link is appropriate because the fitted values are calculated as the antilogs of the log odds, $\log(p/q)$.

By using different link functions, the performance of a variety of models can be compared directly. The total deviance is the same in each case and we can investigate the consequences of altering our assumptions about precisely how a given change in the linear predictor brings about a response in the fitted value of y. The most appropriate link function is the one which produces the minimum residual deviance.

The log link

The log link has many uses, but the most frequent are:

- for count data, where negative fitted values are prohibited

- for explanatory variables that have multiplicative effects, where the log link introduces additivity (remember that the log of times is plus)

The model parameters inspected with summary(model) are in natural logarithms, and the fitted values are the natural antilogs (exp) of the linear predictor.

The logit link

The logit link is the link used for proportion data, and it is generally preferred to the more old fashioned **probit** link. If a fraction p of the insects in a bioassay died, then a fraction $q = (1 - p)$ must have survived out of the original cohort of n animals. The logit link is

$$\text{logit} = \ln\left(\frac{p}{q}\right)$$

This is beautifully simple, and it ensures that the fitted values are bounded both above and below (the predicted proportions may not be greater than 1 or less than 0). The details of how the logit link linearises proportion data are explained in Chapter 28.

The logit link does, however, make it a little tedious to calculate p from the parameter estimates. Suppose the predicted logit at $z = 10$ was -0.328. To get the value of p we find the antilog of the fitted value $x = \exp(-0.328)$ then evaluate

$$p = \frac{1}{1 + 1/x}$$

which gives $p = 0.42$. Note that confidence intervals on p will be asymmetric when back-transformed, and it is good practice to draw bar charts and error bars on the logit scale rather than the proportion (or percentage) scale to avoid this problem.

Other link functions for proportion data

Two other commonly used link functions are the **probit** and the **complementary log-log** links. They are used in bioassay and in dilution analysis respectively, and examples of their use are discussed later.

Use of probits for bioassay is largely traditional, because probit paper used to be available for converting percentage mortality to a linear scale against log dose. Since computers have become widely available, the need for the probit transformation

$$\frac{y_i}{n_i} = \Phi(\eta_i) + \varepsilon_i$$

has declined. The proportion responding (y/n) is linked to the linear predictor by $\Phi(.)$, the unit Normal probability integral. Because the logit is so much simpler to interpret, and because the results of modelling with the two transformations are almost always identical, the logit link function is nowadays recommended for bioassay work, even though probits are based on a reasonable distributional argument for the tolerance levels of individuals.

The complementary log-log link
$$\theta = \ln[-\ln(1 - p)]$$

is not symmetrical about $p = 0.5$ and is often used in simple dilution assay. If the proportion of tubes containing bacteria p is related to dilution x like this:

$$p = 1 - e^{-\lambda x}$$

then the complementary log-log transformation gives

$$\eta = \ln[-\ln(1-p)] = \ln \lambda + \ln x$$

which means that the linear predictor has a slope of 1 when plotted against ln x. We fit the model, therefore, with ln x as an offset, and **glm** estimates the maximum likelihood value of ln λ.

The complementary log-log link should be assessed during model criticism for binary data and for data on proportional responses (Chapters 28 and 30). It will sometimes lead to a lower residual deviance than the symmetrical logit link.

Canonical link functions

The canonical link functions are the default options employed when a particular error structure is specified in the **family** directive in the model formula. Omission of a **link** directive means that the following settings are used.

Error	Canonical link
Normal	identity
Poisson	log
Binomial	logit
Gamma	reciprocal

You should try to memorise these canonical links and to understand why each is appropriate to its associated error distribution. Note that only Gamma errors have a capital initial letter.

Choosing between using a link function (e.g. log link) and transforming the response variable (i.e. having log y as the response variable rather than y) takes a certain amount of experience. The decision is usually based on *whether the variance is constant* on the original scale of measurement. If the variance were constant, you would use a link function. If the variance increased with the mean, you would be more likely to log transform the response.

The likelihood function

The concept of maximum likelihood is unfamiliar to most non-statisticians. Fortunately, the methods that scientists have encountered in linear regression and traditional Anova (i.e. least squares) are the maximum likelihood estimators when the data have Normal errors and the model has an identity link. For other kinds of error structure and different link functions, however, the methods of least squares do not give unbiased parameter estimates, and maximum likelihood methods are preferred. It is easiest to see what maximum likelihood involves by working through two simple examples based on the binomial and Poisson distributions.

The binomial distribution

Suppose we have carried out a single trial, and have found $r = 5$ parasitised animals out of a sample of $n = 9$ insects. Our intuitive estimate of the proportion parasitised is $5/9 = r/n = 0.555$. What is the maximum likelihood estimate of the proportion parasitised? With $n = 9$ and $r = 5$ the formula for the binomial looks like this:

$$P(5) = \left(\frac{9!}{(9-5)! \times 5!}\right)\theta^5(1-\theta)^{(9-5)}$$

Now the likelihood L does not depend upon the combinatorial part of the formula, because θ, the parameter we are trying to estimate, does not appear there. This simplifies the problem, because all we need to do now is to find the value of θ which maximises the likelihood

$$L(\theta) = \theta^5 (1 - \theta)^{(9-5)}$$

To do this we might plot $L(\theta)$ against θ like this:

```
theta<-seq(0,1,.01)
par(mfrow=c(1,1))
plot(theta,theta^5*(1-theta)^4,type="l",ylab="Likelihood")
```

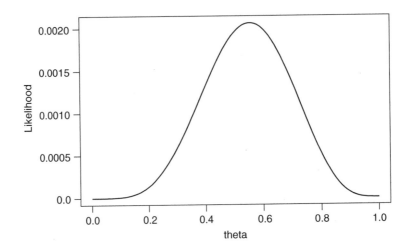

From which it is clear that the maximum likelihood occurs at $= r/n = 5/9 = 0.555$. It is reassuring that our intuitive estimate of the proportion parasitised is r/n as well.

A more general way to find the maximum likelihood estimate of θ is to use calculus. We need to find the derivative of the likelihood with respect to θ, then set this to zero, and solve for θ. In the present case it is easier to work with the log of the likelihood. Obviously, the maximum likelihood and the maximum log likelihood will occur at the same value of θ:

$$\text{log likelihood} = r \ln \theta + (n - r) \ln(1 - \theta)$$

so the derivative of the log likelihood with respect to θ is

$$\frac{dL(\theta)}{d\theta} = \frac{r}{\theta} - \frac{n-r}{1-\theta}$$

remembering that the derivative of $ln\theta$ is $1/\theta$ and of $\ln(1 - \theta)$ is $-1/ln(1 - \theta)$. We set this to zero, rearrange, then take reciprocals to find θ:

$$\frac{r}{\theta} = \frac{n-r}{1-\theta} \quad \text{so} \quad \theta = \frac{r}{n}$$

The maximum likelihood estimate of the binomial parameter is the same as our intuitive estimate.

The Poisson distribution

As a second example, we take the problem of finding the maximum likelihood estimate of μ for a Poisson process in which we observed, say, r lightning strikes per parish in n parishes, giving a total of $\sum r$ lightning strikes in all. The probability density function for the number of strikes per parish is

$$f(r) = \frac{e^{-\mu} \mu^r}{r!}$$

so the initial likelihood is the density function multiplied by itself as many times as there are individual parishes:

$$L(\mu) = \prod_1^n \frac{e^{-\mu} \mu^r}{r!} = e^{-n\mu} \mu^{\sum r}$$

because the constant $r!$ can be ignored. Note that nr is replaced by $\sum r$ the observed total number of lightning strikes. Now it is straightforward to obtain the log likelihood:

$$L(\mu) = -n\mu + \sum r \ln \mu$$

The next step is to find the derivative of the log likelihood with respect to μ:

$$\frac{dL(\mu)}{d\mu} = -n + \frac{\sum r}{\mu}$$

We set this to zero and rearrange to obtain

$$\mu = \frac{\sum r}{n}$$

Again, the maximum likelihood estimator for the single parameter of the Poisson distribution conforms with intuition; it is the mean (in this case the mean number of lightning strikes per parish).

Parameter estimation in generalised linear models

The method of parameter estimation is *iterative, weighted least squares*. You know about least squares methods from Chapters 14 and 15. A GLM is different in that the regression is not carried out on the response variable, y, but on *a linearised version of the link function applied to y*. The *weights* are functions of the fitted values. The procedure is *iterative* because both the adjusted response variable and the weight depend upon the fitted values.

This is how it works; the technical details are on pp. 31–34 in McCullagh and Nelder (1989). Take the data themselves as starting values for estimates of the fitted values. Use this to derive the linear predictor, the derivative of the linear predictor ($d\eta/d\mu$) and the variance function. Then re-estimate the adjusted response variable z and the weight W, as follows:

$$z_0 = \eta_0 + (y - \mu_0) \left(\frac{d\eta}{d\mu} \right)_0$$

where the derivative of the link function is evaluated at μ_0 and

$$W_0^{-1} = \left(\frac{d\eta}{d\mu} \right)_0^2 V_0$$

where V_0 is the variance function of y (see below). Keep repeating the cycle until the changes in the parameter estimates are sufficiently small. It is the difference $(y - \mu_0)$ between the data y and the fitted values μ_0 that lies at the heart of the procedure. The maximum likelihood parameter estimates are given by

$$\sum W(y - \mu)\frac{\mathrm{d}\eta}{\mathrm{d}\mu}x_i = 0$$

for each explanatory variable x_i (summation is over the rows of the data frame). For more detail, see McCullagh and Nelder (1989) and Aitkin *et al.* (1989); a good general introduction to the methods of maximum likelihood is to be found in Edwards (1972).

Deviance: measuring the goodness of fit of a GLM

The fitted values produced by the model are most unlikely to match the values of the data perfectly. The size of the discrepancy between the model and the data is a measure of the inadequacy of the model; a small discrepancy may be tolerable, but a large one will not be. The measure of discrepancy in a GLM to assess the goodness of fit of the model to the data is called the *deviance*. Deviance is defined as -2 times the difference in log likelihood between the current model and a saturated model (i.e. a model that fits the data perfectly). Because the saturated model does not depend on the parameters of the model, minimising the deviance is the same as maximising the likelihood.

Deviance is estimated in different ways for different GLM families. Numerical examples of the calculation of deviance for different GLM families are in Chapter 28 (binomial errors) and Chapter 29 (Poisson errors).

Quasi-likelihood

In some cases we may be uneasy about specifying the precise form of the error distribution. We may know, for example, that it is not Normal (e.g. because the variance increases with the mean), but we don't know with any confidence that the underlying distribution is, say, negative binomial.

There is a very simple and robust alternative known as quasi-likelihood, introduced by Wedderburn in 1974, which uses only the most elementary information about the response variable, namely the variance–mean relationship (see Taylor's power law, p. 56). It is extraordinary that this information alone is often sufficient to retain close to the full efficiency of maximum likelihood estimators.

Suppose that we know that the response is always positive, the data are invariably skew to the right, and the variance increases with the mean. This does not enable us to specify a particular distribution (e.g. it does not discriminate between Poisson or negative binomial errors), hence we cannot use

Table 27.1 Deviance formulas for different GLM families where y is observed data, $\bar{y}$ the mean value of y, μ are the fitted values of y from the maximum likelihood model, and n is the binomial denominator in a binomial GLM. Note that Gamma is the only error structure with a capital initial letter

Family (error structure)	Deviance
Normal	$\sum (y - \bar{y})^2$
Poisson	$2 \sum y \ln(y/\mu) - (y - \mu)$
Binomial	$2 \sum y \ln(y/\mu) + (n - y) \ln[(n - y)/(n - \mu)]$
Gamma	$2 \sum (y - \mu)/y - \ln(y/\mu)$
Inverse gaussian	$\sum (y - \mu)^2/(\mu^2 y)$

techniques like maximum likelihood or likelihood ratio tests. Quasi-likelihood frees us from the need to specify a particular distribution, and requires us only to specify the variance–mean relationship up to a proportionality constant, which can be estimated from the data:

$$\text{var}(y_i) \propto v(\mu_i)$$

An example of the principle at work compares quasi-likelihood with maximum likelihood in the case of Poisson errors; full details are in McCulloch and Searle (2001). The fundamental quantity is

$$E\left[\frac{\partial \log f_{Y_i}(y_i)}{\partial \mu_i}\right] = 0$$

The contribution to the log likelihood from y_i is the integral with respect to μ_i of $\partial \log f_{Y_i}(y_i)/\partial \mu_i$, and we define the log quasi-likelihood via the contribution that y_i makes to it:

$$Q_i = \int_{y_i}^{\mu_i} \frac{y_i - t}{\tau^2 v(t)} \, dt$$

The maximum quasi-likelihood equations are

$$\frac{\partial}{\partial \beta} \sum Q_i = 0$$

which, on evaluating the derivative gives

$$\sum \frac{y_i - \mu_i}{\tau^2 v(\mu_i)} \frac{\partial \mu_i}{\partial \beta} = 0$$

Now suppose that we allow that the variance is equal to the mean, so $v(\mu_i) = \mu_i$ which means that

$$Q_i = \int_{y_i}^{\mu_i} \frac{y_i - t}{\tau^2 t} \, dt = \frac{1}{\tau^2} (y_i \log t - t) \Big|_{y_i}^{\mu_i}$$

$$Q_i = \frac{1}{\tau^2} (y_i \log \mu_i - \mu_i - y_i \log y_i + y_i)$$

This means that the maximum quasi-likelihood (MQL) equations for β are

$$\frac{\partial}{\partial \beta} \sum (y_i \log \mu_i - \mu_i) = 0$$

which you might remember is exactly the same as the maximum likelihood equation for the Poisson (see p. 507). In this case, MQL and ML give precisely the same estimates, and MQL would therefore be fully efficient. Other cases do not work out as elegantly as this, but MQL estimates are generally robust and efficient. Their great virtue is the simplicity of the central premise that $\text{var}(y_i) \propto v(\mu_i)$, and the lack of the need to assume a specific distributional form.

Specifying your own GLM structure

Some data sets are not well described by the standard combinations of link function and variance function. In such cases, you can specify our own functions using the **quasi** option of the family directive. Suppose, for example, that a square root transformation linearised the response, but that the variance increased with the square of the mean. We can incorporate these two attributes like this:

family = quasi(link = sqrt, variance = mu^2)

inside the **glm** call. The model is then refitted, and model checking carried out. If the problem has not been resolved, then a new **quasi** combination can be tried, and model checking reassessed.

The function **robust** generates a new family object, with its component functions suitably modified to perform a robust version of a **glm** or **gam** fit. For instance

family = robust(quasi(link = power(2)))

The value 0 for **power** is a limiting case and results in the log link function; **power** detects this situation.

A GLM with Gamma errors

With continuous response variables, the traditional models of regression and Anova assume constant variance and Normal errors. In many circumstances, however, this structure is inappropriate because:

- variance often increases with the mean

- the distribution of errors is often highly skewed

One solution is to assume lognormal errors in y and to carry out the analysis on the log-transformed response variable, using least squares with Normal errors and the identity link. An alternative is to use a GLM:

- the errors are Gamma (rather than Normal)

- the link is the reciprocal (especially useful with inverse polynomials, p. 383)

The two-parameter Gamma distribution is particularly appropriate as an error structure in cases where the response variable has a constant coefficient of variation rather than a constant variance. The mean of the Gamma distribution is $\alpha\beta$ and the variance is $\alpha\beta^2$. This means that the *variance/mean ratio* is β and the *coefficient of variation* is

$$\mathrm{CV} = \frac{s}{\mu} = \frac{\sqrt{\alpha\beta^2}}{\alpha\beta} = \frac{\sqrt{\alpha}\beta}{\alpha\beta} = \frac{\sqrt{\alpha}}{\alpha} = \frac{1}{\sqrt{\alpha}} = \text{constant}$$

In a plot of log variance against log mean, the graph would be linear with slope $= 2$ (see Taylor's power law, p. 56).

With Gamma errors, the lack of fit is measured by deviance rather than by *SSE*. The value of deviance is calculated for data y with mean μ like this:

$$2\sum \left[\log\left(\frac{\mu}{y}\right) + \frac{(y - \mu)}{\mu} \right]$$

Here are some data on time to death $\{3, 7, 11, 25\}$ from p. 608; they average 11.5 days, so the total deviance is

2*(log(11.5/3)+(3-11.5)/11.5+log(11.5/7)+(7-11.5)/11.5+
          log(11.5/11)+(11-11.5)/11.5+(log(11.5/25)+(25-11.5)/11.5))

```
[1] 2.216189
```

We should check this by comparing it with the output from a GLM with Gamma errors:

```
d<-c(3,7,11,25)
glm(d~1,family=Gamma)
```

```
Degrees of Freedom: 4 Total; 3 Residual
Residual Deviance: 2.216189
```

There you have it. Deviance is not mysterious; it is -2 times the log likelihood.

Further reading

Aitkin, M., Anderson, D. *et al.* (1989) *Statistical Modelling in GLIM*. Oxford, Clarendon Press.

Crawley, M.J. (1993) *GLIM for Ecologists*. Oxford, Blackwell Science.

Dobson, A.J. (1990) *An Introduction to Generalized Linear Models*. London, Chapman & Hall.

McCullagh, P. and Nelder, J.A. (1989) *Generalized Linear Models*. London, Chapman and Hall.

Nelder, J.A. and Wedderburn, R.W.M. (1972) Generalized linear models. *Journal of the Royal Statistical Society, Series A* **135**: 370–384.

28

Proportion data: binomial errors

An important class of problems involves data on proportions:

- studies on percentage mortality

- infection rates of diseases

- proportion responding to clinical treatment

- proportion admitting to particular voting intentions

- data on proportional response to an experimental treatment

What all these have in common is that we know how many of the experimental objects are in one category (say dead, insolvent or infected) and we also know how many are in another (say alive, solvent or uninfected). This contrasts with Poisson count data, where we knew how many times an event occurred, but *not* how many times it did not occur (Chapter 29).

We model processes involving proportional response variables in S-Plus by specifying a GLM with family=binomial. The only complication is that whereas with Poisson errors we could simply say family=poisson, with binomial errors we must specify the number of failures as well as the numbers of successes in a two-vector response variable. To do this we bind together two vectors using **cbind** into a single object, y, comprising the numbers of successes and the number of failures. The *binomial denominator*, n, is the total sample, and

number.of.failures = binomial.denominator − number.of.successes

y <- cbind (number.of.successes, number.of.failures)

The old-fashioned way of modelling this sort of data was to use the percentage mortality as the response variable. There are four problems with this:

- the errors are not normally distributed

- the variance is not constant

- the response is bounded (by 1 above and by 0 below)

- by calculating the percentage, we lose information of the size of the sample, n, from which the proportion was estimated

S-Plus carries out weighted regression, using the individual sample sizes as weights, and the *logit link function* to ensure linearity (as described below).

There are some kinds of proportion data, like **percentage cover**, which are best analysed using conventional models (Normal errors and constant variance) following **arcsine transformation**. The response variable, y, measured in radians, is $\sin^{-1} \sqrt{0.01 \times p}$ where p is cover in percent.

If, however, the response variable takes the form of a **percentage change** in some continuous measurement (such as the percentage change in weight on receiving a particular diet), then rather than arcsine transform the data, it is usually better treated by either of these methods:

- analysis of covariance (Chapter 16), using final weight as the response variable and initial weight as a covariate

- by specifying the response variable as a relative growth rate, measured as log(final weight/initial weight)

Both of these can be analysed with Normal errors without further transformation.

Analyses of data on one and two proportions

For comparisons of one binomial proportion with a constant, use **binom.test** (see p. 180). For comparison of two samples of proportion data, use **prop.test** (see p. 181). The methods of this chapter are required only for more complex models of proportion data, including regression and contingency tables, where generalised linear models are used.

Count data on proportions

The traditional transformations of proportion data were arcsine and probit. The arcsine transformation took care of the error distribution, while the probit transformation was used to linearise the relationship between percentage mortality and log dose in a bioassay. There is nothing wrong with these transformations, and they are available within S-Plus, but a simpler approach is often preferable, and is likely to produce a model that is easier to interpret.

The major difficulty with modelling proportion data is that the responses are *strictly bounded*. There is no way that the percentage dying can be greater than 100% or less than 0%. But if we use simple techniques like regression or analysis of covariance, then the fitted model could quite easily predict negative values or values greater than 100%, especially if the variance was high and many of the data were close to 0 or close to 100%.

The *logistic* curve is commonly used to describe data on proportions, because unlike the straight line model, it asymptotes at 0 and 1 so that negative proportions and responses of more than 100% cannot be predicted. Throughout this discussion we shall use p to describe the proportion of individuals observed to respond in a given way. Because much of their jargon was derived from the theory of gambling, statisticians call these *successes* although, to a demographer measuring death rates this may seem somewhat macabre. The individuals that respond in other ways (the statistician's *failures*) are therefore $(1 - p)$ and we shall call the proportion of failures q. The third variable is the size of the sample, n, from which p was estimated (it is the binomial denominator, and the statistician's *number of attempts*).

An important point about the binomial distribution is that the variance is not constant. In fact, the variance of a binomial distribution with mean np is

$$s^2 = npq$$

so that the variance changes with the mean like this:

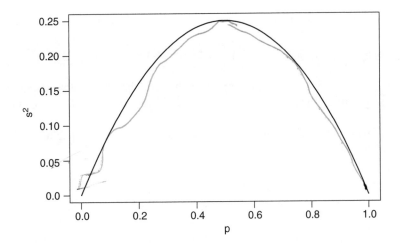

The variance is low when p is very high or very low, and the variance is greatest when $p = q = 0.5$. As p gets smaller, so the binomial distribution gets closer and closer to the Poisson distribution. You can see why this is so by considering the formula for the variance of the binomial (above). Remember that for the Poisson, the variance is equal to the mean, $s^2 = np$. Now, as p gets smaller, q gets closer and closer to 1, so the variance of the binomial converges to the mean:

$$s^2 = npq \approx np \ (q \approx 1)$$

Odds

The logistic model for p as a function of x looks like this:

$$p = \frac{e^{(a+bx)}}{1 + e^{(a+bx)}}$$

and there are no prizes for realising that the model is not linear. But if $x = -\infty$ then $p = 0$ and if $x = +\infty$ then $p = 1$, so the model is strictly bounded. When $x = 0$ then $p = e^a/(1 + e^a)$. The trick of linearising the logistic actually involves a very simple transformation. You may have come across the way in which bookmakers specify probabilities by quoting the *odds* against a particular horse winning a race (they might give odds of 2 to 1 on a reasonably good horse or 25 to 1 on an outsider). This is a rather different way of presenting information on probabilities than scientists are used to dealing with. Thus, where the scientist might state a proportion as 0.666 (2 out of 3), the bookmaker would give odds of 2 to 1 (2 successes to 1 failure). In symbols, this is the difference between the scientist stating the probability p, and the bookmaker stating the odds, p/q. Now if we take the *odds* p/q and substitute this into the formula for the logistic, we get

$$\frac{p}{q} = \frac{e^{(a+bx)}}{1 + e^{(a+bx)}} \left[1 - \frac{e^{(a+bx)}}{1 + e^{(a+bx)}} \right]^{-1}$$

which looks awful. But a little algebra shows that

$$\frac{p}{q} = \frac{e^{(a+bx)}}{1 + e^{(a+bx)}} \left[\frac{1}{1 + e^{(a+bx)}} \right]^{-1} = e^{(a+bx)}$$

Now, taking natural logs, and recalling that $\ln(e^x) = x$ will simplify matters even further, we have

$$\ln\left(\frac{p}{q}\right) = a + bx$$

This gives a *linear predictor*, $a + bx$, not for p but for the *logit* transformation of p, namely $\ln(p/q)$. In the jargon of S-Plus, the logit is the *link function* relating the linear predictor to the value of p.

Here is p as a function of x (left panel) and logit(p) as a function of x (right panel) for the logistic with $a = 0.2$ and $b = 0.1$.

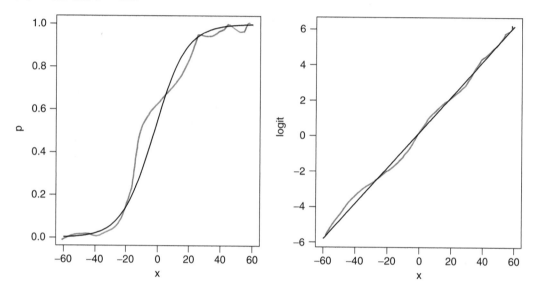

You might ask at this stage, Why not simply do a linear regression of $\ln(p/q)$ against the explanatory x variable? S-Plus has three great advantages here:

- it allows for the non-constant binomial variance

- it deals with the fact that logits are infinite for p near 0 or 1

- it allows for differences between the sample sizes by weighted regression

Deviance with binomial errors

Before we met GLMs we always used *SSE* as the measure of lack of fit (see p. 225). For data on proportions, the maximum likelihood estimate of lack of fit is different from this: if y is the observed count and $\hat{y}$ is the fitted value predicted by the current model, then if n is the binomial denominator (the sample size from which we obtained y successes), the deviance is

$$2\sum y \log\left[\frac{y}{\hat{y}}\right] + (n - y)\log\left[\frac{(n - y)}{(n - \hat{y})}\right]$$

We should try this on a simple numerical example. In a trial with two treatments and two replicates per treatment we got 3 out of 6 and 4 out of 10 in one treatment, and 5 out of 9 and 8 out of 12 in the other treatment. Are the proportions significantly different across the two treatments? The first thing you need to remember is how to calculate the average of two proportions. You don't work out the two proportions, add them up and divide by 2. What you do is count up the total number of successes and divide by the total number of attempts. In our case the overall mean proportion is calculated like this. The total number of successes, y, is $(3 + 4 + 5 + 8) = 20$. The total number of attempts, N, is $(6 + 10 + 9 + 12) = 37$. So the overall average proportion is $20/37 = 0.5405$. Next we need to work out the four expected counts, $\hat{y}$. We multiply each of the sample sizes by the mean proportion, 6×0.5405, etc.:

```
y<-c(3,4,5,8)
N<-c(6,10,9,12)
N*20/37
```

```
[1]   3.243243   5.405405   4.864865   6.486486
```

Now we can calculate the total deviance:

$$2 \times \left\{ 3\log\left(\frac{3}{3.243}\right) + 3\log\left(\frac{3}{6 - 3.243}\right) + 4\log\left(\frac{4}{5.405}\right) + 6\log\left(\frac{6}{10 - 5.405}\right) + 5\log\left(\frac{5}{4.865}\right) \right.$$

$$\left. + 4\log\left(\frac{4}{9 - 4.865}\right) + 8\log\left(\frac{8}{6.486}\right) + 4\log\left(\frac{4}{12 - 6.486}\right) \right\}$$

You wouldn't want to have to do this too often, but it is worth it to demystify deviance:

```
p1<-2*(3*log(3/3.243)+3*log(3/(6-3.243))+4*log(4/5.405)+6*log(6/(10-5.405)))
p2<-2*(5*log(5/4.865)+4*log(4/(9-4.865))+8*log(8/6.486)+4*log(4/(12-6.486)))
```

```
p1+p2
```

```
[1]   1.629673
```

So the total deviance is 1.6297. Let's see what a GLM with binomial errors gives:

```
rv<-cbind(y,N-y)
glm(rv~1,family=binomial)
```

```
Degrees of Freedom: 4 Total; 3 Residual
Residual Deviance: 1.629733
```

Great relief all round. What about the residual deviance? We need to compute a new set of fitted values based on the two individual treatment means. Total successes in treatment A were $3 + 4 = 7$ out of a total sample of $6 + 10 = 16$. So the mean proportion was $7/16 = 0.4375$. In treatment B the equivalent figures were $13/21 = 0.619$. So the fitted values are

```
p<-c(7/16,7/16,13/21,13/21)
p*N
```

```
[1]   2.625000   4.375000   5.571429   7.428571
```

and the residual deviance is

$$2 \times \left\{ 3\log\left(\frac{3}{2.625}\right) + 3\log\left(\frac{3}{6 - 2.625}\right) + 4\log\left(\frac{4}{4.375}\right) + 6\log\left(\frac{6}{10 - 4.375}\right) + 5\log\left(\frac{5}{5.571}\right) \right.$$

$$\left. + 4\log\left(\frac{4}{9 - 5.571}\right) + 8\log\left(\frac{8}{7.429}\right) + 4\log\left(\frac{4}{12 - 7.429}\right) \right\}$$

Again, we calculate this in two parts:

```
p1<-2*(3*log(3/2.625)+3*log(3/(6-2.625))+4*log(4/4.375)+6*log(6/(10-4.375)))
p2<-2*(5*log(5/5.571)+4*log(4/(9-5.571))+8*log(8/7.429)+4*log(4/(12-7.429)))

p1+p2

[1]  0.4201974
```

To see the deviance calculated by **glm** we need to create a two-level factor called treatment to describe the data on which the two different means are based:

```
treatment<-factor(c("A","A","B","B"))
```

Now we can fit the **glm** with treatment as the categorical explanatory variable:

```
glm(rv~treatment,family=binomial)

Degrees of Freedom: 4 Total; 2 Residual
Residual Deviance: 0.4206011
```

This is the value we calculated by hand. So there really is no mystery about the value of deviance. It is just another way of measuring lack of fit. The difference between 0.420 1974 and 0.420 6011 is due to our use of only 3 decimal places in calculating the deviance components. If you were perverse, you could always work it out for yourself. Everything we have done with sums of squares (e.g. F tests and r^2), you can do with deviance.

Overdispersion and hypothesis testing

All the different statistical procedures that we have met in earlier chapters can also be used with data on proportions. Factorial analysis of variance, multiple regression, and a variety of models in which different regression lines are fitted in each of several levels of one or more factors, can be carried out. The only difference is that we assess the significance of terms on the basis of chi-square, the increase in scaled deviance that results from removal of the term from the current model.

Bear in mind that hypothesis testing with binomial errors is less clear-cut than with Normal errors. While the chi-square approximation for changes in scaled deviance is reasonable for large samples (i.e. bigger than about 30), it is poorer with small samples. Most worrisome is the fact that the degree to which the approximation is satisfactory is itself unknown. This means that considerable care must be exercised in the interpretation of tests of hypotheses on parameters, especially when the parameters are marginally significant or when they explain a very small fraction of the total deviance. With binomial or Poisson errors we cannot hope to provide exact p values for our tests of hypotheses.

As with Poisson errors, we need to address the question of overdispersion (Chapter 29). When we have obtained the minimal adequate model, *the residual scaled deviance should be roughly equal to the residual degrees of freedom*. When the residual deviance is larger than the residual degrees of freedom there are two possibilities: either the model is misspecified, or the probability of success, p, is not constant within a given treatment level. The effect of randomly varying p is to increase the binomial variance from npq to

$$s^2 = npq + n(n-1)\sigma^2$$

leading to a large residual deviance. This occurs even for models that would fit well if the random variation were correctly specified.

One simple solution is to assume that the variance is not npq but $npqs$, where s is an unknown *scale parameter* $(s > 1)$. We obtain an estimate of the scale parameter by dividing the Pearson chi-square by the degrees of freedom, and use this estimate of s to compare the resulting scaled deviances for terms in the model using an F test (just as in conventional Anova). While this procedure may work reasonably well for small amounts of overdispersion, it is no substitute for proper model specification. For example, it is not possible to test the goodness of fit of the model.

Model criticism

Next we need to assess whether the standardised residuals are normally distributed and whether there are any trends in the residuals, either with the fitted values or with the explanatory variables. It is necessary to deal with *standardised residuals* because with error distributions like the binomial, Poisson or Gamma distributions, the variance changes with the mean. To obtain plots of the standardised residuals, we need to calculate

$$r_s = \frac{y - \mu}{\sqrt{V}} = \frac{y - \mu}{\sqrt{\mu(1 - \mu/n)}}$$

where V is the formula for the *variance function* of the binomial, μ are the fitted values, and n is the binomial denominator (successes plus failures).

Other kinds of standardised residuals (like Pearson residuals or deviance residuals) are available as options like this:

resid(model.name, type="pearson")

or you could use other types including deviance, working or response (see p. 307). To plot the residuals against the fitted values, we might arcsine transform the x axis (because the fitted data are proportions):

Summary

The most important points to emphasise in modelling with binomial errors are:

- create a two-column object for the response, using **cbind** to join together the two vectors containing the counts of success and failure

- check for overdispersion (residual deviance > residual degrees of freedom), and correct for it by using F tests rather than chi-square if necessary

- remember that you do not obtain exact p values with binomial errors; the chi-square approximations are sound for large samples, but small samples may present a problem

- the fitted values are counts, like the response variable

- the linear predictor is in logits (the log of the odds $= p/q$)

- you can back-transform from logits (z) to proportions (p) by $p = 1/(1 + 1/e^z)$

Applications

You can do as many kinds of modelling in a GLM as in a linear model; here we show examples of:

- regression with binomial errors (continuous explanatory variables)

- analysis of deviance with binomial errors (categorical explanatory variables)

- analysis of covariance with binomial errors (both kinds of explanatory variables)

Regression analysis with proportion data

This is an experiment from insect toxicology, where batches of insects were exposed to different doses of insecticide and the number of dead insects was recorded as the response variable:

```
attach(bioassay)
names(bioassay)
```

```
[1] "dose"    "dead"    "batch"
```

We begin by looking at the data. We need to calculate the proportion of insects killed:

```
pkill<-dead/batch
```

Then we can plot the proportion killed against the log of the dose like this:

```
plot(log(dose),pkill)
```

and fit a straight line through the scatterplot using **abline**:

```
abline(lm(pkill~log(dose)))
```

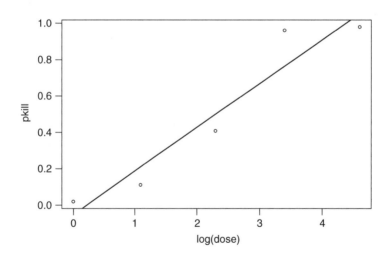

This plot demonstrates why we don't use linear regression with proportion data. At low doses the model predicts negative proportions killed (spontaneous generation) while at high doses it predicts more than 100% mortality. The correct plot is logit(proportion killed) against log(dose). First we calculate the logits, then plot the graph:

```
logit<-log(pkill/(1-pkill))
plot(log(dose),logit)
abline(lm(logit~log(dose)))
```

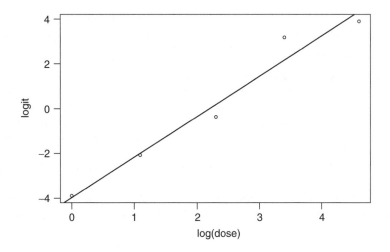

This is the regression that we shall be modelling in **glm**. We begin by binding together the vectors containing the counts of dead and live (= batch − dead) insects to form the response variable, y:

    y<-cbind(dead,batch-dead)

Now fit the model using **glm** with binomial errors:

    model<-glm(y~log(dose),family=binomial)

    summary(model)

    Coefficients:
                    Value Std. Error   t value
    (Intercept) -4.531819 0.4380398 -10.34568
      log(dose) 1.964394 0.1750339 11.22293

As in any regression, the first parameter is the intercept and the second is the slope. The table of coefficients is interpreted as meaning that our equation relating killing power to ln(dose) looks like this:

$$\ln\left(\frac{p}{1-p}\right) = -4.531\,819 + 1.964\,394 \times \ln(\text{dose})$$

The second column of numbers shows the standard errors of the intercept and slope respectively, and the final column shows Student's t test values. Here they are of no interest at all because we would not expect the intercept or the slope to be zero in a bioassay:

    (Dispersion Parameter for Binomial family taken to be 1)

Remember, binomial errors is an assumption, not a fact. We need to check that the dispersion parameter really is equal to 1:

        Null Deviance: 408.3527 on 4 degrees of freedom
    Residual Deviance: 10.82791 on 3 degrees of freedom

In fact, there is evidence of rather serious overdispersion: $10.828/3 = 3.61$. This could be caused by misspecification of the model (e.g. the relationship might not be linear) or by the influence of unmeasured variables. We shall need to take account of this later on, when estimating confidence intervals for the LD50 and LD90 (lethal doses for 50% and 90% of the insects).

```
Number of Fisher Scoring Iterations: 4

Correlation of Coefficients:
            (Intercept)
log(dose)   -0.9374279
```

Note that the estimates of the slope and the intercept are very closely correlated ($r = -0.94$).

We can use **mcheck** for model criticism, or plot(model). With so few points on the graph, the model-checking plots are rather uniformative, so we shall use the simpler output of **mcheck**:

mcheck(model)

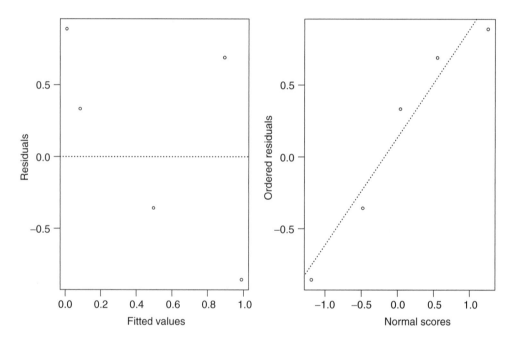

There is a rather worrying U-shape to the residuals, suggesting a misspecified model, but the errors look reasonably close to Normal. Coupled with the overdispersion we calculated earlier, and the very low replication ($n = 5$) these model criticisms suggest that we should be more than usually circumspect in interpreting these results.

A cautionary example

A problem with proportion data arises from unequal weights: 50% based on a sample of 1 out of 2 is a very different thing from 50% based on 200 out of 400. The following example shows the sort of thing that can go wrong.

In a study of density-dependent seed mortality in trees, seeds were placed in six groups at different densities (from 4 to 100). The scientist returned after 48 hours and counted the number of seeds taken away by animals. We are interested to see if the *proportion* of seeds taken is influenced by the size of the cache of seeds. These are the data:

```
taken<-c(3,3,6,16,26,39)
size<-c(4,6,10,40,60,100)
```

Now calculate the proportion removed (p) and the logarithm of pile size (ls):

```
p<-taken/size
ls<-log(size)
plot(ls,p)
abline(lm(p~ls))
```

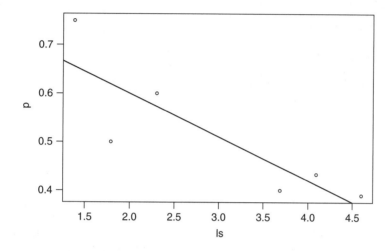

It certainly looks as if the proportion taken declines as density increases. Let's do the regression to test the significance of this:

```
summary(lm(p~ls))
```

```
Coefficients:
             Estimate Std. Error  t value  Pr(>|t|)
(Intercept)   0.77969    0.08939    8.723  0.000952 ***
ls           -0.08981    0.02780   -3.230  0.031973 *
```

So the density-dependent effect is significant ($p = 0.032$). But these were proportion data. Perhaps we should have transformed them before doing the regression? We can repeat the regression using the logit transformation:

```
logit<-log(p/(1-p))
```

```
summary(lm(logit~ls))
```

```
Coefficients:
             Estimate Std. Error  t value  Pr(>|t|)
(Intercept)    1.1941     0.3895    3.065    0.0375 *
ls            -0.3795     0.1212   -3.132    0.0351 *
```

The regression is still significant, suggesting density dependence in removal rate.

Now we shall carry out the analysis properly, taking account of the fact that the very high death rates were actually estimated from very small samples of seeds (4, 6 and 10 individuals respectively).

Using **glm** with binomial errors takes care of this by giving low weight to estimates with small binomial denominators. First we need to construct a response vector (using **cbind**) that contains the number of successes and the number of failures. In our example this means the number of seeds taken away and the number of seeds left behind. We do it like this:

```
left<-size - taken
y<-cbind(taken,left)
```

Now the modelling. We use **glm** instead of **lm**, but otherwise everything else is the same as usual:

```
model<-glm(y~ls,binomial)

summary(model)

Coefficients:
              Estimate Std. Error z value Pr (>|z|)
(Intercept)    0.8815     0.7488    1.177    0.239
ls            -0.2944     0.1817   -1.620    0.105

(Dispersion parameter for binomial family taken  to be 1)

    Null deviance: 3.7341  on 5  degrees of freedom
Residual deviance: 1.0627  on 4  degrees of freedom
AIC: 25.543
```

When we do the analysis properly, using weighted regression in **glm**, we see that there is no significant effect of density on the proportion of seeds removed ($p = 0.105$). This analysis was carried out by a significance test on the parameter values. It is usually better to assess significance by a deletion test. To do this, we fit a simpler model with an intercept but no slope:

```
model2<-glm(y~1,binomial)

anova(model,model2)

Analysis of Deviance Table

Response: y

    Resid. Df Resid. Dev Df Deviance
ls      4       1.0627
1       5       3.7341 -1  -2.6713
```

The thing to remember is that the change in deviance is a chi-square value. In this case the chi-square on 1 d.f. is only 2.671 and this is less than the value in tables (3.841), so we conclude *there is no evidence for density dependence* in the death rate of these seeds. It is usually more convenient to have the chi-square test printed out as part of the analysis of deviance table. This is very straightforward. We just include the option test="Chi" in the **anova** directive:

```
anova(model,model2,test="Chi")

Analysis of Deviance Table
Response: y

    Resid. Df Resid. Dev Df Deviance P(>|Chi|)
ls      4       1.0627
1       5       3.7341 -1  -2.6713    0.1022
```

The moral is clear. Tests of density dependence need to be carried out using **weighted regression**, so that the death rates estimated from small samples do not have undue influence on the value of the regression slope. In this example, using the correct linearisation technique (logit transformation) did not solve the problem; logit regression agreed wrongly with simple regression that there was significant inverse density dependence (i.e. the highest death rates were estimated at the lowest densities). But these high rates were estimated from tiny samples (3 out of 4 and 3 out of 6 individuals). The correct analysis uses weighted regression and binomial errors, in which case these points are given less influence. The result is that the data no longer provided convincing evidence of density dependence. In fact, the data are consistent with the hypothesis that the death rate was the same at all densities.

Analysis of deviance with proportion data

This experiment involves female-biased sex ratios in different genotypes of insect. The question is whether different genotypes produce significantly different proportions of males among their offspring.

```
attach(sexratio)
names(sexratio)
```

```
[1] "male"        "total"       "genotype"
```

The first step is to look at the data. To make the plot easier to understand, we shall calculate the proportion of eggs producing males and do a box plot of this for each genotype:

```
pmale<-male/total
plot(genotype,pmale)
```

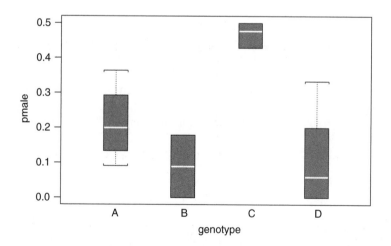

There are apparently substantial differences between the genotypes in mean sex ratio, but the variability (especially in genotype D) is quite large. We begin the statistical modelling by creating the response variable y, as the two variables *male* and *female* (= *total* − *male*) bound together into a single object:

```
y<-cbind(male,total-male)
```

There are no more preliminaries, so we can use **glm** to carry out the analogue of a one-way analysis of variance (called an analysis of deviance) using our single categorical explanatory variable, *genotype*:

```
model<-glm(y~genotype, family=binomial)
summary(model)
```

The only difference is that we write **glm** instead of **aov**, and add the expression family=binomial. The output gives the model formula:

```
Call: glm(formula = y ~ genotype, family = binomial)
```

It is useful to have the model formula printed here when we come back to look at the output days (or months) later, when we have forgotten entirely what the object model contained.

```
Deviance Residuals:
      Min        1Q     Median        3Q       Max
 -1.590747 -1.014794 -0.2068726 0.2412256 1.79567
```

This is a summary of the lack of fit of the model. Deviance is equivalent to sums of squares in linear models, and this shows that the middle 50% of the deviance residuals lay between −1.0148 and 0.2412. It also prints the largest and smallest residuals.

```
Coefficients:
                 Value Std. Error    t value
(Intercept) -1.2074291  0.1867334 -6.4660571
  genotype1 -0.2405267  0.2881459 -0.8347393
  genotype2  0.4285099  0.1353875  3.1650631
  genotype3 -0.2278071  0.1089334 -2.0912512
```

This is the most important part of the summary. It shows the parameter estimates (headed Value) as Helmert contrasts (see p. 340), their standard errors and a t test comparing the values to zero. The most important thing to understand here is that the parameter estimates are **logits**. They are the logs of $p/(1 - p)$. Because the explanatory variable, *genotype*, is categorical the parameter estimates are means and differences between means, expressed on the logit scale. We interpret this table as showing that there *are* significant differences between the genotypes in the mean proportion of males emerging from clutches of eggs, because at least one of the rows labelled genotype has a t value bigger than 2 (in fact, two of the rows look significantly different). Note that the genotype levels are numbered as contrasts (1–3) rather than given their names (A, B, C and D). Generally, factor levels will appear in alphabetical order unless we rank them differently (see p. 251).

```
Dispersion Parameter for Binomial family taken to be 1
```

With binomial, exponential or Poisson errors the dispersion parameter is set at a value of 1. This is because these are one-parameter models where the variance is a known function of the mean, and is not estimated separately from the data. This, of course, is an assumption, and we need to test it. If all is well, the residual deviance should be approximately the same as the residual degrees of freedom. If the residual deviance is substantially larger than the residual deviance, we have a problem known as **overdispersion** (see p. 318). We check for the possibility of overdispersion on the next lines of output:

```
   Null Deviance: 32.24692 on 13 degrees of freedom

Residual Deviance:  14.5101 on 10 degrees of freedom
```

The ratio of residual deviance to residual d.f. is $14.51/10 = 1.451$. This is greater than 1 but not unduly worrying. As a rule of thumb, accounting for overdispersion will increase the size of the standard errors by about the square root of this ratio. In this case they will be increased by $\sqrt{1.451} = 1.2046$. We return to this later (p. 528). The remainder of the output shows the number of times the computer went round the iterative loop to arrive at its parameter estimates:

```
Number of Fisher Scoring Iterations: 4
```

and the correlations between the various estimates:

```
Correlation of Coefficients:
           (Intercept)   genotype1    genotype2
genotype1   0.3364157
genotype2  -0.2772673   -0.3093347
genotype3   0.0179454   -0.1922277    0.1584304
```

To see the sex ratios predicted by the model, we simply type

fitted(model)

and obtain the proportions directly. Note that we get the proportions, not the counts (the response variable) or the logits (the linear predictor):

```
        1          2          3          4          5
0.2372881  0.2372881  0.2372881  0.2372881  0.1612903
        6          7          8          9         10
0.1612903  0.4693878  0.4693878  0.4693878  0.1311475
       11         12         13         14
0.1311475  0.1311475  0.1311475  0.1311475
```

If, for any reason, you want the values of the logits, then they are stored as part of the model object called **model$linear.predictors**. Remember that we have fitted a four-level factor to the data. The first four numbers all came from genotype A, so they all have the same fitted value (0.237 2881), which is the average proportion male for the first genotype. This was computed as the total number of males emerging from genotype A eggs divided by the total number of eggs. We can easily verify this using **tapply** to compute the subtotals:

tapply(male,genotype,sum)

```
A    B    C    D
14   5    23   8
```

tapply(total,genotype,sum)

```
A    B    C    D
59   31   49   61
```

so the mean proportion for genotype A is $14/59 = 0.2372881$ as predicted. With binomial and Poisson errors, we test the significance of model terms using chi-square, and we can specify this as an option in producing our **analysis of deviance** table. The function **anova** produces an analysis of deviance table automatically when it is applied to a **glm** object:

anova(model,test="Chi")

```
            Df Deviance Resid.  Df Resid. Dev      Pr(Chi)
     NULL                       13   32.24692
genotype     3 17.73682         10   14.51010  0.0004983831
```

Adding the four-level factor called *genotype* to the null model reduced the deviance from 32.25 to 14.51 and this reduction in deviance of 17.74 on 3 d.f. is significant at $p = 0.000\,498$ (note that the result will not be quite so significant once we correct for overdispersion). For comparison, the $\alpha = 0.05$ value of chi-square in tables with 3 d.f. is 7.851. So there are significant differences in sex ratio between the genotypes.

The simplest way to correct for overdispersion is to use an F test rather than a chi-square test in producing the analysis of deviance table:

```
anova(model,test="F")
```

```
            Df Deviance Resid. Df Resid. Dev  F Value       Pr(F)
    NULL                      13    32.24692
genotype     3 17.73682       10    14.51010 4.668026 0.02741896
```

Doing this has reduced the significance from 0.000 498 to $p = 0.0274$, but the differences in sex ratio are still significant. One way to do the F test is to calculate an error variance based on the overdispersion ratio, 1.451 (see above), and a treatment variance based on the ratio of explained deviance (17.737) to treatment degrees of freedom ($17.737/3 = 5.912$) just like in a least squares Anova. This method gives $F = 5.912/1.451 = 4.075$. You will see that the F ratio in the analysis of deviance table, above, is slightly different from this (4.668), because the computer has used an alternative way of estimating the error variance, based on Pearson's chi-square rather than residual deviance.

Model simplification with overdispersion

Just as with Poisson errors, logistic analysis with binomial errors assumes that the variance is a known function of the mean. In this case the mean is np and the variance ought to be $np(1 - p)$ as explained above. If, once we have fitted the minimal adequate model, the residual deviance is larger than the residual degrees of freedom then this assumption is unfounded and we have overdispersion. As with Poisson errors, the simplest way to compensate for this is to carry out hypothesis testing using F tests rather than chi-square, and to use an empirical scale parameter as the denominator of the F statistic.

This example concerns the germination of seeds of two genotypes of the parasitic plant *Orobanche* and two plant extracts (bean and cucumber) that were used in an attempt to stimulate germination (Collett, 1991):

```
attach(germination)
names(germination)
```

```
[1] "count"     "sample"     "Orobanche"     "extract"
```

We start by calculating the proportion germinated, so that we can plot the means:

```
p<-count/sample
par(mfrow=c(1,2))
plot(extract,p)
plot(Orobanche,p)
```

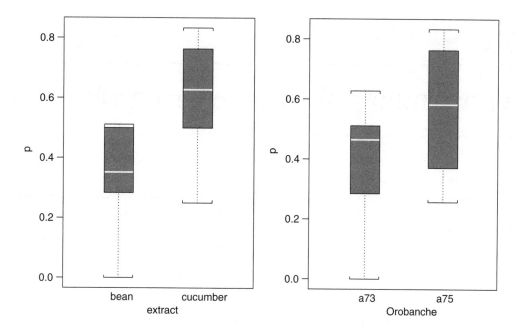

There is a clear effect of extract on germination rate, but the effect of *Orobanche* genotype is much less pronounced. Count is the number of seeds that germinated out of a batch of size = sample. So the number that didn't germinate is sample—count, and we can construct the response vector like this:

> y<-cbind(count, sample-count)

Each of the categorical explanatory variables has two levels:

> levels(Orobanche)

```
[1] "a73"   "a75"
```

> levels(extract)

```
[1] "bean"     "cucumber"
```

We want to test the hypothesis that there is no interaction between *Orobanche* genotype (a73 or a75) and plant extract (bean or cucumber) on the germination rate of the seeds. This requires a factorial analysis using the asterisk operator * like this:

> model<-glm(y ~ Orobanche * extract, binomial)

> summary(model)

```
Coefficients:
                  Estimate Std. Error   z value  Pr(>|z|)
(Intercept)        -0.4122     0.1842    -2.238    0.0252 *
Orobanche          -0.1459     0.2232    -0.654    0.5132
extract             0.5401     0.2498     2.162    0.0306 *
Orobanche.extract   0.7781     0.3064     2.539    0.0111 *

(Dispersion parameter for binomial family taken to be 1)
```

```
   Null deviance: 98.719  on 20  degrees of freedom
Residual deviance: 33.278  on 17  degrees of freedom
AIC: 117.87
```

At first glance, it looks as if there is a highly significant interaction ($p = 0.0111$). But we need to check that the model is sound. The *first thing is to check for overdispersion*. The residual deviance is 33.278 on 17 d.f. so the model is quite badly overdispersed:

```
33.279 / 17
```

```
[1] 1.957588
```

The overdispersion factor is almost 2. The simplest way to take this into account is to use what is called an 'empirical scale parameter' to reflect the fact that the errors are not binomial as we assumed, but were larger than this (i.e. overdispersed) by a factor of 1.9576. We use **update** to remove the interaction term in the normal way:

```
model2<-update(model, ~. - Orobanche:extract)
```

The only difference is that we use an F test instead of a chi-square test to compare the original and simplified models:

```
anova(model,model2,test="F")
```

```
Analysis of Deviance Table
                                    Resid. Df Resid. Dev Df Deviance      F Pr(>F)
Orobanche + extract + Orobanche:extract 17      33.278
Orobanche + extract                     18      39.686 -1   -6.408 3.442 0.08099
```

Now you see that the interaction is not significant ($p = 0.081$). For comparison, here is the wrong deletion test, using chi-square:

```
anova(model,model2,test="Chi")
```

```
Analysis of Deviance Table
                                    Resid. Df Resid. Dev Df Deviance Pr(>|Chi|)
Orobanche + extract + Orobanche:extract 17      33.278
Orobanche + extract                     18      39.686 -1   -6.408      0.011
```

Doing it this way, and failing to correct for overdispersion, suggests that the interaction was significant and that different genotypes of *Orobanche* respond differently to the two plant extracts. Note that we haven't made the overdispersion go away, we have simply taken it into account in carrying out our hypothesis testing.

The next step is to see if any further model simplification is possible:

```
anova(model2,test="F")
```

```
          Df Deviance Resid. Df Resid. Dev      F     Pr(>F)
NULL                     20        98.719
Orobanche  1   2.544    19        96.175  1.195    0.2887
extract    1  56.489    18        39.686 26.542 6.691e-005 ***
```

There is a highly significant difference between the two plant extracts on germination rate, but it is not obvious that we need to keep *Orobanche* genotype in the model. We try removing it:

```
model3<-update(model2, ~. - Orobanche)
anova(model2,model3,test="F")
```

```
Analysis of Deviance Table
                     Resid. Df Resid. Dev Df Deviance      F Pr(>F)
Orobanche + extract        18      39.686
extract                    19      42.751 -1   -3.065  1.440 0.2457
```

There is no justification for retaining *Orobanche* in the model. So the minimal adequate model contains just two parameters:

coef(model3)

```
(Intercept)     extract
 -0.5121761 1.0574031
```

What, exactly, do these two numbers mean? Remember that the coefficients are from the linear predictor. They are on the transformed scale, so because we are using binomial errors, they are in logits, $\ln(p/(1-p))$. To turn them into the germination rates for the two plant extracts requires us to do a little calculation. To go from a logit x to a proportion p, you need to do the following sum:

$$p = \frac{1}{1 + e^{-x}}$$

So our first x value is -0.5122 and we calculate

1/(1+1/(exp(-0.5122)))

```
[1]   0.3746779
```

This says that the mean germination rate of the seeds with the first plant extract was 37%. What about the parameter for extract (1.057)? Remember that with categorical explanatory variables *the parameter values are differences between means*. So to get the second germination rate we *add 1.057 to the intercept* before back-transforming:

1/(1+1/(exp(-0.5122+1.0574)))

```
[1]   0.6330212
```

This says that the germination rate was nearly twice as great (63%) with the second plant extract (cucumber). Obviously we want to generalise this process, and also to speed up the calculations of the estimated mean proportions. We can use **predict** to help here:

tapply(predict(model3),extract,mean)

```
      bean    cucumber
 -0.5121761  0.5452271
```

This gives us the average logits for the two levels of extract, showing that seeds germinated better with cucumber extract than with bean extract. To get the mean proportions we just apply the back transformation to this **tapply** (use Up Arrow to edit):

tapply(1/(1+1/exp(predict(model3))),extract,mean)

```
      bean    cucumber
 0.3746835  0.6330275
```

These are the two proportions we calculated earlier. There is an even easier way to get the proportions, because there is an option for **predict** called type="response" which makes predictions on the back-transformed scale automatically:

```
tapply(predict(model3,type="response"),extract, mean)
```

```
      bean  cucumber
 0.3746835 0.6330275
```

It is interesting to compare these figures with the average of the raw proportions and the overall average. First we need to calculate the proportion germinating, *p*, in each sample:

```
p<-count/sample
```

then we can average it by extract:

```
tapply(p,extract,mean)
```

```
      bean  cucumber
 0.3487189 0.6031824
```

You see that this gives a *different* answer. Not too different in this case, but different nonetheless. The correct way to average proportion data is to add up the total counts for the different levels of extract, and only then to turn them into proportions:

```
tapply(count,extract,sum)
```

```
bean cucumber
 148      276
```

This means that 148 seeds germinated with bean extract and 276 with cucumber extract:

```
tapply(sample,extract,sum)
```

```
bean cucumber
 395      436
```

This means there were 395 seeds treated with bean extract and 436 seeds treated with cucumber. So the answers we want are 148/395 and 276/436. We automate the calculations like this:

```
ct<-as.vector(tapply(count,extract,sum))
```

```
sa<-as.vector(tapply(sample,extract,sum))
```

```
ct/sa
```

```
[1]   0.3746835 0.6330275
```

These are the correct mean proportions that were produced by **glm**. The moral here is that you calculate the average of proportions by using total counts and total samples and *not* by averaging the raw proportions.

Analysis of covariance with proportion data

Here we carry out a logistic analysis of covariance to compare three different insecticides in a toxicology bioassay. Three products (A, B and C) were applied to batches of insects in circular arenas. The initial numbers of insects (*n*) varied between 15 and 24. The insecticides were each applied at 17 different doses, varying on a log scale from 0.1 to 1.7 units of active ingredient per square metre. The number of insects found dead after 24 hours, *c*, was counted.

```
attach(logistic)
names(logistic)
```

```
[1] "logdose" "n"         "dead"      "product"
```

First we calculate the proportion dying:

```
p<-dead/n
```

Now we produce the scatterplot:

```
plot(logdose,p,type="n")
```

We used type="n" to make sure that all the points from all three products fit onto the plotting surface. Now we add the points for each product, one at a time using different plotting symbols:

```
points(logdose[product=="A"],p[product=="A"],pch=16)
points(logdose[product=="B"],p[product=="B"])
points(logdose[product=="C"],p[product=="C"],pch="+")
```

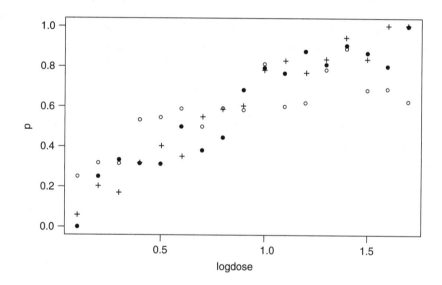

It looks as if the slope of the response curve is shallower for product B (open circles), but there is no obvious difference between products A (solid circles) and C (crosses). We shall test this using analysis of covariance in which we begin by fitting three separate logistic regression lines (one for each product) then ask whether all the parameters need to be retained in the minimum adequate model.

Begin by creating the response vector using **cbind** like this:

```
y<-cbind(dead,n-dead)
```

Now we fit the maximal model using the * notation to get three slopes and three intercepts:

```
model<-glm(y~product*logdose,binomial)
```

```
summary(model)
```

```
Coefficients:
```

	Estimate	Std. Error	z value	Pr(>|z|)	
(Intercept)	-2.0410	0.3028	-6.741	1.57e-011	***
productB	1.2920	0.3841	3.364	0.000768	***

```
productC              -0.3080    0.4301  -0.716  0.473977
logdose                2.8768    0.3423   8.405  < 2e-016 ***
productB.logdose      -1.6457    0.4213  -3.906  9.38e-005 ***
productC.logdose       0.4319    0.4887   0.884  0.376845
```

(Dispersion parameter for binomial family taken to be 1)

```
      Null deviance: 311.241  on 50  degrees of freedom
Residual deviance:  42.305  on 45  degrees of freedom
AIC: 201.31
```

First we check for overdispersion. The residual deviance was 42.305 on 45 degrees of freedom, so that is perfect. There is no evidence of overdispersion at all, so we can use chi-square tests in our model simplification. There looks to be a significant difference in slope for product B (as we suspected from the scatterplot); its slope is shallower by -1.6457 than the slope for product A (labelled `logdose` in the table above). But what about product C? Is it different from product A in its slope or its intercept? The z tests certainly suggest that it is not significantly different. Let's remove it by creating a new factor called *newfac* that combines A and C:

newfac<-factor(1+(product=="B"))

Now we create a new model in which we replace the three-level factor called *product* by the two-level factor called *newfac*:

model2<-update(model , ~. - product*logdose + newfac*logdose)

and then we ask whether the simpler model is significantly less good at describing the data than the more complex model?

anova(model,model2,test="Chi")

```
Analysis of Deviance Table

Response: y

                            Resid. Df Resid. Dev Df Deviance P(>|Chi|)
product+logdose+product:logdo 45       42.305
logdose+newfac+logdose:newfac 47       43.111  -2  -0.805     0.668
```

Evidently, this model simplification was completely justified ($p = 0.668$). So is this the minimal adequate model? We need to look at the coefficients:

summary(model2)

```
Coefficients:
                Estimate Std. Error z value  Pr(>|z|)
(Intercept)      -2.1997    0.2148  -10.240  < 2e-016 ***
logdose           3.0988    0.2438   12.711  < 2e-016 ***
newfac            1.4507    0.3193    4.543  5.54e-006 ***
logdose.newfac   -1.8676    0.3461   -5.397  6.79e-008 ***
```

Yes it is. All of the four coefficients are significant (*** in this case); we need two different slopes and two different intercepts. So now we can draw the fitted lines through our scatterplot. You may need to revise the steps involved in this. We need to make a vector of values for the x axis (we might call it xv) and a vector of values for the two-level factor to decide which graph to draw (we might call it nf):

```
max(logdose)
```

```
[1]   1.7
```

```
min(logdose)
```

```
[1]   0.1
```

So we need to generate a sequence of 100 or so values of xv between 0.1 and 1.7:

```
xv<-seq(0.1,1.7,0.01)
```

```
length(xv)
```

```
[1]   161
```

This tells us that the vector of factor levels in nf needs to be made up of exactly 161 repeats of factor level = 1 then 161 repeats of factor level = 2, like this:

```
nf<-factor(c(rep(1,161),rep(2,161)))
```

Inside the **predict** directive we combine xv and nf into a data frame where we give them *exactly the same names* as the explanatory variables had in model 2 (i.e. *logdose* and *newfac*). Remember that we need to back-transform from **predict** (which is in logits) onto the scale of proportions (see above).We shall draw the graph in two steps, first for factor level 1 (i.e. for products A and C combined), and then for the second factor level (product B), using type="response" to carry out the back transformation for us automatically:

```
lines(xv,predict(model2,type="response",data.frame(logdose=c(xv,xv),newfac=nf))[nf==1])
lines(xv,predict(model2,type="response",data.frame(logdose=c(xv,xv),newfac=nf))[nf==2])
```

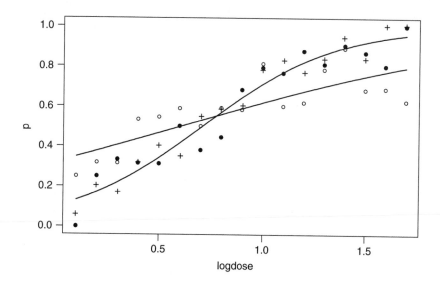

All three products have very similar LD50s (roughly logdose = 0.8), but higher doses of A and C are much better than B at killing insects in large numbers.

Note that it is essential that all the vectors in the data frame are the same length; we have had to include two sets of the x values (length = 161) by writing c(xv,xv) to match the length of the factor levels *nf* (length = 322). Note also the use of subscripting of the **predict** object, to get only the 161 values we want to plot against the 161 values of *xv* for any particular product. It will take a good deal of practice before these multiple fitted plots seem as straightforward as they really are.

Summary: binomial errors

- make a two-column response vector containing the successes and failures

- use **glm** with family=binomial (you don't need to include the family= bit)

- fit the maximal model

- test for overdispersion

- if you find overdispersion then use F tests not chi-square tests of deletion

- begin model simplification by removing the interaction terms

- remove non-significant main effects if there are no significant interactions

- use **plot** to obtain your model-checking diagnostics

- back-transform using **predict** with the option type="response" to obtain means

Further reading

Collett, D. (1991) *Modelling Binary Data*. London, Chapman & Hall.

29

Count data: Poisson errors

Up to Chapter 26 the data were all continuous measurements like weights, heights, lengths, temperatures, growth rates, and so on. A great deal of the data collected by scientists, medical statisticians and economists, however, is in the form of *counts* (whole numbers or integers). The number of individuals that died, the number of firms going bankrupt, the number of days of frost, the number of red blood cells on a microscope slide, or the number of craters in a sector of lunar landscape are all potentially interesting variables for study. With count data, the number 0 often appears as a value of the response variable (consider what a 0 would mean in the context of the examples just listed).

For our present purposes, it is useful to think of count data as coming in four types:

- data on *frequencies*, where we count how many times something happened, but we have no way of knowing how often it did *not* happen (e.g. lightning strikes, bankruptcies, deaths, births)

- data on *proportions*, where both the number doing a particular thing, and the total group size are known (insects dying in an insecticide bioassay, sex ratios at birth, proportions responding in a questionnaire)

- *category* data, in which the response variable is a count distributed across a categorical variable with two or more levels

- *binary* response variables (dead or alive, solvent or insolvent, infected or immune)

We dealt with proportion data in Chapter 28 and we shall deal with binary response variables in Chapter 30. Here we are concerned with pure counts rather than proportions. Straightforward linear regression methods (constant variance, Normal errors) are not appropriate for count data for five main reasons:

- the linear model might lead to the prediction of negative counts

- the variance of the response variable is likely to increase with the mean

- the errors will not be normally distributed

- zeros are difficult to handle in transformations

- some distributions (e.g. lognormal or Gamma) don't allow zeros

S-Plus handles count data very elegantly using **glm** by specifying family=poisson which sets errors=Poisson and link=log. The log link ensures that all the fitted values are positive, while the Poisson errors take account of the fact that the data are integer and have variances that are equal to their means.

The Poisson distribution

The Poisson distribution is widely used for the description of count data that refer to cases where we know how many times something happened (e.g. kicks from cavalry horses, lightning strikes, bomb hits), but we have no way of knowing how many times it did not happen. This is in contrast to the binomial distribution (Chapter 28) where we know how many times something did not happen as well as how often it did happen (e.g. if we got 6 heads out of 10 tosses of a coin, we must have got 4 tails).

The Poisson is a one-parameter distribution, specified entirely by the mean. The variance is identical to the mean, so the variance/mean ratio is equal to one. Suppose we are studying the structure of concrete bridges, and our data consist of the numbers of cracks per $10\,\mathrm{m}$ length of roadway (x). Many sections have no cracks at all, but some may have as many as 5 or 6 cracks. If the mean number of cracks per section is λ, then the probability of observing x cracks per section is given by

$$P(x) = \frac{e^{-\lambda}\lambda^x}{x!}$$

This can be calculated very simply on a hand calculator because

$$P(x) = P(x-1)\frac{\lambda}{x}$$

This means that if you start with the *zero term*

$$P(0) = e^{-\lambda}$$

then each successive probability is obtained simply by multiplying by the mean and dividing by x.

Because the data are whole numbers (integers), it means that the residuals (y − the fitted values) can only take a restricted range of values. If the estimated mean was 0.5, for example, then the residuals for counts of 0, 1, 2 and 3 could only be −0.5, 0.5, 1.5 and 2.5. The Normal distribution assumes the residuals can take any values (it is a continuous distribution).

Similarly, the Normal distribution allows for negative fitted values. Since we cannot have negative counts, this is clearly not appropriate. S-Plus deals with this by using the log link function when Poisson errors are specified:

$$y = \exp\left(\sum \beta_i x_i\right)$$

The fitted values are antilogs and so they can never be negative. Even if the linear predictor were $-\infty$, the fitted value would be $e^{-\infty} = 0$.

A further difference from the examples in previous chapters is that S-Plus uses maximum likelihood methods other than least squares to estimate the parameters values and their standard errors. Given a set of data and a particular model, then the maximum likelihood estimates of the parameter values are *those values that make the observed data most likely*, hence the name (see p. 88).

Because the Poisson is a one-parameter distribution (the mean is equal to the variance), S-Plus does not attempt to estimate a scale parameter (it is set to a default value of 1.0). If the error structure of the data really is Poisson, then the ratio of residual deviance to degrees of freedom after model fitting should be 1.0. If the ratio is substantially greater than 1.0, then the data are said to show overdispersion, and remedial measures may need to be taken (e.g. the use of an empirical scale parameter or the specification of negative binomial errors).

With Poisson errors, the change in deviance attributable to a given factor is distributed asymptotically as chi-square. This makes hypothesis testing extremely straightforward. We simply remove a given factor from the maximal model and note the resulting change in deviance and in the degrees of freedom. If the change in deviance is larger than the critical value of chi-square (**qchisq**) then we retain the term in the model; if the change in deviance is less than the value of chi-square in tables, the factor is insignificant and can be left out of the model.

Thus, the only important differences you will need to remember in modelling with Poisson errors rather than Normal errors are:

- use **glm** rather than **aov** or **lm**

- the family=poisson directive must be specified (but the family= bit is optional)

- hypothesis testing involves deletion followed by chi-square tests

- beware of overdispersion, and correct for it if necessary by using F tests

- do not collapse contingency tables over important explanatory variables (see p. 556)

Deviance with Poisson errors

Up to Chapter 26, lack of fit was measured by *SSE*, which is the residual or error sum of squares:

$$SSE = \sum (y - \hat{y})^2$$

where $\hat{y}$ are the fitted values estimated by the model. With the **glm** function, *SSE* is only the maximum likelihood estimate of lack of fit when the model has Normal errors and the identity link, in which case you have to ask, Why am I using **glm**? Generally, we use *deviance* to measure lack of fit in **glm**. For Poisson errors the deviance is

$$\text{Poisson deviance} = 2 \sum O \ln \left(\frac{O}{E} \right)$$

where O is the observed count, and E is the expected count as predicted by the current model. Let's see how this works. Suppose you have four counts, two from each of two locations. Location A produced counts of 3 and 6. Location B produced counts of 4 and 7. The total deviance (like *SST*) is based on the whole sample of four numbers. The expected count is the overall mean:

```
mean(c(3,6,4,7))
```

```
[1]  5
```

This allows us to calculate the total deviance:

$$2 \times \left\{ 3 \times \ln \left(\frac{3}{5} \right) + 6 \times \ln \left(\frac{6}{5} \right) + 4 \times \ln \left(\frac{4}{5} \right) + 7 \times \ln \left(\frac{7}{5} \right) \right\}$$

Calculating the value gives

```
2*(3*log(3/5)+6*log(6/5)+4*log(4/5)+7*log(7/5))
```

```
[1]   2.048368
```

Now the two different location means are $9/2 = 4.5$ and $11/2 = 5.5$ respectively. The residual deviance after fitting a two-level factor for location should therefore be

$$2 \times \left\{ 3 \times \ln \left(\frac{3}{4.5} \right) + 6 \times \ln \left(\frac{6}{4.5} \right) + 4 \times \ln \left(\frac{4}{5.5} \right) + 7 \times \ln \left(\frac{7}{5.5} \right) \right\}$$

```
2*(3*log(3/4.5)+6*log(6/4.5)+4*log(4/5.5)+7*log(7/5.5))
```

```
[1]   1.848033
```

Given that the total deviance is 2.048 and the residual deviance is 1.848, we calculate the treatment deviance (due to differences between locations) as $2.048 - 1.848 = 0.2$.

Let's see if **glm** with Poisson errors gives the same answers. Data entry first:

```
y<-c(3,6,4,7)
location<-factor(c("A","A","B","B"))
```

Now the statistical modelling:

```
glm(y~location,poisson)

Degrees of Freedom: 4 Total; 2 Residual
Residual Deviance: 1.848033
```

So far so good. The residual deviance (1.848) is as we calculated it to be. How about the total and treatment deviances? We can get the total deviance if we fit the null model y ~ 1:

```
glm(y~1,family=poisson)

Degrees of Freedom: 4 Total; 3 Residual
Residual Deviance: 2.048368
```

So there is no mystery to the values of deviance. We could calculate them if we wanted to, at least for models as simple as this, but note that more complex models require *iterative fits* to estimate the parameter values.

Continuous explanatory variables with count data: log-linear regression

This example involves counts of prostate cancer (cancer 'clusters') and distance from a nuclear processing plant. The response variable contains many zeros and is completely unsuitable for standard regression analysis. The single explanatory variable is continuous: distance to the nuclear plant from the clinic where the diagnosis was made (not from where the patients actually lived or worked). The issue is whether or not the data provide any evidence that the number of cancers increases with proximity to the plant.

```
attach(clusters)
names(clusters)

[1] "Cancers"    "Distance"

plot(Distance,Cancers)
```

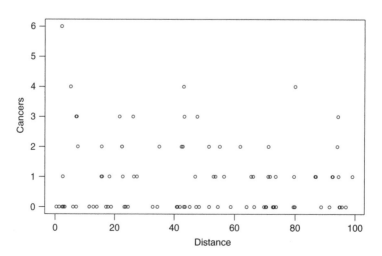

The first thing that you notice about plots of count data is that all the data are in sharp horizontal rows reflecting the integer values of the response. There are lots of zeros at all distances away from the nuclear plant. The largest value (the cluster of six cases) is close to the plant. But is there evidence for any trend in the number of cases as distance from the plant increases?

The modelling is exactly like any other regression except that we replace **lm** by **glm**, and add the phrase family=poisson after the model formula (family= is optional):

> model<-glm(Cancers~Distance,family=poisson)

> summary(model)

```
Coefficients:
                   Value    Std. Error      t value
(Intercept)    0.186933539  0.18777172    0.9955362
   Distance   -0.006139087  0.00365356   -1.6803029
```

There is a negative trend in the data (slope $= -0.006\,14$) but it is not significant under a two-tailed test ($t = 1.68$). If you were desperate, you might say that a one-tailed test is appropriate, because we expected *a priori* that the slope would be negative. I don't buy that. Anyway, we are not finished yet. Poisson errors is an assumption not a fact.

```
(Dispersion Parameter for Poisson family taken to be 1)
```

We need to test for *overdispersion*. This would be evidenced if the residual deviance were larger than the residual degrees of freedom:

```
Null Deviance: 149.4839 on 93 degrees of freedom

Residual Deviance: 146.6431 on 92 degrees of freedom
```

The dispersion parameter is actually $146.64/92 = 1.594$ and we should make allowance for this overdispersion in our analysis. As a rule of thumb, the standard error of the parameters will increase as $\sqrt{\text{dispersion parameter}}$ (i.e. they will be 1.26 times larger in this case). A better way to adjust for overdispersion is to use an F test with an empirical scale parameter instead of a chi-square test. We can accomplish this by deleting distance from the model using **update**, then comparing the model fits using **anova** in which we specify test="F":

> model2<-update(model, ~ . - Distance)

Be careful with the punctuation; it is , ~ . - (comma tilde dot minus).

> anova(model,model2,test="F")

```
Analysis of Deviance Table

Response: Cancers

     Terms Resid. Df Resid. Dev Test Df  Deviance   F Value      Pr(F)
1 Distance        92   146.6431
2        1        93   149.4839      -1 -2.840826 1.841963  0.1780414
```

There is no evidence for a decline in the number of cancers with distance. An F value as large as 1.84 will arise by chance alone with probability $p = 0.18$ when there is no trend in cancers with distance.

Analysis of deviance: categorical explanatory variables with count data

Count data were obtained on the numbers of slugs found beneath 40 tiles placed in a stratified random grid over each of two permanent grasslands. This was part of a pilot study in which the question was simply whether the mean slug density differed significantly between the two grasslands:

```
attach(slugsurvey)
names(slugsurvey)

[1]  "slugs"  "field"
```

These count data seem ideally suited for analysis using **glm** with Poisson errors:

```
plot(field,slugs)
```

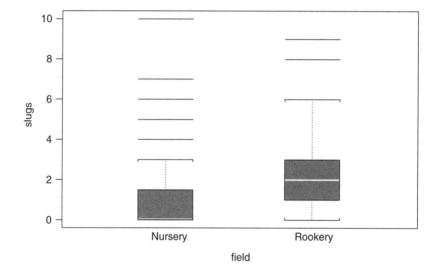

It certainly looks as if median slug numbers are higher in Rookery than in Nursery, but the range of counts is very large in both fields, so the significance of the difference must be in some doubt. We fit the model first as a simple one-way analysis of deviance, using Poisson errors and the log link:

```
model<-glm(slugs~field,poisson)

summary(model)

Coefficients:

              Estimate Std. Error z value Pr(>|z|)
(Intercept)    0.2429     0.1399    1.737 0.082446 .
field          0.5790     0.1748    3.312 0.000925 ***
```

The *t* test for difference between fields is highly significant, so it looks like mean slug numbers really are different. But just a minute. We have not yet checked for overdispersion:

```
(Dispersion parameter for poisson family taken to be 1)

    Null deviance: 224.86  on 79  degrees of freedom
Residual deviance: 213.44  on 78  degrees of freedom
```

The scale parameter (here labelled Dispersion parameter) is $213.44/78 = 2.74$, not 1 as was assumed by the model. The simplest thing we can do to compensate for this is to carry out an F test rather than a chi-square test of deletion. The F test uses the empirical scale parameter as an estimate equivalent to the error variance, and performs a test much harsher than the chi-square test. We can compare the two, chi-square first:

model2<-update(model,~.-field)

anova(model,model2,test="Chi")

```
         Resid. Df Resid. Dev Df Deviance P(>|Chi|)
field         78    213.438
1             79    224.859 -1  -11.422      0.001
```

This suggests that the difference between the fields is highly significant ($p = 0.001$). Now do exactly the same deletion, but use the F test with the empirical scale parameter:

anova(model,model2,test="F")

```
         Resid. Df Resid. Dev Df Deviance       F   Pr(>F)
field         78    213.438
1             79    224.859 -1  -11.422    3.604  0.06134 .
```

A big change. Under the F test the difference in mean slug density is *not* significant.

We could try a parametric transformation, and analyse the data using a linear model (**aov**). For instance:

model3<-aov(log(slugs+1)~field)

summary(model3)

```
            Df  Sum of Sq  Mean Sq   F Value        Pr(F)
    field    1     4.4750  4.475004  8.961176  0.003692791
Residuals   78    38.9514  0.499377
```

This says the difference is highly significant ($p < 0.004$), so clearly the adjustment for overdispersion in the F test is extremely unforgiving.

The alternative parametric transformation for count data is the **square root**. Here we compare a straightforward analysis of variance on the raw count data (Normal errors and constant variance wrongly assumed) with a model based on the square root transformation:

model4<-aov(slugs~field)

anova(model4)

```
            Df Sum of Sq  Mean Sq  F Value       Pr(F)
    field    1     20.00 20.00000 3.900488  0.05181051
Residuals   78    399.95  5.12756
```

The untransformed Anova suggests that the difference is not quite significant. Now for a square root transformation of the counts, assuming (now with much greater justification) that the errors are Normal and the variance constant:

model5<-aov(slugs^0.5~field)

anova(model5)

```
              Df  Sum of Sq   Mean Sq    F Value        Pr(F)
    field  1     7.44812   7.448123   9.312308    0.003110717
Residuals 78    62.38557   0.799815
```

So the linear model with square root transformed counts indicates that the difference in slug density between the two fields is highly significant.

The next option in dealing with overdispersion is to use **quasi** to define a different family of error structures. For instance, we might retain the log link (this is good for constraining the predicted counts to be non-negative), but we might allow that the variance increases with the square of the mean (like a discrete version of a Gamma distribution), rather than as the mean (as assumed by Poisson errors). This is how we write the model:

```
model6<-glm(slugs~field,family=quasi(link=log,variance=mu^2))
```

The software recognises mu as the mean of the distribution. The result of the fit is this:

```
summary(model6)

Coefficients:
                Value Std. Error    t value
(Intercept) 0.5324631   0.1625783   3.275118
      field 0.2895169   0.1625783   1.780785

(Dispersion Parameter for Quasi-likelihood family taken to be
2.114536)
```

Note the estimate of the dispersion parameter, above. The t test does not indicate a significant difference between fields. Model checking plots indicate that the errors are much more nearly Normal under the $\log(\text{slugs} + 1)$ transformation than in the **quasi** model:

```
Null Deviance: 11.16246 on 79 degrees of freedom

Residual Deviance: 4.548488 on 78 degrees of freedom
```

Let's check the deletion test:

```
model7<-update(model6,~.-field)
anova(model6,model7,test="F")

   Terms Resid. Df Resid. Dev Df  Deviance   F Value        Pr(F)
1  field        78    4.54849
2      1        79   11.16246 -1  -6.613969  3.127858   0.08087445
```

Not significant. What about a non-parametric test? We can use the Wilcoxon rank sum test to compare the two (unpaired) samples:

```
wilcox.test(slugs[field=="Nursery"],slugs[field=="Rookery"])
```

Notice that we need to split the response variable into two separate vectors, one for each field, in order to use this function (in general, the two vectors might be of different lengths):

```
        Wilcoxon rank-sum test

data:slugs[field=="Nursery"]and slugs[field=="Rookery"]

rank-sum normal statistic with correction Z = -3.0581,
p-value = 0.0022
alternative hypothesis: true mu is not equal to 0
```

The non-parametric test says that mean slug numbers are significantly different in the two fields ($p = 0.0022$). This p value is not particularly reliable, however, because there are so many ties in the data as a result of all the zeros and ones.

Thus, several of the tests suggest that the difference between the mean slug densities is highly significant, while other tests suggest that the difference is insignificant. What is clear is that there is more to this than mere differences between the fields. Differences between the means explain only 5% of the variation in slug counts from tile to tile (deviance change of 11.42 out of 224.86). Within fields it is clear that the data are highly aggregated. It is very common for field data to have this kind of overdispersed, spatially aggregated structure, and it would be naive of us to assume that simple error structures will always work perfectly.

This kind of problem, where one kind of test says a difference is significant and another test says the same difference is not significant, comes up all the time when dealing with data with a low mean and a high variance. There is no obvious right answer, but the analyses correcting for overdispersion are clearly signalling that the result, so significant with linear models, should be treated with a more than usual degree of circumspection.

Summary of overdispersion with Poisson errors

Overdispersion occurs when the residual deviance is greater than the residual degrees of freedom once the minimal adequate model has been fitted to the data. It means that the errors are not, in fact, Poisson (i.e. equal to the mean) but are actually greater than assumed. This means that the estimated standard errors are too small, and the significance of model terms is more or less severely overestimated. We need to take care of overdispersion, and there are several options we can follow. In order of simplicity, these are

- carry out significance tests using "F" rather than "Chi" in the **anova** directive
- use family=quasi instead of family=poisson and specify a variance function
- use negative binomial errors

Ancova with count data: categorical and continuous explanatory variables

A long-term agricultural experiment had 90 grassland plots, each 25 m × 25 m, differing in biomass, soil pH and species richness (the count of species in the whole plot). It is well known that species richness declines with increasing biomass, but the question addressed here is whether the slope of that relationship differs with soil pH. The plots were classified according to a three-level factor as high, medium or low pH with 30 plots in each level. The response variable is the count of species, so **glm** with Poisson errors is a sensible choice. The continuous explanatory variable is long-term average biomass measured in June, and the categorical explanatory variable is soil pH. With a mixture of continuous and categorical explanatory variables, analysis of covariance is the appropriate method.

```
attach(species)
names(species)

[1]  "pH"   "Biomass"   "Species"
```

We begin by plotting the data, using different symbols for each level of soil pH. It is important in cases like this to ensure that the axes in the first plot directive are scaled appropriately to accommodate all of the data. The trick here is to plot the axes with nothing between them; this is the type="n" option:

```
plot(Biomass,Species,type="n")
```

Now we can add a scatterplot of points for each level of soil pH, using different plotting characters **pch** for each:

```
points(Biomass[pH=="high"],Species[pH=="high"])
points(Biomass[pH=="mid"],Species[pH=="mid"],pch=16)
points(Biomass[pH=="low"],Species[pH=="low"],pch=0)
```

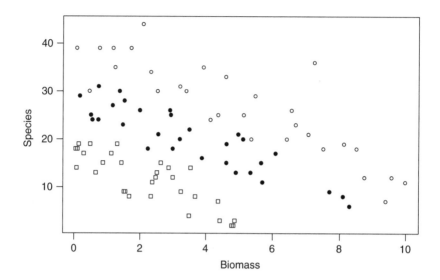

There are two clear patterns: species richness declines with increasing biomass (left to right), and species richness increases with increasing soil pH (bottom to top). The question now arises as to whether the effect of biomass in reducing species richness depends upon the level of soil pH? This is a question that is answered by analysis of covariance, in which we fit different parameters for each level of soil pH. There is no point in fitting a linear model to these data because at low soil pH, high biomass (e.g. biomass > 6) would predict nonsensical *negative* counts of species richness. Instead, we fit a log-linear model, in which species richness is predicted from the antilog of a set of linear functions of *Biomass*:

$$Species = \exp(a + b \times Biomass)$$

with different values of a and b for each level of soil pH.

The procedure is exactly the same as in standard analysis of covariance. We fit a full model with different slopes and intercepts for each factor level, then we simplify the model by assessing whether a common slope would describe the data equally well. The only difference is that we replace **lm** by **glm** and add the phrase family=poisson after the model formula:

```
model <-glm(Species~pH*Biomass,poisson)

summary(model)

Coefficients:
                Value    Std. Error      t value
(Intercept)  3.38577729  0.041205385   82.1683216
       pH1  -0.40778804  0.051410886   -7.9319396
       pH2   0.02544186  0.028580716    0.8901757
```

```
    Biomass  -0.16943387  0.014803648  -11.4454135
 pH1Biomass  -0.07751020  0.019974571   -3.8804437
 pH2Biomass   0.01520604  0.009281891    1.6382482
```

(Dispersion Parameter for Poisson family taken to be 1)

Null Deviance: 452.3459 on 89 degrees of freedom

Residual Deviance: 83.20114 on 84 degrees of freedom

Note that there is no need to correct for overdispersion (residual deviance is fractionally less than the residual degrees of freedom).

```
model2<-update(model,~.-pH:Biomass)
model3<-update(model2,~.-pH)
model4<-update(model3,~.-Biomass)
```

Don't forget the punctuation in **update** (, ~ . -). Now we can get a full Anova table comparing the four different models:

```
anova(model,model2,model3,model4,test="Chi")
```

Terms	Resid. Df	Resid. Dev	Test	Df	Deviance	Pr(Chi)
1 pH * Biomass	84	83.2011				
2 pH + Biomass	86	99.2415	-pH:Biomass	-2	-16.0404	0.0003287607
3 Biomass	88	407.6728	-pH	-2	-308.4313	0.0000000000
4 1	89	452.3459		-1	-44.6731	0.0000000000

Our initial impression that the slopes were the same was completely wrong; in fact, the slopes are very significantly different ($p < 0.0004$). We are not surprised to find that the main effects of *Biomass* and *pH* are significant. So we need to retain a model with different slopes.

Plotting the fitted values as smooth lines, separately for each level of soil pH, demonstrates the boundedness of log-linear models. A linear regression would predict negative values for species richness at high biomass. In contrast, the **glm** predicts species richness declining asymptotically towards low values. In reality of course, it is impossible that plant species richness would ever fall below 1 so long as the plots were vegetated.

The curve fitting requires that we make a data frame to contain values for all of the explanatory variables (the continuous variable *Biomass* for the x axis and the categorical explanatory variable *pH* for the three different graphs). First we generate the values for the x axis:

```
x<-seq(0,10,0.1)
length(x)
```

```
[1]   101
```

This means that we need to generate 101 repeats of each of the factor levels for *pH*:

```
levels(pH)
```

```
[1] "high"   "low"   "mid"
```

We created a new vector of length 3×101 to contain the factor levels:

```
acid<-factor(c(rep("low",101),rep("mid",101),rep("high",101))
```

and now we need to make the vector of x values (*Biomass*) the same length:

```
x<-c(x,x,x)
```

Finally, we use **predict** to draw the three lines, one at a time. We could back-transform the predicted values manually (using the antilog function **exp**) but here we use type="response" to do this for us. We use subscripts (square brackets [ ]) to select subsets of the points for plotting the lines; note that we need to take a subset of the *x* values as well as of the predicted values to ensure the vectors for generating the lines are exactly the same length.

```
lines(x[acid=="high"],predict(model,type="response",data.frame(Biomass=x,pH=acid))[acid=="high"])
lines(x[acid=="mid"],predict(model,type="response",data.frame(Biomass=x,pH=acid))[acid=="mid"])
lines(x[acid=="low"],predict(model,type="response",data.frame(Biomass=x,pH=acid))[acid=="low"])
```

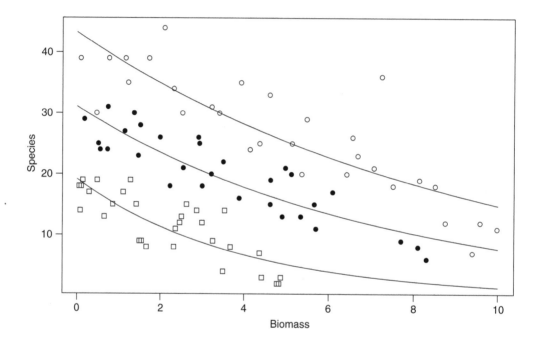

Because of the log link, the model never predicts negative species counts at high biomass (the predicted values go asymptotically to zero for high values of biomass). We have used a slightly different convention here than on p. 535; the vector of *x* values has to be subscripted in the **lines** function, because we made it to be the same length as the vector of factor levels, called *acid* (the sequence was repeated three times). The advantage of this is that we do not need to repeat the *x* values inside the data frame used by **predict**. It takes a lot of practice before this becomes clear.

Contingency tables

Where the response variable is a count, and all of the explanatory variables are categorical, then contingency tables are the appropriate analysis. If sample size is so small that one or more of the expected frequencies is less than 5 or so, then Fisher's exact test is appropriate (Chapter 11). For larger samples, the simplest way to analyse count data is the 2×2 contingency table. The traditional analysis of such tables used Pearson's chi-square to test the null hypothesis that two factors were independent in their effects on the response variable (see p. 181). Here we show the alternative (often called the *G* test) which uses log-linear models to address the same question. As with Pearson, the test statistic is chi-square distributed with one degree of freedom.

Suppose we have the following contingency table. It shows counts of factories classified by their no-smoking policy and their productivity rating.

Productivity	Smoking	No smoking
High	13	44
Low	25	29

The response variable is a count of the number of factories falling into each of four categories: those rated as having higher or lower than average productivity, and those with and without no-smoking policies. The question to be addressed is whether there is any evidence of an association between these two factors. We set up the data like this. There are three variables: the response, which we'll call *count*, a two-level factor for smoking, and a two-level factor for productivity. The data frame is so small, we may as well type in the values directly:

    count<-c(13,25,44,29)

Note that the data are entered column-wise rather than row-wise:

    smoking<-factor(c("yes","yes","no","no"))
    productivity<-factor(c("high","low","high","low"))

Carrying out the log-linear modelling (**glm** with Poisson errors) is straightforward. We give a name to the model object, then specify the model formula in the usual way. The only extra detail is that we need to specify family=poisson:

    model<-glm(count~smoking+productivity,poisson)

We shall not bother with the model summary, because most of the information is superfluous here (the table is dominated by the coefficients for the nuisance variables that are needed to constrain the row and column totals). All we need is the value of the residual deviance:

    model
    Call:
    glm(formula = count ~ smoking + productivity, family =
    poisson)

    Coefficients:
     (Intercept)   smoking productivity
        3.270511   -0.3264358   -0.02703373
    Degrees of Freedom: 4 Total; 1 Residual
    Residual Deviance: 6.878277

The coefficients are of no interest. All we want is the value of the residual deviance (6.878) to compare with chi-square tables with 1 d.f. (3.841). Because the calculated value of chi-square is larger than the value in tables, we reject the null hypothesis of independence in the distribution of these two factors.

Note that the test does not tell us whether there is a positive or a negative correlation between the two variables. For this, we need to look at the data and the fitted values (these are the counts we would have expected if the distributions really had been independent):

```
fitted(model)

        1            2            3            4
  19.51355     18.48652     37.48649     35.51351

count

  [1]     13           25           44           29
```

This tells us that if the distributions had been independent, then we should have expected to find 19 or 20 factories where smoking was associated with high productivity. We actually found only 13 such factories, so the significant association between the two variables is negative. The mechanism is not obvious, but low productivity was associated with smoking at work in these data.

A complex contingency table: Schoener's lizards

It is all too easy to analyse complex contingency tables in the wrong way and to produce answers that are not supported by the data. Problems typically occur because people fail (or forget) to include all the interactions between the nuisance variables that are necessary to constrain the marginal totals. Note that the problems almost never arise if the same analysis can be structured so that it can be carried out as the analysis of proportion data using binomial errors (compare the example explained below with its reanalysis as proportion data later on, where there are no nuisance variables at all).

The example concerns niche differences between two lizard species:

```
attach(lizards)
names(lizards)

[1] "n"    "sun"    "height"    "perch"    "time"    "species"
```

Schoener (1968) collected information on the distribution of two *Anolis* lizard species (*A. opalinus* and *A. grahamii*) to see if their ecological niches were different in terms of where and when they perched to prey on insects. Perches were classified by twig diameter, their height in the bush, whether the perch was in sun or shade when the lizard was counted, and the time of day at which they were foraging. The response variable is a count of the number of times a lizard of each species was seen under each of the contingencies. The modelling looks like this. First we fit a *saturated model* which has a parameter for every data point. There are no degrees of freedom and hence the model has no explanatory power.

```
model1<-glm(n~species*sun*height*perch*time,family=poisson)
```

Prepare yourself for a nasty shock !

```
summary(model1)
```

```
Coefficients:
                        Estimate  Std. Error    z value  Pr(>|z|)
(Intercept)            1.386e+000  5.000e-001      2.773   0.00556  **
species               -1.000e-014  7.071e-001  -7.73e-015   1.00000
sun                    9.163e-001  5.916e-001      1.549   0.12143
height                -1.369e+001  2.847e+002     -0.048   0.96165
perch                 -2.877e-001  7.638e-001     -0.377   0.70642
timeMid.day           -1.386e+000  1.118e+000     -1.240   0.21500
timeMorning           -6.931e-001  8.660e-001     -0.800   0.42349
species.sun            5.878e-001  8.097e-001      0.726   0.46786
species.height         1.479e+001  2.847e+002      0.052   0.95857
```

species.perch	5.108e-001	1.017e+000	0.503	0.61530
species.timeMid.day	2.079e+000	1.275e+000	1.631	0.10284
species.timeMorning	2.303e+000	1.025e+000	2.247	0.02463 *
sun.height	1.248e+001	2.847e+002	0.044	0.96502
sun.perch	6.454e-002	8.991e-001	0.072	0.94277
sun.timeMid.day	2.079e+000	1.183e+000	1.757	0.07884 .
sun.timeMorning	7.885e-001	9.700e-001	0.813	0.41631
height.perch	1.259e+001	2.847e+002	0.044	0.96472
height.timeMid.day	1.386e+000	4.026e+002	0.003	0.99725
height.timeMorning	6.931e-001	4.026e+002	0.002	0.99863
perch.timeMid.day	2.877e-001	1.607e+000	0.179	0.85795
perch.timeMorning	6.931e-001	1.190e+000	0.582	0.56032
species.sun.height	-1.391e+001	2.847e+002	-0.049	0.96103
species.sun.perch	-1.099e+000	1.200e+000	-0.916	0.35974
species.sun.timeMid.day	-1.429e+000	1.358e+000	-1.052	0.29283
species.sun.timeMorning	-1.762e+000	1.151e+000	-1.530	0.12598
species.height.perch	-1.530e+001	2.847e+002	-0.054	0.95714
species.height.timeMid.day	-2.485e+000	4.026e+002	-0.006	0.99508
species.height.timeMorning	-2.223e+000	4.026e+002	-0.006	0.99559
species.perch.timeMid.day	-1.204e+000	1.846e+000	-0.652	0.51431
species.perch.timeMorning	-1.833e+000	1.429e+000	-1.283	0.19965
sun.height.perch	-1.208e+001	2.847e+002	-0.042	0.96615
sun.height.timeMid.day	-1.792e+000	4.026e+002	-0.004	0.99645
sun.height.timeMorning	-2.776e-001	4.026e+002	-0.001	0.99945
sun.perch.timeMid.day	4.055e-001	1.700e+000	0.239	0.81147
sun.perch.timeMorning	-1.598e-001	1.341e+000	-0.119	0.90514
height.perch.timeMid.day	-1.259e+001	4.930e+002	-0.026	0.97963
height.perch.timeMorning	-1.300e+001	4.930e+002	-0.026	0.97897
species.sun.height.perch	1.442e+001	2.847e+002	0.051	0.95960
species.sun.height.timeMid.day	2.989e+000	4.026e+002	0.007	0.99408
species.sun.height.timeMorning	2.040e+000	4.026e+002	0.005	0.99596
species.sun.perch.timeMid.day	1.182e+000	1.981e+000	0.597	0.55083
species.sun.perch.timeMorning	1.417e+000	1.641e+000	0.864	0.38786
species.height.perch.timeMid.day	1.609e+000	5.693e+002	0.003	0.99774
species.height.perch.timeMorning	1.585e+001	4.930e+002	0.032	0.97436
sun.height.perch.timeMid.day	1.183e+001	4.930e+002	0.024	0.98085
sun.height.perch.timeMorning	1.057e+001	4.930e+002	0.021	0.98290
species.sun.height.perch.timeMid.day	-1.307e+000	5.693e+002	-0.002	0.99817
species.sun.height.perch.timeMorning	-1.330e+001	4.931e+002	-0.027	0.97847

(Dispersion parameter for poisson family taken to be 1)

Null deviance: 7.3756e+002 on 47 degrees of freedom
Residual deviance: 5.4480e-005 on 0 degrees of freedom
AIC: 259.25

There are two key things to note here: first, the residual deviance and degrees of freedom are both zero because we have intentionally fitted a *saturated model* which has as many parameters (48) as there are data points; and second, all but 23 of the 48 estimated parameters are nuisance variables. The nuisance variables are of absolutely no scientific interest and are included only to ensure that all of the marginal totals are correctly constrained. The only parameters that are of scientific interest are 23 *interaction terms involving species*.

The only way to model a data set like this successfully is to be fantastically well organised. Terms involving species are deleted stepwise from the current model, starting with the highest-order interactions. Non-significant terms are left out. Significant terms are added back into the model.

We need to keep very accurate track of which terms we have deleted and which terms remain to be deleted. Remember, you can't remove any three-way interactions until the four-way interactions containing the relevant factors have been removed.

The 23 parameters to be tested are involved in the following 15 potentially interesting interaction terms that involve species:

```
species.sun
species.height
species.perch
species.timeMid.day
species.timeMorning
species.sun.height
species.sun.perch
species.sun.timeMid.day
species.sun.timeMorning
species.height.perch
species.height.timeMid.day
species.height.timeMorning
species.perch.timeMid.day
species.perch.timeMorning
species.sun.height.perch
species.sun.height.timeMid.day
species.sun.height.timeMorning
species.sun.perch.timeMid.day
species.sun.perch.timeMorning
species.height.perch.timeMid.day
species.height.perch.timeMorning
species.sun.height.perch.timeMid.day
species.sun.height.perch.timeMorning
```

I suggest that you start at the bottom of this list and tick off the interaction terms as you delete them. Once you have tested, say, all of the four-way interactions involving species and found them to be non-significant, then leave them all out (but leave the high-order interactions involving the nuisance variables in the model).

Using automatic model simplification with **step** is not much use in a case like this because:

- it tends to be too lenient, and to leave non-significant terms in the model

- it is likely to remove interaction terms between nuisance variables and these must stay in the model in order to constrain the marginal totals properly

There is no way round it. We have to do the model simplification the long way, using **update** to delete terms and **anova** to test the significance of deleted terms. I have used cut and paste to make the following table of deletion tests:

```
model2<-update(model1,~.-species:sun:height:perch:time)
anova(model1,model2,test="Chi")

    Resid. Df Resid. Dev Df Deviance P(>|Chi|)
1         0    0.00005
2         2    0.00010 -2 -0.00004   0.99998
```

This means that we can adopt model2 as the base model and systematically delete the various four-way interactions involving species:

```
model3<-update(model2,~.-species:sun:height:perch)
anova(model2,model3,test="Chi")

Resid. Df Resid. Dev  Df Deviance P(>|Chi|)
1         2      0.0001
2         3      2.7089 -1  -2.7088     0.0998
```

And so on.

	Resid. Df	Resid. Dev	Test	Df	Deviance	Pr(Chi)
	2	0.00014458	-species:sun:height:perch:time	-2	-0.00004	0.9999766
	3	2.708911	-species:sun:height:perch	-1	-2.708766	0.09979817
	5	3.111413	-species:sun:height:time	-2	-0.4025021	0.8177071
	7	7.928383	-species:height:perch:time	-2	-4.816971	0.08995144
	9	8.573263	-species:sun:perch:time	-2	-0.6448801	0.7243793
	10	8.587620	-species:sun:perch	-1	-0.01435679	0.9046259
	11	11.97484	-species:sun:height	-1	-3.387218	0.06570373
	13	13.34891	-species:sun:time	-2	-1.374076	0.5030639
	15	13.37456	-species:perch:time	-2	-0.02564741	0.9872582
	16	13.68243	-species:height:perch	-1	-0.3078676	0.5789916
	18	14.20496	-species:height:time	-2	-0.522531	0.7700764
***	20	25.80257	-species:time	-2	-11.59761	0.003031181
***	19	36.27283	-species:height	-1	-22.06787	2.6317e-006
***	19	21.89253	-species:sun	-1	-7.687569	0.005560247
***	19	27.33579	-species:perch	-1	-13.13083	0.000290476

There are no significant interactions between the explanatory variables that have any effect on the distribution of these two lizard species. All of the factors have highly significant main effects, however. The two lizard species utilise different parts of the bush, occupy perches of different diameters, are active at different times of day, and are found in differing levels of shade. The published analyses that report significant interactions between factors all made the mistake of unintentionally leaving out one or more interactions involving the nuisance variables. This is a very easy thing to do, unless a strict regime of *simplifying downwards from the saturated model* is followed religiously. If you always simplify down from a complex model to a simple model, this problem will never arise and you will save yourself from making embarrassing mistakes of interpretation. With this many comparisons being done, I would always work at $p = 0.01$ for significance, rather than $p = 0.05$.

Reanalysis of Schoener's lizards as proportion data: getting rid of the nuisance variables

The analysis of count data in complex contingency tables is difficult, tedious and very easy to get wrong. If the response can be reformulated as a proportion, then the analysis is much more straightforward, because there is no longer any need to include the vast numbers of nuisance parameters necessary to constrain the marginal totals.

The computational part of the exercise here is data compression. We need to reduce the data frame called lizards so that it has the proportion of all lizards that are *Anolis grahamii* as the response, with the same set of explanatory variables (but each with half as many rows as before). It is essential to ensure that all of the cases are in exactly the same sequence for both of the lizard species. To do this, we create a sorted version of the data frame:

```
sorted.lizards<-lizards[order(lizards[,2],lizards[,3],lizards[,4],lizards[,5],lizards[,6]),1:6]
```

So now we can save the ordered *grahamii* and *opalinus* vectors as *ng* and *no* and bind them together to use as the response vector in the binomial analysis

```
ng<-sorted.lizards[,1][sorted.lizards[,6]=="grahamii"]
no<-sorted.lizards[,1][sorted.lizards[,6]=="opalinus"]
y<-cbind(ng,no)
```

The 4 shortened explanatory variables (the factors s, h, p, and t) are now produced:

```
s<-sorted.lizards[,2][sorted.lizards[,6]=="grahamii"]
h<-sorted.lizards[,3][sorted.lizards[,6]=="grahamii"]
p<-sorted.lizards[,4][sorted.lizards[,6]=="grahamii"]
t<-sorted.lizards[,5][sorted.lizards[,6]=="grahamii"]
```

You should check that these have been automatically declared to be factors (hint: use **is.factor**). Now we are ready to fit the saturated model:

```
model<-glm(y ~ s*h*p*t, binomial)
```

Let's see how good **step** is at simplifying this saturated model:

```
step(model)

Start:  AIC = 48.0002
 y ~ s * h * p * t

Single term deletions

          Df    Sum of Sq          RSS          Cp
 <none>                    0.000000000  48.00000
 s:h:p:t  1 0.002745804    0.002745804  46.00275

Step:  AIC= 46.0003
 y ~ s + h + p + t + s:h + s:p + h:p + s:t + h:t + p:t + s:h:p
 + s:h:t + s:p:t + h:p:t
```

(out goes the 4-way interaction s:h:p:t)

```
Single term deletions

Model:
y ~ s + h + p + t + s:h + s:p + h:p + s:t + h:t + p:t + s:h:p
+ s:h:t + s:p:t + h:p:t

         Df Sum of Sq       RSS        Cp
<none>                 0.000000  46.00000
 s:h:p  1  0.022982  0.022982  44.02298
 s:h:t  2  0.012405  0.012405  42.01240
 s:p:t  2  0.802822  0.802822  42.80282
 h:p:t  2  2.797531  2.797531  44.79753

Step: AIC= 42.4418
 y ~ s + h + p + t + s:h + s:p + h:p + s:t + h:t + p:t + s:h:p
 + s:p:t + h:p:t
```

(out goes the 3-way interaction s:h:t)

```
Single term deletions

Model:
y ~ s + h + p + t + s:h + s:p + h:p + s:t + h:t + p:t + s:h:p
+ s:p:t + h:p:t

        Df Sum of Sq      RSS        Cp
<none>                 0.287591 42.28759
 s:h:p   1  0.031109 0.318700 40.31870
 s:p:t   2  0.627395 0.914986 38.91499
 h:p:t   2  3.584907 3.872498 41.87250

Step: AIC = 39.0715
 y ~ s + h + p + t + s:h + s:p + h:p + s:t + h:t + p:t + s:h:p
+ h:p:t                    (out goes the 3-way interaction s:p:t)

Single term deletions

Model:
y ~ s + h + p + t + s:h + s:p + h:p + s:t + h:t + p:t + s:h:p
+ h:p:t

        Df Sum of Sq      RSS        Cp
<none>                 0.966941 38.96694
   s:t   2  2.231904 3.198845 37.19884
 s:h:p   1  0.026440 0.993381 36.99338
 h:p:t   2  4.000991 4.967932 38.96793

Step: AIC= 39.3016
         (AIC has increased, so this deletion was not accepted)

y ~ s + h + p + t + s:h + s:p + h:p + s:t + h:t + p:t + h:p:t
                (it wants to keep the 3-way interaction s:h:p)
```

As often happens, **step** has left a very complicated model with two 3-way interactions and six 2-way interactions in it. Let's see how we get on by hand in simplifying the reduced model (model1) that **step** has bequeathed us:

```
model1<-glm(y ~ s+h+p+t+s:h+s:p+h:p+h:t+p:t+s:h:p+h:p:t,family=binomial)
```

Now we remove the terms in sequence and use **anova** to compare the simpler model with its more complex predecessor:

```
model2<-update(model1,~.-h:p:t)
anova(model1,model2,test="Chi")

Analysis of Deviance Table

  Resid. Df Resid. Dev Df Deviance p(>|Chi|)
1         7     3.3405
2         9     8.5422 -2  -5.2017    0.0742
```

This simplification caused a non-significant increase in deviance ($p = 0.07$) so we make model2 the current model, and remove the next 3-way interaction, s:h:p (see below). We keep removing terms until a significant terms in discovered, as follows:

```
                              Resid. Df Resid. Dev Df Deviance  P(>|Chi|)
model2 <-update(model1,~.-h:p:t)    9      8.5422 -2  -5.2017     0.0742
model3 <-update(model2,~.-s:h:p)   10     10.9032 -1  -2.3610     0.1244
model4 <-update(model3,~.-p:t)     12     10.9090 -2  -0.0058     0.9971
model5 <-update(model4,~.-h:t)     14     11.7667 -2  -0.8577     0.6513
model6 <-update(model5,~.-h:p)     15     11.9789 -1  -0.2122     0.6450
model7 <-update(model6,~.-s:p)     16     11.9843 -1  -0.0054     0.9414
model8 <-update(model7,~.-s:h)     17     14.2046 -1  -2.2203     0.1362
model9 <-update(model8,~.-t)       19     25.802  -2 -11.597      0.003   ***
model10<-update(model8,~.-p)       18     27.3346 -1 -13.1300     0.0003  ****
model11<-update(model8,~.-h)       18     36.271  -1 -22.066   2.634e-06  ****
model12<-update(model8,~.-s)       18     21.8917 -1  -7.6871     0.0056  ***
```

Note that because there was a significant increase in deviance when time was deleted from model8, we use model 8 not model9 in assessing the significance of perch diameter (and of the other two main effects as well). The main effect probabilities are identical to those obtained by contingency table analysis with Poisson errors (above). We need to look at the model summary for model8 to see whether all of the factor levels need to be retained:

```
summary(model8)
```

```
Call: glm(formula = y ~ s + h + p + t, family = binomial)
```

```
Coefficients:

                 Value Std. Error   t value
(Intercept) -1.53482789 0.17218762 -8.913695
          s  0.42363733 0.16111981  2.629331
          h -0.56499535 0.12849678 -4.396961
          p  0.38131710 0.10562126  3.610231
         t1 -0.48196106 0.14077322 -3.423670
         t2 -0.08495018 0.07898896 -1.075469
```

```
(Dispersion Parameter for Binomial family taken to be 1)
        Null Deviance: 70.10183 on 23 degrees of freedom
    Residual Deviance: 14.20457 on 17 degrees of freedom
```

The second contrast for time-of-day (t2) is not significant (its t-value is less than 2), so it might be possible to reduce the number of levels from 3 to 2. The question is which two levels to lump together ? It turns out that lumping morning and afternoon together (t2) and comparing these with mid-day causes a significant increase in deviance and hence is unjustified (chi-square = 6.056, d.f. = 1, p = 0.014). Here is the code for lumping together morning and mid-day and comparing this with afternoon (t3)

```
t3<-t
levels(t3)[c(2,3)]<-"other"
levels(t3)
```

```
[1] "Afternoon" "other"
```

```
model9<-glm(y~s+h+p+t3,binomial)
anova(model8,model9,test="Chi")
```

```
Analysis of Deviance Table

Response: y

            Terms Resid. Df Resid. Dev   Test Df   Deviance    Pr(Chi)
1  s + h + p + t       17   14.20457
2  s + h + p + t3      18   15.02320 1 vs. 2 -1 -0.8186255  0.3655824
```

This simplification was justified (p = 0.37), so we accept model9.

summary(model9)

```
Call: glm(formula = y ~ s + h + p + t3, family = binomial)

Coefficients:

                Value Std. Error    t value
(Intercept) -1.3869466  0.1789906  -7.748714
          s  0.3935751  0.1579030   2.492511
          h -0.5593822  0.1282340  -4.362200
          p  0.3742429  0.1052055   3.557256
         t3 -0.4358541  0.1305493  -3.338616

(Dispersion Parameter for Binomial family taken to be 1)
     Null Deviance: 70.10183 on 23 degrees of freedom
Residual Deviance: 15.0232 on 18 degrees of freedom
```

All the parameters are significant, so this is the minimal adequate model. There are just 5 parameters, and the model contains no nuisance variables (compare this with the massive contingency table model on p. 550). The ecological interpretation is straightforward: the two lizard species differ significantly in their niches on all the niche-axes that were measured. However, there were no significant interactions (nothing subtle was happening like swapping perch sizes at different times of day).

The danger of contingency tables

In observational studies we quantify only a limited number of explanatory variables. It is inevitable that we shall fail to note (or to measure) a number of factors that have an important influence on the behaviour of the system in question. That's life, and given that we make every effort to note the important factors, there is little we can do about it. The problem comes when we ignore factors that have an important influence on behaviour. This difficulty can be particularly acute if we *aggregate data over important explanatory variables*. This is known as 'collapsing a contingency table'. An example should make this clear.

Suppose we are carrying out a study of induced defences in trees. A preliminary trial has suggested that early feeding on a leaf by aphids may cause chemical changes in the leaf which reduce the probability of that same leaf being attacked later in the season by hole-making insects. To this end, we mark a large cohort of leaves then score whether they were infested by aphids early in the season and whether they were holed by insects later in the year. The work was carried out on two different trees and the results were as follows.

Tree	Aphids	Holed	Intact	Total leaves	Proportion holed
Tree1	Without	35	1750	1785	0.0196
	With	23	1146	1169	0.0197
Tree2	Without	146	1642	1788	0.0817
	With	30	333	363	0.0826

There are four variables: the response variable, *Count*, with 8 values (highlighted above), a two-level factor for late season feeding by caterpillars (Holed or Intact), a two-level factor for early season aphid feeding (With or Without aphids) and a two-level factor for tree (the observations come from two separate trees, imaginatively named Tree1 and Tree2):

```
attach(induced)
names(induced)
```

```
[1] "Tree"    "Aphid"    "Caterpillar"    "Count"
```

We call the model id (for induced defences) and fit a **saturated model**. This is a curious thing, which has as many parameters as there are values of the response variable. The fit of the model is perfect, so there are no residual degrees of freedom and no residual deviance. The reason that we fit a saturated model is that this is always the best place to start modelling complex contingency tables. If we fit the saturated model, then there is no risk that we inadvertently leave out important interactions between the nuisance variables:

```
id<-glm(Count~Tree*Aphid*Caterpillar,family=poisson)
```

The asterisk notation ensures that the saturated model is fitted, because all of the main effects and two-way interactions are fitted, along with the three-way interaction *Tree* by *Aphid* by *Caterpillar*. The model fit involves the estimation of $2 \times 2 \times 2 = 8$ parameters, and exactly matches the 8 values of the response variable, *Count*. There is no point looking at the saturated model in any detail, because the reams of information it contains are all superfluous. The first real step in the modelling is to use **update** to remove the three-way interaction from the saturated model, and then to use **anova** to test whether the three-way interaction is significant or not:

```
id2<-update(id, ~ . - Tree:Aphid:Caterpillar)
```

The punctuation here is , ~ . - and it is very important. Note the use of colons rather than asterisks to denote interaction terms rather than main effects plus interaction terms. Now we can see whether the three-way interaction was significant by specifying test="Chi" like this:

```
anova(id,id2,test="Chi")
```

```
      Resid. Df  Resid. Dev Df Deviance  P(>|Chi|)
   1          0 -3.975e-013
   2          1    0.00079 -1 -0.00079    0.97756
```

This shows clearly that the interaction between caterpillar attack and leaf holing does not differ from tree to tree ($p = 0.977\,56$). Note that if this interaction had been significant, then we would have stopped the modelling at this stage. But it wasn't, so we leave it out and continue. What about the main question? Is there an interaction between caterpillar attack and leaf holing? To test this we delete Aphid:Caterpillar interaction from the model, call it id3, and assess the results using **anova**:

```
id3<-update(id2, ~ . - Aphid:Caterpillar)
anova(id2,id3,test="Chi")
```

	Resid. Df	Resid. Dev	Test Df	Deviance	Pr(Chi)
1	1	0.000791372			
2	2	0.004085322	-Aphid:Caterpillar -1	-0.00329395	0.9542322

There is absolutely no hint of an interaction ($p = 0.954$). The interpretation is clear. This work provides no evidence for induced defences caused by early season caterpillar feeding.

Now we shall do the analysis the wrong way, in order to show the danger of collapsing a contingency table over important explanatory variables. Suppose we went straight for the interaction of interest, Aphid:Caterpillar. We might proceed like this:

```
wrong<-glm(Count~Aphid*Caterpillar,family=poisson)
wrong1<-update(wrong,~. - Aphid:Caterpillar)
anova(wrong,wrong1,test="Chi")
```

```
  Resid. Df Resid. Dev             Test Df  Deviance       Pr(Chi)
1         4    550.1917
2         5    556.8511 -Aphid:Caterpillar -1 -6.659372 0.009863566
```

The Aphid:Caterpillar interaction is highly significant ($p < 0.01$) providing strong evidence for induced defences. Wrong ! By failing to include *Tree* in the model, we have omitted an important explanatory variable. As it turns out, the trees differ enormously in their average levels of leaf holing:

```
as.vector(tapply(Count,list(Caterpillar,Tree),sum))[1]/ tapply(Count,Tree,sum) [1]
```

```
   Tree1
 0.01963439
```

```
as.vector(tapply(Count,list(Caterpillar,Tree),sum))[3]/ tapply(Count,Tree,sum) [2]
```

```
   Tree2
 0.08182241
```

Tree2 has more than four times the proportion of its leaves with holes made by caterpillars. We should really have determined this by a more thorough preliminary analysis. If we had been paying attention when we did the modelling the wrong way, then we should have noticed that the model containing only *Aphid* and *Caterpillar* had massive **overdispersion**, and this should have alerted us that all was not well. The moral is simple and clear. Always fit a saturated model first, containing all the variables of interest and all the interactions involving the nuisance variables (*Tree* in this case). Only delete from the model those interactions that involve the variables of interest (*Aphid* and *Caterpillar* in this case). Main effects are meaningless in contingency tables, as are the model summaries. *Always test for overdispersion*. It will never be a problem if you follow the advice of simplifying down from a saturated model, because you only ever leave out non-significant terms.

Bird ring recoveries

Certain kinds of survival analysis involve the periodic recovery of small numbers of dead animals. Bird ringing records are a good example of this kind of data. Out of a reasonably large, known number of ringed birds, a usually small number of dead birds carrying rings are recovered each year. The number of rings recovered each year declines because the pool of ringed birds declines each year as a result of natural mortality. There are two important variables in this case: the probability of a bird dying in a given time period; and the probability that, having died, the ring will be discovered. It is possible that one or both of these parameters varies with the age of the bird (i.e. with the time elapsed since the ring was applied). We can use S-Plus with Poisson errors to fit log-linear models to data like this, and to compare estimated survival and ring recovery rates from different cohorts of animals.

The probability that a bird survives from ringing up to time t_{j-1} is the survivorship s_{j-1}. The number of ringed birds still alive at time t_{j-1} is therefore $N_0 s_{j-1}$, where N_0 is the number of birds initially ringed and released at time t_0. Thus, the number of birds that die in the present time interval D_j is

$$D_j = N_0 s_{j-1} d_j$$

where d_j is the probability of an animal that survived to time t_{j-1} dying during the interval $(t_j - t_{j-1})$. Now, out of these D_j dead animals, we expect to recover a small proportion p_j, so the expected number of recoveries, R_j, is

$$R_j = N_0 s_{j-1} d_j p_j$$

Maximum likelihood estimates of the death and recovery probabilities are possible only if assumptions are made about the way in which the two parameters change with age and with time after ringing (e.g. Seber 1982). Common assumptions are that the probability of death is constant, once a given age has been reached, and that the probability of discovery of a dead bird is constant over time.

In order to analyse ring return data by log-linear modelling, we must combine the two probabilities of death and recovery into a single parameter z_j. This is the *probability of a death being recorded* in year j for an animal that was alive at the beginning of year j.

$$R_j = N_0 s_{j-1} z_j$$

Now, taking logs, rearranging and then taking antilogs we can write this expression as

$$R_j = \exp[\ln N_0 - \lambda_{j-1} t_{j-1} + \ln z_j]$$

replacing the survivorship term S_{j-1} by the proportion dying λt. Since we know the number of birds originally marked, we can use $\ln N_0$ as an offset. Then, using the log link, we are left with a graph of log of recoveries $\ln R_j$ against time t_j in which the intercept is given by the log of the recovery probability ($\ln z_j$) and the slope λ_{j-1} is the survival rate per unit time over the period 0 to t_{j-1}.

S-Plus can now be used to analyse count data on the number of recoveries of dead, ringed birds, and to compare models based on different assumptions about survivorship and recovery:

- different cohorts have constant death rates and recovery rates, and these rates are the same in all cohorts (the null model)

- different cohorts have the same death rates but different recovery probabilities

- different cohorts differ in both their death rates and their recovery rates

- the death rates may be time dependent

- a variety of more complex models

Suppose that tawny owls were ringed in three different kinds of woodland: oak, birch and mixed forest. The numbers marked in the three habitats were 75, 49 and 128 respectively. The numbers of dead, ringed birds recovered in subsequent years are in the following data frame:

```
attach(owlrings)
names(owlrings)

[1] "recovered"    "year"    "wood"    "marked"
```

The model has the number of rings recovered as the response variable (a count with Poisson errors, with the number of birds marked appearing in the model as an offset (on the log scale of the linear predictor, because the log link is the default for Poisson models, hence the name log-linear models).

```
model<-glm(recovered~wood*year+offset(log(marked)),family=poisson)
```

summary(model)

```
Coefficients:
                 Value Std. Error     t value
(Intercept) -2.32685842 0.28990217 -8.0263573
      wood1  0.07171189 0.35164327  0.2039336
      wood2 -0.05259181 0.20694348 -0.2541361
       year -0.40965835 0.09375710 -4.3693581
  wood1year -0.06239019 0.11411776 -0.5467176
  wood2year  0.01110562 0.06670412  0.1664908

(Dispersion Parameter for Poisson family taken to be 1)
Null Deviance: 42.83914 on 20 degrees of freedom
Residual Deviance: 15.67033 on 15 degrees of freedom
```

There does not look to be any justification for keeping different slopes for the different woodlands ($t < 0.6$), so we use **update** to remove the interaction term, then compare the complex and simpler models using **anova**. Note the lack of overdispersion.

model2<-update(model,~.-wood:year)

anova(model,model2,test="Chi")[,-1]

```
  Resid. Df Resid. Dev     Test  Df   Deviance    Pr(Chi)
1        15   15.67033
2        17   16.03535 -wood:year  -2 -0.3650134 0.8331791
```

There is clearly no justification for retaining different slopes in different woodlands. What about differences in the intercepts of different woodlands? We use **update** to simply model2, like this:

model3<-update(model2,~.-wood)

anova(model2,model3,test="Chi")

```
                              Resid. Df Resid. Dev Df Deviance P(>|Chi|)
wood + year + offset(log(marked))    17    16.0353
year + offset(log(marked))           19    16.2502 -2  -0.2148    0.8982
```

Not even close ($p = 0.898$). What about the slope?

model4<-update(model3,~.-year)

anova(model3,model4,test="Chi")

```
                          Resid. Df Resid. Dev Df Deviance  P(>|Chi|)
year + offset(log(marked))       19    16.250
offset(log(marked))              20    42.839 -1  -26.589 2.517e-007
```

No doubt about the significance of the slope. So model3 looks to be minimal adequate:

summary(model3)

```
Coefficients:
            Estimate Std. Error z value  Pr(>|z|)
(Intercept) -2.30548    0.27283  -8.450  < 2e-016 ***
year        -0.42521    0.09088  -4.679 2.89e-006 ***
```

```
(Dispersion parameter for poisson family taken to be 1)

    Null deviance: 42.839 on 20 degrees of freedom
Residual deviance: 16.250 on 19 degrees of freedom
AIC: 62.439
```

The final model is not overdispersed (16.25 on 19 d.f.). The interpretation is that the log of the recovery rate (the intercept) is -2.30548 and that the log of the survival rate (the slope) is -0.42521. Taking antilogs we get

exp(-2.30548)

```
[1] 0.09971093
```

exp(-0.42521)

```
[1] 0.6536325
```

We conclude that the annual survivorship of the owls is about 65% and that survival does not differ significantly from one woodland to the other. The annual probability of recovery of a ring from a dead, ringed bird is estimated to be about 10%.

Error checking by plot(model3) shows a few problems: the deviance declines markedly with the fitted values, and there is substantial non-normality in the error distribution, particularly for the largest negative residuals. But this will not affect the parameter estimates greatly, and parameter estimation is our main purpose here.

Further reading

Agresti, A. (1990) *Categorical Data Analysis*. New York, John Wiley.

Gordon, A.E. (1981) *Classification: Methods for the Exploratory Analysis of Multivariate Data*. New York, Chapman & Hall.

Santer, T.J. and Duffy, D.E. (1990) *The Statistical Analysis of Discrete Data*. New York, Springer-Verlag.

30

Binary response variables

Many statistical problems involve binary response variables. For example, we often classify individuals like this:

- dead or alive

- occupied or empty

- healthy or diseased

- wilted or turgid

- male or female

- literate or illiterate

- mature or immature

- solvent or insolvent

- employed or unemployed

and it is interesting to understand the factors that are associated with an individual being in one class or the other (Cox and Snell 1989). Binary regression is likely to be useful if *one or more of the explanatory variables is continuous*. If all of the explanatory variables are categorical then the response is better aggregated and analysed as proportion data (Chapter 28). In a study of company insolvency, for instance, the data would consist of a list of measurements made on the insolvent companies (their age, size, turnover, location, management experience, workforce training, and so on) and a similar list for the solvent companies. The question then becomes which, if any, of the explanatory variables increase the probability of an individual company being insolvent.

The response variable contains only 0s or 1s; for example, 0 to represent dead individuals and 1 to represent live ones. There is a single column of numbers for the response, in contrast to proportion data (Chapter 28). The way S-Plus treats this kind of data is to assume that the 0s and 1s come from *a binomial trial with sample size 1*. If the probability that an animal is dead is p, then the probability of obtaining y (where y is either dead or alive, 0 or 1) is given by an abbreviated form of the binomial distribution with $n = 1$, known as the Bernoulli distribution:

$$P(y) = p^y (1 - p)^{(1-y)}$$

The random variable y has a mean of p and a variance of $p(1 - p)$, and the object is to determine how the explanatory variables influence the value of p.

The trick to using binary response variables effectively is to know when it is worth using them and when it is better to lump the successes and failures together and analyse the *total counts* of dead individuals, occupied patches, insolvent firms or whatever. The question you need to ask yourself is, Do I have unique values of one or more explanatory variables for each and every individual case?

If the answer is yes, then analysis with a binary response variable is likely to be fruitful. If the answer is no, then there is nothing to be gained and you should reduce your data by aggregating the counts to the resolution at which each count *does* have a unique set of explanatory variables. For example, suppose that all your explanatory variables were categorical, say sex (male or female), employment (employed or unemployed) and region (urban or rural)). In this case there is nothing to be gained from analysis using a binary response variable because none of the individuals in the study have *unique* values of any of the explanatory variables. It might be worthwhile if you had each individual's body weight, for example, then you could ask the question, When I control for sex and region, are heavy people more likely to be unemployed than light people?. In the absence of *unique* values for any explanatory variables, there are two useful options:

- analyse the data as a $2 \times 2 \times 2$ contingency table using Poisson errors, where the count of the total number of individuals in each of the 8 contingencies is the response variable (Chapter 29) in a data frame with just 8 rows.

- decide which of your explanatory variables is the key (perhaps you are interested in gender differences), then express the data as proportions (the number of males and the number of females) and recode the binary response as count of a two-level factor. The analysis is now of proportion data (e.g. the proportion of all individuals that are female) using binomial errors (Chapter 28).

If you *do* have unique measurements of one or more explanatory variables for each individual, these are likely to be continuous variables (that's what makes them unique to the individual in question). They will be things like body weight, income, medical history, distance to the nuclear reprocessing plant, and so on. This being the case, successful analyses of binary response data tend to be multiple regression analyses or complex analyses of covariance, and you should consult the relevant chapters for details on model simplification and model criticism.

In order to carry out linear modelling on a binary response variable, we take the following steps:

- create a vector of 0s and 1s as the response variable

- use **glm** with family=binomial

- you can change the link function from default logit to complementary log-log

- fit the model in the usual way

- test significance by deletion of terms from the maximal model, and compare the change in deviance with chi-square

- note that there is no such thing as overdispersion with a binary response variable, hence no need to change to using F tests (see below)

Choice of link function is generally made by trying both links and selecting the link that gives the lowest deviance. The logit link that we used earlier is symmetric in p and q, but the complementary log-log link is asymmetric.

Deviance with binary responses

Because the response variable can take only two different values, 0 or 1, the residuals after model fitting $(y_i - \hat{y}_i)$ for any given value of y can take only one of two values:

$$y_i - \hat{y}_i = \begin{cases} 1 - \hat{y} & \text{if } y = 1 \\ -\hat{y} & \text{if } y = 0 \end{cases}$$

This means that the residuals are definitely *not* drawn from a normal distribution with zero mean and constant variance σ^2. The Pearson residuals, e_i, are calculated as

$$e_i = \frac{y_i - \hat{y}_i}{\sqrt{\hat{y}_i(1 - \hat{y}_i)}}$$

these have mean 0 and variance roughly equal to 1 (roughly, because of the estimation of β in the linear predictor); they are not normally distributed. The sum of the squares of the Pearson residuals gives the Pearson goodness of fit, χ^2.

Incidence functions

In this example the response variable is called *incidence*; a value of 1 means that an island was occupied by a particular species of bird, and 0 means that the bird did not breed there. The explanatory variables are the area of the island (km^2), the isolation of the island (distance from the mainland, km), a score representing a subjective assessment of the habitat quality of the island for that bird species (*quality*), and estimates of the population densities of natural enemies and competitors for this species of bird. We shall fit a multiple regression model containing all the explanatory variables, then simplify it, and test the simpler model for non-linear and interaction effects:

```
attach(island)
names(island)
```

```
[1] "incidence"     "area"          "isolation"     "quality"
    "enemy"         "competitors"
```

Because this is a binary regression and the response consists of a vector of 0s and 1s, we don't need to use **cbind** to make a two-column vector (as with logistic regression in Chapter 28). We start by fitting a maximal model, and looking at the coefficients:

```
model<-glm(incidence~area+isolation+quality+enemy+competitors, binomial)
summary(model)
```

```
Coefficients:
                 Value  Std. Error    t value
(Intercept)  7.187897805 4.360531865   1.6483993
       area  0.945187182 1.084874602   0.8712410
  isolation -1.515738426 0.537137090  -2.8218838
    quality  0.203854408 0.193622743   1.0528433
      enemy -0.175208716 0.597942623  -0.2930193
competitors  0.001031434 0.001551567   0.6647691

(Dispersion Parameter for Binomial family taken to be 1)
      Null Deviance: 68.0292 on 49 degrees of freedom
  Residual Deviance: 26.86407 on 44 degrees of freedom
```

There is no such thing as overdispersion in a binary regression, so we can go straight to the *t* values. On this quick assessment, it looks as if *isolation* is the only significant explanatory variable. Remember, however, that the explanatory variables may be correlated with one another, so we can only assess significance in a convincing manner by deleting terms from the model. Let's see how far automatic model simplification gets us, using **step**:

```
step(model)
```

```
Start: AIC = 38.8641
```

Deleting *enemy*, *competitors* and *quality* in turn causes stepwise reductions in AIC and is therefore justified:

```
Step:    AIC = 36.9488
Step:    AIC = 35.3334
Step:    AIC = 34.4048
```

However, attempting to delete *isolation* or *area* causes an increase in AIC and is not justified. This is the model that emerges from the **step** simplification:

```
incidence ~ area + isolation

          Df Sum of Sq      RSS       Cp
<none>                  28.63421 34.63421
    area   1  5.495239 34.12945 38.12945
isolation  1  8.287513 36.92173 40.92173

Coefficients:
 (Intercept)        area isolation
    6.640931 0.5806414  -1.37179

Degrees of Freedom: 50 Total; 47 Residual
Residual Deviance: 28.40477
```

Step has removed three of the original five explanatory variables, leaving only *area* and *isolation*. We don't always trust the results of **step** to have given us the minimal adequate model (Chapter 25), so we refit its model as model2 and summarise the results as an Anova table using **summary.aov** like this:

```
model2<-glm(incidence~area+isolation,binomial)
summary.aov(model2)

          Df Sum of Sq  Mean Sq  F Value       Pr(F)
    area   1   3.27707 3.277071  5.37896 0.02477707
isolation  1   8.28751 8.287513 13.60307 0.00058485
Residuals 47  28.63421 0.609239
```

Both variables are significant, with *isolation* having a much bigger effect on the probability of occupancy than *area*. The sum of squares for *area* is different in our Anova table than in **step's**. This is because *order matters*. To see what has happened, try fitting the model *isolation + area* rather than *area + isolation*. Because they are correlated, the deviance explained by both variables together (39.62) is less than the sum of the deviances explained when each variable is fitted separately (17.86 + 31.39 = 49.25). Now we test for non-linear and interaction effects. We calculate quadratic terms for *area* and *isolation* and a product for the interaction effect. These are then added to model2 to make a more complex model3:

```
a2<-area^2
i2<-isolation^2
ai<-area*isolation
model3<-glm(incidence~area+isolation+a2+i2+ai,binomial)
```

Now we need to compare model3 with model2 to see if these new explanatory variables have led to a significant increase in the explanatory power of the model. As usual, we use **anova** to do this:

```
anova(model2,model3,test="Chi")
```

```
                      Terms Resid. Df Resid. Dev      Test Df Deviance    Pr(Chi)
1             area + isolation        47   28.40477
2 area + isolation + a2 + i2 + ai     44   26.55506 +a2+i2+ai  3  1.84971 0.6041776
```

There is no evidence of any non-linear or interaction effects ($p > 0.6$). So it looks as if model2 is our minimal adequate model: *isolation* is most important, *area* is significant, but none of the other variables has a significant effect on the probability that the island is occupied by breeding populations of this bird species.

The last thing to do is plot the fitted model through the data. This is best achieved by fitting new models in which *area* and *isolation* are fitted on their own (we shall see why this is so later, when we have a model that contains a significant interaction):

```
model4<-glm(incidence~area,binomial)
model5<-glm(incidence~isolation,binomial)
```

To draw the graphs we need to find out the maximum and minimum values of the two explanatory variables:

```
min(area)
```

```
[1] 0.1530263
```

```
max(area)
```

```
[1] 9.26902
```

and use these to generate a sequence of x values from which we shall draw the fitted curve (about 100 values produces nice smooth curves). So for *area* we write

```
avals<-seq(0,10,.1)
```

Now we produce the scatterplot of 0s and 1s against *area*:

```
plot(area,incidence)
```

It may take a while before you realise what the graph means. It helps to draw the fitted values of model4, using **lines**:

```
lines(avals,predict(model4,list(area=avals),type="response"))
```

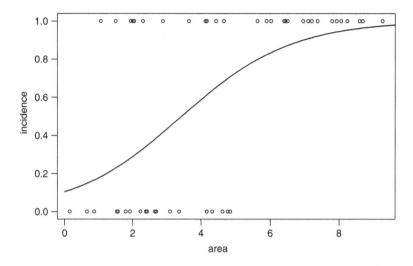

Note the use of `type="response"` to back-transform the logits onto the scale of 0 to 1 (Chapter 28). Remember also that we have to use **list** in order to provide the numerical values for the *x* axis (called *avals*) under exactly the variable name as used in model4 (*area*). Now we repeat the operation for *isolation*:

```
min(isolation)
```
```
[1]  2.022689
```
```
max(isolation)
```
```
[1]  9.577015
```
```
ivals<-seq(2,10,.1)
```

Now draw the scatterplot of *incidence* against *isolation*:

```
plot(isolation,incidence)
```

and add the fitted values as a smooth curve from model5:

```
lines(ivals,predict(model5,list(isolation=ivals),type="response"))
```

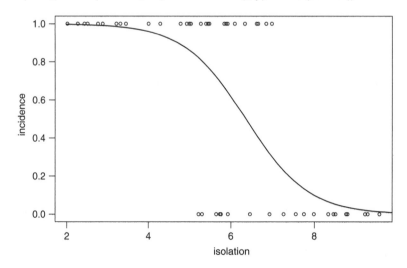

In conclusion, we find that the probability an island is occupied by our species of bird increases with the area of the island and declines with its isolation. There is no evidence for any effects of competitors, enemies or habitat quality on the probability of occupation, and no evidence for any non-linear effects of area or isolation. There was no interaction between area and isolation, which means that small, isolated islands were not disproportionately less likely (or more likely) to be occupied than large, nearby islands.

An ecological trade-off: reproductive effort and death rate

We shall use a binary response variable to analyse some questions about the costs of reproduction in a perennial plant. Suppose that we have one set of measurements on plants that survived through the winter to the next growing season, and another set of measurements on otherwise similar plants that died between one growing season and the next. We are particularly interested to know whether, for a plant of a given size, the probability of death is influenced by the number of seeds it produced during the previous year.

```
attach(flowering)
names(flowering)

[1] "State"    "Flowers"    "Root"
```

We have two explanatory variables in this example, seed production this summer (Flowers) and the size of its rootstock (Root). The expectation is that plants that produce more flowers are more likely to die during the following winter, and plants with bigger roots are less likely to die than plants with smaller roots.

The modelling is straightforward. It is a multiple regression with two continuous explanatory variables (Flowers and Root), where the response variable, State, is binary with values either 0 or 1. We begin by fitting the full model, including an interaction term, using the * operator:

```
model<-glm(State~Flowers*Root, binomial)

summary(model)

Coefficients:
                Estimate Std. Error z value Pr(>|z|)
(Intercept)     -2.96084    1.52124   -1.946  0.05161 .
Flowers         -0.07886    0.04150   -1.901  0.05736 .
Root            25.15152    7.80222    3.224  0.00127 **
Flowers.Root    -0.20911    0.08798   -2.377  0.01746 *

(Dispersion parameter for binomial family taken to be 1)

    Null deviance: 78.903  on 58  degrees of freedom
Residual deviance: 37.128  on 55  degrees of freedom
AIC: 45.128
```

The first question is whether the interaction term is required in the model. It certainly looks as if it is required ($p = 0.01746$), but we need to check this by deletion. We use **update** to compute a simpler model, leaving out the interaction Flowers:Root:

```
model2<-update(model , ~. - Flowers:Root)

anova(model,model2,test="Chi")
```

Resid. Df		Resid.	Resid.	Df	Deviance	P(>|Chi|)
Flowers + Root + Flowers:Ro		55	37.128			
Flowers + Root		56	54.068	-1	-16.940	3.858e-005

This is highly significant, so we conclude that the way that flowering affects survival depends upon the size of the rootstock. Inspection of the model coefficients indicates that a given level of flowering has a bigger impact on reducing survival on plants with small roots than on larger plants. Both explanatory variables and their interaction are important, so we cannot simplify the original model.

Here are separate 2D graphs for the effects of *Flowers* and *Root* on *State*.

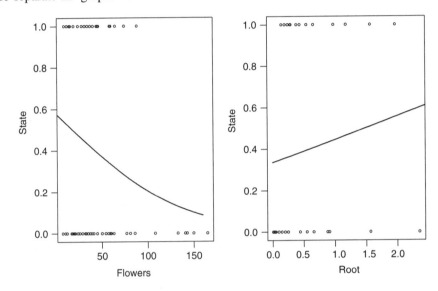

And here is a 3D surface showing the interaction between the two explanatory variables:

```
xyz<-expand.grid(Flowers=seq(0,160,16),Root=seq(0,2.5,.25))
xyz$State<-as.vector(predict(model,xyz))
wireframe(State~Flowers*Root,xyz)
```

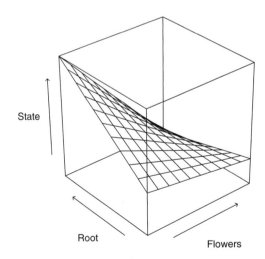

Logistic Ancova

In this example the binary response variable is parasite infection (infected or not) and the explanatory variables are weight and age (continuous) and sex (categorical). We begin with data inspection:

```
attach(Parasite)
names(Parasite)

[1] "infection" "age"        "weight"      "sex"

par(mfrow=c(1,2))
plot(infection,weight)
plot(infection,age)
```

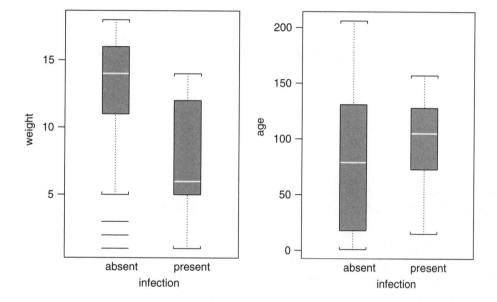

Infected individuals are substantially lighter than uninfected individuals, and occur in a much narrower range of ages. To see the relationship between *infection* and *sex* (both categorical variables) we can use **table**:

```
table(infection,sex)

           female male
  absent       17   47
  present      11    6
```

which indicates that the infection is much more prevalent in females (11/28) than in males (6/53).

We begin, as usual, by fitting a maximal model with different slopes for each level of the categorical variable:

```
model<-glm(infection~age*weight*sex,family=binomial)

summary(model)
```

```
Coefficients:
                 Estimate Std. Error z value Pr(>|z|)
(Intercept)     -0.109124   1.375388  -0.079   0.937
age              0.024128   0.020874   1.156   0.248
weight          -0.074156   0.147678  -0.502   0.616
sex             -5.969133   4.275952  -1.396   0.163
age.weight      -0.001977   0.002006  -0.985   0.325
age.sex          0.038086   0.041310   0.922   0.357
weight.sex       0.213835   0.342825   0.624   0.533
age.weight.sex  -0.001651   0.003417  -0.483   0.629
```

It certainly does not look as if any of the high-order interactions are significant. Instead of using **update** and **anova** for model simplification, we can use **step** to compute the AIC for each term in turn:

```
step(model)

Start:  AIC = 71.7056
  infection ~ age * weight * sex
```

First, it tests whether the three-way interaction is required:

```
                   Df Sum of Sq        RSS       Cp
     <none>                       62.93241 78.93241
age:weight:sex  1 0.2345435 63.16696 77.16696

Step:  AIC = 69.9426
```

This causes a reduction in AIC of just $71.7 - 69.9 = 1.8$ and hence is not significant. Next it looks at the three two-way interactions and decides which to delete first:

```
             Df Sum of Sq      RSS       Cp
   <none>                  66.32119 80.32119
age:weight  1  2.501245 68.82244 80.82244
age:sex     1  1.737103 68.05830 80.05830
weight:sex  1  0.178271 66.49946 78.49946

Step:  AIC = 68.1222

             Df Sum of Sq      RSS       Cp
   <none>                  62.97433 74.97433
age:weight  1  2.652202 65.62653 75.62653
   age:sex  1  1.830635 64.80496 74.80496

Step:  AIC = 68.1425

Coefficients:
  (Intercept)        age      weight        sex
     -2.26311  0.0359926 -0.03648283 -1.871469
   age:weight      age:sex
-0.002220601  0.01022932

Degrees of Freedom: 81 Total; 75 Residual
Residual Deviance: 56.12219
```

The **step** procedure suggests that we retain two two-way interactions: age:weight and age:sex. Let's see if we would have come to the same conclusion using **update** and **anova**:

```
model2<-update(model,~.-age:weight:sex)
anova(model,model2,test="Chi")[,-1]
```

-age:weight:sex

$p = 0.6264$ so no evidence of 3-way interaction. Now for the 2-ways:

```
model3<-update(model2,~.-age:weight)
anova(model2,model3,test="Chi")[,-1]
```

-age:weight

$p = 0.0984166$ so no really persuasive evidence of an age:weight term.

```
model4<-update(model2,~.-age:sex)
anova(model2,model4,test="Chi")[,-1]
```

Note that we are testing all the two-way interactions by deletion from the model that contains all two-way interactions (model2):

-age:sex

$p = 0.1697075$, so nothing there, then.

```
model5<-update(model2,~.-weight:sex)
anova(model2,model5,test="Chi")[,-1]
```

-weight:sex

$p = 0.6717231$, or there. As one often finds, **step** has been relatively generous in leaving terms in the model. It is a good idea to try adding back marginally significant terms to see if they have greater explanatory power when fitted on their own:

```
model6<-update(model2,~.-weight:sex-weight:age-sex:age)
model7<-update(model6,~.+weight:age)
anova(model7,model6,test="Chi")[,-1]
```

-weight:age

$p = 0.19$, so no evidence for this interaction. Now for the main effects:

```
model8<-update(model6,~. -weight)
anova(model6,model8,test="Chi")[,-1]
```

-weight

$p = 0.0003220985$, so weight is highly significant, as we expected from the initial boxplot:

```
model9<-update(model6,~. -sex)
anova(model6,model9,test="Chi")[,-1]
```

-sex

$p = 0.019\,675\,38$, so *sex* is quite significant:

```
model10<-update(model6,~. -age)
anova(model6,model10,test="Chi")[,-1]

-age
```

$p = 0.047\,523\,63$, so *age* is marginally significant. Note that all the main effects were tested by deletion from the model that contained all the main effects (model6).

It is worth establishing whether there is any evidence of non-linearity in the response of *infection* to *weight* or *age*. We might begin by fitting quadratic terms for the two continuous explanatory variables:

```
model11<-glm(infection~sex+age+weight+age^2+weight^2,family=binomial)
```

Then dropping each of the quadratic terms in turn:

```
model12<-glm(infection~sex+age+weight+age^2,family=binomial)
anova(model11,model12,test="Chi")[,-1]

-I(weight^2)
```

$p = 0.034\,474\,23$, so the `weight^2` term is needed.

```
model13<-glm(infection~sex+age+weight+weight^2,family=binomial)
anova(model11,model13,test="Chi")[,-1]

-I(age^2)
```

$p = 0.009\,896\,428$, and so is the age^2 term.

It is worth looking at these non-linearities in more detail to see if we can do better with other kinds of model (e.g. non-parametric smoothers, piecewise linear models or step functions).

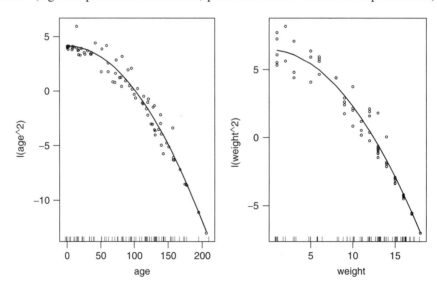

An alternative approach is to use **gam** rather than **glm** for the continuous covariates:

```
gam1<-gam(infection~sex+s(age)+s(weight),family=binomial)
plot.gam(gam1,resid=T)
```

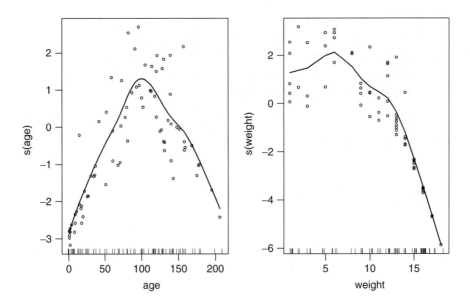

These non-parametric smoothers are excellent at showing the humped relationship between *infection* and *age*, and at highlighting the possibility of a threshold at *weight* ≈ 12 in the relationship between *weight* and *infection*. We can now return to **glm** to incorporate these ideas. We shall fit age and age^2 as before, but try a **piecewise linear fit** for weight, estimating the threshold weight as 12 (see above); we could try a range of values and see which gives the lowest deviance. The piecewise regression is specified by the term

I((weight - 12) * (weight > 12))

The I (as is) is necessary to stop the * being evaluated as an interaction term in the model formula. What this expression says is, regress infection on the value of weight-12, but only do this when weight>12 is true. Otherwise, assume that *infection* is independent of *weight*.

model14<-glm(infection~sex+age+age^2+I((weight-2)*(weight>12)),family=binomial)

The effect of *sex* on *infection* is not quite significant ($p = 0.07$ for a chi-square test on deletion), so we leave it out:

model15<-update(model14,~.-sex)

anova(model14,model15,test="Chi")[,-1]

```
  Resid. Df Resid. Dev Test Df  Deviance     Pr(Chi)
1        76    48.68691
2        77    51.95327 -sex -1 -3.266359 0.0707144
```

summary(model15)

```
Coefficients:
                                       Value    Std. Error    t value
                  (Intercept)   -3.120773865 1.2657194652  -2.465613
                          age    0.076578920 0.0323046648   2.370522
                    I(age^2)   -0.000384259 0.0001843757  -2.084109
I((weight - 12) * (weight > 12)) -1.351012211 0.5072017671  -2.663658
```

```
(Dispersion Parameter for Binomial family taken to be 1)
Null Deviance: 83.23447 on 80 degrees of freedom
Residual Deviance: 51.95327 on 77 degrees of freedom
```

This is our minimal adequate model. All the terms are significant when tested by deletion, the model is parsimonious, and the residuals are well behaved.

model16<-glm(infection~poly(age,2)+I((weight-12)*(weight>12)),family=binomial)

plot.gam(model16,resid=T)

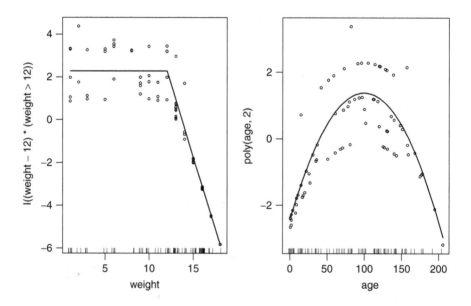

The humped plot for infection against age is obtained by replacing the two terms age + age^2 in model15 by the single expression poly(age,2) in model16.

A 3 × 3 crossover trial

In a crossover experiment, the same set of treatments is applied to different individuals (or locations) but in different sequences *to control for order effects*. So a set of farms might have predator control for 3 years followed by no control for 3 years, while another set may have 3 years of no control followed by 3 years of predator control. Crossover trials are most commonly used on human subjects, as in the present example of a drug trial on a new painkiller using a group of headache sufferers. There were three experimental treatments: a control pill that had no painkiller in it (the placebo), neurofen, and brandA, the new product to be compared with the established standard. The interesting thing about crossover trials is that they involve repeated measures on the same individual, so a number of important questions need to be addressed:

- does the order in which the treatments are administered to subjects matter?

- if so, are there any carryover effects from treatments administered early on?

- are there any time series trends through the trial as a whole?

The response variable, y, is binary: 1 means pain relief was reported and 0 means no pain relief.

```
attach(crossover)
names(crossover)

[1] "y" "drug" "order" "time" "n" "patient" "carryover"
```

In a case like this, data inspection is best carried out by summarising the *totals* of the response variable over the explanatory variables. Your first reaction might be to use **table** to do this, but this just counts the number of cases:

```
table(drug)
```

A	B	C
86	86	86

which shows only that all 86 patients were given each of the three drugs: A is the placebo, B is neurofen and C is brandA. What we want is the sum of the y values (the number of cases of expressed pain relief) for each of the drugs. The tool for this is **tapply**:

```
tapply(y,drug,sum)
```

A	B	C
22	61	69

Before we do any complicated modelling, it is sensible to see if there are any significant differences between these three counts. It looks (as we hoped that it would) as if pain relief is reported less frequently (22 times) for the placebo than for the painkillers (61 and 69 reports respectively). It is much less clear, however, whether brandA (69) is significantly better than neurofen (61). Looking on the bright side, there is absolutely no evidence that it is significantly worse than neurofen! To do a really quick chi-square test on these three counts we just type

```
glm(c(22,61,69)~1,family=poisson)
```

and we see

```
Degrees of Freedom: 3 Total; 2 Residual
Residual Deviance: 28.55763
```

which is a chi-square of 28.558 on just two degrees of freedom, hence highly significant. Now we look to see whether there are any obvious effects of treatment sequence on reports of pain relief. In a three-level crossover like this there are six orders: ABC, ACB, BAC, BCA, CAB and CBA. Patients are allocated to *order* at random. There were different numbers of replications in this trial, which we can see using **table** on *order* and *drug*:

```
table(drug,order)
```

	1	2	3	4	5	6
A	15	16	15	12	15	13
B	15	16	15	12	15	13
C	15	16	15	12	15	13

which shows that replication varied from a low of 12 patients for order 4 (BCA) to a high of 16 for order 2 (ACB). A simple way to get a feel for whether or not order matters is to tabulate the total reports of relief by order and by drug:

```
tapply(y,list(drug,order),sum)
```

```
    1  2  3  4  5  6
A   2  5  5  2  4  4
B  13 13 10 10 10  5
C  12 13 13 10  9 12
```

Nothing obvious here. C (our new drug) did worst in order 5 (CAB) when it was administered first (9 cases), but this could easily be due to chance. We shall see. It is time to begin the modelling:

```
order<-factor(order)
```

With a binary response, we can make the single vector of 0s and 1s (y) into the response variable directly, so we type

```
model<-glm(y~drug*order+time,family=binomial)
```

There are lots of different ways of modelling an experiment like this. Here we are concentrating on the treatment effects and worrying less about the correlation and time series structure of the repeated measurements. We fit the main effects of *drug* (a three-level factor) and *order* (a six-level factor) and their interaction (which has $2 \times 5 = 10$ d.f.). We also fit *time* to see if there is any pattern to the reporting of pain relief through the course of the experiment as a whole. This is what we find:

```
summary(model)
```

```
Coefficients:
                Estimate Std. Error z value Pr (>|z|)
(Intercept)    -1.862803   1.006118  -1.851  0.064101  .
drugB           3.752595   1.006118   3.730  0.000192  ***
drugC           3.276086   0.771058   4.249 2.15e-005  ***
order2          1.083344   0.930765   1.164  0.244453
order3          1.187650   0.858054   1.384  0.166322
order4          0.280355   1.092937   0.257  0.797553
order5          0.869198   0.881368   0.986  0.324038
order6          1.078864   0.771058   1.399  0.161753
time           -0.008997   0.385529  -0.023  0.981383
drugB.order2   -1.479810   1.512818  -0.978  0.327985
drugC.order2   -1.012297   1.159837  -0.873  0.382776
drugB.order3   -2.375300   1.330330  -1.785  0.074181  .
drugC.order3   -0.702144   1.232941  -0.569  0.569026
drugB.order4   -0.551713   1.680933  -0.328  0.742747
drugC.order4   -0.066207   1.428863  -0.046  0.963043
drugB.order5   -2.038855   1.338492  -1.523  0.127697
drugC.order5   -1.868020   1.208188  -1.546  0.122072
drugB.order6   -3.420667   1.222788  -2.797  0.005151  **

(Dispersion parameter for binomial family taken to be 1)

    Null deviance: 349.42  on 257  degrees of freedom
Residual deviance: 268.85  on 240  degrees of freedom
AIC: 304.85
```

Remember that with binary response data we are *not* concerned with overdispersion. Looking up the list of coefficients, starting with the highest-order interactions, it looks as if drugB.order6 with its *t* value of 2.768 might be significant, and drugB.order3, perhaps. We test this by deleting the interaction drug:order from the model:

> model2<-update(model, ~. - drug:order)

> anova(model,model2,test="Chi")

```
Terms Resid. Df Resid. Dev     Test Df  Deviance   Pr(Chi)
1 drug * order + ti    240 268.8494
2 drug + order + time  249 283.3486 -drug:order -9 -14.49911 0.1056456
```

There is no compelling reason to retain this interaction ($p > 10\%$). So we leave it out and test for an effect of *time*:

> model3<-update(model2, ~. - time)

> anova(model2,model3,test="Chi")

```
                Terms Resid. Df Resid. Dev Test Df   Deviance    Pr(Chi)
1 drug + order + time     249    283.3486
2         drug + order    250    283.7556 -time -1 -0.4070609 0.5234651
```

Nothing at all. What about the effect of *order*?

> model4<-update(model3, ~. - order)

> anova(model3,model4,test="Chi")

```
        Terms Resid. Df Resid. Dev    Test Df  Deviance    Pr(Chi)
1 drug + order    250    283.7556
2         drug    255    286.9939 -order -5 -3.238335 0.6632949
```

Nothing. What about the differences between the drugs?

> summary(model4)

```
Coefficients:
                Value Std. Error   t value
(Intercept) -1.067840  0.2470281 -4.322746
      drugB  1.959838  0.3426400  5.719816
      drugC  2.468703  0.3659238  6.746495
```

Notice that, helpfully, the names of the treatments appear in the coefficients list. Now it is clear that our drug C had log odds greater than neurofen by $2.4687 - 1.9598 = 0.5089$. The standard error of the difference between two treatments is about 0.366—taking the larger (less well replicated) of the two values so a *t* test between neurofen and our drug gives $t = 0.5089/0.366 = 1.39$. This is not significant.

Following model simplification, the coefficients are now the logits of our original proportions that we tabulated during data exploration: 22/86, 61/86 and 69/86 (you might like to check this). The analysis tells us that 69/86 is not significantly greater than 61/86. If we had known to begin with that none of the interactions, sequence effects or temporal effects had been significant, we could have analysed these results much more simply, by turning the response into the proportion of patients reporting pain relief, like this:

```
relief<-c(22,61,69)
patients<-c(86,86,86)
treat<-factor(c("A","B","C"))
```

```
model<-glm(cbind(relief,patients-relief)~treat, binomial)
```

```
summary(model)
```

```
Coefficients:
              Estimate Std. Error z value  Pr(>|z|)
(Intercept) -1.0678      0.2471   -4.321 1.55e-005 ***
treatB       1.9598      0.3427    5.718 1.08e-008 ***
treatC       2.4687      0.3666    6.734 1.65e-011 ***

(Dispersion parameter for binomial family taken to be 1)

    Null deviance: 6.2424e+001  on 2  degrees of freedom
Residual deviance: 1.5479e-014  on 0  degrees of freedom
AIC: 19.824
```

Treatments B and C differ by only about 0.5 with a standard error of 0.367, so we may be able to combine them. Note the use of != for 'not equal to':

```
t2<-factor(1+(treat != "A"))
```

```
model2<-glm(cbind(relief,patients-relief)~t2,binomial)
```

```
anova(model,model2,test="Chi")
```

```
        Resid. Df Resid. Dev Df Deviance P(>|Chi|)
treat         0 1.548e-014
t2            1    2.02578 -1 -2.02578   0.15465
```

We can reasonably combine these two factor levels, so the minimum adequate model has only two parameters, and we conclude that the new drug is not significantly better than neurofen. But if the two drugs cost the same, which of them would *you* take if you had a headache?

Further reading

Collett, D. (1991) *Modelling Binary Data*. London, Chapman & Hall.
Cox, D.R. and Snell, E.J. (1989) *Analysis of Binary Data*. London, Chapman & Hall.

31

Tree models

If you have not met tree models before, then prepare to be impressed. Tree models are computationally intensive methods that come into their own in situations where there are many explanatory variables and we don't know which of them to select. Often there are so many explanatory variables that we could not test them all, even if we wanted to invest the huge amount of time that would be necessary. Tree models are particularly good at tasks that might in the past have been regarded as the realm of multivariate statistics (e.g. classification problems).

The great virtues of tree models are:

- they are very simple

- they are excellent for initial data inspection

- they give a very clear picture of the structure of the data

- they provide a highly intuitive insight into the kinds of interactions between variables

It is best to begin by looking at a tree model in action, before thinking about how it works. Here is the air pollution example that we have worked on already as a multiple regression (see p. 442):

```
attach(Pollute)
names(Pollute)
```

```
[1]    "Pollution"   "Temp"   "Industry"   "Population"   "Wind"
[6]    "Rain"        "Wetdays"
```

```
model<-tree(Pollute)
plot(model,type="u")
text(model)
```

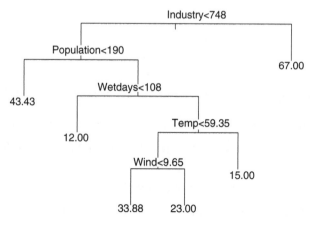

You follow a path from the top of the tree (called, in defiance of gravity, the *root*) and proceed to one of the terminal nodes (called a *leaf*) by following a succession of rules (called *splits*). The numbers at the tips of the leaves are the mean values in that subset of the data (mean SO_2 concentration in this case). The details are explained below.

Background

The model is fitted using *binary recursive partitioning* whereby the data are successively split along coordinate axes of the explanatory variables so that at any node, the split which maximally distinguishes the response variable in the left and the right branches is selected. Splitting continues until nodes are pure or the data are too sparse (fewer than six cases, by default):

- if the response variable is a factor, the tree is called a *classification* tree

- if the response variable is continuous, the tree is called a *regression* tree

Regression trees

This is how a tree models works. Each explanatory variable is assessed in turn, and the variable explaining the greatest amount of the deviance in y is selected. Deviance is calculated on the basis of a threshold in the explanatory variable; this threshold produces two mean values for the response (one mean above the threshold, the other below the threshold).

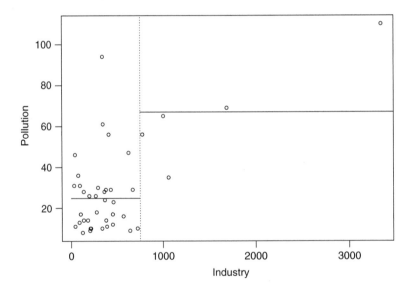

The procedure works like this. For a given explanatory variable (say Industry):

- select a threshold value of the explanatory variable (the vertical dotted line)

- calculate the mean value of the response variable above and below this threshold (the two horizontal solid lines)

- use the two means to calculate the deviance (as with *SSE*, see p. 245).

- look to see which value of the threshold gives the lowest deviance

- split the data into high and low subsets on the basis of the threshold for this variable

- repeat the whole procedure on each subset of the data

- keep going until no further reduction in deviance is obtained, or there are too few data points to merit further subdivision (e.g. the right-hand side of the Industry split is too sparse to allow further subdivision)

The deviance is defined as

$$D = \sum_{\text{cases } j} (y_j - \mu_{[j]})^2$$

where $\mu_{[j]}$ is the mean of all the values of the response variable assigned to node j and this sum of squares is added up over all the nodes. The *value* of any split is defined as the reduction in this residual sum of squares. The probability model used in S-Plus is that the values of the response variable are normally distributed within each leaf of the tree with mean μ_i and variance σ^2. Note that because this assumption applies to the terminal nodes, the interior nodes represent a mixture of different Normal distributions, so the deviance is only appropriate at the terminal nodes (i.e. for the leaves).

If the twigs of the tree are categorical (i.e. levels of a factor like names of particular species) then we have a **classification tree**. On the other hand, if the terminal nodes of the tree are predicted values of a continuous variable, then we have a **regression tree**.

The key questions are these:

- which variables for the division should be used?

- what is the best way to achieve the splits for each selected variable?

Regression trees

In this case the response variable is a continuous measurement, but the explanatory variables can be any mix of continuous and categorical variables. You can think of regression trees as analogous to multiple regression models. The difference is that a regression tree works by forward selection of variables, whereas we have been used to carrying out regression analysis by deletion (backward selection).

```
attach(Pollute)
names(Pollute)

[1]   "Pollution"   "Temp"      "Industry"      "Population"
[5]   "Wind"        "Rain"      "Wet.days"
```

The regression tree is fitted by stating that the continuous response variable Pollution is to be estimated as a function of *all* the explanatory variables in the data frame called Pollute by use of the ~. notation like this:

```
model<-tree(Pollution~., Pollute)
```

To see the regression tree we type

```
plot(model)
```

and to overlay the mean values for each of the leaves and the conditions for the split at each node we type

```
text(model)
```

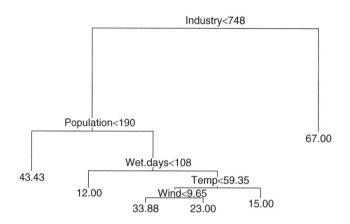

For a regression tree, the **print** method produces the following kind of output:

print(model)

```
node), split, n, deviance, yval
      * denotes terminal node
 1)  root 41 22040 30.05
  2) Industry<748 36   11260  24.92
    4) Population<190    7   4096 43.43 *
    5) Population>190   29   4187 20.45
   10) Wet.days<108    11     96 12.00 *
   11) Wet.days>108    18   2826 25.61
     22) Temp<59.35    13   1895 29.69
       44) Wind<9.65    8   1213 33.88 *
       45) Wind>9.65    5    318 23.00 *
     23) Temp>59.35     5    152 15.00 *
  3) Industry>748 5   3002  67.00 *
```

The terminal nodes (the leaves) are denoted by * (there are six of them). The node number is on the left, labelled by the variable on which the split at that node was made. Next comes the split criterion which shows the threshold value of the variable that was used to create the split. The number of cases going *into* the split (or into the terminal node) comes next. The penultimate figure is the deviance at that node. Notice how the deviance goes down as non-terminal nodes are split. In the root, based on all $n = 41$ data points, the deviance is *SST* (see p. 244) and the y value is the overall mean for Pollution. The last figure on the right is the *mean value* of the response variable within that node or at that leaf. The highest mean pollution (67.00) was in node 3 and the lowest (12.00) was in node 10.

Note how the nodes are nested: within node 2, for example, node 4 is terminal but node 5, is not; within node 5, node 10 is terminal but node 11 is not; within node 11, node 23 is terminal but node 22 is not, and so on.

Tree models lend themselves to circumspect and critical analysis of complex data frames. In the present example, the aim is to understand the causes of variation in air pollution levels from case to case. The interpretation of the regression tree would proceed something like this:

- the five most extreme cases of Industry stand out (mean $= 67.00$) and need to be considered separately

- for the rest, Population is the most important variable but interestingly it is *low* populations that are associated with the highest levels of pollution (mean = 43.43); ask yourself, Which might be cause, and which effect?

- for the rest, Wet.days is a key determinant of pollution; the places with the fewest wet days (less than 108 per year) have the lowest pollution levels of anywhere in the data frame (mean = 12.00)

- for those places with more than 108 wet days, it is Temp that is most important in explaining variation in pollution levels; the warmest places have the lowest air pollution levels (mean = 15.00)

- for the cooler places with lots of wet days, it is Wind speed that matters: the windier places are less polluted than the still places

This kind of complex and contingent explanation is much easier to see, and to understand, in tree models than in the output of a multiple regression.

Tree models as regressions

To see how a tree model works when there is a single, continuous response variable, it is useful to compare the output with a simple linear regression. Take the relationship between mileage and weight in the car.test.frame data:

```
attach(car.test.frame)
names(car.test.frame)

[1]  "Price"     "Country"     "Reliability"     "Mileage"
[5]  "Type"      "Weight"      "Disp."           "HP"

plot(Weight,Mileage)
```

The heavier cars do fewer miles per gallon, but there is a lot of scatter. The tree model starts by finding the weight that splits the mileage data in a way that explains the maximum deviance. This weight turns out to be 2567.5:

```
a<-mean(Mileage[Weight<2567.5])
b<-mean(Mileage[Weight>=2567.5])
lines(c(1500,2567.5,2567.5,4000),c(a,a,b,b))
```

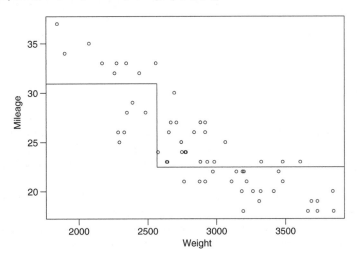

The next thing the tree model does is to work out the threshold weight that would best split the mileage data for the lighter cars; this turns out to be 2280. It then works out the threshold split for the heavier cars; this turns out to be 3087.5. And so the process goes on, until there are too few cars in each split to justify continuation (five or fewer by default). To see the full regression tree as a function plot, we can use the **predict** function with the regression tree object model1 like this:

```
car.model<-tree(Mileage~Weight)

plot(Weight,Mileage)
wt<-seq(1500,4000,10)
y<-predict(car.model,list(Weight=wt))
lines(wt,y)
```

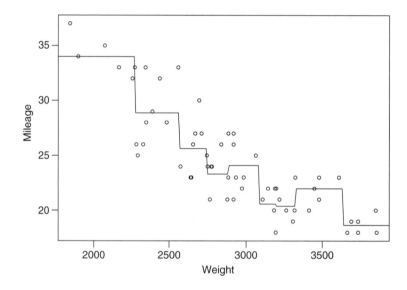

You would not normally do this, of course (and you *could* not do it with more than two explanatory variables) but it is a good way of showing how tree models work with a continuous response variable. It also makes the point that *tree models have a tendency to overinterpret the data*; for instance, the occasional ups in the generally negative correlation probably don't mean anything substantial.

Model simplification

Model simplification in regression trees is based on a *cost-complexity measure*. This reflects the trade-off between fit and explanatory power. A model with a perfect fit would have as many parameters as there were data points, and would consequently have no explanatory power at all. We return to the air pollution example, analysed earlier where we fitted the tree model object called model.

Regression trees can be overelaborate and can respond to random features of the data (the so-called *training set*). To deal with this, S-Plus contains a set of procedures to prune trees on the basis of the cost-complexity measure. The function **prune.tree** determines a nested sequence of subtrees of the supplied tree by recursively 'snipping' off the least important splits, based upon the cost-complexity measure. The **prune.tree** function returns an object of class tree.sequence, which contains the following components:

```
prune.tree(model)

$size:

[1]  6  5  4  3  2  1
```

This shows the number of terminal nodes in each tree in the cost-complexity pruning sequence; the most complex model had six terminal nodes (see above).

```
$dev:

[1]  8876.589  9240.484  10019.992  11284.887  14262.750  22037.902
```

This shows the total deviance of each tree in the cost-complexity pruning sequence.

```
$k:

[1]  -Inf  363.8942  779.5085  1264.8946  2977.8633  7775.1524
```

This is the value of the cost-complexity pruning parameter of each tree in the sequence. If determined algorithmically (as here, k is not specified as an input), its first value defaults to -Inf, its lowest possible bound.

```
plot(prune.tree(model))
```

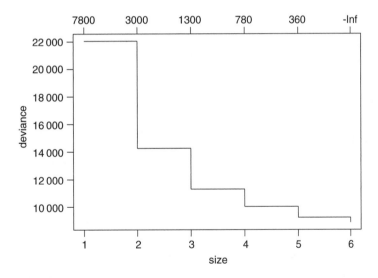

This shows the way that deviance declines as complexity is increased. The total deviance is 22037.902 (size = 1) and this is reduced as the complexity of the tree increases up to six nodes.

An alternative is to specify the number of nodes to which you want the tree to be pruned; this uses the best= option. Suppose we want the best tree with four nodes. Then

```
model2<-prune.tree(model,best=4)
plot(model2,type="u")
text(model2)
```

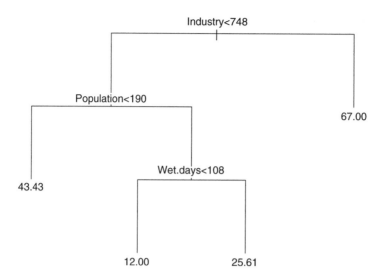

As a printout, this is

model2

```
node),  split,  n,  deviance,  yval
       * denotes terminal node
1) root 41 22040 30.05
  2) Industry<748 36 11260 24.92
    4) Population<190 7 4096 43.43 *
    5) Population>190 29 4187 20.45
      10) Wet.days<108 11 96 12.00 *
      11) Wet.days>108 18 2826 25.61 *
  3) Industry>748 5 3002 67.00 *
```

Subtrees and subscripts

It is straightforward to remove parts of trees, or to select parts of trees, using subscripts; for example:

- a negative subscript [-3] leaves off everything above node 3
- a positive subscript [3] selects only that part of the tree above node 3

Classification trees with categorical explanatory variables

Tree models are a superb tool for helping to write efficient and effective taxonomic keys. Suppose that all of our explanatory variables are categorical, and that we want to use tree models to write a dichotomous key. There is only one entry for each species, so we want the twigs of the tree to be the individual rows of the data frame (i.e. we want to fit a tree perfectly to the data). To do this we need to specify two extra arguments:

- minsize = 2
- mindev = 0

In practice it is better to specify a very small value for the minimum deviance (say 10^{-6}) rather than zero (see below).

The example relates to the nine lowland British species in the genus *Epilobium* (Onagraceae). We have eight categorical explanatory variables and we want to find the optimal dichotomous key. The data frame looks like this:

attach(epilobium)

species	stigma	stem.hairs	gland.hairs	seeds	pappilose	stolons	petals	base
hirsutum	lobed	spreading	absent	none	uniform	absent	>9mm	rounded
parviflorum	lobed	spreading	absent	none	uniform	absent	<10mm	rounded
montanum	lobed	spreading	present	none	uniform	absent	<10mm	rounded
lanceolatum	lobed	spreading	present	none	uniform	absent	<10mm	cuneate
tetragonum	clavate	appressed	present	none	uniform	absent	<10mm	rounded
obscurum	clavate	appressed	present	none	uniform	stolons	<10mm	rounded
roseum	clavate	spreading	present	none	uniform	absent	<10mm	cuneate
palustre	clavate	spreading	present	appendage	uniform	absent	<10mm	rounded
ciliatum	clavate	spreading	present	appendage	ridged	absent	<10mm	rounded

Producing the key could not be easier:

```
model<-tree(species~.,epilobium,mindev=1e-6,minsize=2)
plot(model)
text(model)
```

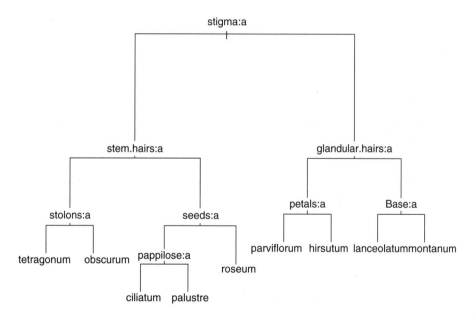

Here is the tree written as a dichotomous key:

1.	Stigma entire and club-shaped	2
1.	Stigma four-lobed	6
2.	Stem hairs all appressed	3
2.	At least some stem hairs spreading	4

3.	Glandular hairs present on hypanthium	*E. obscurum*
3.	No glandular hairs on hypanthium	*E. tetragonum*
4.	Seeds with a terminal appendage	5
4.	Seeds without terminal appendage	*E. roseum*
5.	Surface of seed with longitudinal papillose ridges	*E. ciliatum*
5.	Surface of seed uniformly papillose	*E. palustre*
6.	At least some spreading hairs non-glandular	7
6.	Spreading hairs all glandular	8
7.	Petals large (>9 mm)	*E. hirsutum*
7.	Petals small (<10 mm)	*E. parviflorum*
8.	Leaf base cuneate	*E. lanceolatum*
8.	Leaf base rounded	*E. montanum*

The computer has produced a working key to a difficult group of plants. The result stands as testimony to the power and usefulness of tree models. The same principle underlies good key-writing as is used in tree models: find the characters that explain most of the variation, and use them to split the cases into roughly equal-sized groups at each dichotomy.

Classification trees for replicated data

In this example from plant taxonomy, the response variable is a four-level, categorical variable called Taxon (it is a label expressed as Roman numerals I to IV). The aim is to use the measurements from the seven morphological explanatory variables to construct the best key to separate these four taxa (the 'best' key is the one with the lowest error rate, the key that misclassifies the smallest possible number of cases).

```
attach(taxonomy)
names(taxonomy)
```
```
[1]   "Taxon"   "Petals"   "Internode"   "Sepal"
[5]   "Bract"   "Petiole"  "Leaf"        "Fruit"
```

Using the tree model for classification could not be simpler:

```
model1<-tree(Taxon~.,taxonomy)
```

We begin by looking at the plot of the tree:

```
plot(model1)
text(model1)
```

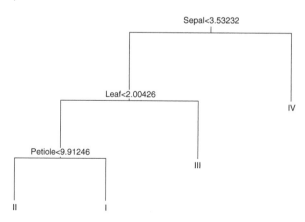

With only a small degree of rounding on the suggested break points, the tree model suggests a simple (and for these 120 plants, completely error-free) key for distinguishing the four taxa:

1.	Sepal length > 4.0	*Taxon IV*
1.	Sepal length <= 4.0	2
2.	Leaf width > 2.0	*Taxon III*
2.	Leaf width <= 2.0	3
3.	Petiole length < 10	*Taxon II*
3.	Petiole length >= 10	*Taxon I*

The **summary** option for classification trees produces the following:

```
summary(model1)

Classification tree:
tree(formula = Taxon ~ ., data = taxonomy)
Variables actually used in tree construction:
[1] "Sepal"  "Leaf" "Petiole"
Number of terminal nodes: 4
Residual mean deviance: 0 = 0 / 116
Misclassification error rate: 0 = 0 / 120
```

Three of the seven variables were chosen for use (Sepal, Leaf and Petiole); four variables were assessed and rejected (Petals, Internode, Bract and Fruit). The key has four nodes and hence three dichotomies. As you see, the misclassification error rate was an impressive 0 out of 120. It is noteworthy that this classification tree does much better than some multivariate classification methods.

For classification trees, the **print** method produces a great deal of information:

```
print(model1)

node), split, n, deviance, yval, (yprob)
      * denotes terminal node
1) root 120 332.70 I (0.2500 0.2500 0.2500 0.25)
  2) Sepal<3.53232 90 197.80 I (0.3333 0.3333 0.3333 0.00)
    4) Leaf<2.00426 60 83.18 I (0.5000 0.5000 0.0000 0.00)
      8) Petiole<9.91246 30 0.00 II (0.0000 1.0000 0.0000 0.00) *
      9) Petiole>9.91246 30 0.00 I (1.0000 0.0000 0.0000 0.00) *
    5) Leaf>2.00426 30 0.00 III (0.0000 0.0000 1.0000 0.00) *
  3) Sepal>3.53232 30 0.00 IV (0.0000 0.0000 0.0000 1.00) *
```

The **node** number is followed by the **split criterion** (e.g. Sepal <3.53 at node 2). Then comes the number of cases passed through that node (90 in this case, versus 30 going into node 3, which is the terminal node for Taxon IV). The remaining deviance within this node is 197.8 (compared with zero in node 3, where all the individuals are alike; they are all Taxon IV). Next is the name of the factor levels left in the split (I, II and III in this case, with the convention that the first in the alphabet is listed), then a list of the empirical probabilities—the fractions of all the cases at that node that are associated with each of the levels of the response variable (in this case the 90 cases are equally split between I, II and III and there are no individuals of Taxon IV at this node, giving 0.33, 0.33, 0.33 and 0 as the four probabilities).

There is quite a useful plotting function for classification trees called **partition.tree** but it is only sensible to use it when the model has two explanatory variables. Its use is illustrated here by taking the two most important explanatory variables, Sepal and Leaf:

```
model2<-tree(Taxon~Sepal+Leaf,taxonomy)
partition.tree(model2)
```

This shows how the phase space defined by sepal length and leaf width has been divided up between the four taxa, but it does not show where the data fall. We could use points(Sepal,Leaf) to overlay the points, but for illustration we shall use **text.** We create a vector called label that has a for Taxon I, b for Taxon II, and so on:

```
label<-ifelse(Taxon=="I", "a", ifelse(Taxon=="II","b",ifelse(Taxon=="III","c","d")))
```

then we use these letters as a **text** overlay on the **partition.tree** like this:

```
text(Sepal,Leaf,label)
```

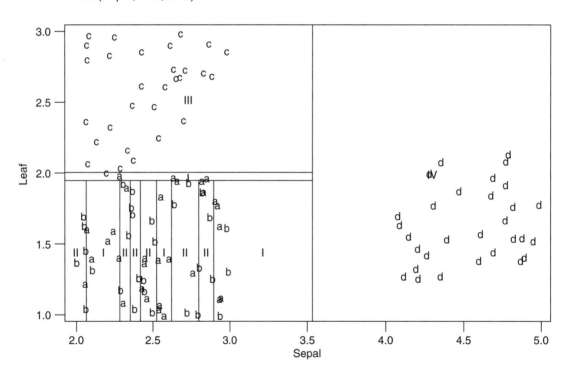

You see that Taxon III and Taxon IV (c and d) are beautifully separated on the basis of sepal length and leaf width, but Taxon I and Taxon II are all jumbled up (recall that they are separated from one another on the basis of petiole length).

Further reading

An article by Clark and Pregibon gives a detailed explanation of tree model simplification by growing, displaying and challenging trees: See Chambers and Hastie (1992, pp. 377–419). These procedures involve inspecting nodes, identifying observations, snipping branches, selecting subtrees, and so on.

Breiman, L., Friedman, L.H. *et al.* (1984) *Classification and Regression Trees*. Belmont, California, Wadsworth International Group.

Chambers, J.M. and Hastie, T.J. (1992). *Statistical Models in S*. Pacific Grove, California, Wadsworth and Brooks Cole.

Venables, W.N. and Ripley, B.D. (1997). *Modern Applied Statistics with S-Plus*. New York, Springer-Verlag.

32

Non-parametric smoothing

Local regression fits regression surfaces to data as a single smooth function of all the explanatory variables. It is like multiple regression analysis but the fitted surface is non-parametric. Loess is particularly useful when the curvature is complex and there is no well-established theory to suggest a suitable parametric form for the relationship. The method works by assuming that within any neighbourhood of a point defined by the set of explanatory variables, the regression surface is well approximated by a function from a particular parametric class (e.g. polynomials of degree 1 or 2). The response is smoothed as a function of all the explanatory variables using non-parametric regression procedures to determine the fit. Unlike **gam** (additive models with non-parametric smoothers), local regression models allow the possibility of interactions between the explanatory variables in determining the shape of the surface.

The local regression model is

$$y_i = g(x_i) + \varepsilon_i$$

where $g(x_i)$ is the smoothed expected value of y at x_i and the errors ε_i are independent random variables with mean 0 that are either Normal or skew (as specified later). The size of the neighbourhood over which the linear or quadratic polynomial smoothing is carried out is determined by a parameter α. The smaller the interval, the rougher the surface and the more degrees of freedom that are used up (see p. 597). For linear smoothing and two explanatory variables, three parameters are used (a constant and two slopes); for quadratic, six are involved (a constant, two slopes, two quadratic terms and an interaction term). When there are categorical explanatory variables in the model, the smoothing is carried out independently for each combination of factor levels. The decisions that need to be made are these:

- neighbourhood size (to give a smooth or rough surface)

- Normal errors or skew errors

- constant variance or *a priori* weights

- locally linear in the predictors or locally quadratic

Other optional choices include whether or not to normalise the scales of the explanatory variables, whether to drop squared terms and retain only the interaction term, and whether to fit some explanatory variables parametrically. Coupled with assiduous model checking, the method of local regression allows for the calculation of confidence intervals and robust prediction. Locally quadratic models may have at most 4 predictor variables; locally linear models may have at most 15. Note that the computer memory needed by **loess** increases exponentially with the number of variables.

Testing for curvature: non-parametric smoothing

The human eye is so good at pattern recognition that it is prone to overinterpret the complexity of functional relationship in a scatterplot. It is useful, therefore, to have some independent assessment of the shape of a relationship, before a great deal of effort is expended in modelling the data with a particular model form. The tool for the job is non-parametric smoothing.

Suppose we have the following scatterplot of Yield against Irrigation. The question is whether or not there is any hint of a hump in the curve (i.e. whether yield declines with irrigation at the highest levels of water input).

```
attach(irrigation)
names(irrigation)

[1]  "Irrigation" "Yield"

plot(Irrigation,Yield)
```

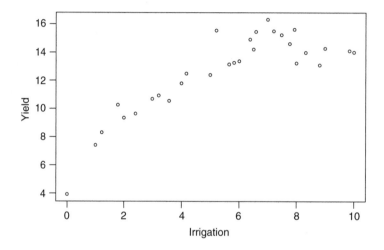

It looks possible that the curve has a hump, but it's far from convincing. Let's see what a non-parametric smoother (**lowess**) produces:

```
lines(lowess(Irrigation,Yield))
```

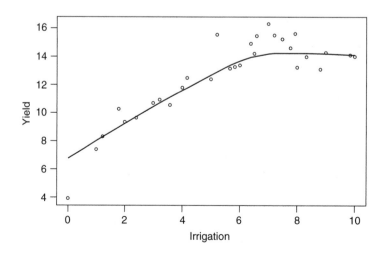

There is a reasonably clear asymptote, but no real hint of a hump. Curve smoothing represents a classic example of a *trade-off* in statistics. At one end of the continuum is a two-parameter straight line—the ultimate in smoothness. At the other extreme is a line which joins every point to every other along the vector of ordered x values—the ultimate in roughness. To draw the joined-up graph, it is important that you understand the difference between the three directives **sort, rank** and **order** (see p. 24).

```
ord<-order(Irrigation)
par(mfrow=c(1,2))
plot(Irrigation, Yield)
abline(lm(Yield~Irrigation))

plot(Irrigation,Yield)
lines(Irrigation[ord],Yield[ord])
```

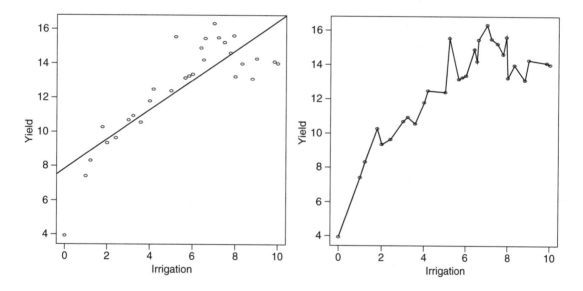

What we want to discover is the optimal compromise between these two. There are several different smoothers that we could use, but the same trade-off underlies them all, and they all deal with the trade-off in essentially similar ways. The straight line uses 2 d.f. for the two parameters estimated from the data (the slope and the intercept). The line going through every point uses up as many parameters as there are data points, $p =$ d.f. $= n$. Use of more degrees of freedom is seen as a penalty. We want the smoothest graph that makes a reasonable job of describing the relationship between y and x—the graph that pays the lowest penalty in terms of degrees of freedom. Let's see what this means in terms of the smoothing function **smooth.spline**.

With 3 d.f. there is no evidence of a hump:

```
par(mfrow=c(1,1))
plot(Irrigation,Yield)
lines(smooth.spline(Irrigation,Yield,df=3),lty=1)
```

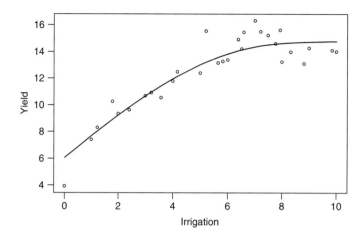

With 4 d.f. there is a hint of a himp.

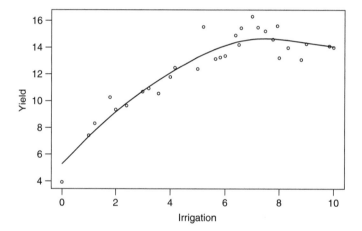

With 6 d.f. there is distinct evidence of a hump.

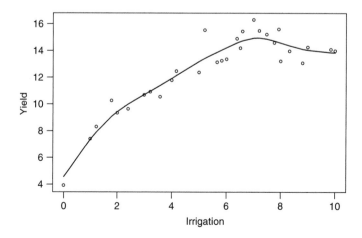

The principle of parsimony would always favour an asymptote over a hump, but these things are always going to be difficult to call, because the data are bound to be sparse at the extremes of the x axis. The way to find out whether the curve has a hump is to do an experiment: increase x above its current maximum value (about 7 here), and see if y decreases significantly.

Non-linear patterns

Smoothing functions like **loess** are excellent for highlighting non-linear patterns in data. Here is a panel plot of the ethanol data showing oxides of nitrogen (NO_x) as a function of E for five levels of C:

```
attach(ethanol)
names(ethanol)
```

```
[1]  "NOx"  "C"  "E"
```

In this panel plot we specify a grid on both the x and y axes (with two horizontal reference lines, $h = 2$), plot the data as a scatterplot using black (col = 1) solid circles (pch = 16), and fit a **loess** with span = 2/3 and degree = 1 (you can experiment to see what happens when span and degree are altered):

```
xyplot(NOx~E|C,
       panel = function(x, y) {
       panel.grid(h = 2)
       panel.xyplot(x, y, col = 1, pch = 16)
       panel.loess(x, y, span = 2/3, degree=1)   })
```

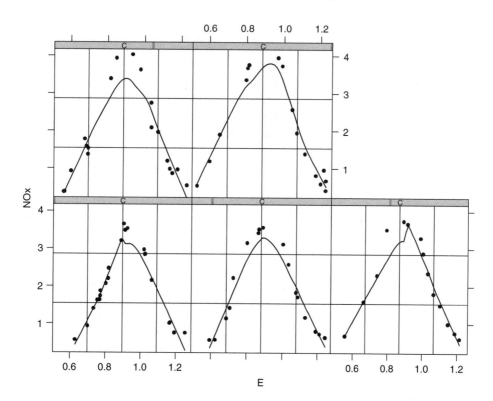

The panels are ranked from lower left to upper right on the basis of the relative values of C. Because C is a continuous variable, it has been treated as an ordered factor for the purposes of creating the panels (the relative values of C show up as small red marks on your colour screen). The non-linearity is very pronounced, but the maximum occurs at roughly the same value of E for each level of C. The peak value of NO_x increases with C from lower left to upper right.

Multiple regression with loess: surface-fitting

This example concerns spatial data in which we have measured a response variable z at several (x, y) locations. The object here is simply to investigate the shape of the landscape of z values; statistical analysis of z in terms of other explanatory variables is considered in Chapter 36.

```
attach(lodata)
names(lodata)
```

```
[1] "x" "y" "z"
```

A sensible way to begin graphical analysis of data like these is to use local regression like this:

```
model<-loess(z~x*y)
```

and **summary** produces the following output:

```
summary(model)
```

```
Call:
loess (formula = z ~ x * y)
  Number of Observations:              121
  Equivalent Number of Parameters:     9.7
  Residual Standard Error:             0.295

  Multiple R-squared:                  1
  Residuals:
      min    1st Q     median   3rd Q       max
  -0.7092  -0.1942  -0.006058  0.1662    0.8698
```

The smoother has used the equivalent of 10 degrees of freedom (parameters $= 9.7$) in producing an excellent fit to the data.

```
plot(model)
```

The **plot** command produces this conditioning plot of the fitted values (without the data).

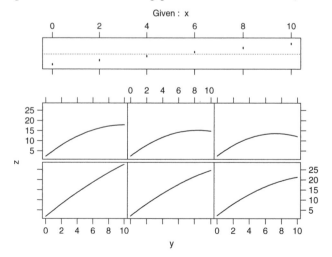

To plot the surface in 3D we can utilise the fact that both x and y are already on a regular grid (so we don't need to use **interp** first):

```
fit<-predict(model)
wireframe(fit~x+y)
```

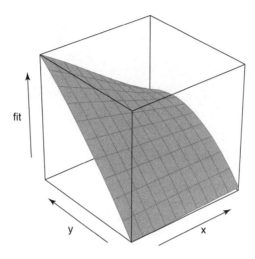

This 3D plot makes it very clear that the hump in the relationship between z and y occurs only at the highest values of x. The function **contourplot** works exactly like **wireframe**:

```
contourplot(fit~x+y)
```

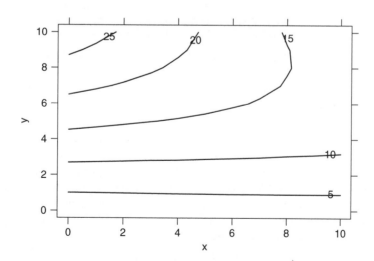

The function **levelplot** produces colour intensities instead of contour lines on a 2D map of y against x; the syntax is the same as **wireframe** and **contourplot**:

```
levelplot(fit~x+y)
```

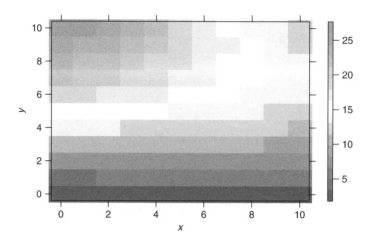

The much smoother **image** plot requires interpolation before we can use it:

image(interp(x,y,fit))

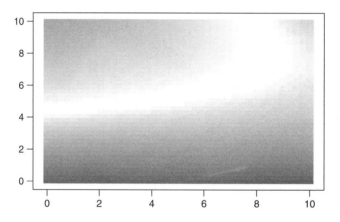

Generalised additive models (gam)

Prior to Chapter 32, continuous explanatory variables were added to models as linear functions, linearised parametric transformations, or through various link functions. In all cases, an explicit or implicit assumption was made about the parametric form of the function to be fitted to the data (whether quadratic, logarithmic, exponential, reciprocal or whatever). As we have seen, however, non-parametric smoothers can capture the shape of a relationship between y and x without us prejudging the issue.

Generalised additive models (GAMs) extend the range of application of generalised linear models (GLMs) by allowing non-parametric smoothers in addition to parametric forms combined with a range of link functions. Two smoothing functions are available, **s** and **lo**. These can be used on their own, or mixed with parametric functions like this:

$$y \sim s(x) + s(w) + s(z)$$

$$y \sim x + lo(w) + z$$

The first model has smoothed functions for all three of its continuous explanatory variables, using the **s** smoother, while the second fits parametric linear models for x and z and a non-parametric smooth function for w using the **lo** smoother. Both **lo** and **s** can be mixed in the same formula, but it is more efficient and less memory intensive to use one smoothing method or the other in any one GAM. The error structure of a GAM can be specified in the same way as a GLM (families binomial, Poisson and Gamma can be used). The **gam** function has many of the attributes of both **glm** and **lm**, and the output can be modified using **update**.

The model is fitted by iteratively smoothing partial residuals in a process known as backfitting. The algorithm separates the parametric and non-parametric parts of the fit. Since the parametric part is estimated using weighted linear least squares within the backfitting algorithm, the object returned has all the components of a GLM. Because of this, you can use all the familiar methods like **print, plot, summary, anova, predict**, and **fitted** after a GAM has been fitted to data.

Degrees of freedom are approximated as penalties for the complexity of the non-parametric curve fitted; the more complex the curve, the higher the penalty and the more d.f. are lost. Note that in **gam** the d.f. are real numbers rather than integers.

The ozone data

This data frame contains measurements of radiation, temperature, wind speed and ozone concentration. We want to model ozone concentration as a function of the three continuous explanatory variables using non-parametric smoothers rather than specified non-linear functions:

```
attach(ozone.data)
names(ozone.data)
```
```
[1] "rad" "temp" "wind" "ozone"
```

For data inspection we use **pairs** with a non-parametric smoother, **lowess**:

```
pairs(ozone.data, panel=function(x,y) {points(x,y); lines(lowess(x,y))} )
```

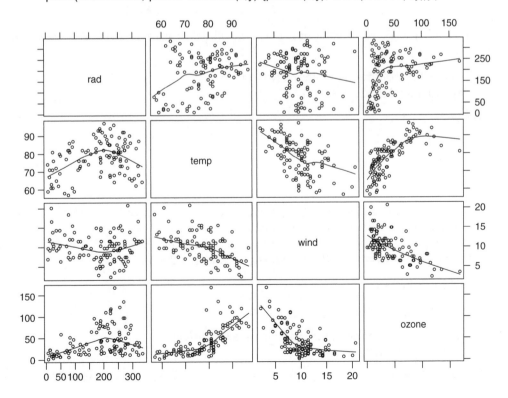

Now fit all three explanatory variables using the **s** smoother:

```
model<-gam(ozone~s(rad)+s(temp)+s(wind))
summary(model)

Call: gam(formula = ozone ~ s(rad) + s(temp) + s(wind))
Deviance Residuals:
      Min       1Q     Median       3Q       Max
-45.70703 -10.0524 -3.357995 7.959326 73.13056

(Dispersion Parameter for Gaussian family taken to be
303.0611 )

    Null Deviance: 121801.9 on 110 degrees of freedom

Residual Deviance: 29699.72 on 97.99912 degrees of
freedom

Number of Local Scoring Iterations: 1

DF for Terms and F-values for Nonparametric Effects

              Df Npar DF   Npar F      Pr(F)
(Intercept)    1
     s(rad)    1         3 2.050060 0.1117955
    s(temp)    1         3 6.677994 0.0003792
    s(wind)    1         3 9.246513 0.0000192
```

Radiation appears to be non-significant, so we should delete it (model2) and compare the two models:

```
model2<-gam(ozone~s(temp)+s(wind))
anova(model,model2,test="F")

Analysis of Deviance Table

Response: ozone

                Terms Resid. Df Resid. Dev     Test       Df  Deviance  F Value       Pr (F)
1 s (rad) + s (temp) + s (wind)  97.9991   29699.72
2 s (temp) + s (wind) 102.0018   34496.71 - s (rad) -4.002692 -4796.984 3.954448 0.005100486
```

The deletion of radiation caused a highly significant increase in deviance ($p = 0.005$), emphasising the fact that deletion is a better test than inspection of parameters (above).

We should investigate the possibility that there is an interaction between wind and temperature. We cannot fit interaction terms directly in **gam**, so we calculate the product

```
wt<-wind*temp
```

then add this to the model:

```
model3<-gam(ozone~s(temp)+s(wind)+s(rad)+s(wt))
summary(model3)

Call: gam (formula = ozone ~ s (temp) + s (wind) + s (rad) + s (wt))
```

```
Deviance Residuals:
      Min        1Q      Median        3Q         Max
 -47.74943 -9.30572 -2.497074 7.06667 67.54792
```

(Dispersion Parameter for Gaussian family taken to be 295.5693)

    Null Deviance: 121801.9 on 110 degrees of freedom

Residual Deviance: 27782.91 on 93.99798 degrees of freedom

Number of Local Scoring Iterations: 1

DF for Terms and F-values for Nonparametric Effects

```
             Df Npar Df   Npar F       Pr (F)
(Intercept)  1
   s(temp)   1        3 7.170581    0.0002192
   s(wind)   1        3 5.571937    0.0014584
    s(rad)   1        3 2.025007    0.1156243
     s(wt)   1        3 2.500044    0.0652082
```

The interaction appears to be of borderline significance ($p = 0.064$), so we try **anova**, comparing model (without the interaction) with model3 (including it):

    anova(model,model3,test="F")

This gives $p = 0.175\,4421$, nowhere close to significance, reinforcing the view that model simplification is preferable. We can inspect the fit of model3 like this:

    par(mfrow-c(2,2))
    plot.gam(model3,resid=T,rugplot=F)

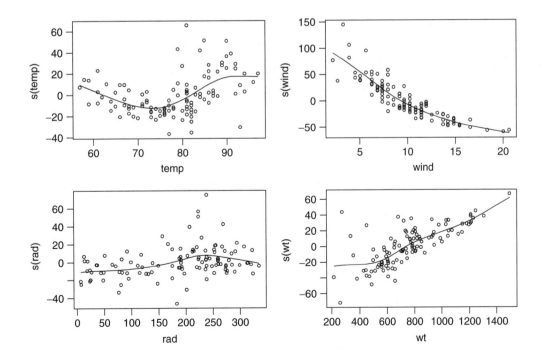

For comparison, here is the regression tree for these ozone data:

```
model<-tree(ozone~.,ozone.data)
plot(model, type="u")
text(model)
```

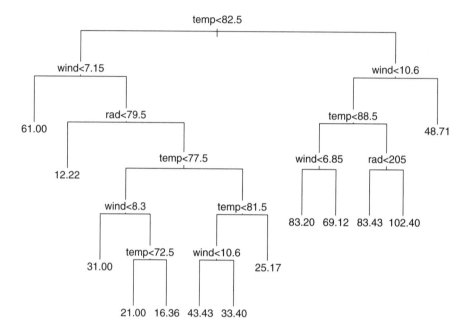

This shows that radiation is important, but only during periods of high wind at relatively low temperatures; then low levels of radiation are associated with very low levels of ozone (mean = 12.22). Likewise, at high temperatures on relatively still days, the level of ozone is positively correlated with radiation, up to a mean of 102.4 at the highest levels of radiation (>205). This comparison of **gam** and **tree** reinforces the view that regression trees are excellent tools for exposing high-order interaction effects in complex data sets.

Further reading

Chambers, J.M., Cleveland, W.S. *et al.* (1983) *Graphical Methods for Data Analysis.* Belmont, California, Wadsworth.

Chambers, J.M. and Hastie, T.J. (1992) *Statistical Models in S.* Pacific Grove, California, Wadsworth and Brooks Cole.

Hastie, T. and Tibshirani, R. (1990) *Generalized Additive Models.* London, Chapman & Hall.

33

Survival analysis

A great many studies in statistics deal with deaths or with failures of components; they involve the numbers of deaths, the timing of death, or the risks of death to which different classes of individuals are exposed. The analysis of survival data is a major focus of the statistics business (Kalbfleisch and Prentice 1980; Miller 1981; Fleming and Harrington 1991), and S-Plus supports a wide range of tools for the analysis of survival data. The main theme of this chapter is the analysis of data that take the form of measurements of the *time to death*, or the *time to failure* of a component. Up to now we have dealt with mortality data by considering the proportion of individuals that were dead *at a given time*. In this chapter each individual is followed until it dies, then the time of death is recorded (this will be the response variable). Individuals that survive to the end of the experiment will die at an unknown time in the future; they are said to be *censored* (see below).

A Monte Carlo experiment

With data on time to death, the most important decision to be made concerns the error distribution. The key point to understand is that the variance in age at death is almost certain to increase with the mean, hence standard models (assuming constant variance and Normal errors) will be inappropriate. You can see this at once with a simple Monte Carlo experiment. Suppose that the per week probability of failure of a component is 0.1 from one factory but 0.2 from another. We can simulate the fate of an individual component in a given week by generating a uniformly distributed random number between 0 and 1. If the value of the random number is less than or equal to 0.1 (or 0.2 for the second factory), the component fails during that week and its lifetime can be calculated. If the random number is larger than 0.1, the component survives to the next week. The lifetime of the component is simply the number of the week in which it finally failed. Thus a component that failed in the first week has an age at failure of 1 (this convention means there are no zeros in the data frame).

The simulation is very simple. We create a vector of random numbers, *rnos*, long enough to be almost certain to contain a value that is less than our failure probabilities of 0.1 and 0.2. Remember that the mean life expectancy is the reciprocal of the failure rate, so our mean lifetimes will be $1/0.1 = 10$ and $1/0.2 = 5$ weeks respectively. A length of 100 should be more than sufficient:

```
rnos<-runif(100)
```

The trick is to find the week number in which the component failed; this is the smallest subscript for which rnos <= 0.1 for factory 1. We can do this very efficiently using the **which** function; **which** returns *a vector of subscripts* for which the specified logical condition is true. So for factory 1 we would write

```
which(rnos<= 0.1)
```

```
[1]   5   8   9  19  29  33  48  51  54  63  68  74  80  83  94  95
```

This means that for my first set of 100 random numbers, 16 of them were less than or equal to 0.1. The important point is that the *first* such number occurred in week 5. So the simulated value of the age of failure of this first component is 5 and is obtained from the vector of failure ages using the subscript [1]:

```
which(rnos<= 0.1)[1]
```

```
[1]  5
```

All we need to do to simulate the lifespans of a sample of 20 components, called death1, is to repeat the above procedure 20 times:

```
death1<-numeric(20)
```

```
for (i in 1:20) {
rnos<-runif(100)
death1[i]<- which(rnos<= 0.1)[1]
}
```

```
death1
```

```
[1] 5  8 12 23 11 3  8  3 12 13  1  5  9  1  7  9 11  1  2 8
```

The fourth component survived for a massive 23 weeks but the eleventh component failed during its first week. The simulation has roughly the right average weekly failure rate:

```
1/mean(death1)
```

```
[1]  0.1315789
```

which is as close to 0.1 as we could reasonably expect from a sample of only 20 components. Now we do the same for the second factory with its failure rate of 0.2:

```
death2<-numeric(20)
```

```
for (i in 1:20) {
rnos<-runif(100)
death2[i]<- which(rnos<= 0.2)[1]
}
```

The sample mean is again quite reasonable (if a little on the low side):

```
1/mean(death2)
```

```
[1]  0.1538462
```

We now have the simulated raw data to carry out a comparison in age at failure between factories 1 and 2. We combine the two vectors into one, and generate a vector to represent the factory identities:

```
death<-c(death1,death2)
factory<-factor(c(rep(1,20),rep(2,20)))
```

We get a visual assessment of the data using **plot**:

```
plot(factory,death)
```

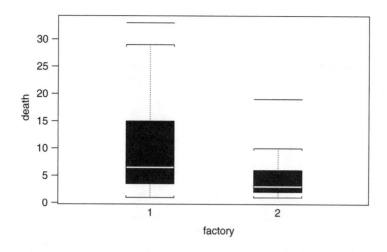

The median age at failure for factory 1 is somewhat greater, but the variance in age at failure is much higher in factory 1 than in factory 2. For data like this we expect the variance to be proportional to the square of the mean, so an appropriate error structure is the Gamma (as explained below). We model the data very simply as a one-way analysis of deviance using **glm** with family=Gamma (note the upper case G).

Modelling death rate as a GLM with Gamma errors

```
model1<-glm(death~factory,Gamma)
summary(model1)

Coefficients:
            Estimate Std. Error t value  Pr(>|t|)
(Intercept)  0.08474    0.01861   4.554 5.29e-005 ***
factory      0.06911    0.03884   1.779    0.0832 .

(Dispersion parameter for Gamma family taken to be 0.983125)

    Null deviance: 37.175  on 39  degrees of freedom
Residual deviance: 33.670  on 38  degrees of freedom
```

We conclude that the factories are not significantly different in mean age at failure of these components ($p = 0.0832$). So, even with a twofold difference in the true failure rate, we are unable to detect a significant difference in mean age at failure with samples of size $n = 20$. But the **glm** with Gamma errors comes closer to detecting the difference than did a conventional analysis with Normal errors and constant variance. See if you can work out how to demonstrate this: for the data above, $p = 0.113$ with the inappropriate Anova compared with $p = 0.0832$ here. *The moral is that, for data like this on age at death, you are going to need really large sample sizes in order to find significant differences.*

Survival analysis

We are unlikely to know much about the error distribution in advance of the study, except that it will certainly not be Normal! In S-Plus we are offered several choices for the analysis of survival data:

- Gamma

- exponential

- piecewise exponential

- extreme value

- log-logistic

- lognormal

- Weibull

and, in practice, it is often difficult to choose between them. In general, the best solution is to try several distributions and to pick the error structure that produces the minimum error deviance.

Parametric survival models are used in circumstances where prediction is the object of the exercise (e.g. in analyses where extreme conditions are used to generate accelerated failure times). Alternatively, we could use non-parametric methods, the most important of which are Kaplan–Meier and Cox proportional hazards, which are excellent for comparing the effects of different treatments on survival, but they do not predict beyond the last observation and hence cannot be used for extrapolation.

Background

Since everything dies eventually, it is often not possible to analyse the results of survival experiments in terms of the proportion that were killed (as we did in Chapter 28); in due course, they *all* die. Look at the following figure.

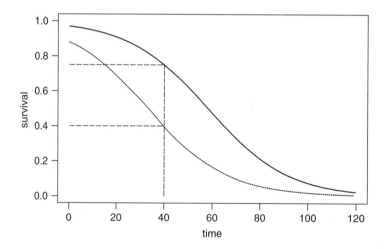

It is clear that the two treatments caused different patterns of mortality, but both start out with 100% survival and both end up with zero. We could pick some arbitrary point in the middle of the distribution at which to compare the percentage survival (say at time = 40), but this may be difficult in practice, because one or both of the treatments might have few observations at the same location. Also, the choice of when to measure the difference is entirely subjective and hence open to bias. It is much better to use S-Plus's powerful facilities for the analysis of survival data than it is to pick an *arbitrary* time at which to compare two proportions.

Demographers, actuaries and ecologists use three interchangeable concepts when dealing with data on the timing of death:

- survivorship

- age at death

- instantaneous risk of death

There are three broad patterns of survivorship. Type I: most of the mortality occurs late in life (e.g. humans). Type II: mortality occurs at a roughly constant rate throughout life. Type III: most of the mortality occurs early in life (e.g. salmonid fishes). On a log scale, the numbers surviving from an initial cohort of 1000, say, would look like this.

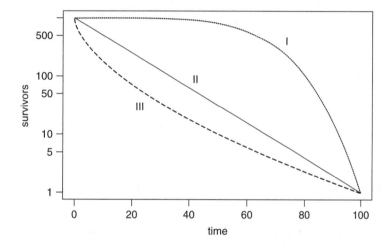

The survivor function

The survivorship curve plots the natural log of the proportion of a cohort of individuals that started out at time 0 that is still alive at time t. For the so-called Type II survivorship curve, there is a linear decline in log numbers with time (see above). This means that a constant proportion of the individuals alive at the beginning of a time interval will die during that time interval (i.e. the proportion dying is density independent and constant for all ages). When the death rate is highest for the younger age classes then we get a steeply descending, Type III survivorship curve. When it is the oldest animals that have the highest risk of death then we obtain the Type I curve (characteristic of human populations in affluent societies where there is low infant mortality).

The density function

The density function describes the fraction of all deaths from our initial cohort that are likely to occur in a given instant of time. For the Type II curve this is a negative exponential. Because the fraction of individuals dying is constant with age, the number dying declines exponentially as the number of survivors (the number of individuals at risk of death) declines exponentially with the passage of time. The density function declines more steeply than exponentially for Type III survivorship curves. In the case of Type I curves, however, the density function has a maximum at the time when the product of the risk of death and the number of survivors is greatest (see below).

The hazard function

The hazard is the instantaneous risk of death, i.e. the derivative of the survivorship curve. It is the instantaneous rate of change in the log of the number of survivors per unit time. Thus, for the Type II survivorship, the hazard function is a horizontal line, because the risk of death is constant with age. Although this sounds highly unrealistic, it is a remarkably robust assumption in many applications. It also has the substantial advantage of parsimony. In some cases, however, it is clear that the risk of death changes substantially with the age of the individuals, and we need to be able to take this into account in carrying out our statistical analysis. In the case of Type III survivorship, the risk of death declines with age, while for Type I survivorship (as in humans) the risk of death increases with age.

The exponential distribution

The exponential distribution is a one-parameter distribution in which the hazard function is independent of age (i.e. it describes a Type II survivorship curve). The exponential is a special case of the Gamma distribution in which the shape parameter α is equal to 1.

Density function

The density function is the probability of dying in the small interval of time between t and $t + \mathrm{d}t$; a plot of the number dying in the interval around time t as a function of t (i.e. the proportion of the original cohort dying at a given age) declines exponentially:

$$f(t) = \frac{\mathrm{e}^{-t/\mu}}{\mu}$$

where both μ and $t > 0$. Note that the density function has an intercept of $1/\mu$ (remember that e^0 is 1). The number from the initial cohort dying per unit time declines exponentially with time, and a fraction $1/\mu$ dies during the first time interval (and during every subsequent time interval).

Survivor function

The survivor function shows the proportion of individuals from the initial cohort that are still alive at time t:

$$S(t) = \mathrm{e}^{-t/\mu}$$

The survivor function has an intercept of 1 (i.e. all the cohort is alive at time 0), and shows the probability of surviving at least as long as t.

Hazard function

The hazard function is the statistician's equivalent of the ecologist's *instantaneous death rate*. It is defined as the ratio between the density function and the survivor function, and is the conditional density function at time t, given survival up to time t. In the case of Type II curves this has an extremely simple form:

$$h(t) = \frac{f(t)}{S(t)} = \frac{e^{-t/\mu}}{\mu e^{-t/\mu}} = \frac{1}{\mu}$$

because the exponential terms cancel out. Thus, with the exponential distribution the *hazard is the reciprocal of the mean time to death*, and vice versa. For example, if the mean time to death is 3.8 weeks, then the hazard is 0.2632; if the hazard were to increase to 0.32, then the mean time to death would decline to 3.125 weeks. The **survivor**, **density** and **hazard** functions of the exponential distribution are as follows; note the changes of scale on the y axes.

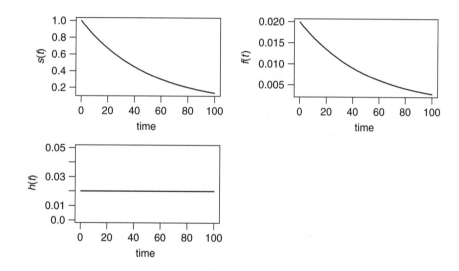

Of course, the death rate may not be a linear function of age. For example, the death rate may be high for very young animals and for very old animals, in which case the survivorship curve is like an S-shape on its side.

Kaplan–Meier survival distributions

The Kaplan–Meier survival distribution is a discrete-stepped survivorship curve that adds information as each death occurs. Suppose we had $n = 5$ individuals and that the times at death were 12, 17, 29, 35 and 42 weeks after the beginning of a trial. Survivorship is 1 at the outset, and stays at 1 until time 12, when it steps down to $4/5 = 0.8$. It stays at 0.8 until time 17, when it drops to $0.8 \times 3/4 = 0.6$. It stays at 0.6 until time 29, when it drops to $0.6 \times 2/3 = 0.4$, then drops at time 35 to $0.4 \times 1/2 = 0.2$ then to zero at time 42.

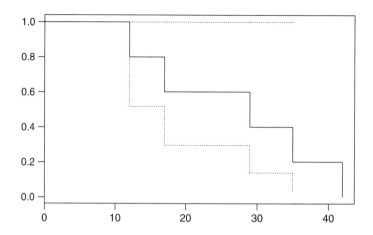

The solid line shows the survival distribution and the dotted lines show the confidence intervals (see below). In general, therefore, we have two groups at any one time: the number of deaths $d(t_i)$ and the number at risk $r(t_i)$; represents those that have not yet died—the survivors. The Kaplan–Meier survivor function is

$$\hat{S}_{KM} = \prod_{t_i < t} \frac{r(t_i) - d(t_i)}{r(t_i)}$$

which, as we have seen, produces a step at every time that one or more deaths occur. The censored individuals that survive beyond the end of the study are shown by a + on the plot or after their age in the data frame (thus 65 means died at time 65, but 65+ means still alive when last seen at age 65).

Age-specific hazard models

In many circumstances the risk of death increases with age. There are many models to chose from.

Distribution	Hazard
Exponential	constant $= 1/\mu$
Weibull	$\alpha\lambda(\lambda t)^{\alpha-1}$
Gompertz	be^{ct}
Makeham	$a + be^{ct}$
Extreme value	$\sigma^{-1}e^{(t-\eta)/\sigma}$
Rayleigh	$a + bt$

The Rayleigh distribution is obviously the simplest model in which hazard increases with time, but the Makeham is widely regarded as the best description of hazard for human subjects. Post infancy, there is a constant hazard (a) which is due to age-independent accidents, murder, suicides, etc., with an exponentially increasing hazard in later life. The Gompertz assumption was that 'the average exhaustion of a man's power to avoid death is such that at the end of equal infinitely small intervals of time he has lost equal portions of his remaining power to oppose destruction which he had at the commencement of these intervals'. Note that the Gompertz differs from the Makeham only by the omission of the extra background hazard (*a*), and this becomes negligible in old age.

These plots show how hazard changes with age for the following distributions: from top left to bottom right: exponential, Weibull, Gompertz, Makeham, extreme value and Rayleigh.

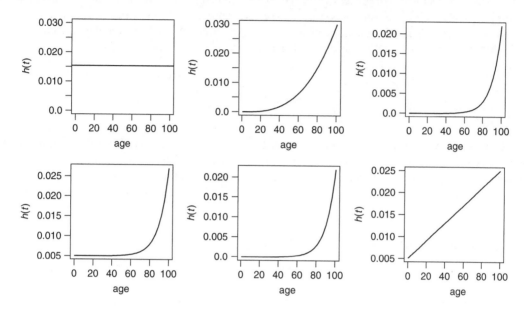

Survival analysis in S-Plus

There are three cases that concern us here:

- constant hazard and no censoring
- constant hazard with censoring
- age-specific hazard, with or without censoring

The first case is dealt with in S-Plus by specifying a **glm** with exponential errors. This involves using Gamma errors with the scale factor fixed at 1.

The second case involves the use of a **glm** with Poisson errors and a log link, where *the censoring indicator is the response variable*, and log(time of death) is an offset (see below and p. 537).

The third case is the one that concerns us mainly in this chapter. We can choose to use **parametric** models, based around the **Weibull** distribution, or **non-parametric** techniques, based around the **Cox proportional hazard** model.

Cox proportional hazards model

The Cox proportional hazards model is the most widely used regression model for survival data. It assumes that the hazard is of this form

$$\lambda(t; Z_i) = \lambda_0(t) r_i(t)$$

where $Z_i(t)$ is the set of explanatory variables for individual i at time t. The *risk score* for subject i is

$$r_i(t) = e^{\beta Z_i(t)}$$

in which β is a vector of parameters from the linear predictor and $\lambda_0(t)$ is an *unspecified baseline hazard function* that will cancel out in due course. The antilog guarantees that λ is positive for any regression model $\beta Z_i(t)$. If a death occurs at time t^*, then conditional on this death occurring, the likelihood that it is individual i that dies rather than any other individual at risk, is

$$L_i(\beta) = \frac{\lambda_0(t^*)r_i(t^*)}{\sum_j Y_j(t^*)\lambda_0(t^*)r_j(t^*)} = \frac{r_i(t^*)}{\sum_j Y_j(t^*)r_j(t^*)}$$

The product of these terms over all times of death $L(\beta) = \prod L_i(\beta)$ was christened a partial likelihood by Cox (1972). This is clever, because maximising $\log(L(\beta))$ allows an estimate of β without knowing anything about the baseline hazard function ($\lambda_0(t)$ is a nuisance variable in this context). The proportional hazards model is non-parametric in the sense that it depends only on the **ranks** of the survival times.

An example of survival analysis without censoring

To see how the exponential distribution is used in modelling, we take an example from plant ecology in which individual seedlings were followed from germination until death. We have the times to death measured in weeks for two cohorts, each of 30 seedlings. The plants were germinated at two times (cohorts), in early September (treatment 1) and mid October (treatment 2). We also have data on the size of the gap into which each seed was sown (a covariate x). The questions are these:

- is an exponential distribution suitable to describe these data?

- was survivorship different between the two planting dates?

- did gap size affect the time to death of a given seedling?

```
attach(seedlings)
names(seedlings)

[1] "cohort"    "death"    "gapsize"
```

There are several important functions for plotting and analysing survival data. The function **Surv** (note the capital S) takes two vectors. The first contains the time (or age) at which the individual was last seen, and the second indicates the status of that individual (i.e. whether it was dead or alive when it was last seen: 1 = dead, 0 = alive). Individuals are said to be censored if they were alive at the time they were last seen. All the seedlings died in this example, so status = 1 for all the cases. The function **survfit** calculates the survivorship curve on the basis of the age at death and censoring information in **Surv**. Use of **plot** with **survift** as its argument produces a graph of the survivorship curve:

```
status<-1*(death>0)

plot(survfit(Surv(death,status)),ylab="Survivorship",xlab="Weeks")
```

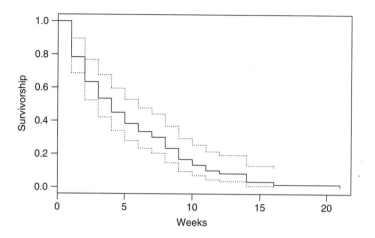

This shows the survivorship of seedlings over the 20 weeks of the study period until the last of the seedlings died in week 21. The dotted lines show the confidence limits and are the default when only one survivorship curve is plotted. No axis labels are plotted unless we provide them (as here).

Statistical modelling is extremely straightforward, but somewhat limited, because interaction effects and continuous explanatory variables are not allowed with **survfit** (but see below). We begin with a simple model for the effect of cohort on its own:

model1<-survfit(Surv(death,status)~cohort)

Just typing the name of the model object produces a useful summary of the two survivorship curves, their means, standard errors and 95% confidence intervals for the age at death:

model1

```
Call: survfit(formula = Surv(death, status) ~ cohort)

                     n  events   mean  se(mean)  median  0.95LCL  0.95UCL
  cohort=October    30      30   5.83     0.903        5        3        9
cohort=September    30      30   4.90     0.719        4        2        7
```

Survival in the two cohorts is clearly not significantly different (just look at the overlap in the confidence intervals for median age at death). Using the **summary** function produces fully documented survival schedules for each of the two cohorts:

summary(model1)

```
                         cohort=October
time n.risk n.event survival std.err lower 95% CI upper 95% CI
   1     30       7   0.7667  0.0772      0.62932        0.934
   2     23       3   0.6667  0.0861      0.51763        0.859
   3     20       3   0.5667  0.0905      0.41441        0.775
   4     17       2   0.5000  0.0913      0.34959        0.715
   5     15       3   0.4000  0.0894      0.25806        0.620
   6     12       1   0.3667  0.0880      0.22910        0.587
   8     11       2   0.3000  0.0837      0.17367        0.518
   9      9       4   0.1667  0.0680      0.07488        0.371
  10      5       1   0.1333  0.0621      0.05355        0.332
```

11	4	1	0.1000	0.0548	0.03418	0.293
14	3	1	0.0667	0.0455	0.01748	0.254
16	2	1	0.0333	0.0328	0.00485	0.229
21	1	1	0.0000	NA	NA	NA

```
                        cohort=September
 time n.risk n.event survival std.err lower 95% CI upper 95% CI
    1     30      6     0.8000  0.0730       0.6689        0.957
    2     24      6     0.6000  0.0894       0.4480        0.804
    3     18      3     0.5000  0.0913       0.3496        0.715
    4     15      3     0.4000  0.0894       0.2581        0.620
    5     12      1     0.3667  0.0880       0.2291        0.587
    6     11      2     0.3000  0.0837       0.1737        0.518
    7      9      2     0.2333  0.0772       0.1220        0.446
    8      7      2     0.1667  0.0680       0.0749        0.371
   10      5      1     0.1333  0.0621       0.0535        0.332
   11      4      1     0.1000  0.0548       0.0342        0.293
   12      3      1     0.0667  0.0455       0.0175        0.254
   14      2      2     0.0000      NA           NA           NA
```

Using the **plot** function produces a set of survivorship curves. Note the use of the vector of different line types lty=c(1,3) for plotting:

```
plot(model1,lty=c(1,3))
```

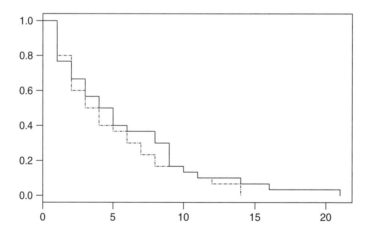

This produces a plot of the two survivorship curves using line types 1 and 3 for the October and September cohorts respectively (the factor levels are in alphabetical order, as usual). Note that the axes are unlabelled unless you specify xlab and ylab.

To investigate the effects of a continuous explanatory variable like gap size, we use Cox proportional hazards to give the survivorship curve at the average gap size:

```
model2<-survfit(coxph(Surv(death,status)~gapsize))
```

Typing the model name alone gives this:

```
model2
```

```
Call: survfit.coxph(object = coxph(Surv(death, status) ~ gapsize))

 n events mean se(mean) median 0.95LCL 0.95UCL
60      60 5.49    0.619      4       3       6
```

Using the **summary** function gives the full survival table and confidence intervals:

summary(model2)

```
Call: survfit.coxph(object = coxph(Surv(death, status) ~ gapsize))
time n.risk n.event survival std.err lower 95% CI     upper 95% CI
   1     60      13  0.78600  0.0527     0.689271            0.896
   2     47       9  0.63769  0.0618     0.527386            0.771
   3     38       6  0.53812  0.0641     0.426011            0.680
   4     32       5  0.45467  0.0641     0.344883            0.599
   5     27       4  0.38793  0.0628     0.282503            0.533
   6     23       3  0.33807  0.0609     0.237442            0.481
   7     20       2  0.30474  0.0593     0.208086            0.446
   8     18       4  0.23773  0.0549     0.151129            0.374
   9     14       4  0.17098  0.0487     0.097862            0.299
  10     10       2  0.13785  0.0446     0.073147            0.260
  11      8       2  0.10501  0.0396     0.050169            0.220
  12      6       1  0.08885  0.0366     0.039606            0.199
  14      5       3  0.04132  0.0252     0.012487            0.137
  16      2       1  0.02565  0.0199     0.005607            0.117
  21      1       1  0.00935  0.0119     0.000771            0.113
```

plot(model2)

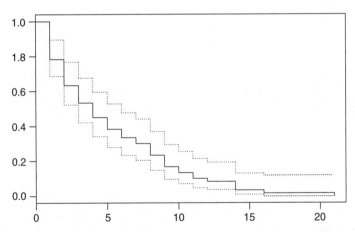

To test for differences in baseline survival for each cohort, we use **survdiff** like this:

survdiff(Surv(death,status)~cohort)

	N	Observed	Expected	(O-E)^2/E	(O-E)^2/V
cohort=October	30	30	32.9	0.259	0.722
cohort=September	30	30	27.1	0.315	0.722

Chisq=0.7 on 1 degrees of freedom, p=0.395

The baseline survival is not significantly different for the two cohorts.

For a full analysis of covariance, fitting gap size separately for each cohort, we use the **strata** option in the model formula with Cox proportional hazards:

coxph(Surv(death,status)~strata(cohort)*gapsize)

```
                        coef exp(coef) se(coef)     z    p
              gapsize 0.357      1.43     0.43 0.829 0.41
strata(cohort):gapsize 0.359     1.43     0.43 0.833 0.40
```

Gap size has no effect on survival in either cohort. To test whether the coefficients are a function of time, we use the **cox.zph** function:

model3<-cox.zph(coxph(Surv(death,status)~strata(cohort)*gapsize))

plot(model3)

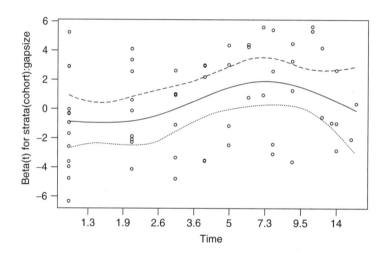

There is no evidence of temporal changes in the parameters. This is a truly dull data set. Nothing is happening at all. We tidy up by removing the vectors **status** and **death**:

rm(status, death)

Censoring

Censoring occurs when we do not know the time to death for all of the individuals. This comes about principally because some individuals outlive the experiment. We can say they survived for the duration of the study but we have no way of knowing at what age they will die. These individuals contribute something to our knowledge of the survivor function, but nothing to our knowledge of the age at death. Another reason for censoring occurs when individuals are lost from the study; they may be killed in accidents, they may emigrate, or they may lose their identity tags.

In general, then, our survival data may be a mixture of times at death and times after which we have no more information on the individual. We deal with this by setting up an extra vector called the *censoring indicator* to distinguish between the two kinds of numbers. If a time really is a time to death, then the censoring indicator takes the value 1. If a time is just the last time we saw an individual alive, then the censoring indicator is set to 0. Thus, if we had the time data T and censoring indicator W:

```
T   4  7  8  8  12  15  22
W   1  1  0  1  1   0   1
```

this would mean that all the data were genuine times at death except for two cases, one at time 8 and another at time 15, when individuals were seen alive but never seen again.

With repeated sampling in survivorship studies, it is usual for the degree of censoring to decline as the study progresses. Early on, many of the individuals are alive at the end of each sampling interval, whereas few if any survive to the end of the last study period.

An example with censoring and non-constant hazard

This study involved four cohorts of cancer patients, each of 30 individuals. They were allocated to Drug A, Drug B, Drug C or a placebo at random. The year in which they died (recorded as time after treatment began) is the response variable. Some patients left the study before their age at death was known (these are the censored individuals with status = 0).

```
attach(cancer)
names(cancer)
```

```
[1] "death"     "treatment"     "status"
```

We start by plotting the survivorship curves for patients in the four different treatments:

```
plot(survfit(Surv(death,status)~treatment),lty=c(1:4))
```

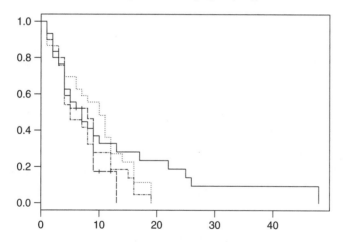

The mean age at death for the four treatments (ignoring censoring) was

```
tapply(death[status==1],treatment[status==1],mean)
```

```
DrugA    DrugB    DrugC   placebo
 9.48     8.36     6.80      5.24
```

The patients receiving Drug A lived an average of more than 4 years longer than those receiving the placebo. We need to test the significance of these differences, remembering that the variance in age at death is very high:

```
tapply(death[status==1],treatment[status==1],var)
```

```
DrugA    DrugB    DrugC placebo
117.5     32.7     27.8     11.4
```

We start with the simplest model, assuming exponential errors and constancy in the risk of death:

```
model<-survReg(Surv(death,status)~treatment,dist="exponential")
summary(model)
                     Value Std. Error      z        p
       (Intercept)   2.448      0.200 12.238 1.95e-034
    treatmentDrugB  -0.125      0.283 -0.442 6.58e-001
    treatmentDrugC  -0.430      0.283 -1.520 1.28e-001
  treatmentplacebo  -0.333      0.296 -1.125 2.61e-001

Scale fixed at 1

Exponential distribution
Loglik(model)=-310.1    Loglik(intercept only)=-311.5
      Chisq=2.8 on 3 degrees of freedom, p= 0.42
Number of Newton-Raphson Iterations: 3
n=120
```

The key column in the output shows the p values of the four parameters. This analysis does not distinguish between the four treatments, despite the large differences in mean age at death (the smallest p value is 0.128). The placebo is not significantly different from Drug A. Next we try a model with a different error assumption, the extreme value distribution:

```
model<-survReg(Surv(death,status)~treatment,dist="extreme")

summary(model)
                     Value  Std. Error      z        p
       (Intercept)  22.91       2.0686 11.07 1.69e-028
    treatmentDrugB -11.16       2.7548 -4.05 5.13e-005
    treatmentDrugC -13.38       2.7487 -4.87 1.12e-006
  treatmentplacebo -13.29       2.9357 -4.53 5.97e-006
        Log(scale)   2.21       0.0717 30.76 8.32e-208

Scale= 9.08

Extreme value distribution
Loglik(model)=-371.7    Loglik(intercept only)=-383.7
      Chisq=23.94 on 3 degrees of freedom, p=0.000026
Number of Newton-Raphson Iterations: 4
n=120
```

This model indicates highly significant effects of treatment on death rate (p values <0.0001). Drug A gave improved survival compared to the placebo and the other two drugs, but the differences between Drugs B and C and the placebo were not significant. The scale parameter of 9.08 indicates that mortality was strongly age dependent, so the earlier exponential distribution (scale = 1), assuming that mortality was not age dependent, was not appropriate. A full analysis of these data would fit more covariates (details about each patient) and test for time dependency in the model parameters. The present point is that if as here the death risk is a function of age, then assuming the simpler exponential model does *not* allow us to detect the significant differences that exist between the treatments.

```
      rm(status)      <- remove the variable called status from last model
```

An example with censoring

The next example comes from a study of mortality in 150 adult cockroaches. There were three experimental *group*s, and the animals were followed for 50 days. The groups were treated with three different insecticidal Bt toxins added to their diet. The initial body mass of each insect (*weight*) was recorded as a covariate. The day on which each animal died (*death*) was recorded, and animals which survived up to the 50th day were recorded as being censored (for them, the censoring indicator *status* = 0).

```
attach(roaches)
names(roaches)
[1] "death"   "status"   "weight"   "group"
plot(group,weight)
```

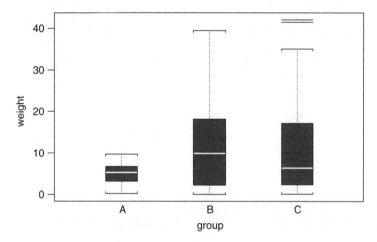

The insects in batches B and C are much more variable than those in batch A. The overall survivorship curves for the three groups are obtained as before:

```
plot(survfit(Surv(death,status)~group),lty=c(1,3,5))
```

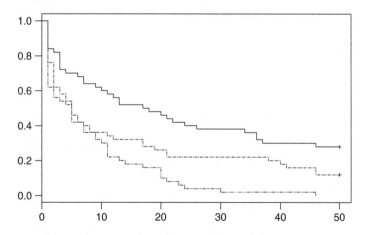

The + symbols at the end of the survivorship curves for groups A and B indicate that there was censoring in these groups (not all of the individuals were dead at the end of the experiment).

Parametric regression in survival models uses the **survReg** function, for which you can specify a wide range of different error distributions. Here we use the exponential distribution for the purposes of demonstration (we can chose dist from extreme, logistic, gaussian or exponential and we can choose link = "log" or link = "identity") and fit the full analysis of covariance model to begin with:

```
model<-survReg(Surv(death,status)~weight*group,dist="exponential")

summary(model)
```

```
                  Value Std. Error      z        p
   (Intercept)   3.8702     0.3854 10.041 1.00e-023
        weight  -0.0803     0.0659 -1.219 2.23e-001
        groupB  -0.8853     0.4508 -1.964 4.95e-002
        groupC  -1.7804     0.4386 -4.059 4.92e-005
   weightgroupB  0.0643     0.0674  0.954 3.40e-001
   weightgroupC  0.0796     0.0674  1.180 2.38e-001

Scale fixed at 1

Exponential distribution
Loglik(model)=-480.6   Loglik(intercept only)=-502.1
      Chisq=43.11 on 5 degrees of freedom, p=3.5e-008
Number of Newton-Raphson Iterations: 4
n=150
```

Model simplification proceeds in the normal way. You could use **update**, but here (for variety only) we refit progressively simpler models and test them using **anova**. First we take out the different slopes for each group:

```
model2<-survReg(Surv(death,status)~weight+group,dist="exponential")

anova(model,model2,test="Chi")

Analysis of Deviance Table
Response: Surv(death, status)
```

	Resid. Df	Resid. Dev	Df	Deviance	P(>|Chi|)
weight + group + weight:group	144	301.6			
weight + group	146	303.4	-2	-1.8	0.4

The interaction is not significant so we leave it out and try deleting weight:

```
model3<-survReg(Surv(death,status)~group,dist="exponential")
anova(model2,model3,test="Chi")

Analysis of Deviance Table
Response: Surv(death, status)
```

	Resid. Df	Resid. Dev	Df	Deviance	P(>|Chi|)
weight + group	146	303.4			
group	147	304.4	-1	-1.0	0.3

This is not significant, so we leave it out and try deleting group:

```
model4<-survReg(Surv(death,status)~1,dist="exponential")

anova(model3,model4,test="Chi")

Analysis of Deviance Table
Response: Surv(death, status)

        Resid. Df Resid. Dev Df Deviance P(>|Chi|)
group         147        304
1             149        345 -2      -40  1.7e-009
```

This is highly significant, so we add it back. The minimal adequate model is model3 with the three-level factor group, but there is no evidence that initial body weight had any influence on survival.

```
summary(model3)

survReg(formula = Surv(death, status) ~ group, dist =
"exponential")
               Value Std. Error       z          p
(Intercept)    3.467      0.167   20.80 3.91e-096
    groupB    -0.671      0.225   -2.99 2.83e-003
    groupC    -1.386      0.219   -6.34 2.32e-010

Scale fixed at 1

Exponential distribution
Loglik(model)= -482    Loglik(intercept only)= -502.1
    Chisq= 40.35 on 2 degrees of freedom, p= 1.7e-009
Number of Newton-Raphson Iterations: 4
n= 150
```

You can immediately see the advantage of doing proper survival analysis when you compare the predicted mean ages at death for model3 with the crude arithmetic averages of the raw data on age at death:

```
tapply(predict(model3,type="response"),group,mean)

        A        B       C
 32.05555 16.38635 8.02

tapply(death,group,mean)

    A     B     C
 23.08 14.42 8.02
```

If there is no censoring (all the individuals died, as in Group C) then the estimated mean ages at death are identical. But when there is censoring, the arithmetic mean underestimates the age at death, and when the censoring is substantial (as in Group A) this underestimate is very large (23.08 vs. 32.06).

Box 33.1 The likelihood function for censored data

A given individual contributes to the likelihood function depending upon whether it is alive or dead at time t. If it has died, then the censoring indicator is 1 and we learn more about $f(t)$; if it is still alive, then w_i is 0 and we learn only about $S(t)$:

$$L(\beta) = \prod_{i=1}^{n} [f(t_i)]^{w_i} [S(t_i)]^{1-w_i}$$

Now recall that the hazard function $h(t)$ is given by $f(t)/S(t)$. Thus we have $[S(t)^w]$ in the denominator and $[S(t)^{1-w}]$ in the numerator, so the likelihood function becomes

$$L(\beta) = \prod_{i=1}^{n} [h(t_i)]^{w_i} S(t_i)$$

which involves data only from the hazard function of the uncensored individuals. If we replace $1/\mu_i$ by λ_i, then we can write the likelihood function for the exponential distribution as follows:

$$L(\beta) = \prod_{i=1}^{n} \lambda_i^{w_i} e^{-\lambda t_i}$$

For reasons that will become clear in a moment, it is convenient to multiply both the numerator and the denominator by $\prod t_i^{w_i}$. This gives

$$L(\beta) = \frac{\displaystyle\prod_{i=1}^{n} (\lambda_i t_i)^{w_i} e^{-\lambda t_i}}{\displaystyle\prod_{i=1}^{n} t_i^{w_i}}$$

Because the denominator is not a function of the estimated parameters, β, it can be omitted from the likelihood formula, leaving only a term for the likelihood of a set of n observations, w_i, having independent Poisson distributions with means $\lambda_i t_i$, where w_i is either 0 or 1. The model can be fitted to the hazard rate λ in one of two ways. Let θ_i represent the Poisson mean $\lambda_i t_i$. Using the *linear hazard model*, we have

$$\lambda_i = \beta' x_i$$

and

$$\theta_i = \beta'(t_i x_i)$$

This is somewhat long-winded, and the fitting is easier if the *log-linear hazard model* is employed, because

$$-\log \mu_i = \log \lambda = \beta' x_i$$

and so

$$\log \theta_i = \log \lambda_i + \log t_i = \beta' x_i + \log t_i$$

This is easier to fit, because we do not need to multiply through all the explanatory vectors by t_i. Instead, we use log t_i as an offset, and proceed as follows:

- use the censoring indicator w_i as the response variable
- declare the errors as Poisson
- declare the link function as the log link
- declare log t_i as an offset
- fit the model as usual

Censored survival data analysed using glm with Poisson errors

The rather curious procedure of using the censoring indicator full of 0s and 1s as the response variable (Box 33.1) should become clearer with an example. Suppose that in a GM trial we have two groups of caterpillars, each comprising 21 larvae of the same initial size and age. They are fed on leaves from two kinds of plants: the first group get leaves from a plant that has been genetically engineered to express Bt toxin in its foliage, while the second group get leaves from an otherwise identical but non-transgenic strain. The data consist of the time at death in days (when $w = 1$) or the time when the animal was lost to the study ($w = 0$). The survival analysis is very simple. The response variable is the censoring indicator (*status*) with Poisson errors and log(*time*) as an offset:

```
attach(transgenic)
names(transgenic)
```

```
[1] "time"   "status" "diet"
```

where *time* is age at death (days), *status* is the censoring indicator ($1 =$ dead, $0 =$ alive at the end of the experiment), and *diet* is a two-level factor. Preliminary data inspection involves calculating the mean age at death for insects fed on control and transgenic Bt-expressing leaves using **tapply**:

```
tapply(time,diet,mean)
```

```
  control    transgenic
 17.9524      8.666667
```

Evidently the control insects lived much longer on average. It is useful to know how the censoring is distributed across individuals in the different treatments. We use **table** to count the cases:

```
table(status,diet)
```

```
      control transgenic
  0        12          0
  1         9         21
```

This is very revealing. All of the censoring (12 cases of *status* $= 0$) occurred in the control group of insects. None of the insects fed on Bt-expressing leaves survived until the end of the experiment. Nine control insects died compared with all 21 of the insects fed a diet of transgenic leaves. The model has the censoring indicator as the response variable, and the variation to be explained by the model is introduced by the offset (log time of death):

```
model<-glm(status~diet+offset(log(time)),family=poisson)
```

Note how **offset** appears as an additive part of the model formula.

```
summary(model)
```

```
Coefficients:
              Estimate Std.  Error   z value   Pr(>|z|)
(Intercept)   -3.6861        0.3333  -11.060   < 2e-016 ***
diet           1.5266        0.3984    3.832   0.000127 ***

(Dispersion parameter for poisson family taken to be 1)

    Null deviance: 54.503  on 41  degrees of freedom
Residual deviance: 38.017  on 40  degrees of freedom
AIC: 102.02
```

Diet looks highly significant, but we prefer to test by deletion, using chi-square:

```
model2<-update(model,~.-diet)
```

```
anova(model,model2,test="Chi")
```

```
Analysis of Deviance Table
Response: status

                         Resid. Df Resid. Dev Df Deviance  P(>|Chi|)
diet + offset(log(time))      40      38.017
offset(log(time))             41      54.503 -1  -16.485 4.903e-005
```

So, no doubt there then. Diet had an enormously significant effect on mean time at death. It is useful to know how to back-transform these coefficient tables when there are offsets.

The antilogs of the estimates give the hazard. The mean age at death is the reciprocal of the hazard. So, for the control insects, mean age at death is

```
1/exp(tapply(predict(model),diet,mean))
```

```
  control transgenic
39.888888   8.666667
```

Note that where there is a lot of censoring (as in the case of the control insects) the estimated mean age at death is substantially greater than the arithmetic mean age at death (39.89 vs. 17.10) whereas the non-censored means are identical (8.67 vs. 8.67):

```
tapply(time,diet,mean)
```

```
  control transgenic
17.095238   8.666667
```

Further reading

Cox, D.R. and Oakes, D. (1984) *Analysis of Survival Data*. London, Chapman & Hall.

Fleming, T. and Harrington, D. (1991) *Counting Processes and Survival Analysis*. New York, John Wiley.

Kalbfleisch, J. and Prentice, R.L. (1980) *The Statistical Analysis of Failure Time Data*. New York, John Wiley.

Miller, R.G. (1981) *Survival Analysis*. New York, John Wiley.

34

Time series analysis

Time series data are vectors of numbers, typically regularly spaced in time. Yearly counts of animals, daily prices of shares, monthly means of temperature, minute by minute details of blood pressure are all examples of time series, but they are measured on different timescales. Sometimes the interest is in the time series itself (e.g. whether or not it is cyclic), or how well the data fit a particular theoretical model (e.g. non-linear curve fitting), and sometimes the time series is incidental to a designed experiment (e.g. repeated measures). We cover each of these cases in turn.

The three key concepts in time series analysis are:

- trend

- serial dependence

- stationarity

Most time series analyses assume the data are untrended. If they do show a consistent upward or downward trend, then they can be detrended before analysis (e.g. by differencing). Serial dependence arises because the values of adjacent members of a time series may well be correlated. Stationarity is a technical concept but it can be thought of simply as meaning that the time series has the same properties wherever you start looking at it (e.g. white noise is a sequence of mutually independent random variables each with mean zero and variance $\sigma^2 > 0$).

Nicholson's blowflies

The Australian ecologist A.J. Nicholson reared blowfly larvae on pieces of liver in laboratory cultures that his technicians kept running continuously for almost 7 years (361 weeks to be exact). The time series for numbers of adult flies looks like this:

```
attach(blowfly)
names(blowfly)

[1] "flies"
```

First, make the variable *flies* into a time series object:

```
flies<-ts(flies)
```

and plot it:

```
tsplot(flies)
```

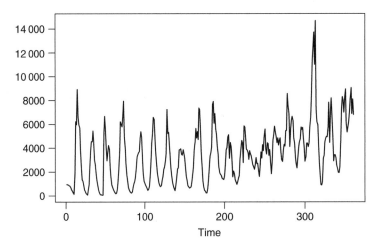

This classic time series has two clear features:

- for the first 200 weeks the system exhibits beautifully regular cycles

- after week 200 things change (perhaps a genetic mutation had arisen); the cycles become much less clear-cut, and the population begins a pronounced upward trend

There are two key concepts to understand in time series analysis: **autocorrelation** and **partial autocorrelation**. The first describes how this week's population is related to last week's population. This is the autocorrelation at lag 1. Then it calculates how this week's population is related to the population size in the week before last. This is the autocorrelation at lag 2. And so on. This should become clear if we draw the scatterplots from which the first four autocorrelation terms are calculated. We need to make a copy of the time series which is lagged by 1 week. This is easy using subscripts to remove the first element of the vector called *flies* like this:

```
f1<-flies[-1]
```

We want to plot *f*1 against *flies*. But we can't do this because the two vectors have different lengths:

```
length(f1)
```

```
[1] 360
```

```
length(flies)
```

```
[1] 361
```

To compensate for having removed the first element of flies to create *f*1, we must remove the last element of flies in order to plot one against the other. Because we intend eventually to produce four scatterplots, we set up the plotting region to take four graphs in a 2 × 2 array:

```
par(mfrow=c(2,2))
```

Now we can look at how this week's population is correlated with next week's:

```
plot(flies[-361],f1)
```

The correlation is very strong, but notice how the variance increases with population size: small populations this week are invariably correlated with small populations next week, but large populations this week may be associated with large or small populations next week. Now we calculate the population 2 weeks ahead, and 3 and 4 weeks ahead by dropping successively more data from the top of the vector called *flies*:

```
f2<-flies[-c(1,2)]
f3<-flies[-c(1,2,3)]
f4<-flies[-c(1,2,3,4)]
```

The remaining three scatterplots can now be completed:

```
plot(flies[-c(361,360)],f2)
plot(flies[-c(361,360,359)],f3)
plot(flies[-c(361,360,359,358)],f4)
```

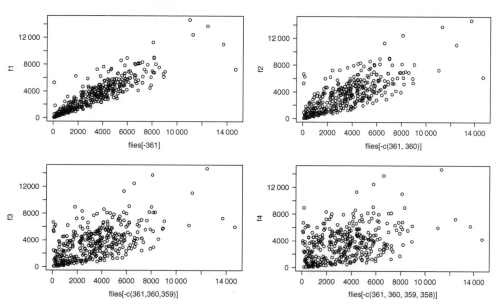

The striking pattern here is the way that the correlation fades away as the size of the lag increases. Because the population is cyclic, the correlation goes to zero, then becomes weakly negative and then becomes strongly negative. This occurs at lags that are half the cycle length. Looking back at the time series, the cycles look to be about 20 weeks in length. So let's repeat the exercise by producing scatterplots at lags of 7, 8, 9 and 10 weeks:

```
f7<-flies[-c(1:7)]
f8<-flies[-c(1:8)]
f9<-flies[-c(1:9)]
f10<-flies[-c(1:10)]
plot(flies[-c(355:361)],f7)
plot(flies[-c(354:361)],f8)
plot(flies[-c(353:361)],f9)
plot(flies[-c(352:361)],f10)
```

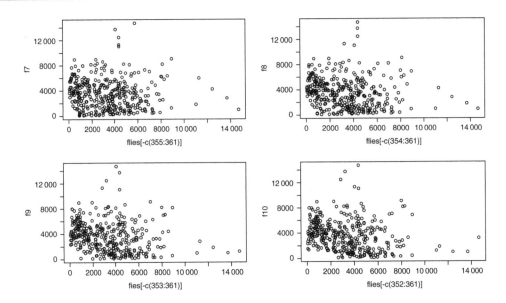

The negative correlation at lag 10 gradually emerges from the fog of no correlation at lag 7.

Partial autocorrelation is the relationship between this week's population and the population at lag t *when we have controlled for the correlations between all of the successive weeks between this week and week t*.

More formally, the autocorrelation function $\rho(k)$ at lag k is

$$\rho(k) = \frac{\gamma(k)}{\gamma(0)}$$

where $\gamma(k)$ is the autocovariance function at lag k of a stationary random function $Y(t)$ given by

$$\gamma(k) = \text{cov}\,\{Y(t), Y(t-k)\}$$

The most important properties of the autocorrelation coefficient are these:

- they are symmetric backwards and forwards, so $\rho(k) = \rho(-k)$

- the limits are $-1 \le \rho(k) \le 1$

- when $Y(t)$ and $Y(t-k)$ are independent, then $\rho(k) = 0$

- the converse of this is not true, so that $\rho(k) = 0$ does not imply that $Y(t)$ and $Y(t-k)$ are independent (look at the scatterplot for $k = 7$ in the scatterplots above)

In a first-order autoregressive process,

$$Y_t = \alpha Y_{t-1} + Z_t$$

This says 'this week's population is α times last week's population plus a random term Z_t'. The randomness is **white noise**; the values of Z are **serially independent**, they have a **mean of zero** and they have **finite variance** σ^2.

In a stationary times series $-1 < \alpha < 1$. In general, then, the autocorrelation function of $Y(t)$ is

$$\rho_k = \alpha^k \qquad (k = 0, 1, 2, \ldots)$$

The **partial autocorrelation** is the correlation between $Y(t)$ and $Y(t+k)$ after regression of $Y(t)$ on $Y(t+1)$, $Y(t+2)$, $Y(t+3), \ldots, Y(t+k-1)$. It is obtained by solving the Yule–Walker equation

$$\rho_k = \sum_1^p \alpha_i \rho_{k-i} \qquad (k > 0)$$

with the ρ replaced by r (correlation coefficients estimated from the data). Partial correlation coefficients are explained on p. 191 but suppose we want the partial autocorrelation between time 1 and time 3. To calculate this, we need the three ordinary correlation coefficients $r_{1,2}$ $r_{1,3}$ and $r_{2,3}$. The partial $r_{13,2}$ is then

$$r_{13,2} = \frac{(r_{13} - r_{12}r_{23})}{\sqrt{(1 - r_{12}^2)(1 - r_{23}^2)}}$$

Let's look at the correlation structure of the blowfly data. First, the autocorrelation plot **acf** to look for evidence of cyclic behaviour:

acf(flies)

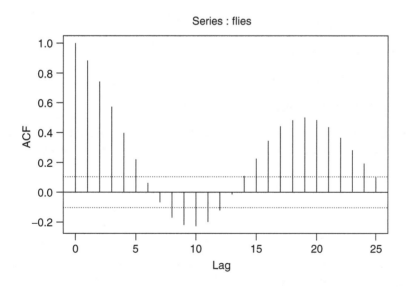

You will not see more convincing evidence of cycles than this. The blowflies exhibit highly significant, regular cycles with a period of 19 weeks. What kind of time lags are involved in the generation of these cycles? We use partial autocorrelation (type="p") to find this out:

acf(flies,type="p")

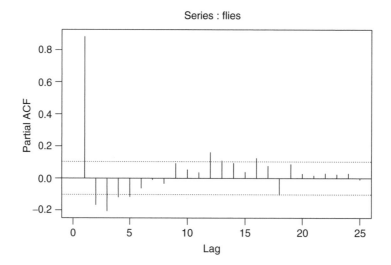

Series : flies

The significant density-dependent effects are manifested at lags of 2 and 3 weeks, with other, marginally significant negative effects at lags of 4 and 5 weeks. These lags reflect the duration of the larval and pupal period (1 and 2 periods respectively). The cycles are clearly caused by overcompensating density dependence, resulting from intraspecific competition between the larvae for food (what Nicholson christened 'scramble competition'). There is a curious positive feedback at lag 12 weeks (12–16 weeks, in fact). Perhaps you can think of a possible cause for this?

We should investigate the behaviour of the second half of the time series separately. Let's say it is from week 201 onwards:

```
second<-flies[201:361]
```

Now test for a linear trend in mean fly numbers:

```
summary(lm(second~I(1:length(second))))
```

Note the use of I in the model formula (I stands for 'as is') to tell S-Plus that the colon we have used is to generate a sequence of x values for the regression (and not an interaction term as it would otherwise have assumed).

```
Coefficients:
                       Value  Std. Error   t value    Pr(>|t|)
        (Intercept) 2827.5305    336.6606    8.3988     0.0000
I(1:length(second))   21.9448      3.6050    6.0873     0.0000

Residual standard error: 2126 on 159 degrees of freedom
Multiple R-Squared: 0.189
```

This shows that there is a highly significant upward trend of about 22 extra flies on average each week in the second half of the time series. We can detrend the data by subtracting the linear regression:

```
detrended<-second - predict(lm(second~I(1:length(second))))
tsplot(detrended)
```

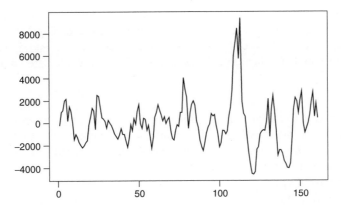

There are still cycles but they are weaker and less regular. We repeat the correlation analysis on the detrended data:

    acf(detrended)

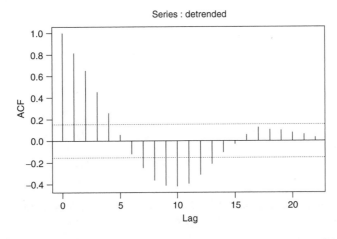

These look more like damped oscillations than repeated cycles. What about the partials?

    acf(detrended,type="p")

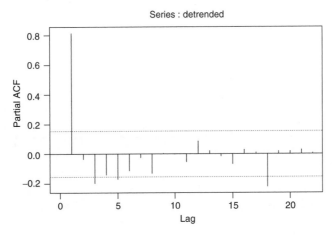

There are still significant negative partials at lags 3 and 5, but now there is a curious extra negative partial at lag 18. It looks, therefore, as if main features of the ecology are the same (scramble contest for food between the larvae, leading to negative partials at 3 and 5 weeks after 1 and 2 generation lags), but population size is drifting upwards and the cycles are showing a tendency to dampen out.

Soay sheep

This example concerns a free-living, food-limited population of feral sheep living on the isolated island of St Kilda in the North Atlantic. The animals are interacting with a complex mosaic of vegetation, but they have no competitors and no natural enemies. It was previously thought that the population was cyclic. Let's have a look at the data:

```
attach(soaysheep)
names(soaysheep)
[1] "Year"      "Population"     "Delta"
tsplot(Population)
```

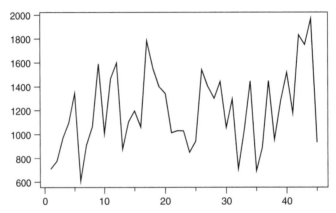

The population certainly fluctuates widely, but does it cycle? We look at the autocorrelation structure to find out:

```
acf(Population)
```

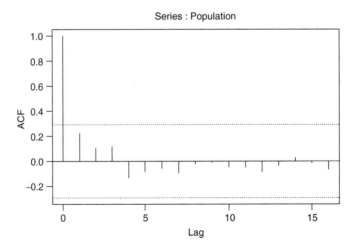

Nothing at all. What about the partials, to look for evidence of time-lagged density dependence?

acf(Population,type="p")

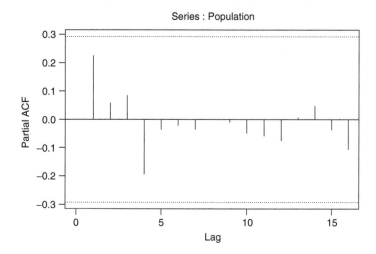

Again, nothing even close to significance. The population clearly does not cycle. We can investigate the regulation of the population by looking at the pattern of density dependence. How do the direction and rate of population change depend upon population size? We define population change as delta, which is the difference in log population size in successive years: positive values of delta reflect population increase, negative values reflect population decline, while stationary populations would have delta = 0. There is a very useful function in S-Plus called **diff** which works out the difference between values in a vector. The default (which we want here) is to work out the difference between successive values (i.e. with lag = 1) but you can specify different lags if you need to. It is easy to calculate the vector of deltas like this:

delta<-diff(log(Population))

The only complication that we need to bear in mind is that **diff** produces a shorter vector because with a lag of 1 there is one less difference than there are numbers in the original series. We can see this using the **length** function:

length(delta)

[1] 44

length(Population)

[1] 45

This means that we need to create a new, shorter vector if we want to analyse density dependence in population change (you might try plotting delta against Population to see what happens). It is the last value of Population that is redundant, so we create the shorter vector like this:

numbers<-Population[-45]

Now we can begin to study the density dependence by plotting population change against the log of population size:

plot(log(numbers),delta)

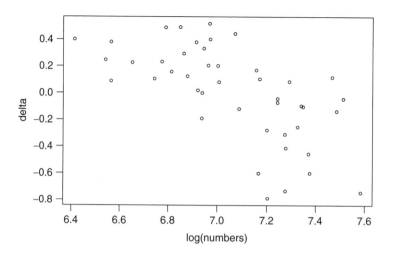

The picture would be clearer, perhaps, if we added a non-parametric smoothed line to help draw out the shape of the relationship. We begin by adding a dashed line to show delta = 0 (population stasis):

```
abline(0,lty=2)
```

To fit a smooth line we need first to generate a series of x values. On a log scale these need to go from 6.4 up to 7.6:

```
xv<-seq(6.4,7.6,0.01)
```

Now we use **predict** with the **loess** function to generate the fitted line. Note that we have to antilog the x values in the data frame supplied to **predict**, because the **loess** model worked with *numbers* not the *logs* of the numbers:

```
lines(xv,predict(loess(delta~log(numbers)),data.frame(numbers=exp(xv))))
```

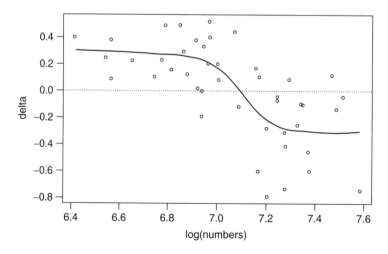

The dashed line shows population stasis (delta = 0). There appears to be a threshold population of about 7.1 (on this log scale) below which the population tends to increase at about delta = 0.3 per year, and above which it tends to decline at about delta $= -0.3$ per year. The density dependence is very clear-cut, but rather unusual in form. Most theoretical models of population dynamics assume that the intensity of density dependence increases as population size goes up (here delta is density *independent* over the range 7.2 to 7.6). Although delta is highly variable, and some of the highest population densities were followed by population *increases*, the sheep population is regulated about an equilibrium population of about $\exp(7.1) = 1212$ animals.

Canadian lynx

Nicholson's blowflies were a single species interacting with a regular food supply in a controlled environment, and the Soay sheep were a relatively simple plant–herbivore interaction. The Canadian lynx, however, represent a free-ranging, large-scale carnivore population feeding mainly on snowshoe hares, which in turn are interacting with a food supply that shows pronounced changes in quantity and quality though time (e.g. feeding by hares may cause the induction of chemical defences in some of the plant species).

This famous data set was first compiled by Charles Elton in 1942 from the trading records of the Hudson's Bay Company. It is a classic example of a cyclic population. The important question as to what causes the cycle is still unresolved, more than 60 years later. Is it a predator–prey cycle, caused by top-down processes with the crashes caused by overexploitation of hares by lynx? Or is it a plant–herbivore cycle, with lynx numbers driven into a cycle by a bottom-up plant–herbivore interaction? Or is it both of these?

Here we carry out a basic time series analysis, and address three broad questions:

- is the series trended up or down in the long term?

- is there evidence of regular cyclic behaviour?

- what kinds of periodic phenomena appear to be operating and at what timescales (i.e. what are the lags in the system)?

First we look at the data in a time series plot, **tsplot**, with lynx numbers against years:

```
attach(Lynx)
names(Lynx)
```
```
[1] "Numbers"
```
```
tsplot(Numbers,ylab="Lynx numbers")
```

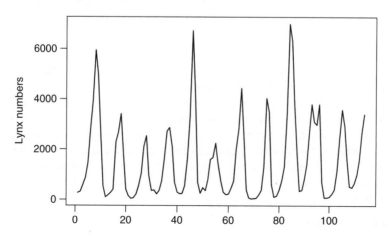

It certainly looks to be untrended and it is clearly cyclic. We test for a trend by a simple linear regression of lynx numbers against years. The data frame only contains the lynx numbers, so we need to create a sequence of x values:

```
year<-1:length(Numbers)
```

Now we regress Numbers on year and ask whether or not there is a significant trend?

```
summary(lm(Numbers ~ year))
```

```
Coefficients:
                  Value    Std. Error    t value    Pr(>|t|)
(Intercept)  1349.1158      299.6479     4.5023      0.0000
      year      3.2852        4.5229     0.7264      0.4691
```

There is a slight upward trend of just over three extra lynx per year, but this is nowhere near significant ($p = 0.47$), so we conclude that the time series is untrended.

Next we test for the existence of significant cyclic behaviour. There are two important elements to this: the autocorrelation plot and the partial autocorrelation plot. The autocorrelation plot is obtained like this:

```
acf(Numbers)
```

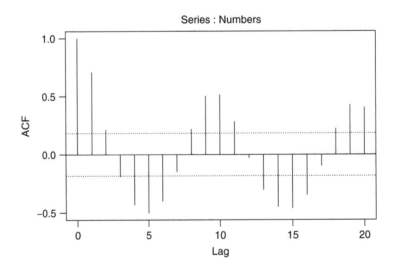

The plot shows clear cycles, with a period of roughly 10 years. What the y axis shows is the correlation between lynx numbers taken different numbers of years apart (lag). There is obviously a perfect correlation between lynx numbers and themselves (lag $= 0$), and there is a strong positive correlation between lynx numbers this year and lynx numbers last year (lag $= 1$). There is a strong negative correlation between lynx numbers this year and lynx numbers 5 years ago (roughly -0.5) and a reasonably strong and significant correlation between lynx numbers now and lynx numbers 9 and 10 years ago. The horizontal dotted lines show correlations that would be significant at 5%. So we conclude that there are significant cycles and their period is about 10 years.

The next plot shows the partial autocorrelations:

```
acf(Numbers,type="p")
```

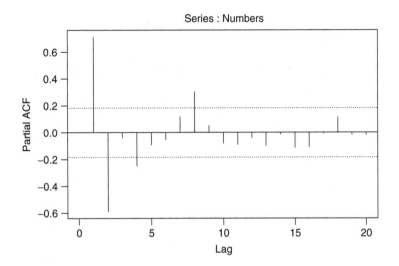

Series : Numbers

The partial autocorrelation plot gives us insights into the kind of ecological processes that might be generating the cycles. In general, *cycles are caused by time-lagged density dependence*. Here we see significant evidence for strong negative feedback on lynx numbers at a lag of 2 years, and weaker but still significant evidence for a further feedback with a lag of 4 years. There is a significant positive feedback with a lag of 8 years. It is impossible to prove the mechanisms that lie behind these negative and positive feedbacks by time series analysis, but they are extremely interesting and may reflect density dependence caused by predator–prey and plant–herbivore interactions. I have no idea what (if anything) lies behind the positive partial correlation that acts with an 8 year lag, but it might be genetic. These data are analysed further, using a time series model, on p. 666.

Moving average

The simplest way of seeing pattern in time series data is to plot the moving average. A useful summary statistic is the three-point moving average

$$y_i' = \frac{y_{i-1} + y_i + y_{i+1}}{3}$$

The function **ma3** will compute the three-point moving average for any input vector x:

```
ma3<-function (x){
  y<-numeric(length(x)-1)
  for (i in 2:length(x)-1) {
      y[i]<-(x[i-1]+x[i]+x[i+1])/3
  }
y }
```

The time series of mean monthly temperatures will illustrate the use of moving average:

```
attach(temp)
tm<-ma3(temp)
plot(temp)
lines(tm)
```

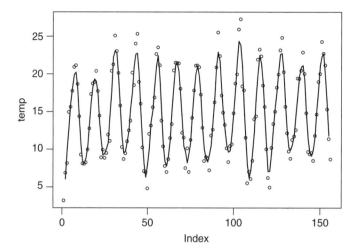

The seasonal pattern of temperature change over the 13 years of the data is clear. Note that a moving average can never capture the maxima or minima of a series (because they are averaged away). Note also that the three-point moving average is undefined for the first and last points in the series.

Seasonal data

Many time series applications involve data that exhibit seasonal cycles. The commonest applications involve weather data. Here are daily maximum and minimum temperatures from Silwood Park in south-east England over the period 1987–2000 inclusive:

```
attach(SilwoodWeather)
names(SilwoodWeather)

[1] "upper"   "lower"    "rain"   "month"   "yr"

plot(upper,type="l")
```

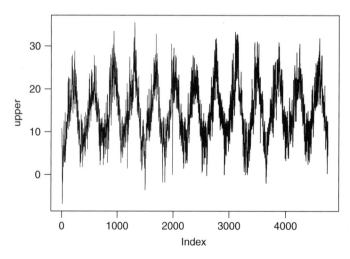

The seasonal pattern of temperature change over the 13 year period is clear, but there is no clear trend (e.g. warming). Note that the x axis is labelled by the day number of the time series (Index).

We start by modelling the seasonal component. The simplest models for cycles are scaled so that a complete annual cycle is of length 1.0 (rather than 365 days). Our series consists of 4748 days over a 13 year span, so we write

```
index<-1:4748
4748/13
```

```
[1] 365.2308
```

```
time<-index/365.2308
```

The equation for the seasonal cycle is

$$y = \alpha + \beta \sin(2\pi t) + \gamma \cos(2\pi t) + \varepsilon$$

This is a linear model, so we can estimate its three parameters very simply:

```
model<-lm(upper~sin(time*2*pi)+cos(time*2*pi))
```

To investigate the fit of this model, we need to create the scatterplot using very small symbols (otherwise the fitted line will be completely obscured). The smallest useful plotting symbol is the dot ".":

```
plot(time, upper, pch=".")
lines(time, predict(model))
```

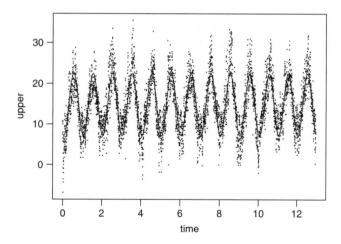

The three parameters of the model are all highly significant:

```
summary(model)
```

```
Coefficients:
```

	Value	Std. Error	t value	Pr(>|t|)
(Intercept)	14.8034	0.0506	292.5076	0.0000
sin(time * 2 * pi)	-2.5276	0.0716	-35.3156	0.0000
cos(time * 2 * pi)	-7.2237	0.0716	-100.9302	0.0000

```
Residual standard error: 3.487 on 4745 degrees of freedom
Multiple R-Squared: 0.7067
F-statistic:5717 on 2 and 4745 degrees of freedom, the p-value
is 0
```

We can investigate the residuals to look for patterns (e.g. trends in the mean, or autocorrelation structure). Remember that the residuals are stored as part of the model object $resid:

```
plot(model$resid,pch=".")
```

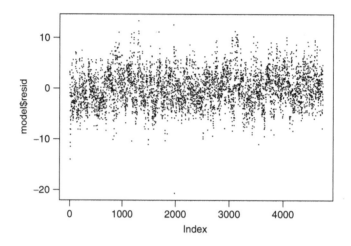

There looks to be some periodicity in the residuals, but no obvious trends. To look for serial correlation in the residuals, we use the **acf** function like this:

```
par(mfrow=c(1,2))
acf(model$resid)
acf(model$resid,type="p")
```

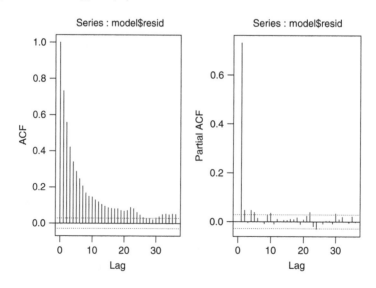

There is very strong serial correlation in the residuals, and this drops off roughly exponentially with increasing lag (left graph). The partial autocorrelation at lag $= 1$ is very large (0.7317), but the correlations at higher lags are much smaller. This suggests that an AR(1) model might be appropriate. This is the statistical justification behind the old joke about the weather forecaster who was asked what tomorrow's weather would be. And his reply? 'Like today's.'

Pattern in the monthly means

The monthly average upper temperatures show a beautiful seasonal pattern when analysed by **acf**:

```
temp<-ts(as.vector(tapply(upper,list(month,yr),mean)))
acf(temp)
```

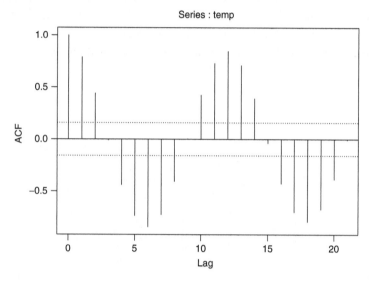

There is a perfect cycle with period $= 12$ (as expected). What about patterns across years?

```
ytemp<-ts(as.vector(tapply(upper,yr,mean)))
acf(ytemp)
```

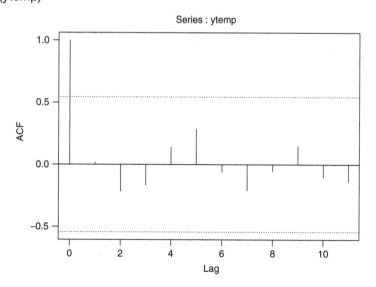

Nothing! The pattern you may (or may not) see depends upon the scale at which you look for it. As for spatial patterns (Chapter 36), so it is with temporal patterns. There is strong pattern between days within months (tomorrow will be like today). There is very strong pattern from month to month within years (January is cold, July is warm). But there is no pattern at all from year to year (there may be progressive global warming, but it is not apparent within this recent time series, and there is absolutely no evidence for untrended serial correlation).

Built-in time series functions

The analysis is simpler, and the graphics are better labelled if we convert the temperature data into a regular time series object using **rts**. We need to specify the first date (January 1987) as start=c(1987,1), and the number of data points per year as frequency=365:

```
high<-rts(upper,start=c(1987,1),frequency=365)
```

Now use **ts.plot** to see a plot of the time series, correctly labelled by years:

```
ts.plot(high)
```

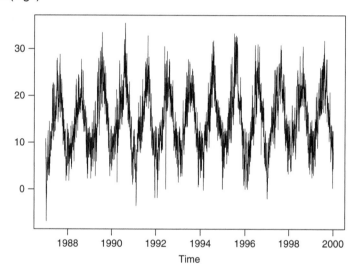

Decompositions

It is useful to be able to turn a time series into components:

- trend

- seasonal fluctuation

- residuals

The function **stl** (letter *l* not figure 1) performs seasonal decomposition. First, we make a time series object, specifying the start date and the frequency:

```
up<-ts(upper,start=c(1987,1),frequency=365.25)
```

Now use **stl** to decompose the series:

```
high<-stl(up,"period")
```

The plot function produces the series, the trend and the residuals in a single frame:

```
plot(high)
```

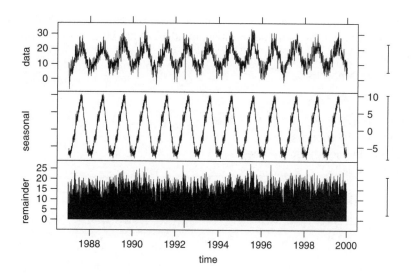

The **stl** object, high, has the two components (seasonal and residuals) as attributes that can be addressed using $sea and $rem respectively.

First, the data need to be converted to monthly means before we can use this function:

```
highmonths<-as.vector(tapply(upper,list(month,yr),mean))
```

and highmonths needs to be turned into a regular time series object:

```
highs<-rts(highmonths,start=c(1987,1),frequency=12)
```

The time series highs can now be decomposed using the **stl** function:

```
highdec<-stl(highs,"periodic")
```

The decomposed time series object highdec has components for the residual component $rem and the seasonal component $sea. We can plot these on the same axes like this:

```
ts.plot(highdec$rem,highdec$sea)
```

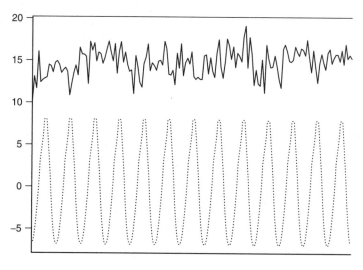

The seasonal component (lower dotted curve) has a mean of zero. The residual component (upper solid curve) is our main interest. Are the residuals trended? What kind of serial correlation structure do the residuals exhibit?

```
length(highmonths)
```

```
[1] 156
```

```
time<-1:156
plot(time,highmonths)
model<-lm(highmonths~time)
abline(model)
```

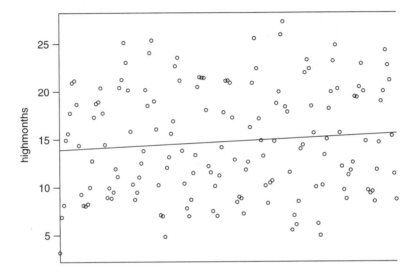

There is an upward trend. But is it significant?

```
summary(model)
```

```
Coefficients:
                Value   Std. Error   t value   Pr(>|t|)
(Intercept)   13.9065      0.9079    15.3169     0.0000
       time    0.0110      0.0100     1.0939     0.2757
```

No it isn't.

Spectral analysis

There is an alternative approach to time series analysis, which is based on the analysis of **frequencies** rather than fluctuations of numbers. Frequency is the reciprocal of cycle period. Ten-year cycles would have a frequency 0.1 per year (see the lynx example on p. 639).

The fundamental tool of spectral analysis is the **periodogram**. This is based on the squared correlation between the time series and sine/cosine waves of frequency ω, and conveys exactly the same information as the autocovariance function. It may (or may not) make the information easier to interpret. Using the function is straightforward; we employ the **spectrum** directive like this:

```
spectrum(Numbers)
```

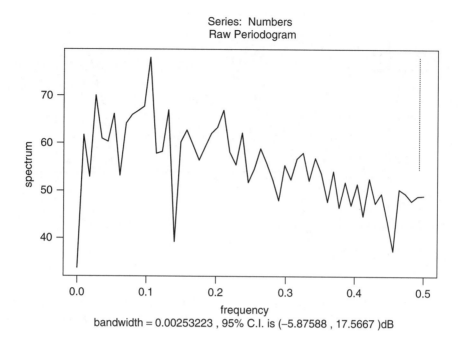

Series: Numbers
Raw Periodogram

bandwidth = 0.00253223 , 95% C.I. is (−5.87588 , 17.5667)dB

The plot is on a log scale, in **units of decibels**, and the x legend shows the bandwidth and the 95% confidence interval in decibels (the vertical bar in the top right corner). The figure is interpreted as showing strong cycles with a frequency of about 0.1 where the maximum value of spectrum occurs. That is to say, it indicates cycles with a period of $1/0.1 = 10$ years. There is a hint of longer-period cycles (the local peak at frequency 0.033 would produce cycles of length $1/0.033 = 30$ years) but no real suggestion of any shorter-period cycles.

The analysis of *wavelets* is a major growth industry in the time series analysis (see Bruce and Gao 1996 for details).

Multiple time series

When we have two or more time series measured over the same period, the question naturally arises as to whether or not the ups and downs of the different series are correlated. It may be that we suspect that change in one of the variables causes changes in the other (e.g. changes in the number of predators may cause changes in the number of prey, because more predators means more prey eaten). We need to be careful, of course, because it will not always be obvious which way round the causal relationship might work (e.g. predator numbers may go up because prey numbers are higher; ecologists call this a numerical response). Suppose we have the following sets of counts:

```
attach(twoseries)
names(twoseries)
```

```
[1] "x"   "y"
```

We start by inspecting the two time series on the same time axis:

```
tsplot(x,y)
```

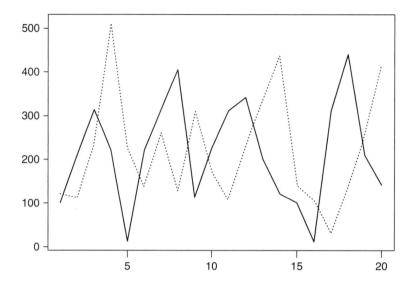

There is some evidence of periodicity (at least in x) and it looks as if y lags behind x by roughly two periods (sometimes one). First, we carry out straightforward analyses on each time series separately:

```
par(mfrow=c(1,2))
acf(x,type="p")
acf(y,type="p")
```

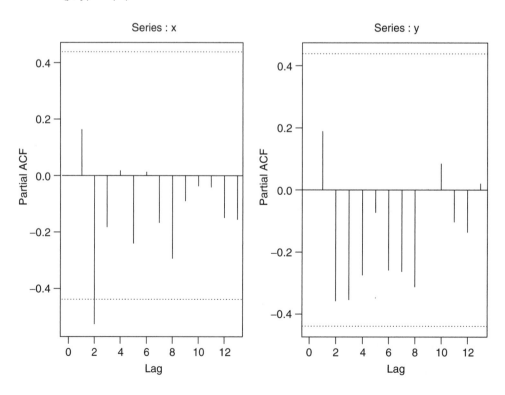

As we suspected, the evidence for periodicity is stronger in x than in y: the partial autocorrelation is significant at lag $= 2$ for x, but none of the partials are significant for y.

To look at the cross-correlation between the two series we use the same **acf** function, but we provide it with a matrix containing both x and y as its argument:

```
par(mfrow=c(1,1))
acf(cbind(x,y))
```

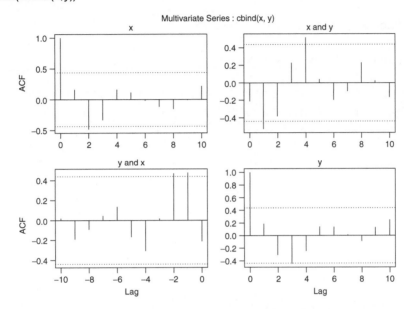

The four plots show the results for x, x and y, y and x, and y. The plots for x and y are exactly the same as in the single-series case. Note the reversal of the time lag scale on the plot of y vs. x. In the case of partial autocorrelations, things are a little different, because we now see in the xx column the values of the autocorrelation of x controlling for x *and for* y (not just for x as in the earlier figure):

```
acf(cbind(x,y),type="p")
```

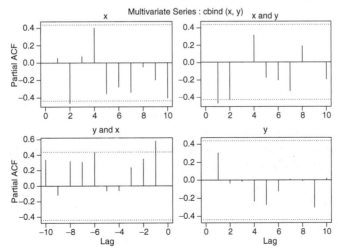

This indicates a hint (not quite significant) of a positive partial autocorrelation for x at lag $= 4$ that we did not see in the analysis of the single time series of x. There are significant negative cross-correlations of x and y at lags of 1 and 2, strongly suggestive of some sort of relationship between the two time series.

To obtain the matrix of **autocovariances** rather than autocorrelations, just use the option type="covariance":

```
acf(cbind(x,y),type="cov")
```

```
Autocovariances matrix:
      lag          x.x           x.y           y.y
 1     0    13724.5098    -3124.7251    15719.0879
 2     1     2255.7205    -7732.7710     2980.3345
 3     2    -6659.2241    -5627.2852    -4864.6938
 4     3    -4624.3433     3350.2915    -6997.0342
 5     4     2239.8669     7567.2954    -3843.0750
 6     5     1570.4475      614.8588     2201.1594
 7     6     -317.4420    -2896.0400     2183.4688
 8     7    -1643.5115    -1434.8738      215.0656
 9     8    -2158.9961     3370.3152    -1434.8875
10     9     -144.9305      353.8112     2078.0469
11    10     3009.0051    -2441.4126     3940.8687

      lag          y.x
 1     0    -3124.7251
 2    -1     7009.0415
 3    -2     6897.1577
 4    -3      259.7162
 5    -4    -4532.7573
 6    -5    -2467.2913
 7    -6     1993.2224
 8    -7      664.6262
 9    -8    -1355.3250
10    -9    -2825.9238
11   -10      280.9874
```

The first and third columns are the same as would be obtained from the univariate time series (the autocovariance of x, or the autocovariance of y).

To get a visual assessment of the relationship between x and y, we can plot the change in y against the change in x using the **diff** function like this:

```
plot(diff(x),diff(y))
```

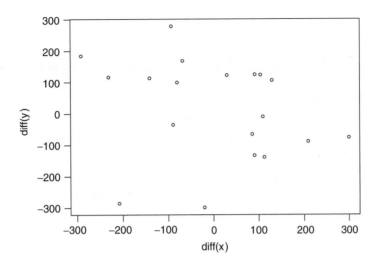

The negative correlation is clear enough, but there are two conspicuous outliers, making it clear that not all negative changes in x are associated with large positive changes in y.

Coupled time series: ragwort and cinnabar moth

In ecological work, we are often interested in the way in which the population dynamics of species in different trophic levels are related. In this example we study a 19 year time series of a plant called ragwort (*Senecio jacobaea*, Asteraceae) and its principal insect herbivore, the cinnabar moth (*Tyria jacobaeae*, Lepidoptera, Arctiidae) in mesic acid grasslands in Silwood Park, Berkshire. You might expect that the plant would have a positive impact on the population dynamics of the moth and that the moth might have a negative impact on the population dynamics of the plant (much as in a Lotka–Volterra predator–prey interaction):

$$\frac{\mathrm{d}V}{\mathrm{d}t} = rV\left[\frac{K-V}{K}\right] - \alpha VN$$

$$\frac{\mathrm{d}N}{\mathrm{d}t} = \beta VN - \gamma N$$

where V is the biomass of the plant and N is the number of cinnabar moth caterpillars. This model has a stable equilibrium in which plant numbers V^* are determined entirely by the attributes of the food-limited herbivore:

$$V^* = \frac{\gamma}{\beta}$$

We use time series analysis to see what light the field data throw on the applicability of this simple theoretical model:

```
attach(interaction)
names(interaction)
[1] "year"  "cinnabar"  "ragwort"
```

You should begin by convincing yourself that there is no hint of periodicity in either of the separate time series (cinnabar or ragwort). Hint: use **acf** and type="p" on both series.

To work on the time series in more detail, we need to calculate the differences in log population size from year to year (this gives a measure of relative population change). The question is whether cinnabar affects population change in ragwort (and vice versa) and whether there are any significant time lags in the system. (For example, does ragwort population change this year depend upon cinnabar numbers last year? First we difference the log time series:

```
cd<-diff(log(cinnabar))
rd<-diff(log(ragwort))
```

then use **cbind** to bind the two vectors together in a data frame called data:

```
data<-cbind(cd,rd)
```

To measure the strength of interaction between the plant and the herbivore, we use the logged differences to fit what is called a **multivariate autoregressive model**. The model (**ar** stands for autoregressive) is very easy to use:

```
ar(data)
```

This simple command produces reams of output. You need to scroll back to find $ar:

```
$ar:
, , 1
                [,1]                    [,2]
[1,]  -0.5287896          0.01722965
, , 2
                [,1]                    [,2]
[1,]  -0.02249318        -0.4034163
```

This 2×2 matrix gives the strengths of the four first-order interactions between the plant and the herbivore. [1,1] is the impact of cinnabar on cinnabar (self-regulating intraspecific density dependence $= -0.5288$) and [2,2] is the impact of ragwort on ragwort population change (self-regulating intraspecific density dependence $= -0.4034$). We conclude that both populations are very strongly regulated by immediate (non-lagged) density dependence. Surprisingly, however, the other two interaction strengths are not significantly different from zero. There is no significant impact of cinnabar moth on ragwort population change ([2, 1] $= -0.022\,49$), nor of plant abundance of cinnabar moth population change ([1, 2] $= 0.017\,22$).

Next we investigate the population regulation of the insects in more detail by plotting the log of population change against population density, like this:

```
plot(cinnabar[-19],cd)
```

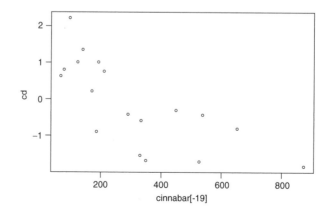

This shows the strong density dependence in cinnabar moth dynamics, with an equilibrium population density (cd = 0) of about 200 insects per unit area.

There are various ways of plotting both time series on the same axes. The most obvious way turns out to be the least satisfactory:

tsplot(cinnabar,ragwort)

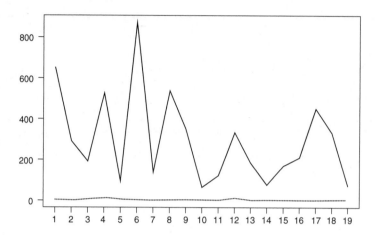

With this method you get a very good picture of population fluctuation in the cinnabar moth (solid line), but a very poor indication of what is going on with the ragwort plants (dotted line). This is because the scales of measurement of the two species are so different:

mean(ragwort)

[1] 4.075226

mean(cinnabar)

[1] 298.5789

One option is to plot time series of the log population sizes:

tsplot(log(cinnabar),log(ragwort))

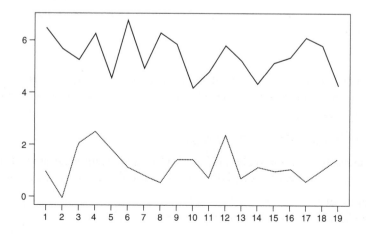

This is better but still not ideal. You could plot them in separate frames, one above the other like this:

```
par(mfrow=c(2,1))
tsplot(cinnabar)
tsplot(ragwort)
```

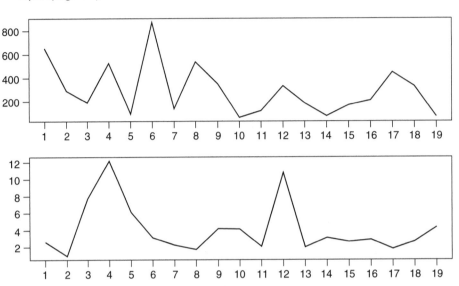

Alternatively, you could use the graphical overlay facility par(new=T), which effectively allows different scaling on the left- and right-hand *y* axes. Start by rescaling the margins so there is enough room to put a separate axis with tick marks and a label on the right-hand *y* axis of the plot:

```
par(mfrow=c(1,1))
par(mar=c(5,4,4,5)+0.1)
```

Now plot the first graph in the normal way:

```
tsplot(cinnabar,ylab="cinnabar")
```

Next we use new=T to override the clearance of the first graph when the next plot directive is issued, and xaxs="d" to specify direct *x* axes in the next plot:

```
par(new=T,xaxs="d")
```

Now the next plot is drawn in the normal way, except that we specify axes=F (we don't want to overwrite the cinnabar numbers axis) and ylab="" (we don't want to write 'ragwort' on top of 'cinnabar'):

```
tsplot(ragwort,axes=F,ylab="",lty=2)
```

To finish off the graph, we want to put a different axis on the right-hand side (this is side = 4), to show the density of ragwort plants, using line = 3.8 to space the text away from the right-hand axis:

```
axis(side=4)
mtext(side=4,line=3.8, "ragwort")
```

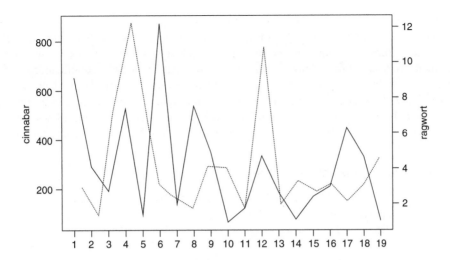

Repeated measures Anova

Designed experiments sometimes involve temporal pseudoreplication; that is to say, repeated measurements taken from the same individuals (patients, quadrats, plants, etc.). Survey work often involves repeated assessment of the same individuals. We need to be careful when analysing data like these, because if we fail to take account of the pseudoreplication our error variance will have too many degrees of freedom. This will mean that we reject the null hypothesis when it is true far more frequently than α (we shall commit too many Type I errors, in other words).

The data consist of heights measured on four occasions in three replicate blocks using four different seed types.

```
attach(repmeasures)
names(repmeasures)

[1] "time"   "seed"   "height"   "rep"
```

The main interest is in differences in the height growth of plants from the different seeds (a simple one-way Anova), and so the easiest way to get rid of the temporal pseudoreplication is to analyse the data only for the *final* heights (at time==4) like this:

```
summary(aov(height~seed,subset=(time==4)))
```

	Df	Sum of Sq	Mean Sq	F Value	Pr(F)
seed	3	14.87932	4.959772	42.63857	0.00002883387
Residuals	8	0.93057	0.116321		

There is a highly significant difference between the mean heights of plants from different seeds. We can see how this comes about by tabulating the means:

```
tapply(height,list(time,seed),mean)
```

	Barrymore	Conway	Culrain	Desire
1	1.000000	1.911940	1.294316	3.236525
2	1.533333	2.514312	2.109192	3.563512
3	2.133333	3.233956	2.518726	4.400554
4	3.000000	4.594047	3.152776	5.715371

This table shows how the mean heights differ at sample time number 4 but it also shows how height increases (almost linearly) with time for each of the seed types. It is clear that Desire has grown taller than Barrymore and Culrain, but it is not clear whether Conway is significantly taller than the smaller two seed types. This would require contrasts or model simplification.

An alternative analysis is to use Ancova and to fit regression lines to describe growth separately for each seed type. If it turns out that the slopes are not significantly different, then the hypothesis involves a comparison of the intercepts of the lines for the four seed types. If the lines are of different slopes, then we need to specify a time for the comparison (say time == 4) and adjust the standard errors accordingly:

summary(aov(height~seed*time))

	Df	Sum of Sq	Mean Sq	F Value	Pr(F)
seed	3	37.85597	12.61866	76.7330	0.0000000
time	1	32.91032	32.91032	200.1249	0.0000000
seed:time	3	0.79070	0.26357	1.6027	0.2038792
Residuals	40	6.57796	0.16445		

The slopes are not significantly different, so we can fit a common slope to the model. To see the coefficients we use **lm** rather than **aov**:

summary(lm(height~seed+time))

Coefficients:

	Value	Std. Error	t value	Pr($>$|t|)
(Intercept)	0.0651	0.1793	0.3634	0.7181
seedConway	1.1469	0.1690	6.7864	0.0000
seedCulrain	0.3521	0.1690	2.0834	0.0432
seedDesire	2.3123	0.1690	13.6825	0.0000
time	0.7406	0.0534	13.8582	0.0000

Conway is significantly taller than Barrymore but Culrain is only marginally taller ($p = 0.0432$). Desire is significantly taller than any of the other varieties ($2.3123 - 1.1469$ is much more than 2 standard errors).

Another way to analyse these data is to use an error term in the Anova to show the nesting and the pseudoreplication. We need to make rep and time into factors:

rep<-factor(rep)
tt<-factor(time)
summary(aov(height~seed+Error(rep/seed/tt)))

Error: rep

	Df	Sum of Sq	Mean Sq	F Value	Pr(F)
Residuals	2	2.125869	1.062934		

Error: seed %in% rep

	Df	Sum of Sq	Mean Sq	F Value	Pr(F)
seed	3	37.85597	12.61866	79.73624	0.00003175197
Residuals	6	0.94953	0.15825		

Error: tt %in% (rep/seed)

	Df	Sum of Sq	Mean Sq	F Value	Pr(F)
Residuals	36	37.20358	1.033433		

You would test for the seed effect by dividing the seed mean square by the rep-in-seed mean square on 3 and 6 d.f. (compare this with our first analysis, using only the final plant heights; the error d.f. are lower (6 rather than 8) and the error variance is somewhat higher (0.158 rather than 0.116). Finally, we could use mixed effects models.

Linear mixed effects models

Here we take advantage of the fact that mixed effects models can deal with random effects (like rep and time above) in an efficient way (see Chapter 35 for details). In an **lme** we need to specify:

- the fixed effects (height ~ seed)

- the temporal random effects (random = ~ time)

- the spatial random effects (rep)

- the nesting structure for the repeated measures (time | rep/seed)

The simplest way to do the initial data inspection for the mixed effects model is to create a **groupedData** object; we call it structure:

> structure<-groupedData(height~time | rep/seed)

This defines the time series with height as a the response variable (height~time), indicates that the subjects on which the repeated measures are made are seed varieties nested within rep (|rep/seed). First we obtain a plot of each of the 12 time series:

> plot(structure)

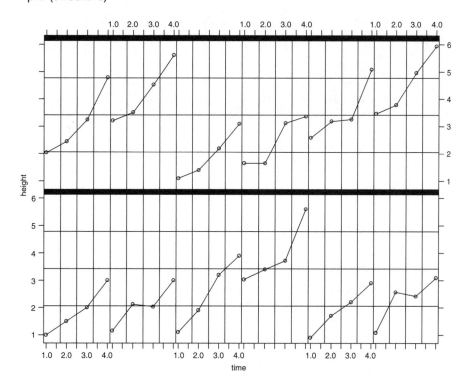

Then we obtain a plot of the time series grouped by seed type (outer=~seed):

plot(structure,outer=~seed)

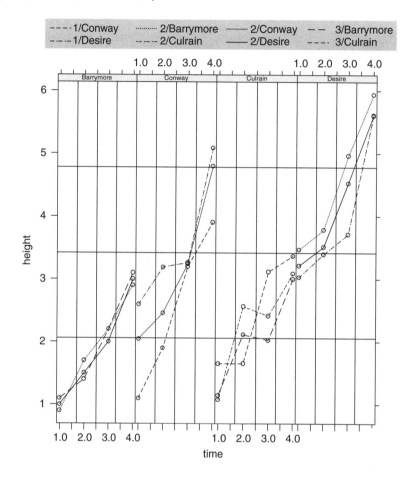

This shows separate lines for each rep of each seed. The separate linear regressions of height against time are easily obtained using **lmList**:

lmList(structure)

```
Call:
  Model: height ~ time | rep
   Data: structure

Coefficients:
       (Intercept)          time
  2      0.3000000      0.6500000
  1      0.2500000      0.6500000
  3      0.2500000      0.6800000
  7      0.1000000      0.9700000
  8      0.8743493      0.9039787
```

```
 9       1.6418689      0.7558104
 4       0.7059286      0.5469925
 5       0.8184280      0.5873083
 6       0.7932158      0.6611735
10       1.9268839      0.8031883
11       2.1633857      0.8203924
12       2.3915161      0.8584937

Degrees of freedom: 48 total; 24 residual
Residual standard error: 0.372922
```

The linear mixed effects model is specified like this:

```
summary(lme(fixed = height~seed,random = ~time|rep/seed))
```

We are interested mainly in the fixed effects of seed on height (height ~ seed), but we have spatial random effects (rep/seed) and temporal pseudoreplication (~time|rep/seed) to take care of:

```
Linear mixed-effects model fit by REML
Data: NULL
      AIC      BIC     logLik
 85.2965 104.9226 -31.64825

Random effects:
Formula:  ~ time | rep
Structure:  General positive-definite
             StdDev     Corr
(Intercept) 4.0587833 (Inter
       time 0.7376433 -1

Formula:  ~ time | seed %in% rep
Structure:  General positive-definite
             StdDev     Corr
(Intercept) 0.27148977 (Inter
       time 0.06082398 -1
   Residual 0.34859050

Fixed effects: height ~ seed
                Value  Std.Error DF t-value  p-value
(Intercept) 4.092899  0.1825308 36 22.42306  <.0001
 seedConway 1.203085  0.1688350  6  7.12580  0.0004
seedCulrain 0.336130  0.1688350  6  1.99088  0.0936
seedDesire  2.355738  0.1688350  6 13.95290  <.0001

Correlation:
            (Intr)  sdCnwy  sdClrn
seedConway  -0.462
seedCulrain -0.462   0.500
seedDesire  -0.462   0.500   0.500

Standardized Within-Group Residuals:
       Min         Q1        Med        Q3        Max
 -1.962299 -0.6380027 0.01606774 0.5312365 2.168464
```

```
Number of Observations: 48
Number of Groups:
 rep seed %in% rep
   3              12
```

Because we are using treatment contrasts, the main effects of seed on height are differences from variety Barrymore (because that name comes first in the alphabet). The standard errors are a little higher than they were in the one-way using only the final heights. Culrain is not significantly taller than Barrymore but both the other two varieties are significantly taller. Finally, Desire is significantly taller than Conway; you have to do this t test in your head: $t = (2.35-1.20)/0.169 = 6.8$. The regression slopes (all varieties and replicates) are in the coefficients:

```
coef(lme(fixed = height~seed,random = ~time|rep/seed))
```

	(Intercept)	seedConway	seedCulrain	seedDesire	time
1/Barrymore	-0.15088808	1.203085	0.33613	2.355738	0.7587862
1/Conway	-0.64896085	1.203085	0.33613	2.355738	0.8703733
1/Culrain	-0.19836877	1.203085	0.33613	2.355738	0.7694237
1/Desire	-0.40979983	1.203085	0.33613	2.355738	0.8167922
2/Barrymore	0.14033832	1.203085	0.33613	2.355738	0.7171767
2/Conway	0.03801650	1.203085	0.33613	2.355738	0.7401007
2/Culrain	0.18343191	1.203085	0.33613	2.355738	0.7075221
2/Desire	0.02526746	1.203085	0.33613	2.355738	0.7429569
3/Barrymore	0.34707455	1.203085	0.33613	2.355738	0.6894272
3/Conway	0.56428991	1.203085	0.33613	2.355738	0.6407627
3/Culrain	0.46027611	1.203085	0.33613	2.355738	0.6640657
3/Desire	0.42498500	1.203085	0.33613	2.355738	0.6719723

This shows the spatial structure of the experiment, with seed varieties nested within replicates. The 12 slopes are in the right-hand column (headed time). For a more in-depth discussion of temporal autocorrelation in mixed effects models, see Chapters 35.

Simulated time series

To see how the correlation structure of a first-order autoregressive process depends on the value of α, we can simulate the process over, say, 250 time periods using different values of α. We generate the white noise Z_t using the random number generator rnorm(n,0,s), which gives n random numbers with a mean of 0 and a standard deviation of s. To simulate the time series we evaluate

$$Y_t = \alpha Y_{t-1} + Z_t$$

multiplying last year's population by α then adding the relevant random number from Z_t. We begin with the special case of $\alpha = 0$, so $Y_t = Z_t$ and the process is pure white noise:

```
Y<-rnorm(250,0,2)

par(mfrow=c(1,2))
tsplot(Y)
acf(Y)
```

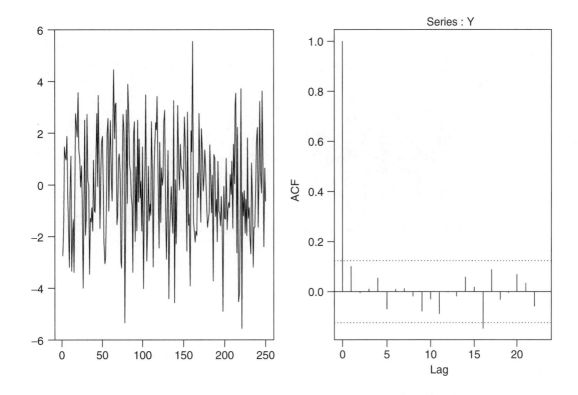

The time series is bound to be stationary because each value of Z is independent of the value before it. The correlation at lag 0 is 1 (of course), but there is absolutely no hint of any correlations at higher lags.

To generate the time series for non-zero values of α, we need to use recursion: this year's population is last year's population times α plus the white noise. We begin with a negative value of $\alpha = -0.5$. First we generate all the noise values (by definition, these *don't* depend on population size):

```
Z<-rnorm(250,0,2)
```

Now the initial population at time 0 is set to 0 (remember that the population is stationary, so we can think of the Y values as departures from the long-term mean population size). This means that $Y_1 = Z_1$. Thus, Y_2 will be whatever Y_1 was, times -0.5, plus Z_2. And so on.

```
Z<-rnorm(250,0,2)
Y<-numeric(250)
Y[1]<-Z[1]
for (i in 2:250) Y[i]<- - 0.5*Y[i-1]+Z[i]
tsplot(Y)
acf(Y)
```

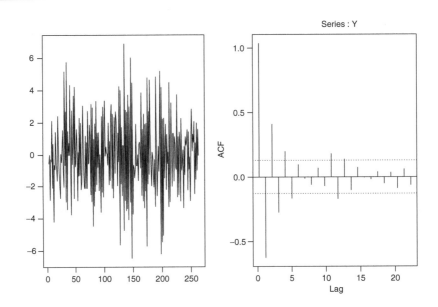

The time series shows rapid return to equilibrium following random departures from it. There is a highly significant negative autocorrelation at lag 1, significant positive autocorrelation at lag 2, and so on, with the size of the correlation gradually damping away. Let's simulate a time series with a positive value of, say, $\alpha = 0.5$:

```
Z<-rnorm(250,0,2)
Y[1]<-Z[1]
for (i in 2:250) Y[i]<- 0.5*Y[i-1]+Z[i]
tsplot(Y)
acf(Y)
```

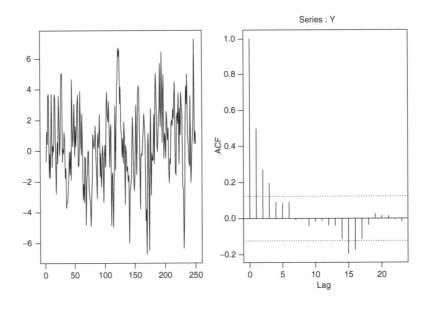

Now the time series plot looks very different, with protracted periods spent drifting away from the long-term average. The autocorrelation shows significant positive correlations for the first three lags.

Finally, we look at the special case of $\alpha = 1$. This means that the time series is a classic **random walk**:

$$Y_t = Y_{t-1} + Z_t$$

```
Z<-rnorm(250,0,2)
Y[1]<-Z[1]
for (i in 2:250) Y[i]<- Y[i-1]+Z[i]
tsplot(Y)
acf(Y)
```

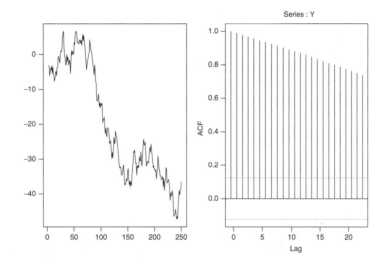

The time series wanders about and strays far away from the long-term average. Using **acf** shows positive correlations dying away very slowly, and still highly significant at lags of more than 20. Of course, if you do another realisation of the process, the time series will look very different, but the out put of **acf** will be similar.

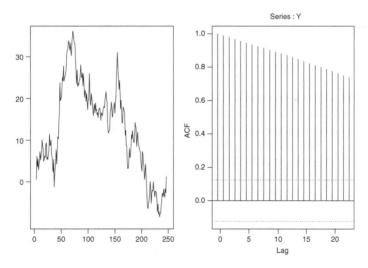

Time series models

Time series models come in three kinds (Box and Jenkins 1976):

- moving average (MA) models where

$$X_t = \sum_0^q \beta_j \varepsilon_{t-j}$$

- autoregressive (AR) models where

$$X_t = \sum_1^p \alpha_i X_{t-i} + \varepsilon_t$$

- autoregressive moving average (ARMA) models where

$$X_t = \sum_1^p \alpha_i X_{t-i} + \sum_0^q \beta_j \varepsilon_{t-j}$$

A moving average of order q averages the random variation over the last q time periods. An autoregressive model of order p computes X_t as a function of the last p values of X, so for a second-order process we would use

$$X_t = \alpha_1 X_{t-1} + \alpha_2 X_{t-2} + \varepsilon_t$$

Typically, we would use the partial autocorrelation plot (above) to determine the order. So, for the lynx data we would use order 2 or 4 depending on taste. Other things being equal, parsimony suggests the use of order 2. The fundamental difference is that a *set* of random components (ε_{t-j}) influence the current value of an MA process, whereas only the *current* random effect (ε_t) affects an AR process. Both kinds of effects are at work in an ARMA processes. Ecological models of population dynamics are typically AR models. For instance,

$$N_t = \lambda N_{t-1}$$

is the discrete-time version of exponential growth $(\lambda > 1)$ or decay $(\lambda < 1)$. It looks just like a first-order AR process with the random effects missing. This is somewhat misleading, however, since *time series are supposed to be stationary*, which would imply a long-term average value of $\lambda = 1$. But in the absence of density dependence (as here) this is impossible. The α of the AR model is *not* the λ of the population model.

Models are fitted using the **arima.mle** directive, and their performances are compared using AIC (see p. 315). The most important component of the model is *order*. This is a vector of length 3 specifying the order of the autoregressive and moving average operators and the number of differences. Thus order=c(1,3,2) is based on a first-order autoregressive process, a third-order moving average and two differences. We give the models names to reflect their AR and MA order, so model21 has second-order AR and first-order MA, then we fit them like this (if you have not done the lynx example, you will need to attach the lynx data frame, see p. 639):

```
model10<-arima.mle(Numbers,model=list(order=c(1,0,0)))
model21<-arima.mle(Numbers,model=list(order=c(2,1,0)))
```

and so on. We can extract the AICs of each fit like this:

model22$aic

[1] 1886.287

model21$aic

[1] 1861.248

model41$aic

[1] 1825.284

model40$aic

[1] 1838.663

model20$aic

[1] 1871.256

model12$aic

[1] 1913.785

model11$aic

[1] 1889.229

model10$aic

[1] 1915.253

The lowest AIC is 1825.284, which suggests that a model with an AR lag of 4 and a moving average term of 1 is best. This implies that a more complex ecological model is required which takes account of both significant partial correlations (at lags of 2 and 4 years) and not just the 2 year lag. It is not clear why the moving average term is important but it may be an age-structure effect.

Further reading

Box, G.E.P. and Jenkins, G.M. (1976) *Time Series Analysis: Forecasting and Control*. Oakland, California, Holden-Day.
Chatfield, C. (1989) *The Analysis of Time Series: An Introduction*. London, Chapman Hall.
Diggle, P.J., Liang, K.-Y. *et al.* (1994). *Analysis of Longitudinal Data*. Oxford, Clarendon Press.
Priestley, M.B. (1981) *Spectral Analysis and Time Series*. London, Academic Press.
Shumway, R.H. (1988) *Applied Statistical Time Series Analysis*. Englewood Cliffs, New Jersey, Prentice Hall.

35

Mixed effects models

Mixed effects models are so called because the explanatory variables are a mixture of fixed effects and random effects (see p. 361):

- fixed effects influence only the *mean* of *y*
- random effects influence only the *variance* of *y*

A random effect should be thought of as coming from a population of effects; the existence of this population is an extra assumption. We speak of *prediction of random effects*, rather than estimation; we *estimate* fixed effects from data, but we intend to make predictions about the population from which our random effects were sampled. Fixed effects are unknown constants to be estimated from the data. Random effects govern the variance–covariance structure of the response variable.

The fixed effects are often experimental treatments that were applied under our direction, and the random effects are either categorical or continuous variables that are distinguished by the fact that we are typically not interested in their parameter values, but only in the variance they explain.

One or more of the explanatory variables represents *grouping* in time or in space. Random effects that come from the same group will be correlated, and this contravenes one of the fundamental assumptions of standard statistical models: *independence of errors*. Mixed effects models take care of this non-independence of errors by modelling the covariance structure introduced by the grouping of the data. The covariance between observations in the same group is σ_a^2, which corresponds to a correlation of $\sigma_a^2/(\sigma_a^2 + \sigma^2)$.

A major benefit of a random effects model is that it economises on the number of degrees of freedom used up by the factor levels; the model is written

$$y_{ij} = \overline{\beta} + (\beta_i - \overline{\beta}) + \varepsilon_{ij}$$

where $\overline{\beta}$ is the overall average ($\overline{\beta} = \sum \beta_i/k$). The random effects model replaces $\overline{\beta}$ by the mean of the response variable over the population of factor levels, and replaces the deviations $(\beta_i - \overline{\beta})$ by random variables whose distribution is to be estimated.

Mixed effects models are particularly useful in cases where there is temporal pseudoreplication (repeated measurements) and/or spatial pseudoreplication (e.g. nested designs or split-plot experiments). The data are arranged in a **groupedData** object, and the **lme** models can allow for:

- spatial autocorrelation between neighbours
- temporal autocorrelation across repeated measures on the same individuals
- differences in the mean response between blocks in a field experiment
- differences between subjects in a medical trial involving repeated measures

The point is that we really do not want to waste precious degrees of freedom in estimating parameters for each of the separate levels of the categorical random variables. On the other hand, we do want to use all measurements we have taken, but because of the pseudoreplication we want to take account of the correlation structure and the variance function:

- the correlation structure is used to model within-group correlation associated with temporal and spatial dependencies, using **correlation**

- the variance function is used to model non-constant variance in the within-group errors using **weights**

Fixed or random effects?

It is difficult without lots of experience to know when to use categorical explanatory variables as fixed effects and when to use them as random effects. Here are some guidelines:

- Am I interested in the effect sizes? Yes, means fixed effects.

- Is it reasonable to suppose that the factor levels come from a population of levels? Yes, means random effects.

- Are there enough levels of the factor in the data on which to base an estimate of the variance of the population of effects? No, means fixed effects.

- Are the factor levels informative? Yes, means fixed effects.

- Are the factor levels just numeric labels? Yes, means random effects.

- Am I mostly interested in making inferences about the distribution of effects, based on the random sample of effects represented in the data frame? Yes, means random effects.

- Is there hierarchical structure? Yes, means you need to ask whether the data are experimental or observations.

- Is it a hierarchical experiment, where the factor levels are experimental manipulations? Yes, means fixed effects in a split-plot design (see p. 345).

- Is it a hierarchical observational study? Yes, means random effects, perhaps in a variance components analysis (see p. 361).

- When your model contains both fixed and random effects, use mixed effects models.

- If your model structure is linear, use linear mixed effects, **lme**.

- Otherwise, specify the model equation and use non-linear mixed effects, **nlme**.

Removing the pseudoreplication

The extreme response to pseudoreplication in a data set is simply to eliminate it. Spatial pseudoreplication can be averaged away (see p. 58) and temporal pseudoreplication can be dealt with by carrying out separate Anovas, one at each time. This approach has two major weaknesses:

- it cannot address questions about treatment effects that relate to the longitudinal development of the mean response profiles (e.g. differences in growth rates between successive times).

- inferences made with each of the separate analyses are not independent, and it is not always clear how they should be combined.

The best option is to make use of linear mixed effects models, **lme**. These are capable of dealing with a wide variety of nested and pseudoreplicated data, as explained below.

Mixed effects models: an introduction

The aim is to work through some very simple examples to see exactly how **lme** differs from **aov** in terms of the output it produces. We start with an example with a single categorical explanatory variable, factory (a factor called fac with five levels), from which we have measures of total annual costs per employee, y:

```
attach(lmedata1)
names(lmedata1)
```

```
[1] "y" "fac"
```

```
fac<-factor(fac)
plot(fac,y)
```

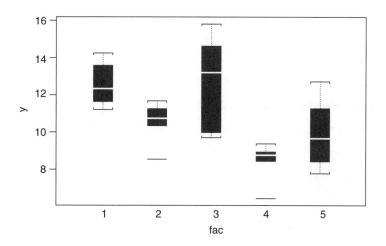

There is considerable variation in the means, but the key feature of the data is that the variance differs substantially from level to level. The overall mean and standard deviation are

```
mean(y)
```

```
[1] 10.85089
```

```
sqrt(var(y))
```

```
[1] 2.247646
```

We start by carrying out the simplest possible analysis of these data, a one-way Anova with Normal errors and constant variance:

```
model1<-aov(y~fac)
summary(model1)
```

```
           Df    Sum of Sq     Mean Sq     F Value          Pr(F)
     fac    4     79.24995    19.81249    7.364627    0.0004606644
Residuals   25    67.25557     2.69022
```

Because there is equal replication within each factor level, we can calculate the *variance compo-nents* directly from the Anova table (see p. 361). The residual variance is obtained directly as the error mean square 2.690 22, and the variance within each factor level is just

(19.81249-2.69022)/6

[1] 2.853712

Make sure you understand where all of these output figures come from, then interpretation of the mixed effects model will be straightforward. If you don't do this, the mixed effects output will seem totally baffling.

Much the most difficult thing about mixed effects modelling is learning how to specify the model formulas for the fixed and random effects correctly. In this case there is no fixed effect; we need only to estimate the overall mean (10.850 89) which, in terms of the model, is the intercept (parameter 1). So the fixed effect, which comes first in the list, is y ~ 1. The random effects are *differences in the intercepts* of the different factories. We are interested principally in the distribution (specifically the variance) of these differences, rather than in the sizes of the differences (hence the use of random effects rather than fixed effects). To achieve this, the random effects model formula is random=~1 | fac, which is read as 'the intercepts given the factor levels'. Note that for the random effects you need to write random= in front of the model formula. Note also, that the model formula has no response variable; there is nothing to the left of the tilde ~. The vertical bar | is read as 'given'. Now we can analyse the same data as a mixed effects model using **lme** instead of **aov** like this:

model2<-lme(y~1,random=~1|fac)

We shall go through the output line by line, to see what everything means:

summary(model2)

```
Linear mixed-effects model fit by REML
 Data: NULL
        AIC        BIC      logLik
   128.3855   132.4874   -61.19274
```

When you are doing a mixed effects model in earnest, you are typically most interested in the random effects (if it was only the fixed effects you wanted, you would save yourself the trouble, and carry out a straightforward analysis of variance). The table for random effects shows their structure (the model formula ~1 | fac), followed by the variance components, expressed as standard deviations:

```
Random effects:
 Formula: ~ 1 | fac
         (Intercept)     Residual
StdDev:     1.689293      1.64019
```

We should check that these are the same values that we obtained earlier, by hand. We need to square the standard deviations to get the variances:

1.64019 ^2

[1] 2.690223

1.689293 ^2

[1] 2.853711

so that's okay. The fixed effects appear after the random effects. In the present example, the only fixed effect is the overall mean, 10.850 89 (see above). Next to this fixed effect size are its standard error (0.812 66), degrees of freedom, and the *t* values and *p* values for a two-tailed test of the null hypothesis that the value of the fixed effect is equal to zero (which, of course, we know it isn't):

```
Fixed effects: y ~ 1
                Value   Std.Error  DF  t-value   p-value
(Intercept) 10.85089     0.81266  25 13.35231   <.0001
```

Next comes a list of the residuals, showing their range and quartiles:

```
Standardized Within-Group Residuals:
      Min          Q1          Med          Q3        Max
-1.700859 -0.4972231 -0.009183497 0.5999522 2.024215
```

Finally, there is a summary of the total number of observations, and the number of groups into which they were divided:

```
Number of Observations: 30
Number of Groups: 5
```

It looks as if there is about as much variation *within* factor levels ($\sigma^2 = 2.69$) as there is *between* the means of the different factor levels ($\sigma_a^2 = 2.69$). At this stage, it is useful to ask whether the mixed effects model is an improvement over the null model containing only variation around the grand mean. We fit a new model using **update**, leaving out the different factor levels, and replacing the random effects by random=~1:

```
model3<-update(model2,random=~1)
anova(model2,model3)
```

The output shows the Akaike information criterion, $-2 \log \text{Lik} + 2p$ (see p. 315), the Bayesian information criterion

$$\text{BIC} = -2 \log \text{Likelihood} + p \log n$$

and the log likelihood for the simpler model 3 and the more complex model 2:

	Model	df	AIC	BIC	logLik
model2	1	3	128.3855	132.4874	-61.19274
model3	2	3	138.6729	142.7748	-66.33644

Because the AIC and BIC are *lower*, we prefer the more complicated model 2. A significant amount of the total variation in costs is explained by differences between the factories.

In summary, the methods of model fitting and model simplification are exactly the same in mixed effects and analysis of variance. All that has changed is that we are interested in the variance explained by the random effects, rather than their effect sizes. This information appears (as standard deviations rather than variances) in the middle of the output table, under the heading Random effects.

Nested designs in mixed effects models

This example involves a study of six different machines and five different operators. The size of the metal component (in mm) produced by the machine on each run is recorded. We are interested in the following questions:

- do the six machines differ?
- do the five operators differ ?
- is there an interaction between operators and machines?

```
attach(lmedata2)
names(lmedata2)
```

```
[1]  "y"            "machine"   "operator"
```

```
interaction.plot(machine,operator,y)
```

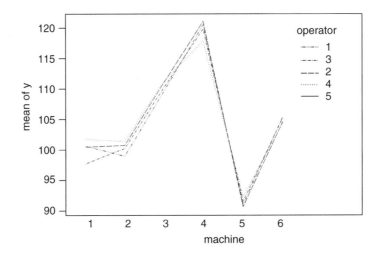

In answer to our three questions, it appears there is a big difference between machines (especially between machines 4 and 5), there may be differences between operators, but there is little obvious evidence of any interaction (the five lines are roughly *parallel* with each other). We begin with a familiar nested analysis of variance using **aov** (see p. 368):

```
operator<-factor(operator)
machine<-factor(machine)
model1<-aov(y~machine/operator)
summary(model1)
```

	Df	Sum of Sq	Mean Sq	F Value	Pr(F)
machine	5	11841.63	2368.327	797.0344	0.0000000
operator %in% machine	24	133.47	5.561	1.8716	0.0146613
Residuals	120	356.57	2.971		

The variance components analysis can be done by hand, because the analysis is balanced: for machine/operator we have $\sigma_a^2 = (5.561 - 2.971)/5 = 0.518$, and for machine we have $\sigma_b^2 = (2368.327 - 5.561)/25 = 94.51$. Now we repeat the analysis using a mixed effects model, **lme**. We are not interested in the mean values for different operators or machines, so we have fixed = y ~ 1. The random effects cause changes in the intercept (~1) and operator is nested within machine, so we put

model2<-lme(y~1,random =~1 | machine/operator)
summary(model2)

Working through the output step by step, we find the AIC, BIC and log likelihood:

```
Linear mixed-effects model fit by REML
 Data: NULL
        AIC       BIC      logLik
   646.5693  658.5851  -319.2846
```

Because the model is nested, the random effects are presented in a new way:

```
Random effects:
 Formula:  ~ 1 | machine
         (Intercept)
 StdDev:    9.721657
```

The variation from machine to machine has a standard deviation of 9.92, so to compare the result with our familiar linear model we need to square this to obtain the variance component, $\sigma_b^2 = 9.721\,657^2 = 94.51061$ (as we found earlier).

```
 Formula:  ~ 1 | operator %in% machine
         (Intercept) Residual
 StdDev:   0.7197165 1.723782
```

Operators within machines vary far less; the variance component is $\sigma_a^2 = 0.719\,7165^2 = 0.517\,9918$ (again, as we found earlier). The fixed effect is just the overall mean:

```
Fixed effects: y ~ 1
                 Value Std.Error  DF  t-value p-value
(Intercept) 104.3221   3.973518 120 26.25434  <.0001
```

```
Standardized Within-Group Residuals:
       Min          Q1        Med         Q3        Max
 -2.408289  -0.6667915  -0.1278863  0.4851294  2.498192
```

```
Number of Observations: 150
Number of Groups:
 machine operator  %in% machine
    6                   30
```

Because the operators do not appear to vary much from one another, it is worth trying model simplification to see if a model with only machine as a random effect is just as good as a description of the data:

model3<-lme(y~1,random=~1|machine)

Then we compare the two models in the usual way, using **anova**:

anova(model2,model3)

```
        Model df      AIC      BIC    logLik   Test L.Ratio p-value
model2      1  4 646.5693 658.5851 -319.2846
model3      2  3 649.0582 658.0701 -321.5291 1 vs 2 4.488937  0.0341
```

The more complicated model 2 with random effects for operators as well as machines is significantly better than the simpler model containing only a random effect for differences between machines ($p = 0.0341$).

Models with both fixed effects and random effects

Mixed effects models only come into their own once we have a mixture of fixed effects and random effects. If all the explanatory variables are random effects, we might as well use *variance components analysis* (p. 361), as the last two examples showed. In this example we have a continuous explanatory variable as our fixed effect. The regression could be fitted separately for each level of the categorical variable (using Ancova) but this would involve the estimation of far too many parameters (a slope and an intercept for every level of the factor). Instead, we might allow the slope or the intercept, or both, to be estimated as random effects:

```
attach(lmedata3)
names(lmedata3)

[1]    "Weight"     "Age"         "Genotype"

plot(Age,Weight,type="n")
for (i in 1:60)
    points(Age[Genotype==Genotype[i]],Weight[Genotype==Genotype[i]],
    pch=ceiling(i/10))
```

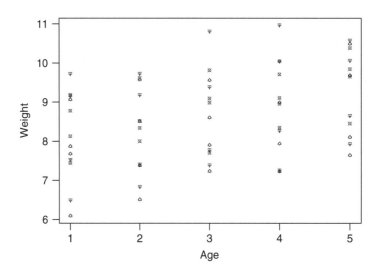

There is a clear upward trend in weight with age, but there is too much overlap of the points to see what is really going on. In a case like this, it is much better to use panel plots:

```
xyplot(Weight~Age|Genotype,pch=16)
```

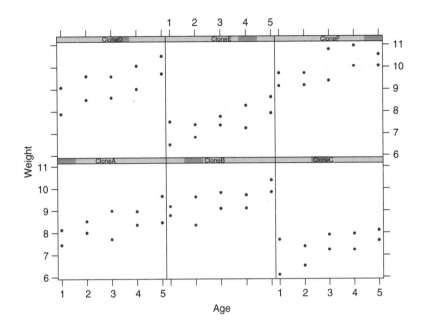

There is a clear upward trend for each genotype, but the slope seems to differ from genotype to genotype. Here is the simple analysis of covariance:

```
model1<-lm(Weight~Age*Genotype)
summary(model1)

Coefficients:
                   Value    Std. Error    t value    Pr(>|t|)
 (Intercept)      7.7558       0.1573    49.3097      0.0000
        Age       0.2996       0.0474     6.3171      0.0000
   Genotype1      0.5201       0.2724     1.9090      0.0623
   Genotype2     -0.4896       0.1573    -3.1127      0.0031
   Genotype3      0.1394       0.1112     1.2538      0.2160
   Genotype4     -0.2257       0.0861    -2.6202      0.0117
   Genotype5      0.2709       0.0703     3.8506      0.0003
AgeGenotype1     -0.0121       0.0821    -0.1468      0.8839
AgeGenotype2     -0.0065       0.0474    -0.1377      0.8911
AgeGenotype3      0.0243       0.0335     0.7243      0.4724
AgeGenotype4      0.0044       0.0260     0.1699      0.8658
AgeGenotype5     -0.0036       0.0212    -0.1704      0.8654

Residual standard error: 0.5195 on 48 degrees of freedom
Multiple R-Squared: 0.8357
```

This indicates that contrary to our initial impression, there are no significant interaction terms, so a model with a common slope should be sufficient. But there are highly significant differences between the intercepts for different genotypes. Our first mixed effects model therefore has the intercept (~1) as a random effect and age as a fixed effect:

```
model2<-lme(Weight~Age,random=~1|Genotype)
summary(model2)

Linear mixed-effects model fit by REML
 Data: NULL
       AIC        BIC      logLik
  119.4781   127.7198  -55.73903

Random effects:
Formula: ~ 1 | Genotype
        (Intercept)      Residual
StdDev:   1.036537     0.4975872

Fixed effects: Weight ~ Age
              Value Std.Error DF  t-value  p-value
(Intercept) 7.755760 0.4491819 53 17.26641  <.0001
       Age 0.299583 0.0454233 53  6.59535   <.0001
 Correlation:
    (Intr)
Age -0.303

Standardized Within-Group Residuals:
       Min         Q1        Med         Q3         Max
 -1.521588 -0.8940179 0.04649235  0.8295459   1.767093

Number of Observations: 60
Number of Groups: 6
```

Note that the slope and intercept of the graph (the fixed effects for Age and Intercept) are exactly the same as in the analysis of covariance (0.2996 and 7.7558 respectively). The variance due to different intercepts (Genotypes) $\sigma_a^2 = 1.036\,537^2 = 1.074\,409$ is large compared with the residual variance $\sigma^2 = 0.497\,5872^2 = 0.247\,593$.

Let's fit a model with random slopes as well as random intercepts. This means that we need to modify the random model formula to contain Age as well as Genotype:

```
model3<-lme(Weight~Age,random=~Age|Genotype)
summary(model3)

Linear mixed-effects model fit by REML
 Data: NULL
       AIC       BIC      logLik
  123.4703   135.833  -55.73515

Random effects:
 Formula:  ~ Age | Genotype
 Structure: General positive-definite
```

```
                      StdDev            Corr
(Intercept) 1.023384289         (Inter
        Age 0.004386592         1
   Residual 0.497549975

Fixed effects: Weight ~ Age
                Value Std.Error DF  t-value p-value
(Intercept) 7.755760 0.4441230 53 17.46309  <.0001
        Age 0.299583 0.0454552 53  6.59073  <.0001
 Correlation:
     (Intr)
Age  -0.27

Standardized Within-Group Residuals:
      Min        Q1        Med        Q3       Max
-1.513056 -0.898523 0.04720394 0.8240707 1.756553

Number of Observations: 60
Number of Groups: 6
```

This shows that the variance attributable to differences in slope (Age s.d. $= 0.004\,38$) is tiny compared with the variance attributable to differences in intercept (Intercept s.d. $= 1.023\,38$). Comparison of models 2 and 3 shows that the simpler model 2 with a single, common slope is much to be preferred:

anova(model3,model2)

```
       Model df     AIC      BIC    logLik   Test      L.Ratio p-value
model3     1  6 123.4703 135.8330 -55.73515
model2     2  4 119.4781 127.7198 -55.73903 1 vs 2 0.007755208  0.9961
```

Model 2 has the lower AIC, and the more complex model 3 has no greater explanatory power ($p = 0.9961$). Note that the mixed effects model uses only 4 degrees of freedom, compared with the 7 d.f. used by analysis of covariance with common slope and six different intercepts.

Analysis of longitudinal data

The key feature of longitudinal data is that individuals are measured repeatedly through time. This would represent temporal pseudoreplication if the data were used uncritically in regression or Anova. The set of observations on one individual subject will tend to be intercorrelated and this correlation needs to be taken into account in carrying out the analysis. The alternative is a cross-sectional study, with all the data gathered at a single point in time, in which each individual contributes a single data point. The advantage of longitudinal studies is that they are capable of separating *age effects* from *cohort effects*; these are inextricably confounded in cross-sectional studies. This is particularly important when differences between years mean that cohorts originating at different times experience different conditions, so that individuals of the same age in different cohorts would be expected to differ. There are two extreme cases in longitudinal studies:

- a few measurements on a large number of individuals
- a large number of measurements on a few individuals

In the first case it is difficult to fit an accurate model for change within individuals, but treatment effects are likely to be tested effectively. In the second case it is possible to get an accurate model of the way that individuals change though time, but there is less power for testing the significance of treatment effects, especially if variation from individual to individual is large. In the first case, less attention will be paid to estimating the correlation structure, while in the second case the covariance model will be the principal focus of attention. The aims are:

- to estimate the average time course of a process

- to characterise the degree of heterogeneity from individual to individual in the rate of the process

- to identify the factors associated with both of these, including possible cohort effects

The response is not the individual measurement, but the *sequence of measurements* on an individual subject. This enables us to distinguish between age effects and year effects; see Diggle *et al.* (1994) for details.

Crossover trials

Crossover trials involve administering treatments to individuals in randomised sequences. Different individuals get their treatments in different orders. In a drug trial, for instance, there may be three treatments: a placebo, drug A and drug B. There are six possible sequences in which they can be administered: PAB, PBA, ABP, APB, BAP and BPA. Each of these should be replicated on different subjects. The order of administering treatments is important because of the possibility of *carryover effects*. Further difficulties arise if there are treatment by time interactions (see p. 576).

Derived variable analysis

The idea here is to get rid of the pseudoreplication by reducing the repeated measures into a set of summary statistics (slopes, intercepts, means), then *analyse these summary statistics* using standard parametric techniques like Anova or regression. The technique is not strong when the values of the explanatory variables change through time.

Derived variable analysis makes most sense when it is based on the parameters of scientifically interpretable non-linear models from each time sequence. However, the best model from a theoretical perspective may not be the best model from the statistical viewpoint.

Transitional models focus on the conditional expectation of Y_{ij} given past outcomes Y_{ij-1}, Y_{ij-2}, Y_{ij-3}, ..., Y_{i1}. They combine the assumptions about the dependence of Y on the explanatory variables and the correlation among repeated measurements of Y into a single equation.

Consider the three models

$$Y_{ij} = \beta_0 + \beta x_{i1} + \beta(x_{ij} - x_{i1}) + \varepsilon_{ij}$$

$$Y_{ij} = \beta_{0i} + \beta(x_{ij} - x_{i1}) + \varepsilon_{ij}$$

$$Y_{ij} = \beta_0 + \beta_C x_{i1} + \beta_L(x_{ij} - x_{i1}) + \varepsilon_{ij}$$

In the first and third cases, the same intercept is assumed for all individuals; in the second case, different individuals have different intercepts. In the first case the coefficient β for the cross-sectional effect is the same as the coefficient β for the longitudinal effect (a strong assumption that will often be wrong in practice). In the third case there are separate coefficients β_C for the cross-sectional effect and β_L the longitudinal effect. In the last model we can test whether the cross-sectional effect of a given explanatory variable is the same as the longitudinal effect. The third model is the best when covariates change through time and vary across subjects at the outset.

The correlation structure is explored by first removing the influence of the explanatory variables and calculating the residuals

$$r_{ij} = y_{ij} - x_{ij}'\hat{\beta}$$

then produce scatterplots of r_{ij} against r_{ik} for all $j < k = 1, 2, \ldots, n$. When the residuals have constant mean and variance and when $\text{Corr}(y_{ij}, y_{ik})$ depends only on $|t_{ij} - t_{ik}|$ the process is said to be weakly stationary (Box and Jenkins 1970).

Model fitting is by weighted least squares, where the weights are the reciprocal of the variance. Computing the variance matrix, however, requires knowledge of the complete correlation structure of the data. We shall never know this in practice, so it important to understand how much loss of efficiency might result from using different weights:

$$\hat{\beta}(V_0) = (\mathbf{X}'\mathbf{V}^{-1}\mathbf{X})^{-1}\mathbf{X}'\mathbf{V}^{-1}\mathbf{y}$$

You should compare this weighted least squares estimator with the standard least squares estimator described on p. 437. The working covariance matrix is often substantially simpler than the true covariance matrix $\mathbf{V}$ (e.g. a block diagonal $\mathbf{W}^{-1}$ with the non-zero elements of the form $\exp(-c|t_j - t_k|)$ where c is a positive constant chosen to reflect the rate of decay anticipated for the actual covariances of the data. According to Diggle *et al.* (1994), more elaborate choices for $\mathbf{W}^{-1}$ are often unnecessary.

A full model of treatments is fitted first. This is then crossed with time. There are three qualitatively different sources of random variation:

- **random effects**: experimental units differ (e.g. genotype, history, size, physiological condition) so that there are intrinsically high responders and other low responders

- **serial correlation**: there may be time-varying stochastic variation within a unit (e.g. physiology, ecological succession) so that correlation depends on the time separation of pairs of measurements on the same individual, with correlation weakening with the passage of time

- **measurement error**: the assay technique may introduce an element of correlation (e.g. shared bioassay of closely spaced samples, different assay of later specimens).

These issues are dealt with by expanding the error term, ε_{ij}, to include random effects, serial correlation and measurement error like this:

$$\varepsilon_{ij} = \mathbf{d}_{ij}'\mathbf{U}_i + W_i(t_{ij}) + Z_{ij}$$

The Z_{ij} are a set of N mutually independent Normal random variables, each with zero mean and variance τ^2. The $\mathbf{U}_i$ are a set of m mutually independent r-element Normal random vectors with zero mean and covariance matrix (say, $\mathbf{G}$) and the $\mathbf{d}_{ij}$ are r-element vectors of explanatory variables attached to the individual measurements. The $W_i(t_{ij})$ are sampled from m independent copies of a stationary Gaussian process with mean zero, variance σ^2 and correlation function $\rho(u)$. The additive structure may be improved by transformation (e.g. by log transformation if the errors are thought to be multiplicative).

The covariance structure is estimated from the residuals after fitting the maximal model (crossed by time if possible). Time series plots, scatterplot matrices at different lags, and empirical variogram plots are all potentially useful (see p. 726). Alternatively, the mean response surface can be described using non-parametric smoothers (see p. 595), with higher weights being give to measurements that were taken closer together in time.

Mixed effects models for time series data

The data on the growth of 48 pigs over a 9 week period come from McCloud in Diggle *et al.* (1994). The data are presented in 9 columns and 48 rows:

```
attach(pig)
names(pig)
```

```
[1]  "Pig"  "t1"  "t2"  "t3"  "t4"  "t5"  "t6"  "t7"  "t8"  "t9"
```

We need to make a grouped data structure. First, we put all of the weight data into a single vector:

```
pig.wt<-c(t1,t2,t3,t4,t5,t6,t7,t8,t9)
```

Next we create a vector for pig identity which is the number 1 to 48 repeated 9 times:

```
pig.id<-c(rep(Pig,9))
```

Now we generate a vector for the week number: 48 ones then 48 twos etc.:

```
pig.time<-c(rep(c(1:9),each=48))
```

Next make them into a data frame:

```
pigs<-data.frame(cbind(pig.time,pig.id,pig.wt))
```

Finally convert the data frame into a grouped data object using **groupedData**:

```
pig.growth<-groupedData(pig.wt~pig.time|pig.id,data=pigs)
```

The formula says that weight is to be modelled as a function of time, with the repeated measurements nested within subjects as indicated by the variable called pig.id. To plot all of the pig time series on the same axes, suppress the panel plot by setting outer= ~1:

```
plot(pig.growth,outer=~1)
```

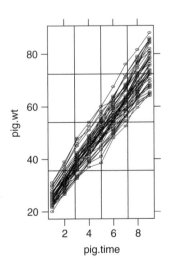

This plot is useful in showing three things:

- growth of the pigs is reasonably linear

- the pigs that were largest at the beginning were largest at the end, and vice versa, so the lines do not cross over one another; this is known as 'tracking')

- the variance in weight is larger at the end of the period than at the beginning; this indicates variation in growth rates from animal to animal

To see the trajectories for each pig more clearly, set outer=~pig.id or just use the default:

    plot(pig.growth)

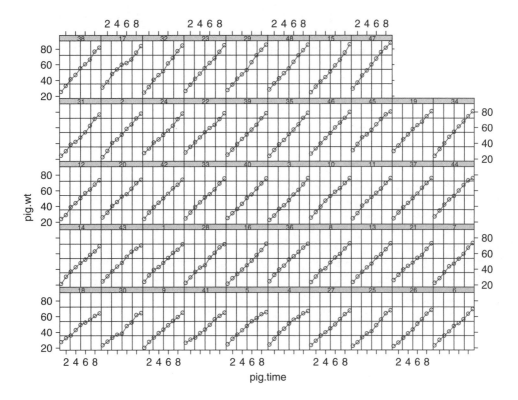

This shows that the linearity assumption is reasonable for most of the animals, although there is a hint that the growth of pig 5 and pig 4 is slowing down towards weeks 8 and 9. The panels are ordered such that the slopes of the growth curves increase from bottom left to top right.

Once the **groupedData** structure has been defined, fitting the linear mixed effects model **lme** could not be more straightforward:

    model<-lme(pig.growth)

The **lme** model produces a summary like this:

    summary(model)

```
Linear mixed-effects model fit by REML
 Data: pig.growth
        AIC      BIC     logLik
   1752.871 1777.254 -870.4356

Random effects:
 Formula: ~ pig.time | pig.id
 Structure: General positive-definite
              StdDev    Corr
(Intercept) 2.6431920 (Inter
   pig.time 0.6164379 -0.063
   Residual 1.2636572

Fixed effects: pig.wt ~ pig.time
              Value Std.Error  DF  t-value p-value
(Intercept) 19.35561 0.4038676 383 47.92564  <.0001
   pig.time  6.20990 0.0920382 383 67.47085  <.0001

 Correlation:
        (Intr)
pig.time -0.133

Standardized Within-Group Residuals:
      Min        Q1       Med        Q3       Max
 -3.620188 -0.5473595 0.01503617 0.5485512 2.993914

Number of Observations: 432
Number of Groups: 48
```

It shows the overall intercept (19.355 61) and slope (6.2099), and the standard deviations of the random effects: the 48 slopes (0.616) and intercepts (2.643) in this case.

To inspect the intercepts and slopes for each of the 48 individual pigs, we look at the coefficients of the model:

coef(model)

```
      (Intercept) pig.time
18      22.65132 4.832617
30      16.68013 5.278909
 9      16.16319 5.473553
41      19.34825 5.120788
17      26.11287 6.276209
...
...
...
32      17.25229 7.369769
23      20.11090 7.050581
29      19.56766 7.283897
48      21.06462 7.121540
15      16.61926 7.288607
47      22.86369 7.363328
```

Now we plot the model, which shows that the standardised residuals are well behaved:

```
plot(model)
```

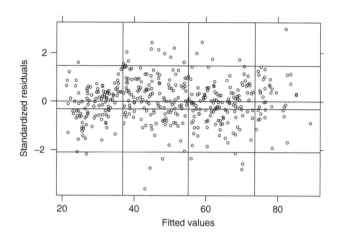

For comparison, analysis of covariance with 96 estimated parameters (48 slopes and 48 intercepts) looks like this:

```
time<-as.numeric(pig.time)
model.ancova<-lm(pig.wt~pig.id*time)
summary(model.ancova)
```

```
Residual standard error: 1.264 on 336 degrees of freedom
```

The analysis of covariance gives exactly the same overall intercept and slope, but is highly over-parameterised compared with the mixed effects model. The residual standard errors are virtually identical (see `Residual` in the random effects table, above).

Treatment effects with time series data

We shall work through an example that involves time series data on the height of plants grown in individual containers with one of two soil types measured over a series of days. The fixed effect is a categorical variable (soil) with two levels (sand and loam). The random effects are time nested within individuals (subjects). The trick with mixed effects modelling is in understanding how to classify all the variables in the data frame:

```
attach(soil)
names(soil)
```

```
[1]  "height"   "days"   "subject"   "mix"
```

There are four variables in the data frame called soil. They are as follows:

- *height* is the response variable (we want to understand the causes of variation in height in terms of the three explanatory variables)

- *days* is a continuous variable and will go on the *x* axis of the graph

- *subject* is a random effect; it is a categorical variable containing the label of the pot in which the plant was grown and is a unique label for each individual plant, hence the name 'subject'

- *mix* is a fixed effect; it is a categorical variable with two levels showing the experimentally allocated soil type

We shall call the grouped data object *growth* and define it as follows. The first element specifies how the response (*height*) is related to the continuous explanatory variable (*days*), and the way that each graph is to be grouped (one for each individual plant, labelled by *subject*). This is achieved in standard model specification format using | 'grouped by':

height ~ days | subject

The second element specifies the fixed effects (*mix*); these are called 'outer' in the jargon of mixed effects modelling:

outer = ~ mix

The elements are combined like this in the **groupedData** function:

growth<-groupedData(height~days|subject,outer=~mix)

We can now do some quite sophisticated tasks extremely simply. For instance

plot(growth)

produces a separate panel plot of *height* against *days* for each individual plant.

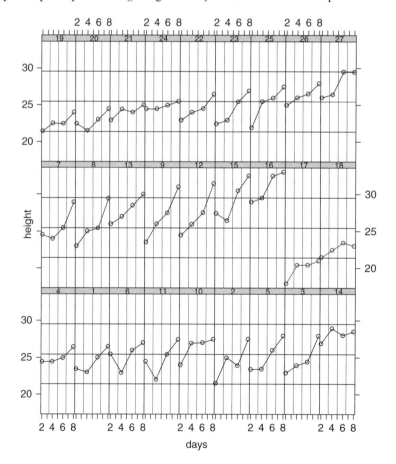

The graphs have been produced by 'joining the dots' for each plant separately (type="b"). Note that the subjects have been ordered by their mean heights: the first 16 graphs are for the plants grown on loam, the next 11 graphs are for the plants grown on sand. They show two things:

- different plants have performed differently (as we expected they might)
- there does appear to be an effect of soil mixture (plants on loam are taller)

An alternative way of plotting grouped data is to have separate panels for each level of the fixed effects, one for sand and one for loam in this case. This is achieved by the option outer=T like this:

plot(growth,outer=T)

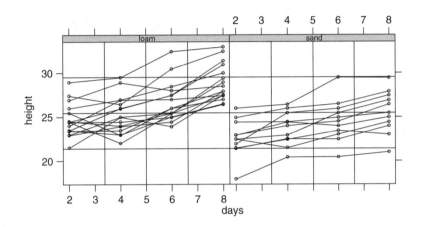

Here the growth curves for all 16 individuals grown on loam are shown in the left panel, and the curves for the 11 individuals grown on sand are in the right panel. It looks as if the slopes are steeper on loam than on sand, but there is little evidence of substantial variation in slope from individual to individual within each treatment (there is very little 'crossing over' in either panel. There is, however, clear evidence of differences in mean initial height between the two soil treatments. Plants were allocated to soil treatments at random initially, so this probably reflects differences in growth before measurements began, 2 days after the plants were moved to their positions in the growth chamber (defined as day 0).

As usual, we begin the modelling by fitting the most complicated model first: this has different slopes for plant growth on each soil type, specified by days*mix:

model1<-lme(fixed=height~days*mix,random= ~days|subject,data=growth)
summary(model1)

```
Random effects:
 Formula: ~ days | subject
 Structure: General positive-definite
               StdDev    Corr
(Intercept) 1.7045889 (Inter
       days 0.1344057 0.125
   Residual 0.9759611
```

The random effects refer to variance introduced by differences between the intercepts of different plants (s.d. = 1.705) and variance introduced by differences in the slopes of the time series (s.d. = 0.1344). The residual s.d. was 0.976.

```
Fixed effects:  height ~ days * mix
                  Value Std.Error DF  t-value p-value
(Intercept)  22.24887 0.4077169 79 54.56941  <.0001
       days   0.59916 0.0501932 79 11.93702  <.0001
        mix  -0.56772 0.4077169 25 -1.39245  0.1760
   days:mix  -0.11612 0.0501932 79 -2.31349  0.0233
```

The fixed effects table is exactly the same as you are used to from linear modelling. The columns show the parameter estimates, their standard errors, degrees of freedom, t tests and p values. We start interpretation by looking at the most complicated term; this is the slope by soil interaction (days:mix). It is significant ($p = 0.0233$), so we should retain it in the model.

```
Number of Observations: 108
Number of Groups: 27
```

Model checking is carried out by diagnostic plots in the usual way:

```
plot(model1)
```

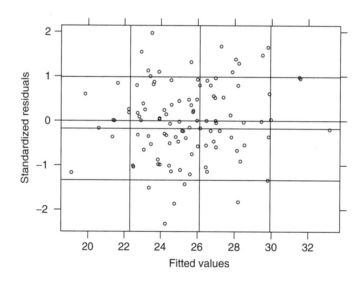

This shows that the residuals are reasonably well behaved (e.g. there is no tendency for the residuals to increase with larger fitted values). There are a great many options for controlling aspects of the output. For example, we might want to look at the residuals for each soil type separately. This is achieved like this:

```
plot(model1, resid(.)~fitted(.) | mix)
```

Note the use of the dot notation to specify the residuals and fitted values to be plotted, and the | symbol to specify the panels to be plotted.

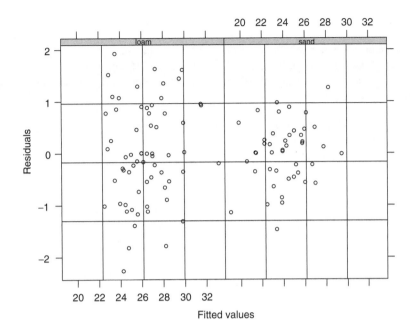

The scatter of the residuals is greater for loam than for sand, but there do not appear to be systematic changes in the residuals on either soil mixture.

If we want them (and it is not obvious that we do) we can get the slopes and intercepts for each of the 27 separate plants using **coef** or, more usefully perhaps, the differences from mean slope and intercept using the random effects function **ranef**:

ranef(model1)

```
     (Intercept)            days
 4  -0.68550571  -0.121539649
 1  -1.38825819  -0.082622992
 6  -0.51494150  -0.102337717
11  -1.18076884  -0.060373870
10   0.15788685  -0.047397651
. . . . . . . . . . . . . .  .
. . . . . . . . . . . .  . .
25   0.50341400   0.112518588
26   1.83784421   0.049671813
27   2.86104304   0.131919853
```

Thus, plant number 4 had an intercept 0.686 lower than average and a slope 0.122 shallower than average, and plant number 27 had an intercept 2.86 higher than average and a slope 0.132 steeper than average. We can plot these very simply:

plot(ranef(model1))

Variance functions

The first thing we might want to take care of is the fact that the variance looked to be bigger for loam than for sand. We can choose from a wide selection of different variance functions; in this case the appropriate specification is **varIdent** (different variances per level of a factor). This enters in the weights option of the model fit:

```
model2<-update(model1,weights=varIdent(form= ~1|mix))
```

Now we can use **anova** in the normal way to compare the two models:

```
anova(model1,model2)
```

```
        Model df      AIC      BIC    logLik    Test   L.Ratio p-value
model1     1  8  404.3475 425.5027 -194.1738
model2     2  9  398.4026 422.2021 -190.2013 1 vs 2 7.944958  0.0048
```

There is strong evidence ($p = 0.0048$) that the variance in height on loam was greater than on sand, hence that model2 is superior to model1.

We might want to specify that the variance is an arbitrary power of the fitted values; this uses the **varPower** function:

```
model3<-update(model1,weights=varPower(form= ~ fitted(.)))
```

```
anova(model1,model3)
```

```
        Model  df      AIC      BIC    logLik    Test   L.Ratio p-value
model1     1   8  404.3475 425.5027 -194.1738
model3     2   9  405.8436 429.6431 -193.9218 1 vs 2 0.5039138  0.4778
```

Thus there is no evidence ($p = 0.478$) that this is an improvement over the initial model, so we should stick with model2.

Temporal autocorrelation

Next you might want to ask questions about the correlation structure of the data. For example, what is the best model for describing the time series structure of the data in terms of the temporal autocorrelation of the heights within individual plants? You might want to specify an AR (autoregressive) or an ARMA (autoregressive moving average) model. Let's test for the presence of autocorrelation of lag 1 by using function **corAR1**:

```
model4<-update(model1, corr=corAR1())
```

```
anova(model1,model4)
```

```
        Model df      AIC      BIC    logLik    Test     L.Ratio p-value
model1     1  8  404.3475 425.5027 -194.1738
model4     2  9  406.3413 430.1409 -194.1707 1 vs 2 0.006197443  0.9373
```

There is absolutely no evidence ($p = 0.937$) of autocorrelation in this case.

Covariance structure of the random effects

Finally, we might want to investigate a different kind of random effects model by altering the structure of the positive-definite matrix structure (**pdMat**). Suppose we wanted to specify independent slope and intercept random effects. This involves the use of a diagonal (**pdDiag**) matrix like this:

```
model5<-update(model1,random=pdDiag(~days))
```

```
anova(model1,model5)
```

```
       Model df      AIC      BIC   logLik   Test   L.Ratio p-value
model1     1  8 404.3475 425.5027 -194.1738
model5     2  7 402.3986 420.9093 -194.1993 1 vs 2 0.0510282  0.8213
```

Again, there is no evidence ($p = 0.82$) that this model is an improvement over model1.

More complex time series structure

This example comes from a study of ovarian cycles in a group of mares, where the question addressed by mixed effects modelling relates to differences in the nature of the variation in cycles from mare to mare.

```
attach(Ovary)
names(Ovary)
```

```
[1]   "Mare"      "Time"        "follicles"
```

```
tsplot(follicles)
```

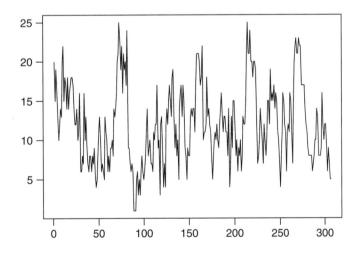

The periodicity in the data is clear; the question is whether (and how) the cycle differs from one mare to another. First we create a grouped data object:

```
ovaries<-groupedData(follicles~Time|Mare)
plot(ovaries)
```

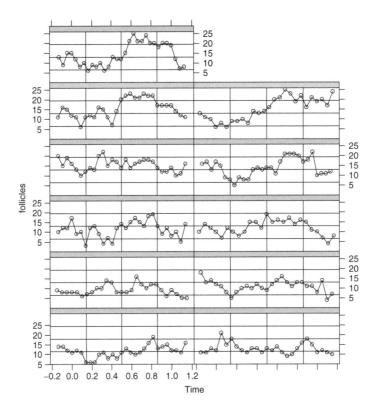

The time data have been scaled so that ovulations occur at times 0 and 1. We want to find the best time series model to describe these data. The fixed effects are to be modelled as a simple trigonometric function (see p. 643)

$$\text{follicles} \sim \sin(2\pi t) + \cos(2\pi t)$$

which looks like this:

```
t<-seq(-0.1,1.1,.01)
plot(t,sin(2*pi*t)+cos(2*pi*t),type="l")
```

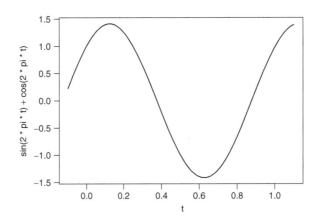

The random effects refer to differences between mares. Rather than estimate several parameters for each individual horse, we intend to estimate the variance in parameter values from mare to mare (a much more parsimonious model). As a first stab at the analysis, we fit a simple linear mixed effects model like this:

```
model1<-lme(follicles~sin(2*pi*Time)+cos(2*pi*Time),data=ovaries)
model1
```

```
Linear mixed-effects model fit by REML
  Data: ovaries
  Log-restricted-likelihood: -805.0166
  Fixed: follicles ~ sin(2*pi*Time) + cos(2*pi*Time)
(Intercept) sin(2 * pi * Time) cos(2 * pi * Time)
   12.18591          -3.296678         -0.8731409

Random effects:
 Formula:  ~ sin(2*pi*Time) + cos(2*pi*Time) | Mare
 Structure: General positive-definite
                      StdDev    Corr
       (Intercept) 3.229273 (Intr) s(2*p*T)
sin(2 * pi * Time) 2.092883 -0.570
cos(2 * pi * Time) 1.067061 -0.801 0.178
          Residual 3.019463

Number of Observations: 308
Number of Groups: 11
```

The fixed effects output shows the three parameters (intercept and two slopes) of the overall fit, which looks like this:

```
plot(t,12.186-3.296678*sin(2*pi*t)-0.8731409*cos(2*pi*t),ylab="follicles",type="l")
```

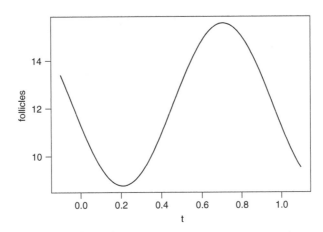

The random effects output shows the variation in these three parameters from mare to mare:

- the intercept has a standard deviation of 3.229

- the sine term has a standard deviation of 2.093

- the cosine term has a standard deviation of 1.067

- residual standard deviation of 3.019

There are several issues to be investigated. What is the best time series model to describe the data, e.g. autoregressive or ARIMA? What is the best structure for the variance–covariance matrices for the random effects, e.g. general positive-definite (as here), diagonal, compound symmetry (see Table 35.1)? The fit of the present model is measured by its AIC:

```
AIC(model1)
```

```
[1] 1630.033
```

Table 35.1 Variance functions for use in mixed effects models

varExp	exponential of a variance covariate
varPower	power of a variance covariate
varConstPower	constant plus power of a variance covariate
varIdent	constant variance(s), allows different variances by factor level
varFixed	fixed weights, determined by a variance covariate
varComb	combination of variance functions

We try a second model with a different variance–covariance structure (**pdDiag**) like this:

```
model2<-lme(follicles~sin(2*pi*Time)+cos(2*pi*Time),data=ovaries,
    random=pdDiag(~sin(2*pi*Time)))
AIC(model1,model2)
```

```
       df      AIC
model1 10 1630.033
model2  6 1638.082
```

The first model has the lower AIC and is therefore preferred.

```
plot(ACF(model1))
```

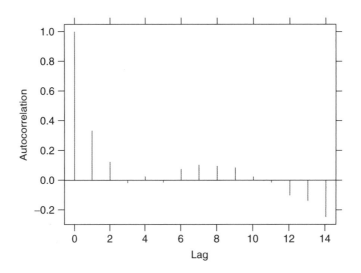

The empirical autocorrelations decline exponentially with increasing lag. They are significant at lag 1 (0.333) and marginally significant at lag 2 (0.124). This suggests an AR(1) model may be reasonable for the within-group correlation. We can update model1 like this:

```
model3<-update(model1,correlation=corAR1( ))
AIC(model1,model3)

        df       AIC
model1 10 1630.033
model3 11 1566.093
```

This is a substantial improvement, so modelling the within-group correlation was much better than our original assumption of independent errors. But have we found the best possible model for the within-group correlation? Instead of an AR model, we can try an ARMA model (autoregressive moving average):

```
model4<-update(model3,correlation=corARMA(p=1,q=1))
AIC(model3,model4)

        df       AIC
model3 11 1566.093
model4 12 1563.418
```

This is a slight improvement, but the mechanistic explanation of the ARMA model is more complex, so it is not clear which model would be selected.

The range of options for specifying correlation structure is shown in Table 35.2.

Different fixed effects models and REML

Generally in mixed effects models we prefer to use restricted maximum likelihood estimation (REML) instead of maximum likelihood (ML), because ML underestimates the sizes of the variance components. The reason for this is that *ML ignores the degrees of freedom used up by the fixed effects*. A general feature of REML estimation of variance components is the fact that the degrees of freedom for the fixed effects are taken into account. For instance, the ML estimate of σ^2 for a single sample based on n observations is SSE/n (as we saw on p. 120). In contrast, the REML estimate takes

Table 35.2 Correlation structures available in mixed effects models

Time series	
corAR1	autoregressive process of order 1
corARMA	autoregressive moving average process
corBand	banded correlation structure
corCAR1	continuous autoregressive process (AR(1) process for a continuous time covariate)
corCompSymm	compound symmetry structure corresponding to a constant correlation
Spatial	
corExp	exponential spatial correlation
corGaus	Gaussian spatial correlation
corLin	linear spatial correlation
corRatio	rational quadratic spatial correlation
corSpher	spherical spatial correlation
corStrat	stratified correlation structure
corSymm	general correlation matrix, with no additional structure

account of the fact that the overall mean used in calculating $SSE = \sum (y - \overline{y})^2$ was a fixed effect estimated from the data, so the REML estimate of σ^2 is $SSE/(n - 1)$.

The best linear unbiased predictor

In a model containing only fixed effects, the best estimate of *effect size* is simply

$$a_i = \overline{y}_i - \mu$$

In a model with random effects, however, there is *shrinkage*; the extent of the shrinkage depends on the size of the between-group variance, σ_a^2, and the correlation between the pseudoreplicates that is introduced by this. The best linear unbiased predictor (BLUP) is smaller than the difference between the means, and is given by

$$a_i = (\overline{y}_i - \mu) \left(\frac{\sigma_a^2}{\sigma_a^2 + \sigma^2/n} \right)$$

which of course is less than $\overline{y}_i - \mu$.

Comparing mixed effects models with different fixed effects

It makes no sense to use **anova** to compare REML models that have *different fixed effects structures*. We were all right in the example on p. 690, because model1 to model5 all had the *same* fixed effects structure days*mix. During model simplification, however, you often want to compare models with different fixed effects structures. Before doing this, you must change from REML to maximum likelihood (ML). Suppose we wanted to use **anova** to compare model1 with a model containing only main effects for *days* and *mix* as days+mix in order to assess the significance of the interaction term. We would proceed like this:

model6<-lme(fixed=height~days*mix,random=~days|subject,data=growth,method="ML")
model7<-lme(fixed=height~days+mix,random=~days|subject,data=growth,method="ML")

anova(model6,model7)

```
        Model df      AIC      BIC    logLik   Test  L.Ratio p-value
model6      1  8 395.5823 417.0394 -189.7912
model7      2  7 398.8201 417.5950 -192.4101 1 vs 2 5.237821  0.0221
```

There is strong evidence ($p = 0.0221$) for retaining the interaction term in the model (i.e. the slope of height versus days is significantly different on the two soil types).

Analysing repeated measures data the wrong way

We have a large field experiment on grass yields from pastures treated with two pesticides (insecticide and molluscicide), in factorial combinations. There are five replicates (blocks), and data were gathered over six successive years from each plot:

attach(repeated)
names(repeated)

```
[1] "Block"  "Time"  "Insect"   "Mollusc"   "Grass"
```

This is what the data look like; **coplot** draws Grass as a function of Time, *given* Block:

coplot(Grass ~ Time | Block)

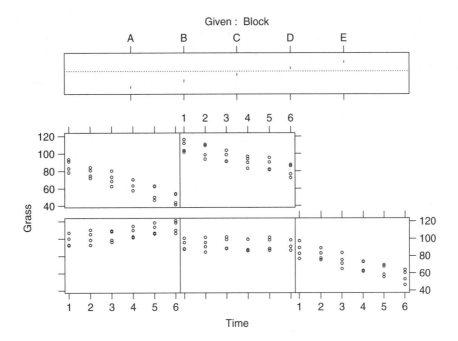

The most obvious way of doing the **wrong** analysis is to fit Block and the 2×2 factorial Insect*Mollusc to the whole data set:

```
model<-aov(Grass~Insect*Mollusc+Error(Block))
```

```
summary(model)
```

```
Error: Block
          Df  Sum Sq Mean Sq F value Pr(>F)
Residuals  4 25905.9  6476.5
```

```
Error: Within
              Df Sum Sq Mean Sq F value     Pr(>F)
Insect         1 3826.1  3826.1 43.9838 1.218e-009 ***
Mollusc        1  314.6   314.6  3.6169    0.05976 .
Insect:Mollusc 1    0.1     0.1  0.0010    0.97429
Residuals    112 9742.6    87.0
```

The conclusion is clear: there is no interaction, but insect exclusion has a highly significant main effect on grass yield, and mollusc exclusion has a close to significant effect. However, you can see at once that the analysis is wrong, because there are far too many degrees of freedom for error (d.f. = 112). There are only five replicates in the whole experiment, so this clearly cannot be right. Both effects are in the expected direction (herbivore exclusion increases grass yield).

```
tapply(Grass,list(Insect,Mollusc), mean)
```

```
             Absent   Present
Sprayed    93.71961  90.42613
Unsprayed  82.37145  79.18800
```

A sensible way to see what is going on is to do a regression of Grass against Time separately for each treatment in each block, to check whether there are temporal trends, and if so, whether the temporal trends are the same in each block. This is an analysis of covariance (Chapter 16) with different slopes (Time) estimated for every combination of Block:Insect:Mollusc:

```
model<-aov(Grass~Block*Insect*Mollusc*Time)
```

```
summary(model)
```

	Df	Sum of Sq	Mean Sq	F Value	Pr(F)
Block	4	25905.93	6476.483	2288.045	0.0000000
Insect	1	3826.05	3826.051	1351.687	0.0000000
Mollusc	1	314.63	314.629	111.154	0.0000000
Time	1	3555.40	3555.400	1256.070	0.0000000
Block:Insect	4	1.59	0.397	0.140	0.9667569
Block:Mollusc	4	10.74	2.685	0.949	0.4404798
Insect:Mollusc	1	0.09	0.091	0.032	0.8583185
Block:Time	4	5898.01	1474.502	520.920	0.0000000
Insect:Time	1	0.39	0.389	0.137	0.7118010
Mollusc:Time	1	13.90	13.900	4.911	0.0295350
Block:Insect:Mollusc	4	11.21	2.802	0.990	0.4179119
Block:Insect:Time	4	3.65	0.914	0.323	0.8619207
Block:Mollusc:Time	4	2.33	0.583	0.206	0.9343920
Insect:Mollusc:Time	1	16.38	16.381	5.787	0.0184497
Block:Insect:Mollusc:Time	4	2.58	0.644	0.227	0.9222823
Residuals	80	226.45	2.831		

This shows some interesting features of the data. Starting with the highest-order interaction and working upwards, we see that there is a significant interaction between Insect, Mollusc and Time. That is to say, the slope of the graph of grass against time is significantly different for different combinations of insecticide and molluscicide. The effect is not massively significant ($p = 0.018$) but it *is* there (another component of this effect appears as a Mollusc by Time interaction). There is an extremely significant interaction between Block and Time, with different slopes in different Blocks (as we saw using **coplot**, earlier). At this stage, we need to consider whether the suggestion of an Insect by Mollusc by Time interaction is worth following up. It is plausible that insects had more impact on grass yield in one kind of year, and molluscs in another. The problem is that this analysis does not take account of the fact that the measurements through time are correlated because they were taken from the same location (i.e. a plot within a block).

One simple way around this is to carry out separate analyses of variance for each time period. We use the **subset** directive to restrict the analysis, and put the whole thing in a loop for time $i = 1$ to 6:

```
for (i in 1:6) print(summary(aov(Grass~Insect*Mollusc+Error(Block),subset=(Time==i))))
```

```
Error: Block
          Df  Sum Sq Mean Sq F value Pr(>F)
Residuals  4 1321.30  330.32

Error: Within
          Df Sum Sq Mean Sq  F value     Pr(>F)
Insect     1 652.50  652.50 257.4725 1.794e-009 ***
Mollusc    1  77.76   77.76  30.6823  0.0001280 ***
```

```
Insect:Mollusc  1    8.47      8.47    3.3435   0.0924206  .
Residuals       12   30.41      2.53
```

Error: Block

	Df	Sum Sq	Mean Sq	F value	Pr(>F)
Residuals	4	2119.29	529.82		

Error: Within

	Df	Sum Sq	Mean Sq	F value	Pr(>F)	
Insect	1	620.38	620.38	253.1254	1.978e-009	***
Mollusc	1	94.88	94.88	38.7122	4.438e-005	***
Insect:Mollusc	1	0.01	0.01	0.0034	0.9545	
Residuals	12	29.41	2.45			

Error: Block

	Df	Sum Sq	Mean Sq	F value	Pr(>F)
Residuals	4	3308.8	827.2		

Error: Within

	Df	Sum Sq	Mean Sq	F value	Pr(>F)	
Insect	1	628.26	628.26	183.8736	1.225e-008	***
Mollusc	1	71.41	71.41	20.9002	0.0006416	***
Insect:Mollusc	1	4.88	4.88	1.4273	0.2552897	
Residuals	12	41.00	3.42			

Error: Block

	Df	Sum Sq	Mean Sq	F value	Pr(>F)
Residuals	4	5063.6	1265.9		

Error: Within

	Df	Sum Sq	Mean Sq	F value	Pr(>F)	
Insect	1	550.43	550.43	194.9970	8.784e-009	***
Mollusc	1	26.50	26.50	9.3878	0.009827	**
Insect:Mollusc	1	3.63	3.63	1.2876	0.278649	
Residuals	12	33.87	2.82			

Error: Block

	Df	Sum Sq	Mean Sq	F value	Pr(>F)
Residuals	4	8327.2	2081.8		

Error: Within

	Df	Sum Sq	Mean Sq	F value	Pr(>F)	
Insect	1	713.18	713.18	301.8366	7.168e-010	***
Mollusc	1	30.45	30.45	12.8863	0.003715	**
Insect:Mollusc	1	1.09	1.09	0.4625	0.509353	
Residuals	12	28.35	2.36			

Error: Block

	Df	Sum Sq	Mean Sq	F value	Pr(>F)
Residuals	4	11710.4	2927.6		

Error: Within

	Df	Sum Sq	Mean Sq	F value	Pr(>F)	
Insect	1	667.19	667.19	459.7610	6.178e-011	***
Mollusc	1	33.34	33.34	22.9714	0.0004388	***
Insect:Mollusc	1	13.04	13.04	8.9868	0.0111120	*
Residuals	12	17.41	1.45			

Each one of these tables has the correct degrees of freedom for error (d.f. = 12) and they all tell a reasonably consistent story. All show significant main effects for Insect and Mollusc—Insect is generally more significant than Mollusc—and there was a significant interaction only at Time 6 (this is the Insect:Mollusc:Time interaction we discovered earlier).

The simplest way to get rid of the temporal pseudoreplication is just to average it away. That would produce a single number per treatment per block (i.e. 20 numbers in all). We need to produce new, shortened vectors (length 20 instead of length 120) for each of the explanatory variables, and for the response variable. The shorter vectors will all need new names. It is easy to deal with the response variable:

```
g<-as.vector(tapply(Grass,list(Insect,Mollusc,Block),mean))
```

You can not use **tapply** directly to make shortened vectors for the categorical variables, because S-Plus does not understand what you mean by 'the mean of a set of character strings' (the factor levels are text). All that is needed, however, is to use **as.numeric** to convert the factor levels into numbers. Then use these numbers as subscripts with *levels* to get back the original names for the factor levels in the shortened vectors (see p. 456):

```
b<-as.vector(tapply(as.numeric(Block),list(Insect,Mollusc,Block),mean))
b<-levels(Block)[b]
m<-as.vector(tapply(as.numeric(Mollusc),list(Insect,Mollusc,Block),mean))
m<-levels(Mollusc)[m]
i<-as.vector(tapply(as.numeric(Insect),list(Insect,Mollusc,Block),mean))
i<-levels(Insect)[i]
```

We should check that all of the factors have been shortened properly:

```
b
[1]  "A" "A" "A" "A" "B" "B" "B" "B" "C" "C" "C" "C" "D" "D" "D" "D" "E"
[18] "E" "E" "E"
i
 [1] "Sprayed"   "Unsprayed" "Sprayed"   "Unsprayed" "Sprayed"
 [6] "Unsprayed" "Sprayed"   "Unsprayed" "Sprayed"   "Unsprayed"
[11] "Sprayed"   "Unsprayed" "Sprayed"   "Unsprayed" "Sprayed"
[16] "Unsprayed" "Sprayed"   "Unsprayed" "Sprayed"   "Unsprayed"
m
 [1] "Absent"  "Absent"  "Present" "Present" "Absent"  "Absent"
 [7] "Present" "Present" "Absent"  "Absent"  "Present" "Present"
[13] "Absent"  "Absent"  "Present" "Present" "Absent"  "Absent"
[19] "Present" "Present"
```

Now we can carry out the Anova fitting block and the 2 × 2 factorial:

```
model<-aov(g~i*m+Error(b))
summary(model)
```

```
Error: b
          Df Sum of Sq  Mean Sq F Value Pr(F)
Residuals  4  4317.655 1079.414

Error: Within
          Df Sum of Sq  Mean Sq   F Value        Pr(F)
    i  1   637.6751 637.6751 1950.628 0.0000000
    m  1    52.4382  52.4382  160.407 0.0000000
  i:m  1     0.0151   0.0151    0.046 0.8332699
Residuals 12    3.9229   0.3269
```

Now we have the right number of degrees of freedom for error (d.f. $= 12$). The analysis shows highly significant main effects for both treatments, but no interaction (recall that it was only significant at Time $= 6$). Note that the error variance $s^2 = 0.327$ is larger than s^2 was in five of the six separate analyses, because of the large Block:Time interaction.

An alternative is to detrend the data by regressing Grass against Time for each treatment combination (Code $= 1$ to 20), then use the slopes and the intercepts of these regressions in two separate Anovas in what is called a *derived variables analysis*.

Here is one way to extract the 20 regression intercepts, a, which are parameter [1] within the **coef** of the fitted model:

```
a<-1:20
Code<-gl(20,6)
for (k in 1:20) a[k]<-coef(lm(Grass~Time,subset=(Code==k)))[1]
```

```
a

 [1]  97.27997 103.22584  87.71079  90.27373  95.41965
 [6] 101.91840  87.22158  88.60817  94.41061 103.25222
[11]  85.07103  89.28192  96.10313 100.96869  86.18559
[16]  91.19490 116.83651 121.69893 106.40919 108.55959
```

This shortened vector now becomes the response variable in a simple analysis of variance, using the shortened factors we calculated earlier:

```
model<-aov(a~i*m+Error(b))
summary(model)
```

```
Error: b
          Df  Sum Sq Mean Sq F value Pr(>F)
Residuals  4 1253.30  313.33

Error: Within
          Df Sum Sq Mean Sq F value      Pr(>F)
i          1 611.59  611.59 594.933 1.359e-011 ***
m          1 107.34  107.34 104.420 2.834e-007 ***
i:m        1  12.32   12.32  11.980   0.004707 **
Residuals 12  12.34    1.03
```

Everything in the model has a significant effect on the intercept, including a highly significant interaction between Insects and Molluscs ($p = 0.0047$). What about the slopes of the Grass against Time graphs, coef [2] of the fitted object?

```
aa<-1:20
for (k in 1:20) aa[k]<-coef(lm(Grass~Time,subset=(Code==k)))[2]

aa

 [1]   3.3885561   2.8508475   3.2423677   3.1764151
 [5]   0.6762409  -0.3742587  -0.1142285   0.1536275
 [9]  -5.5199901  -6.7429909  -6.1245833  -6.1412692
[13]  -6.9687946  -7.7624054  -7.6693108  -7.9243495
[17]  -5.2672184  -5.8186351  -5.5222919  -5.2818614
```

As we saw from the initial graphs, there is great variation in slope from Block to Block, but is there any effect of treatments?

```
model<-aov(aa~i*m+Error(b))
summary(model)

Error: b
            Df Sum Sq Mean Sq F value Pr(>F)
Residuals    4 337.03   84.26

Error: Within
            Df  Sum Sq Mean Sq F value    Pr(>F)
i            1 0.02223 0.02223  0.5453 0.4744357
m            1 0.79426 0.79426 19.4805 0.0008449 ***
i:m          1 0.93608 0.93608 22.9588 0.0004398 ***
Residuals   12 0.48927 0.04077
```

Yes there is. Perhaps surprisingly, there is a highly significant interaction between Insect and Mollusc in their effects on the slope of the graph of Grass biomass against Time. We need to see the values involved:

```
tapply(aa,list(m,i),mean)

          Sprayed    Unsprayed
Absent   -3.569489    -3.203488
Present  -2.738241    -3.237609
```

All the average slopes are negative, but insect exclusion reduces the slope under one mollusc treatment yet increases it under another. The mechanisms underlying this interaction would clearly repay further investigation.

The statistical lesson is that derived variable analysis provided many more insights than either the six separate analyses or the analysis of the time-averaged data.

The same analysis using mixed effects models

This requires a clear understanding of which of our four factors are fixed effects and which are random effects. Insecticide and molluscicide are experimental treatments so these are clearly *fixed* effects. Blocks are different, but we didn't make them different. They are assumed to come from a large population of different locations, hence they are random effects. Things differ through time, but mainly because of the weather, and the passage of the seasons, rather than because of anything we do. So time is also a *random* effect. One of our random effects is spatial (Block) and one temporal (Time). The factorial experiment (Insect*Mollusc) is nested within each block, and Time codes for the repeated measures made within each of the differently treated plots within each block. We need to reflect all this information about the structure of the data frame in the **groupedData** function. The key thing to understand is where, exactly, each of the five variables is placed in a mixed effects model.

The response variable is easy—it is Grass. This goes on the left of the ~ in the model formula. The manipulated parts of the experiment (the fixed effects) are Insect and Mollusc, two-level factors representing plus and minus each kind of herbivore. These factors that apply to the whole experiment are called the **outer** factors. They appear in a model formula (linked in this case by the factorial operator *) to the right of the ~ in the outer statement. But what about Block and Time? You can easily imagine a graph of Grass against Time. This is a time series plot, and each treatment in each block could produce one such time series. Thus, Time goes on the right of the ~ in the model formula. Block is the spatial unit in which the pseudoreplication has occurred, so it is the factor that goes on the right of the conditioning operator | after the model formula. In general, the conditioning factors can be nested, using the / operator. In medical studies the conditioning factor is often the subject from which repeated measures were taken. We call the grouped data object yield like this:

yield<-groupedData(Grass~Time | Block/Insect/Mollusc,data=repeated,outer=~Insect*Mollusc)

With grouped data, some previously complicated things become very simple. For instance, panel plots are produced automatically:

plot(yield)

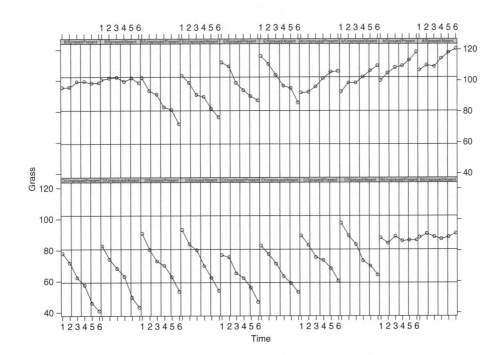

This shows the data for the whole factorial experiment (two levels of Insecticide and two levels of Molluscicide) in a single group of panels (one for each block). What it shows very clearly, however, are the different time series exhibited in the five different blocks. Note that **groupedData** has ordered the blocks from lowest to highest mean grass biomass (D, C, B, A and E) and that grass biomass increases through time on Block E, is roughly constant on Block B and declines on Blocks D, C and A. To see the time series for the four combinations of fixed effects, we use the outer option within the **plot** directive:

plot(yield, outer = ~ Insect*Mollusc)

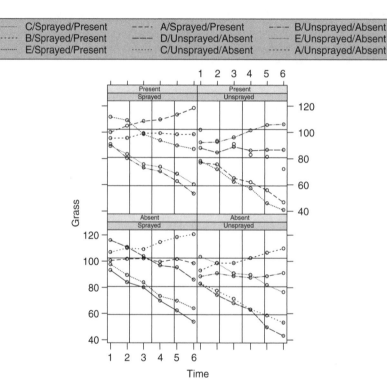

This gives a good impression of the complexity of the outcome: there was a mixture of positive and negative time trends in different blocks for all four treatment combinations.

Now for the mixed effects modelling:

```
model<-lme(Grass~Insect*Mollusc,random=~Time|Block)
summary(model)

Linear mixed-effects model fit by REML
 Data: NULL
```

```
        AIC       BIC       logLik
   538.3928 560.4215  -261.1964

Random effects:
 Formula: ~ Time | Block
 Structure: General positive-definite
             StdDev    Corr
(Intercept) 8.885701 (Inter
       Time 5.193374 -0.261
   Residual 1.644065

Fixed effects: Grass ~ Insect * Mollusc
                 Value Std.Error  DF   t-value p-value
    (Intercept) 96.14375  3.848244 112  24.98380 <.0001
         Insect -5.64657  0.150082 112 -37.62326 <.0001
        Mollusc -1.61923  0.150082 112 -10.78899 <.0001
 Insect:Mollusc  0.02751  0.150082 112   0.18327 0.8549

Number of Observations: 120
Number of Groups: 5
```

The interpretation is unequivocal: the mixed effects model gives no indication of an interaction between insect and mollusc exclusion ($p = 0.8549$). The effects that we saw in the earlier analyses were artefacts of confounding effects associated with blocks. There are variance components for blocks ($8.8857^2 = 78.96$) and for time ($5.193374^2 = 26.97$).

Model criticism in lme

We need to be just as conscious of testing the assumptions of the model with mixed effects models as we are with any other. The key assumptions are these:

- within-group errors are independent and normally distributed with mean zero and variance σ^2

- within-group errors are independent of the random effects

- random effects are normally distributed with mean zero

- random effects have a covariance matrix Ψ, not depending on the group

- random effects are independent for different groups, except as specified by nesting

Most of these can be assessed by model plots, in the usual way. The most informative plots are the residuals, fitted values and estimated random effects. We continue the last example. We plot the residuals against the fitted values separately for each block like this:

```
plot(model, resid(.,type="p")~fitted(.) | Block)
```

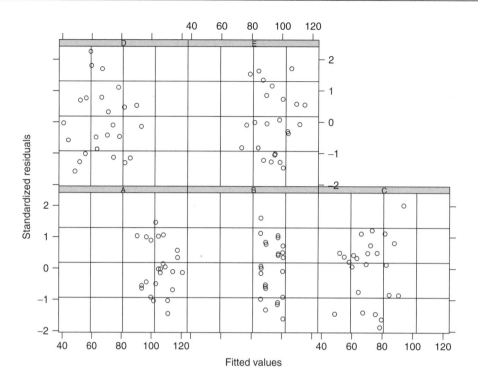

There is no strong evidence of heteroscedasticity (the variance is not a function of the fitted values, nor does it vary from block to block).

Model simplification in lme

During model simplification we would want to remove the non-significant interaction term. Suppose we try to do this in the most obvious way:

```
model<-lme(fixed=Grass~Insect*Mollusc,random=~Time | Block)
model2<-lme(fixed=Grass~Insect+Mollusc,random=~Time | Block)
anova(model,model2)
```

```
 Model df       AIC       BIC    logLik  Test  L.Ratio p-value
 model    1   8 538.3928 560.4215 -261.1964
model2    2   7 534.4668 553.8020 -260.2334 1 vs 2 1.926033 0.1652
```

```
Warning messages:
 Fitted objects with different fixed effects. REML comparisons are not
meaningful. in: anova.lme(model, model2)
```

We get a warning message to remind us that we should not use REML when we want to use **anova** to compare models with different fixed effects. The correct way to proceed is to use the maximum likelihood method like this:

```
model3<-update(model, method="ML")
model4<-update(model2, method="ML")
anova(model3,model4)
```

```
        Model df      AIC      BIC    logLik    Test    L.Ratio p-value
model3      1  8 536.9027 559.2026 -260.4514
model4      2  7 534.9372 554.4497 -260.4686 1 vs 2 0.03452457  0.8526
```

Note how much larger the p value is when we do it the right way (0.8526 vs. 0.1652).

Further reading

Crowder, M.J. and Hand, D.J. (1990) *Analysis of Repeated Measures*. London, Chapman & Hall.

Diggle, P.J., Liang, K.-Y. *et al.* (1994) *Analysis of Longitudinal Data*. Oxford, Clarendon Press.

Khuri, A.I., Mathew, T. *et al.* (1998) *Statistical Tests for Mixed Linear Models*. John Wiley, New York.

McCulloch, C.E. and Searle, S.R. (2001) *Generalized, Linear and Mixed Models*. New York, John Wiley.

Pinheiro, J.C. and Bates, D.M. (2000). *Mixed-effects Models in S and S-Plus*. New York, Springer-Verlag.

36

Spatial statistics

S-Plus is not a geographical information system (GIS), so if it's map-making you are after, then you have come to the wrong place. On the other hand, if you want to do statistical modelling with spatially explicit data, then there are several useful functions in S-Plus. There are three kinds of problems that we tackle with spatial statistics:

- point processes (locations and spatial patterns of individuals)

- maps of a continuous response variable

- spatially explicit dynamics affected by the identity, size and proximity of neighbours

Point processes

There are three broad classes of spatial pattern on a continuum from complete regularity (evenly spaced hexagons) to complete aggregation (all the individuals clustered into a single clump). We call these regular, random and aggregated patterns and they look like this.

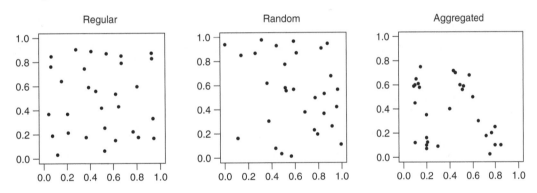

In their simplest form, the data consist of sets of x and y coordinates within some sampling frame like a square or a circle in which all the individuals have been mapped. The first question is often whether there is any evidence to allow rejection of the null hypothesis of *complete spatial randomness* (CSR). In a random pattern the distribution of every individual is completely independent of the distribution of the others. Individuals neither inhibit nor promote one another. In a regular pattern, individuals are more spaced out than random, presumably because of some mechanism (like intraspecific competition) that eliminates individuals that are too close together. In an aggregated pattern, individuals are more clumped than random, presumably because of some process like reproduction with limited dispersal, or because of underlying spatial heterogeneity (good patches and bad patches).

Counts of individuals within sample areas (quadrats) can be analysed by comparing the frequency distribution of counts with a Poisson distribution with the same mean. Aggregated spatial patterns (variance > mean) are often well described by a negative binomial distribution with aggregation parameter k (see p. 485). The main problem with quadrat-based counts is that they are highly *scale dependent*. The same spatial pattern could appear to be regular when analysed with small quadrats, aggregated when analysed with medium-sized quadrats, yet random when analysed with large quadrats.

Distance measures are of two broad types: measures from individuals to their nearest neighbours, and measures from random points to the nearest individual. Recall that the nearest individual to a random point is *not* a randomly selected individual (see p. 53); this protocol favours isolated individuals and individuals on the edge of clumps.

In other circumstances you might be willing to take the existence of patchiness for granted, and to carry out a more sophisticated analysis of the spatial attributes of the patches themselves: their mean size and the variance in size, spatial variation in the spacing of patches of different sizes, and so on.

Nearest neighbours

Suppose that we have been set the problem of drawing lines to join the nearest neighbour pairs of any given set of points (x, y) that are mapped in two dimensions. There are three steps to the computing:

- compute the distance to each neighbour
- identify the smallest neighbour distance for each individual
- use these minimal distances to identify all the nearest neighbours

We start by generating a random spatial distribution of 100 individuals by simulating their x and y coordinates from a uniform probability distribution:

```
x<-runif(100)
y<-runif(100)
```

The graphics parameter pty="s" makes the plotting area square, as we would want for a map like this:

```
par(pty="s")
plot(x,y,pch=16)
```

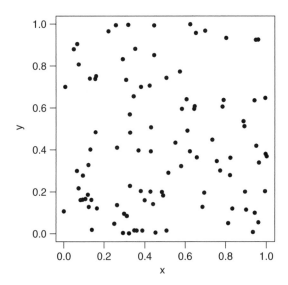

Computing the distances is straightforward; for each individual we use Pythagoras's theorem to calculate the distance to every other plant. The distance between two points with coordinates (x_1, y_1) and (x_2, y_2) is d.

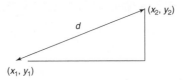

The square on the hypotenuse (d^2) is the sum of the squares on the two adjacent sides: $(x_2 - x_1)^2$ plus $(y_2 - y_1)^2$ so the distance d is given by

$$d = \sqrt{(y_2 - y_1)^2 + (x_2 - x_1)^2}$$

We write a function for this as follows:

```
dist<-function(x1, y1, x2, y2) sqrt((x2 - x1)^2 + (y2 - y1)^2)
```

Now we loop through each individual i and calculate a vector of distances, d, from every other individual. The nearest neighbour distance is the minimum value of d, and the identity of the nearest neighbour, nn, is found using which(d==min(d[-i])) to find the subscript of the minimum value of d (the [-i] is necessary to exclude the distance 0 which results from the ith plant's distance from itself). Here is the complete code to compute nearest neighbour distances, r, and identities, nn, for all 100 individuals on the map:

```
r<-numeric(100)
nn<-numeric(100)
d<-numeric(100)
for (i in 1:100) {
d<-0
for (k in 1:100) d[k]<-dist(x[i],y[i],x[k],y[k])
r[i]<-min(d[-i])
nn[i]<-which(d==min(d[-i]))
}
```

Now we can fulfil the brief and draw lines to join each individual to its nearest neighbour, like this:

```
for (i in 1:100) lines(c(x[i],x[nn[i]]),c(y[i],y[nn[i]]))
```

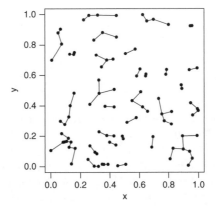

Tests for spatial randomness

Clark and Evans (1954) give a very simple test of spatial randomness. Making the very strong assumption that you know the population density of the individuals, ρ (generally you do not know this, and would need to find it out independently), then the expected mean distance to the nearest neighbour is

$$E(r) = \frac{\sqrt{\rho}}{2}$$

In our example we have 100 individuals in a unit square so $\rho = 0.01$ and $E(r) = 0.05$. The actual mean nearest neighbour distance was

    mean(r)

    [1] 0.05409647

which is very close to expectation. This is clearly a random distribution of individuals (as we constructed it to be). An index of randomness is given by the ratio $\bar{r}/E(r) = 2\bar{r}\sqrt{\rho}$. This takes the value 1 for random patterns, more than 1 for regular (spaced-out) patterns and less than 1 for aggregated patterns.

One problem with such first-order estimates of spatial pattern (including measures like the variance/mean ratio) is that they can give no feel for the way spatial distribution changes *within* an area.

Ripley's K

The *second-order properties* of a spatial point process describe the way that spatial interactions *change* through space. These are computationally intensive measures that take a range of distances within the area, calculate a pattern measure, then plot a graph of the function against distance, to show how the pattern measure changes with scale. The most widely used second-order measure is the K-function, which is defined like this:

$$K(d) = \frac{1}{\lambda} E[\text{number of points} \leq \text{distance } d \text{ of an arbitrary point}]$$

where λ is the mean number of points per unit area (the *intensity* of the pattern). If there is *no spatial dependence*, then the expected number of points that are within a distance d of an arbitrary point is πd^2 times the mean density. So, if the mean density is 2 points per square metre ($\lambda = 2$), then the expected number of points within a 5 m radius is $\lambda \pi d^2 = 2 \times \pi \times 5^2 = 50\pi = 157.1$ If there is clustering, then we expect an excess of points at short distances (i.e. $K(d) > \pi d^2$ for small d). Likewise, for a regularly spaced pattern, we expect an excess of long distances, hence few individuals at short distances (i.e. $K(d) < \pi d^2$). Ripley's K (published in 1976) is calculated as follows:

$$\hat{K}(d) = \frac{1}{n^2} |A| \sum \sum_{i \neq j} \frac{I_d(d_{i,j})}{w_{i,j}}$$

where n is the number of points in region A with area $|A|$, $d_{i,j}$ are the distances between points (the distance between the ith and jth points to be precise). To account for edge effects, the model includes the term $w_{i,j}$ which is the fraction of the area, centred on i and passing through j, that lies within region A (all the $w_{i,j}$ are 1 for points that lie well away from the edges of the region). $I_d(d_{i,j})$ is an indicator function to show which points are to be counted as neighbours at this value of d; it takes the value 1 if $d_{i,j} \leq d$ and 0 otherwise (points with $d_{i,j} > d$ are omitted from the summation). The pattern measure is obtained by plotting $\hat{K}(d)$ against d. This is then compared with the curve that would be observed under complete spatial randomness (i.e. a plot of πd^2 against d). When clustering occurs, $K(d) > \pi d^2$ and the curve lies *above* the CSR curve.

You can see *why* you need the edge correction, from this simple simulation experiment. For individual number 1 with coordinates x[1], y[1], calculate the distances to all the other individuals, using the function **dist** that we wrote earlier (p. 711):

```
distances<-numeric(100)

for(i in 1:100) distances[i]<-dist(x[1],y[1],x[i],y[i])
```

Now find out how many other individuals are within a distance d of this individual. Take as an example $d = 0.1$:

```
sum(distances<0.1)-1
```

```
[1] 6
```

There were six other individuals within a distance $d = 0.1$ of the first individual (the distance 0 from itself is included in the sum so we have to correct for this by subtracting 1). The next step is to generalise the procedure from this one individual to all the individuals. We make a two-dimensional matrix called dd to contain all the distances from every individual (rows) to every other individual (columns):

```
dd<-numeric(10000)
dd<-matrix(dd,nrow=100)
```

The matrix of distances is computed within loops for both individual (j) and neighbour (i) like this:

```
for (j in 1:100) {for(i in 1:100) dd[j,i]<-dist(x[j],y[j],x[i],y[i])}
```

We should check that the number of individuals within 0.1 of individual 1 is still 6 under this new notation. Note the use of blank subscripts [1,] to mean 'all the individuals in row number 1':

```
sum(dd[1,]<0.1)-1
```

```
[1] 6
```

So that's okay. We want to calculate the sum of this quantity over all individuals, not just individual number 1:

```
sum(dd<0.1)-100
```

```
[1] 272
```

This means there are 272 cases in which other individuals are counted within $d = 0.1$ of focal individuals. Next create a vector containing a range of different distances, d, over which we want to calculate $K(d)$ by counting the number of individuals within distance d, summed over all individuals:

```
d<-seq(0.01,1,0.01)
```

For each of these distances we need to work out the total number of neighbours of all individuals. So, in place of 0.1 (in sum above) we need to put each of the d values in turn. The count of individuals is going to be a vector of length 100 (one for each d):

```
count<-numeric(100)
```

Calculate the count for each distance d:

```
for (i in 1:100) count[i]<-sum(dd<d[i])-100
```

The expected count increases with d as πd^2, so we scale our count by dividing by the square of the total number of individuals $n^2 = 100^2 = 10\,000$:

```
K<-count/10000
```

Finally, plot a graph of K against d:

```
plot(d,K,type="l")
```

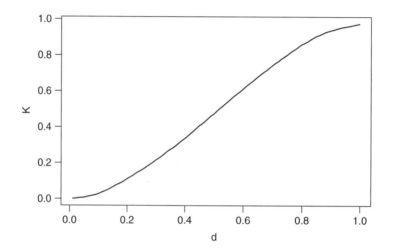

Not surprisingly, when we sample the whole area ($d = 1$), we count all of the individuals in every neighbourhood ($K = 1$). For CSR the graph should follow πd^2 so we add a line to show this:

```
lines(d,pi*d^2)
```

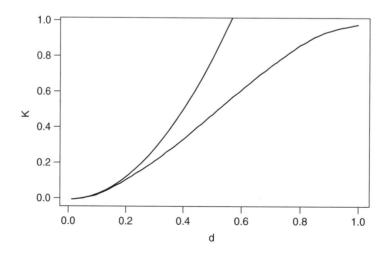

Up to about $d = 0.2$ the agreement between the two lines is reasonably good, but for longer distances our algorithm is counting far too few neighbours. This is because much of the area scanned around marginal individuals is invisible, since it lies outside the study area (there may well be individuals out there but we shall never know). This demonstrates that the *edge correction* is a fundamental part of Ripley's K.

Fortunately, we don't have to write a function to work out a corrected value for K; it is available as **Kfn** in the spatial library of Venables and Ripley. Here we use it to analyse the pattern of trees in their data frame called pines:

```
library(spatial)
pines<-ppinit("pines.dat")
```

First, set up the plotting area with two square frames:

```
par(mfrow=c(1,2),pty="s")
```

On the left, make a map using the x and y locations of the trees, and on the right make a plot of $L(t)$ the pattern measure against distance:

```
plot(pines,pch=16)
plot(Kfn(pines,5),type="s",xlab="distance",ylab="L(t)")
```

Recall that if there were CSR then the expected value of K would be πd^2; to linearise this we could divide by π then take the square root. This is the measure used by Venables and Ripley in their function **Kfn**; they call it $L(t) = \sqrt{K(t)/\pi}$. Now for the simulated upper and lower bounds. The first argument in **Kenvl** (calculating 'envelopes' for K) is the maximum distance, the second is the number of simulations, and the third is the number of individuals within the mapped area (71 in this case):

```
lims<-Kenvl(5,100,Psim(71))
lines(lims$x,lims$lower,lty=2)
lines(lims$x,lims$upper,lty=2)
```

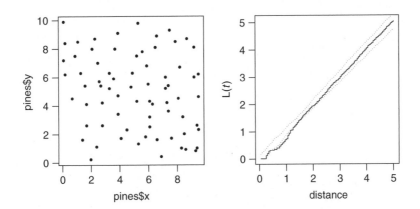

There is a suggestion that at relatively small distances (around 1 or so), the trees are rather *regularly* distributed (more spaced out than random). This shows up as curvature in the plot of $L(t)$ against distance (it should be straight for its whole length under CSR) with the actual line *below* the CSR line. The mechanism underlying this spatial regularity (e.g. non-random recruitment or mortality, or underlying non-randomness in the substrate) would need to be investigated in detail.

Ripley's *K* with an aggregated spatial pattern

The file trees contains a map of the locations of 257 mature Scots pine trees in a permanent 50 m × 50 m plot in heathland in southern England:

```
attach(trees)
names(trees)
```

```
[1] "x" "y"
```

To use the function **Kfn** you need to make a data frame that contains information on the plotting area; this is defined by four quantities: the lower and upper values for *x* (xl and xu) and the lower and upper values for *y* (yl and yu). Here is how we do it:

```
aa<-data.frame(xl=0,xu=1,yl=0,yu=1)
```

Now we need to combine this information with our *x* and *y* coordinates of the individuals into a new data frame called, say, map:

```
map<-NULL
map$y<-y
map$x<-x
map$area<-aa
```

The aim is to produce a map of the individuals on the left, with a plot of $L(t) = \sqrt{K(t)/\pi}$ against distance (*t*) on the right:

```
par(mfrow=c(1,2),pty="s")
plot(x,y,pch=16)
plot(Kfn(map,1),type="s",xlab="distance",ylab="L(t)")
```

The limits are obtained by simulation (on the assumption that the distribution was random), and these are added next:

```
lims<-Kenvl(1,100,Psim(257))
lines(lims$x,lims$upper,lty=2)
lines(lims$x,lims$lower,lty=2)
```

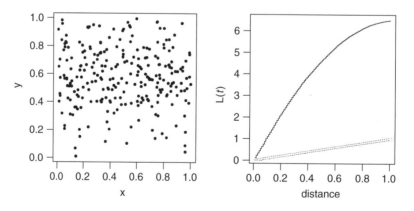

The individuals are highly aggregated, even at the smallest scales measured. It looks as if there is a gradient of *intensity* in the *y* direction (up and down), with higher densities of individuals just north of the centre of the plot. We can test this by counting up the numbers of individuals in different zones (say in strips of width 0.1) across the map, using **cut** with **table** like this:

```
yt<-cut(y,seq(0,1,0.1))
barplot(as.vector(table(yt)),
        names=as.character(seq(.1,1,.1)),xlab="y-distance",ylab="intensity")
```

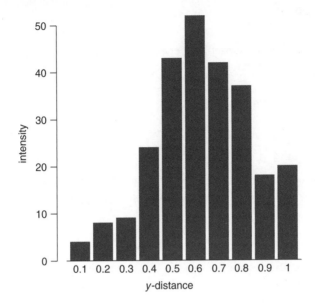

There is an extremely pronounced pattern, with intensity peaking in the strip between $y = 0.5$ and $y = 0.6$. This suggests the existence of an important within-plot gradient (perhaps topographic, perhaps in soil nutrients) running from north to south across the mapped area. This would repay further investigation.

Quadrat-based methods

Another approach to testing for spatial randomness is to count the number of individuals in quadrats of different sizes. Here the quadrats have an area of 0.01, so the expected number per quadrat $= 1$.

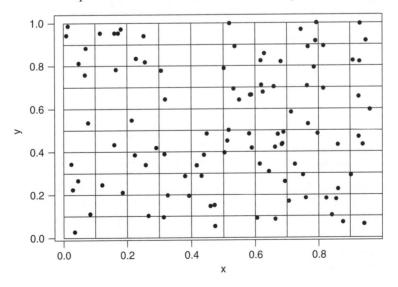

The trick here is to use **cut** to convert the x and y coordinates of the map into bin numbers (between 1 and 10 for the quadrat size we have drawn here). To achieve this, the break points are generated by the sequence $(0,1,0.1)$:

```
xt<-cut(x,seq(0,1,0.1))
yt<-cut(y,seq(0,1,0.1))
```

which creates vectors of integer subscripts between 1 and 10 for xt and yt. Now all we need to do is use **table** to count up the number of individuals in every cell (i.e. in every combination of xt and yt):

```
count<-as.vector(table(xt,yt))
table(count)
```

```
 0  1  2 3 4 5
36 39 18 4 2 1
```

This shows that 36 cells are empty and one cell contained five individuals. Now we need to see what this distribution would look like under a particular null hypotheses. For a Poisson process (see p. 480) we have

$$P(x) = \frac{e^{-\lambda}\lambda^x}{x!}$$

Note that the mean depends upon the quadrat size we have chosen. With 100 individuals in the whole area, the expected number in any one of our 100 cells, λ, is 1.0. The expected frequencies of counts between 0 and 5 are therefore given by

```
expected<-100*exp(-1)/factorial(0:5)
expected
```

```
[1] 36.7879441 36.7879441 18.3939721 6.1313240 1.5328310
[6]  0.3065662
```

The fit between observed and expected is almost perfect (as we should expect, of course, having generated the random pattern ourselves).

Spatial correlation

It is easy to work out the distance of every individual from every other. Then we can work out the distance to the nearest neighbour, second nearest neighbour, third nearest neighbour, and so on. Now we can ask how the value of the response variable depends on the distance to the nearest neighbour, second nearest neighbour, third nearest neighbour, and so on. Finally, we can ask whether the response depends upon the *size* or the *identity* of the neighbour, as well as on the distance of that neighbour.

For these calculations, it is useful to have these two-dimensional matrices:

- a matrix of distances of the ith individual from every other

- a matrix containing the identities of the 1st, 2nd, 3rd, $\ldots$, nth nearest neighbours

We call these matrices ds and *map* respectively. Suppose that the locations of 200 mapped measurements are in x and y:

```
d<-numeric(200)
map<-matrix(0,200,200)
```

```
ds<-matrix(0,200,200)
for (i in 1:200){
d<-0
for (k in 1:200) d[k]<-dist(x[i],y[i],x[k],y[k])
map[i,]<-order(d)
ds[i,]<-d
}
```

The sizes are

```
attach(sizes)
names(sizes)

[1] "x" "y"   "z"
```

We start by producing an interpolated object (called *plan*) on a regular grid of *x* and *y* values for initial graphical data inspection:

```
plan<-interp(x,y,z)
```

Now we can map the individuals and look at size, *z*, as a function of location (x, y) in different styles:

```
par(mfrow=c(2,2))
plot(x,y,pch=16)
persp(plan)
contour(plan)
image(plan)
```

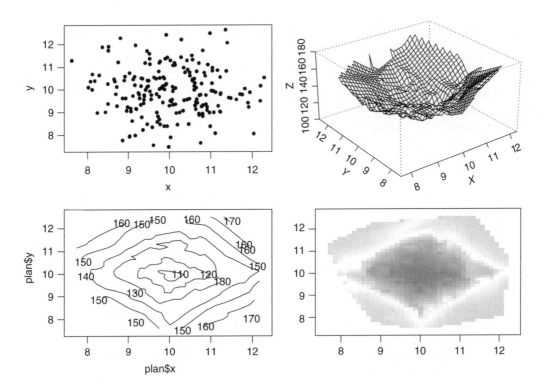

The top left panel shows a map of the individuals. The perspective plot shows the size of the individuals on the vertical (z) axis. The bottom two panels show the same data as a contour plot (left) and as an intensity image (right). There is clear evidence of strong spatial correlation: the individuals in the middle of the map are much smaller than the individuals on the edges. The map shows that the density of individuals is also greater in the central regions.

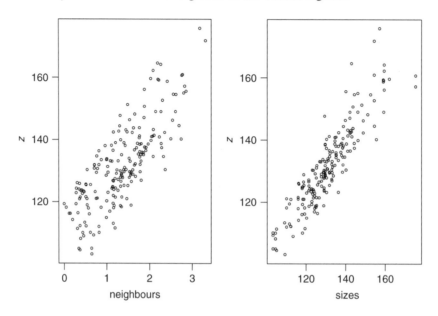

The spatial effect on size appears to be much more important in this case than the effect of the size of the nearest neighbour. Had there been strong competition between neighbouring individuals, then we would have expected the slope of the right-hand plot to be negative. Between them, the two plots indicate that location is the cause of both small size and close neighbour distances, and that the size of the nearest neighbour is of lesser importance. We try a model containing both spatial information (gradients in x and y) and neighbour information (distance and size).

We compute the two explanatory variables like this:

```
neighbours<-ds[,2]
sizes<-z[map[,2]]
```

These two vectors give the distance to the nearest neighbour, and the sizes of the nearest neighbours. Note the use of the subscripts [,2] to extract the distance and the location of the neighbours, and to do this for every individual (the blank before the comma means 'for every row'). We start by looking at whether neighbour distance and size affect the size of individuals:

```
model1<-lm(z~neighbours+sizes)
summary(model1)
```

```
Coefficients:
              Value  Std. Error  t value  Pr(>|t|)
(Intercept) 31.4722     4.9974    6.2977    0.0000
 neighbours  4.7294     0.8675    5.4519    0.0000
      sizes  0.7161     0.0444   16.1148    0.0000
```

```
Residual standard error: 5.424 on 197 degrees of freedom
Multiple R-Squared: 0.8405
F-statistic: 519.2 on 2 and 197 degrees of freedom, the p-
value is 0
```

Both terms are highly significant. Larger individuals have larger nearest neighbour distances. Also, and more surprisingly, bigger individuals are likely to have *bigger* nearest neighbours. We need to look at this more closely. Let's fit geographic coordinates to the model to test for spatial trends:

model2<-lm(z~neighbours+sizes+x+y)
summary(model2)

```
Coefficients:
              Value   Std. Error  t value  Pr(>|t|)
(Intercept)  95.9575   10.6505     9.0097   0.0000
neighbours   11.8872    1.2303     9.6623   0.0000
      sizes   0.4362    0.0545     8.0049   0.0000
          x  -4.2106    0.5696    -7.3928   0.0000
          y   0.4759    0.3321     1.4328   0.1535

Residual standard error: 4.808 on 195 degrees of freedom
Multiple R-Squared: 0.876
F-statistic: 344.3 on 4 and 195 degrees of freedom, the p-
value is 0
```

This says that there is a significant trend in the x direction but not in the y direction. This is not what an inspection of the intensity and contour plots suggest. They both indicate U-shaped responses, so let's test this by fitting quadratic terms in x and y:

model3<-lm(z~neighbours+sizes+x+I(x^2)+y+I(y^2))
summary(model3)

```
Coefficients:
               Value    Std. Error  t value  Pr(>|t|)
(Intercept)  655.5214   113.6284    5.7690    0.0000
 neighbours    5.4555     1.7741    3.0751    0.0024
      sizes    0.3672     0.0533    6.8864    0.0000
          x  -64.0243    11.6921   -5.4759    0.0000
     I(x^2)    3.1003     0.6051    5.1241    0.0000
          y  -50.8449    11.9089   -4.2695    0.0000
     I(y^2)    2.5391     0.5902    4.3023    0.0000

Residual standard error: 4.534 on 193 degrees of freedom
Multiple R-Squared: 0.8908
F-statistic: 262.5 on 6 and 193 degrees of freedom, the p-
value is 0
```

That's more like it. The predicted size of individuals is indeed U-shaped across both the x and y directions (note the negative first terms and positive quadratic terms for both x and y). Apparently, small-scale spatial heterogeneity in substrate quality means there is a *positive* correlation between neighbour sizes, and this effect is greater than any putative competition effects (e.g. you might have expected that large individuals would have small nearest neighbours, and vice versa). We have analysed this example entirely in terms of fixed effects, and we have not taken explicit account of the spatial autocorrelation in dealing with the error structure. We can do better than this.

Geostatistical data

Mapped data commonly show the value of a continuous response variable (e.g. the concentration of a mineral ore) at different spatial locations. The fundamental problem with this kind of data is spatial pseudoreplication. Hot spots tend to generate lots of data, and these data tend to be rather similar because they come from essentially the same place. Cold spots are poorly represented and typically widely separated. Large areas between the cold spots have no data at all.

Spatial statistics take account of this spatial autocorrelation in various ways. The fundamental tool of spatial statistics is the *variogram* (or semivariogram). This measures how quickly spatial autocorrelation $\gamma(h)$ falls off with increasing distance:

$$\gamma(h) = \frac{1}{2\,|N(h)|} \sum_{N(h)} (z_i - z_j)^2$$

where $N(h)$ is the set of all pairwise Euclidean distances $i - j = h$, $|N(h)|$ is the number of distinct pairs within $N(h)$, and z_i and z_j are values of the response variable at spatial locations i and j. There are two important rules of thumb: the distance of reliability of the variogram is less than half the maximum distance over the entire field of data; and you should only consider producing an empirical variogram when you have more than 30 or so pairs of data.

Plots of the empirical variogram against distance are characterised by some quaintly named features which give away its origin in geological prospecting:

- **nugget**: small-scale variation plus measurement error

- **sill**: the asymptotic value of $\gamma(h)$ as $h \to \infty$ representing the variance of the random field

- **range**: the threshold distance (if such exists) beyond which the data are no longer autocorrelated

Variogram plots that do not asymptote may be symptomatic of trended data or a non-stationary stochastic process. The *covariogram* $C(h)$ is the covariance of z values at separation h, for all i and $i + h$ within the maximum distance over the whole field of data:

$$\text{cov}(Z(i + h), Z(i)) = C(h)$$

The *correlogram* is a ratio of covariances:

$$\rho(h) = \frac{C(h)}{C(0)} = 1 - \frac{\gamma(h)}{C(0)}$$

where $C(0)$ is the variance of the random field and $\gamma(h)$ is the variogram. Where the variogram increases with distance, the corellogram and covariogram decline with distance. All three functions, variogram, covariogram and correlogram, have plot methods.

The variogram assumes that the data are untrended. If there are trends, then one option is median polishing; this involves modelling row and column effects from the map like this:

y ~ overall mean + row effect + column effect + residual

This *two-way* model assumes additive effects and would not work if there was an interaction between the rows and columns of the map. An alternative would be to use a **gam** (p. 595) with non-parameteric smoothers for latitude and longitude.

Anisotropy occurs when *spatial autocorrelation changes with direction*. If the sill changes with direction, this is called zonal anisotropy. When it is the range that changes with direction the process is called geometric anisotropy.

Kriging

Geographers have a wonderful knack of making the simplest ideas sound complicated. Kriging is nothing more than linear interpolation through space. Ordinary kriging uses a random function model of spatial correlation to calculate a weighted linear combination of the available samples to predict the response for an unmeasured location. Universal kriging is a modification of ordinary kriging that allows for spatial trends.

The **krige** function models the response variable as a function of spatial locations and a theoretical covariance function in which the range, sill and nugget are specified. The resulting model is then used for prediction in the usual way. The same function performs universal as well as ordinary kriging by specifying a polynomial trend surface as an option in the model formula.

We say no more about models for spatial *prediction* here; details can be found in Kaluzny *et al.* (1998). Our concern is with using spatial information in the interpretation of experimental or observational studies that have a single response variable. The emphasis is on using location-specific measurements to model the spatial autocorrelation structure of the data.

Regression models with spatially correlated errors: generalised least squares

In Chapter 35 we looked at the use of linear mixed effects models for dealing with random effects and temporal pseudoreplication. Here we look at generalised least squares for regression modelling where there are one- or two-dimensional spatial data as explanatory variables, and where we would expect neighbouring values of the response variable to be correlated. This example is a simple trial to compare the yields of 56 different varieties of wheat. What makes the analysis more challenging is that the farms carrying out the trial were spread out over a wide range of latitudes and longitudes.

```
attach(spatialdata)
names(spatialdata)

[1] "Block" "variety" "yield" "latitude" "longitude"
```

We begin with graphical data inspection to see the effect of location on yield:

```
par(mfrow=c(1,2))
plot(latitude,yield)
plot(longitude,yield)
```

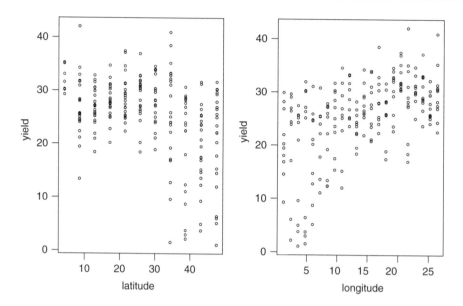

There are clearly big effects of both latitude and longitude on yield. The latitude effect looks like a threshold effect, with little impact for latitudes less than 30. The longitude effect looks more continuous but with a hint of non-linearity (perhaps even a hump). The varieties differ substantially in their mean yields:

```
par(mfrow=c(1,1))
barplot(sort(tapply(yield,variety,mean)))
```

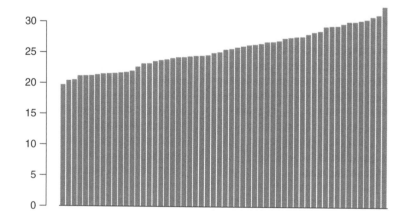

The lowest-yielding varieties are producing about 20 kg and the highest about 30 kg of grain per unit area. There are also substantial block effects on yield:

```
tapply(yield,Block,mean)

      1        2        3        4
27.575  28.81091 24.42589 21.42807
```

Here is the simplest possible analysis, a one-way analysis of variance using variety as the only explanatory variable:

```
model1<-aov(yield~variety)
summary(model1)
```

```
            Df Sum of Sq  Mean Sq   F Value      Pr(F)
  variety   55   2387.487 43.40886 0.7299866 0.9119316
Residuals  168   9990.167 59.46528
```

This says that there are *no significant differences* between the yields of the 56 varieties. We can try a split-plot analysis (see p. 345) using varieties nested within blocks:

```
Block<-factor(Block)
model2<-aov(yield~Block+variety+Error(Block))
summary(model2)
```

```
Error: Block
        Df Sum of Sq  Mean Sq
Block    3  1853.571 617.8569
Error: Within
            Df Sum of Sq  Mean Sq   F Value      Pr(F)
  variety   55   2388.766 43.43211 0.8808874 0.7024903
Residuals 165   8135.317 49.30495
```

This has made no difference to our interpretation. We could fit latitude and longitude as covariates:

```
model3<-aov(yield~Block+variety+latitude+longitude)
summary(model3)
```

```
            Df  Sum of Sq  Mean Sq  F Value       Pr(F)
    Block    3   1853.571  617.857 19.85762 0.00000000
  variety   55   2388.766   43.432  1.39589 0.05651984
 latitude    1    686.117  686.117 22.05145 0.00000560
longitude    1   2377.562 2377.562 76.41368 0.00000000
Residuals  163   5071.639   31.114
```

This makes an enormous difference. Now the differences between varieties are close to significance ($p = 0.0565$).

Finally, we could use mixed effects models to introduce the spatial covariance between yields from locations that are close together. We begin by making a grouped data object:

```
space<-groupedData(yield~variety-1|Block)
```

Now fit a linear model using generalised least squares using **gls** (this allows the errors to be correlated and/or have unequal variances; we add these sophistications later):

```
model4<-gls(yield~variety-1,space)
summary(model4)
```

```
Generalized least squares fit by REML
  Model: yield ~ variety -1
  Data: space
      AIC       BIC        logLik
 1354.742  1532.808     -620.3709
```

```
Coefficients:
                   Value  Std.Error  t-value  p-value
varietyARAPAHOE  29.4375  3.855687  7.634827  <.0001
  varietyBRULE   26.0750  3.855687  6.762738  <.0001
varietyBUCKSKIN  25.5625  3.855687  6.629818  <.0001
```

And so on, for all 56 varieties. The variety means are given, rather than differences between means, because we removed the intercept from the model by using yield~variety-1 rather than yield~variety in the model formula (see p. 214). Now we want to include the spatial covariance. The **Variogram** function is applied to model4 like this:

plot(Variogram(model4,form=~latitude+longitude))

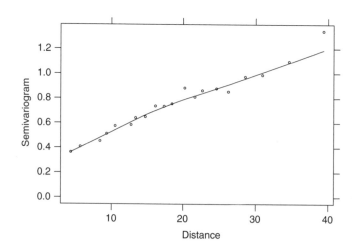

The sample variogram increases with distance, showing the expected spatial correlation. Extrapolating back to zero distance, there appears to be a *nugget* of about 0.2.

There are several assumptions we could make about the spatial correlation in these data. For instance, we could try a spherical correlation structure, using the **corSpher** class (the range of options for spatial correlation structure is shown in Table 35.2). We need to specify the distance at which the semivariogram first reaches 1. Inspection shows this distance to be about 28. We can update model4 to include this information:

model5<-update(model4,
corr=corSpher(c(28,0.2),form=~latitude+longitude,nugget=T))
summary(model5)

```
Generalized least squares fit by REML
  Model: yield ~ variety - 1
  Data: space
       AIC        BIC          logLik
   1185.863  1370.177      -533.9315

Correlation Structure: Spherical spatial correlation
  Parameter estimate(s):
    range      nugget
  27.45743  0.2093148
```

```
Coefficients:
                  Value Std.Error  t-value p-value
varietyARAPAHOE 26.65898  3.437348  7.75568  <.0001
  varietyBRULE  25.84957  3.441787  7.51051  <.0001
varietyBUCKSKIN 34.84837  3.478286 10.01883  <.0001
 varietyCENTURA 25.09473  3.458862  7.25520  <.0001
```

This is a big improvement, and AIC has dropped from 1354.742 to 1185.863. The range (27.46) and nugget (0.209) are very close to our visual estimates. There are other kinds of spatial model, of course. We might try a rational quadratic model (**corRatio**); this needs an estimate of the distance at which the semivariogram $= (1 + \text{nugget})/2 = 1.2/2 = 0.6$, as well as an estimate of the nugget. Inspection gives a distance of about 12.5, so we write

    model6<-update(model4,
    corr=corRatio(c(12.5,0.2),form=~latitude+longitude,nugget=T))

We can use **anova** to compare the two spatial models:

    anova(model5,model6)

```
        Model df      AIC       BIC      logLik
model5     1 59 1185.863 1370.177 -533.9315
model6     2 59 1183.278 1367.592 -532.6389
```

The rational quadratic model (model6) has the lower AIC and is therefore preferred to the spherical model. To test for the significance of the spatial correlation parameters, we need to compare the preferred spatial model6 with the non-spatial model4 (which assumed spatially independent errors):

    anova(model4,model6)

```
      Model df      AIC      BIC     logLik   Test L.Ratio p-value
model4   1 57 1354.742 1532.808 -620.3709
model6   2 59 1183.278 1367.592 -532.6389 1 vs 2 175.464  <.0001
```

The two extra degrees of freedom used up in accounting for the spatial structure are clearly justified. We need to check the adequacy of the **corRatio** model. This is done by inspection of the sample variogram for the normalised residuals of model6:

    plot(Variogram(model6,resType="n"))

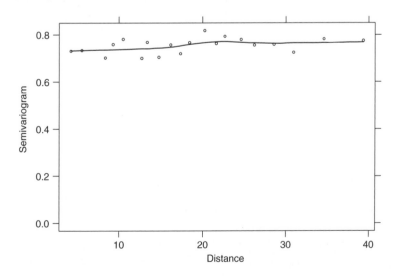

There is no pattern in the plot of the sample variogram, so we conclude that the rational quadratic is adequate. To check for constancy of variance, we can plot the normalised residuals against the fitted values like this:

```
plot(model6,resid (.,type="n")~fitted(.),abline=0)
```

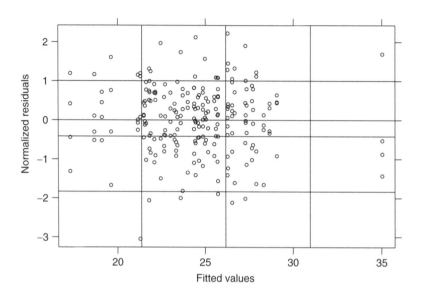

And the normal plot is obtained in the usual way:

```
qqnorm(model6,~resid(.,type="n"))
```

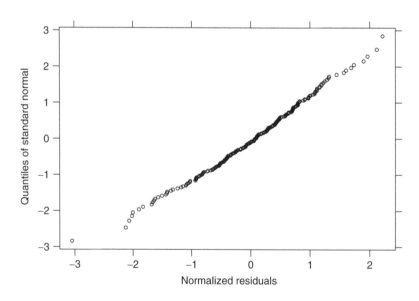

The model looks fine. The next step is to investigate the significance of any differences between the varieties. Use **update** to change the structure of the model from yield~variety-1 to yield~variety:

```
model7<-update(model6,model=yield~variety)
anova(model7)
```

```
Denom. DF: 168
              numDF   F-value   p-value
(Intercept)       1  30.40490   <.0001
    variety      55   1.85092   0.0015
```

The differences between the varieties now appear to he highly significant (recall that they were only marginally significant with our linear model3 using analysis of covariance to take account of the latitude and longitude effects). Specific contrasts between varieties can be carried out using the L argument to **anova**. Suppose that we want to compare the mean yields of the first and third varieties. To do this, we set up a vector of contrast coefficients c(-1,0,1) and apply the contrast like this:

```
anova(model6,L=c(-1,0,1))
```

```
Denom. DF: 168
 F-test for linear combination(s)
 varietyARAPAHOE varietyBUCKSKIN
              -1               1
   numDF   F-value  p-value
1      1  7.696578   0.0062
```

Note that we use model6 (with all the variety means) not model7 (with an intercept plus Helmert contrasts). The varieties Arapahoe and Buckskin exhibit highly significant differences in mean yield.

Further reading

Cressie, N.A.C. (1991) *Statistics for Spatial Data*. New York, John Wiley.
Diggle, P.J. (1983) *Statistical Analysis of Spatial Point Patterns*. London, Academic Press.
Kaluzny, S.P., Vega, S.C. *et al.* (1998) *S + Spatial Stats*. New York, Springer-Verlag.
Mandelbrot, B.B. (1977) *Fractals, Form, Chance and Dimension*. San Fransisco, W.H. Freeman.
Pinheiro, J.C. and Bates, D.M. (2000) *Mixed-effects Models in S and S-Plus*. New York, Springer-Verlag.
Upton, G. and Fingleton, B. (1985) *Spatial Data Analysis by Example*. Chichester, John Wiley.

Bibliography

Abramowitz, M. and Stegun, I. (1964) *Handbook of Mathematical Functions*. Washington, DC, National Bureau of Standards.

Agresti, A. (1990) *Categorical Data Analysis*. New York, John Wiley.

Aitkin, M., Anderson, D., Francis, B. and Hinde, J. (1989). *Statistical Modelling in GLIM*. Oxford, Clarendon Press.

Atkinson, A.C. (1985) *Plots, Transformations, and Regression*. Oxford, Clarendon Press.

Belsley, D.A., Kuh, E. and Welsch, R.E. (1980) *Regression Diagnostics: Identifying Influential Data and Sources of Collinearity*. New York, John Wiley.

Bender, E.A., Case, T.J. and Gilpin, M.E. (1984) Perturbation experiments in community ecology: theory and practice. *Ecology* **65**: 1–13.

Berg, H.C. (1983) *Random Walks in Biology*. Princeton, New Jersey, Princeton University Press.

Bishop, Y.M.M., Fienberg, S.J. and Holland, P.W. (1980) *Discrete Multivariate Analysis: Theory and Practice*. New York, John Wiley.

Box, G.E.P. and Cox, D.R. (1964) An analysis of transformations. *Journal of the Royal Statistical Society, Series B* **26**: 211–246.

Box, G.E.P. and Jenkins, G.M. (1976) *Time Series Analysis: Forecasting and Control*. Oakland, California, Holden-Day.

Box, G.E.P., Hunter, W.G. and Hunter, J.S. (1978) *Statistics for Experimenters: An Introduction to Design, Data Analysis and Model Building*. New York, John Wiley.

Breiman, L., Friedman, L.H., Olshen, R.A. and Stone, C.J. (1984) *Classification and Regression Trees*. Belmont, California, Wadsworth International Group.

Bruce, A. and Gao, H.-Y. (1996) *Applied Wavelet Analysis with S-Plus*. New York, Springer-Verlag.

Burns, P.J. (1992) *Winsorized REML estimates of variance components*, Technical Report, Statistical Sciences, Inc.

Caroll, R.J. and Ruppert, D. (1988) *Transformation and Weighting in Regression*. New York, Chapman & Hall.

Casella, G. and Berger, R.L. (1990) *Statistical Inference. Pacific Grove*, California, Wadsworth and Brooks Cole.

Chambers, J.M. and Hastie, T.J. (1992) *Statistical Models in S. Pacific Grove*, California, Wadsworth and Brooks Cole.

Chambers, J.M., Cleveland, W.S., Kleiner, B. and Tukey, P.A. (1983) *Graphical Methods for Data Analysis*. Belmont, California, Wadsworth.

Chatfield, C. (1989) *The Analysis of Time Series: An Introduction*. London, Chapman & Hall.

Clark, P.J. and Evans, F.C. (1954) Distance to nearest neighbor as a measure of spatial relationships in populations. *Ecology* **35**: 445–453.

Cochran, W.G. and Cox, G.M. (1957) *Experimental Designs*. New York, John Wiley.

Collett, D. (1991) *Modelling Binary Data*. London, Chapman & Hall.

Connolly, J. (1987) On the use of response models in mixture experiments. *Oecologia* **72**: 95–113.

Conover, W.J. (1980) *Practical Nonparametric Statistics*. New York, John Wiley.

Cook, R.D. and Weisberg, S. (1982) *Residuals and Influence in Regression*. New York, Chapman & Hall.

Cox, D.R. (1958) *Planning of Experiments*. New York, John Wiley.

Cox, D.R. (1972) Regression models and life tables. *Journal of the Royal Statistical Society, Series B* **34**: 187–220.

Cox, D.R. and Hinkley, D.V. (1974) *Theoretical Statistics*. London, Chapman & Hall.

Cox, D.R. and Oakes, D. (1984) *Analysis of Survival Data*. London, Chapman & Hall.

Cox, D.R. and Snell, E.J. (1989) *Analysis of Binary Data*. London, Chapman & Hall.

Crawley, M.J. (1993) *GLIM for Ecologists*. Oxford, Blackwell Science.

Cressie, N.A.C. (1991) *Statistics for Spatial Data*. New York, John Wiley.

Crowder, M.J. and Hand, D.J. (1990) *Analysis of Repeated Measures*. London, Chapman & Hall.

Davidian, M. and Giltinan, D.M. (1995) *Nonlinear Models for Repeated Measurement Data*. London, Chapman & Hall.

Diggle, P.J. (1983) *Statistical Analysis of Spatial Point Patterns*. London, Academic Press.

Diggle, P.J., Liang, K.-Y. and Zeger, S.L. (1994) *Analysis of Longitudinal Data*. Oxford, Clarendon Press.

Dobson, A.J. (1990) *An Introduction to Generalized Linear Models*. London, Chapman & Hall.

Draper, N.R. and Smith, H. (1981) *Applied Regression Analysis*. New York, John Wiley.

Edwards, A.W.F. (1972) *Likelihood*. Cambridge, Cambridge University Press.

Efron, B. and Tibshirani, R.J. (1993) *An Introduction to the Bootstrap*. San Francisco, Chapman & Hall.

Eisenhart, C. (1947) The assumptions underlying the analysis of variance. *Biometrics* **3**: 1–21.

Everitt, B.S. (1994) *Handbook of Statistical Analyses Using S-PLUS*. New York, Chapman & Hall / CRC.

Ferguson, T.S. (1996) *A Course in Large Sample Theory*. London, Chapman & Hall.

Fisher, L.D. and Van Belle, G. (1993) *Biostatistics*. New York, John Wiley.

Fisher, R.A. (1925) *Statistical Methods for Research Workers*. Edinburgh, Oliver and Boyd.

Fisher, R.A. (1954) *Design of Experiments*. Edinburgh, Oliver and Boyd.

Fleming, T. and Harrington, D. (1991) *Counting Processes and Survival Analysis*. New York, John Wiley.

Gordon, A.E. (1981) *Classification: Methods for the Exploratory Analysis of Multivariate Data*. New York, Chapman & Hall.

Grimmett, G.R. and Stirzaker, D.R. (1992) *Probability and Random Processes*. Oxford, Clarendon Press.

Hairston, N.G. (1989) *Ecological Experiments: Purpose, Design and Execution*. Cambridge, Cambridge University Press.

Hampel, F.R., Ronchetti, E.M., Rousseeuw, P.J. and Stahel, W.A. (1986) *Robust Statistics: The Approach Based on Influence Functions*. New York, John Wiley.

Harman, H.H. (1976) *Modern Factor Analysis*. Chicago, University of Chicago Press.

Hastie, T. and Tibshirani, R. (1990) *Generalized Additive Models*. London, Chapman & Hall.

Hicks, C.R. (1973) *Fundamental Concepts in the Design of Experiments*. New York, Holt, Rinehart and Winston.

Hoaglin, D.C., Mosteller, F. and Tukey, J.W. (1983) *Understanding Robust and Exploratory Data Analysis*. New York, John Wiley.

Hochberg, Y. and Tamhane, A.C. (1987) *Multiple Comparison Procedures*. New York, John Wiley.

Hsu, J.C. (1996) *Multiple Comparisons: Theory and Methods*. London, Chapman & Hall.

Huber, P.J. (1981) *Robust Statistics*. New York, John Wiley.

Huitema, B.E. (1980) *The Analysis of Covariance and Alternatives*. New York, John Wiley.

Hurlbert, S.H. (1984) Pseudoreplication and the design of ecological field experiments. *Ecological Monographs* **54**: 187–211.

Johnson, N.L. and Kotz, S. (1970) *Continuous Univariate Distributions. Volume 2*. New York, John Wiley.

Kalbfleisch, J. and Prentice, R.L. (1980) *The Statistical Analysis of Failure Time Data*. New York, John Wiley.

Kaluzny, S.P., Vega, S.C., Cardoso, T.P. and Shelly, A.A. (1998) *S+ Spatial Stats*. New York, Springer-Verlag.

Kendall, M.G. and Stewart, A. (1979) *The Advanced Theory of Statistics*. Oxford, Oxford University Press.

Keppel, G. (1991) *Design and Analysis: A Researcher's Handbook*. Upper Saddle River, New Jersey, Prentice Hall.

Khuri, A.I., Mathew, T. and Sinha, B.K. (1998) *Statistical Tests for Mixed Linear Models*. John Wiley, New York.

Krause, A. and Olson, M. (2000) *The Basics of S and S-Plus*. New York, Springer-Verlag.

Lee, P.M. (1997) *Bayesian Statistics: An Introduction*. London, Edward Arnold.

Lehmann, E.L. (1986) *Testing Statistical Hypotheses*. New York, John Wiley.

Little, R.J.A. and Rubin, D.R. (1987) *Statistical Analysis with Missing Data*. New York, John Wiley.

Mandelbrot, B.B. (1977) *Fractals, Form, Chance and Dimension*. San Francisco, W.H. Freeman.

Mardia, K.V., Kent, J.T. and Bibby, J.M. (1979) *Multivariate Statistics*. London, Academic Press.

May, R.M. (1974) Biological populations with non-overlapping generations: stable points, stable cycles and chaos. *Science* **186**: 647–647.

McCullagh, P. and Nelder, J.A. (1989) *Generalized Linear Models*. London, Chapman & Hall.

McCulloch, C.E. and Searle, S.R. (2001) *Generalized, Linear and Mixed Models*. New York, John Wiley.

Mead, R. (1989) *The Design of Experiments*. Cambridge, Cambridge University Press.

Millard, S.P. and Krause, A. (2001) *Using S-PLUS in the Pharmaceutical Industry*. New York, Springer-Verlag.

Miller, R.G. (1981) *Survival Analysis*. New York, John Wiley.

Miller, R.G. (1997) *Beyond ANOVA: Basics of Applied Statistics*. London, Chapman & Hall.

Mosteller, F. and Tukey, J.W. (1977) *Data Analysis and Regression*. Reading, Mass., Addison-Wesley.

Nelder, J.A. and Wedderburn, R.W.M. (1972) Generalized Linear Models. *Journal of the Royal Statistical Society, Series A* **135**: 37–384.

Neter, J., Wasserman, W. and Kutner, M.H. (1985) *Applied Linear Regression Models*. Homewood, Illinois, Irwin.

Neter, J., Kutner, M., Nachstheim, C. and Wasserman, W. (1996) *Applied Linear Statistical Models*. New York, McGraw-Hill.

O'Hagen, A. (1988) *Probability: Methods and Measurement*. London, Chapman & Hall.

Pinheiro, J.C. and Bates, D.M. (2000) *Mixed-effects Models in S and S-Plus*. New York, Springer-Verlag.

Platt, J.R. (1964) Strong inference. *Science* **146**: 347–353.

Priestley, M.B. (1981) *Spectral Analysis and Time Series*. London, Academic Press.

Rao, P.S.R.S. (1997) *Variance Components Estimation: Mixed Models, Methodologies and Applications*. London, Chapman & Hall.

Riordan, J. (1978) *An Introduction to Combinatorial Analysis*. Princeton, New Jersey, Princeton University Press.

Ripley, B.D. (1996) *Pattern Recognition and Neural Networks*. Cambridge University Press, Cambridge.

Robert, C.P. and Casella, G. (1999) *Monte Carlo Statistical Methods*. New York, Springer-Verlag.

Rosner, B. (1990) *Fundamentals of Biostatistics*. Boston, PWS-Kent.

Ross, G.J.S. (1990) *Nonlinear Estimation*. New York, Springer-Verlag.

Santer, T.J. and Duffy, D.E. (1990) *The Statistical Analysis of Discrete Data*. New York, Springer-Verlag.

Schafer, J.L. (1996) *Analysis of Incomplete Multivariate Data*. London, Chapman & Hall.

Schoener, T.W. (1968) The *Anolis* lizards of Bimini: resource partitioning in a complex fauna. *Ecology* **49**, 704–726.

Searle, S.R., Casella, G. and McCulloch, C.E. (1992) *Variance Components*. New York, John Wiley.

Seber, G.A.F. (1982) *Estimation of Animal Abundance and Related Parameters*. London, Griffin.

Shao, J. and Tu, D. (1995) *The Jackknife and Bootstrap*. New York, Springer-Verlag.

Shumway, R.H. (1988) *Applied Statistical Time Series Analysis*. Englewood Cliffs, New Jersey, Prentice Hall.

Silvey, S.D. (1970) *Statistical Inference*. London, Chapman & Hall.

Snedecor, G.W. and Cochran, W.G. (1980) *Statistical Methods*. Ames, Iowa, Iowa State University Press.

Sokal, R.R. and Rohlf, F.J. (1995) *Biometry: The Principles and Practice of Statistics in Biological Research*. San Francisco, W.H. Freeman.

Sprent, P. (1989) *Applied Nonparametric Statistical Methods*. London, Chapman & Hall.

Taylor, L.R. (1961) Aggregation, variance and the mean. *Nature* **189**: 732–735.

Upton, G. and Fingleton, B. (1985) *Spatial Data Analysis by Example*. Chichester, John Wiley.

Venables, W.N. and Ripley, B.D. (1997) *Modern Applied Statistics with S-Plus*. New York, Springer-Verlag.

Weisberg, S. (1985) *Applied Linear Regression*. New York, John Wiley.

Wetherill, G.B., Duncombe, P., Kenward, M., Kollerstrom, J., Paul, S.R. and Vowden, B.J. (1986). *Regression Analysis with Applications*. London, Chapman & Hall.

Winer, B.J., Brown, D.R. and Michels, K.M. (1991) *Statistical Principles in Experimental Design*. New York, McGraw-Hill.

Index

Entries in bold are S-Plus functions